Fish Defenses

Fish Defenses

Volume 1: Immunology

Editors

Giacomo Zaccone
Department of Animal Biology and Marine Ecology
University of Messina
Messina, Italy

J. Meseguer
Department of Cell Biology
University of Murcia
Murcia, Spain

A. García-Ayala
Department of Cell Biology
University of Murcia
Murcia, Spain

B.G. Kapoor
Formerly Professor of Zoology
The University of Jodhpur
Jodhpur, India

Science Publishers
Enfield (NH) Jersey Plymouth

Science Publishers www.scipub.net
234 May Street
Post Office Box 699
Enfield, New Hampshire 03748
United States of America

General enquiries : *info@scipub.net*
Editorial enquiries : *editor@scipub.net*
Sales enquiries : *sales@scipub.net*

Published by Science Publishers, Enfield, NH, USA
An imprint of Edenbridge Ltd., British Channel Islands
Printed in India

ISBN 978-1-57808-327-5

Cover illustration: Reproduced from Chapter 1 by C.J. Secombes, J. Zou and S. Bird with kind permission of the authors.

Library of Congress Cataloging-in-Publication Data

Fish defenses/editors, Giacomo Zaccone ...[et al.].
v. cm.
Includes bibliographical references.
Contents: v. 1. Immunology
ISBN 978-1-57808-327-5 (hardcover)
1. Fishes--Defenses. I. Zaccone, Giacomo.
QL639.3.F578 2008
571.9'617--dc22

2008016632

"...in **Mozart** the Nature has produced exceptional, an unrepeatable one, in any case never more repeated, masterpiece."

Wolfgang Hildesheimer

Preface

In its widest sense the term defense refers to all the mechanisms used by living organisms as protection against foreign environmental agents such as microorganisms and their products, chemicals, drugs, animal actions, etc. Among these, immunity is the main endogenous mechanism of defense which helps to distinguish between self and non self. This recognition mechanism originated with the formation of cell markers probably involving cell surface molecules that were able to specifically bind and adhere to other molecules present on opposing cell surfaces. This simple method seems to have evolved into the full complexity of what we call an immune response.

The greatest complexity of the immune response is shown by vertebrates which are endowed with innate and acquired immunity, although, increasingly, evidence for types of acquired immunity is coming to light within the invertebrates. Immunological studies performed mostly in mammals have been the reference for studies in other vertebrates. However, the efforts of research scientists around the world have now produced findings that allow us to identify significant differences among the immune system of the major vertebrate groups. Fish immunity, particularly, shows striking differences from that observed in homoeothermic animals.

The study of immunological fish defenses has advanced considerably in recent decades. This has been due to the key position of fish in terms of the evolution of acquired immunity and to the rapid expansion of aquaculture over this period, where disease control is of prime concern. In addition, some fish species are seen as powerful scientific models for different field of study.

The objective of this book is to present a compilation of some of the main findings that reflect current thinking on fish immune defenses.

Two chapters are devoted to fish innate immunity: the antimicrobial peptides and the cellular processes involved in macrophage-mediated host defense. Also, two chapters look at adaptive immunity in fish and review current knowledge on the molecular organization of antibody genes, the structural and functional features of the antibody molecule, the development of antibody-producing cells and the organization and function of the system which leads to an antibody response. The discovery of a new immunoglobulin class and the characterization of teleost IGH loci are discussed. Another chapter is dedicated to the fish immune response to eukaryotic parasites (immune responses to surface and internal parasites) as well as the evasion and suppression of the immune response by such pathogens (elusive parasite).

The fish cytokine network, immune regulatory peptides coordinating innate and adaptive responses, is analysed in a further chapter, outlining their discovery, activities and potential application. Two chapters are devoted to immune-endocrine interactions in fish, namely, the effects of estrogens as fish immunoregulators. Linked to this, the morpho-functional features of leukocytes and cytokines present in the fish testis are also described.

Finally, two chapters give up-to-date reviews on applied aspects of manipulating fish immune defenses in aquaculture. Current knowledge of the immune system of sea bass (*Dicentrarchus labrax* L.), a teleost species of immense commercial interest for Mediterranean aquaculture, is reported, and the potential use of CpG ODNs as immunostimulants in aquaculture is presented, together with their possible use to improve future vaccine formulations.

Acknowledgements

We especially thank Professor Chris Secombes (University of Aberdeen, Scotland, UK) for his encouragement, suggestions and collaboration during the preparation of this book.

Giacomo Zaccone
Jose Meseguer Penalver
Alfonsa García-Ayala
B.G. Kapoor

Contents

List of Contributors

Belosevic Miodrag

Department of Biological Sciences, University of Alberta, Edmonton, Alberta, Canada.
E-mail: mike.belosevic@ualberta.ca

Bird S.

Scottish Fish Immunology Research Centre, University of Aberdeen, Aberdeen, AB24 2TZ, Scotland.

Bromage E.

Department of Environmental and Aquatic Animal Health, Virginia Institute of Marine Science, 1208 Greate Rd., College of William and Mary, Gloucester Point, VA 23062, USA.

Brown G.

Department of Environmental and Aquatic Animal Health, Virginia Institute of Marine Science, 1208 Greate Rd., College of William and Mary, Gloucester Point, VA 23062, USA.

Buonocore Francesco

Laboratory of Animal Biotechnology, Department of Environmental Sciences, University of Tuscia, Largo dell' Università, I-01100 Viterbo, Italy.

Chaves-Pozo Elena

Department of Cell Biology, Faculty of Biology, University of Murcia, 30100 Murcia, Spain.
E-mail: echaves@um.es

Cuesta A.

Department of Cell Biology, University of Murcia, 30100 Murcia, Spain.

Esteban M.A.

Department of Cell Biology, University of Murcia, 30100 Murcia, Spain.

Fernandes Jorge M.O.

Marine Molecular Biology and Genomics Research Group, Faculty of Biosciences and Aquaculture, Bodø University College, N-8049 Bodø, Norway.
E-mail: jorge.fernandes@hibo.no

García-Ayala Alfonsa

Department of Cell Biology, Faculty of Biology, University of Murcia, 30100 Murcia, Spain.
E-mail: agayala@um.es

Grayfer Leon

Department of Biological Sciences, University of Alberta, Edmonton, Alberta, Canada.

Haddad George

Department of Biological Sciences, University of Alberta, Edmonton, Alberta, Canada.

Haines A.

Department of Environmental and Aquatic Animal Health, Virginia Institute of Marine Science, 1208 Greate Rd., College of William and Mary, Gloucester Point, VA 23062, USA.

Hanington Patrick C.

Department of Biological Sciences, University of Alberta, Edmonton, Alberta, Canada.

Hoole Dave

Keele University, Keele, Staffordshire, ST5 5BG, UK.
E-mail: d.hoole@biol.keele.ac.uk; d.hoole@keele.ac.uk

Iwanowicz Luke R.
USGS, Leetown Science Center, Aquatic Ecology Branch, Kearneysville, WV 25430, USA.
E-mail: liwanowicz@usgs.gov

Kaattari I.
Department of Environmental and Aquatic Animal Health, Virginia Institute of Marine Science, 1208 Greate Rd., College of William and Mary, Gloucester Point, VA 23062, USA.

Kaattari S.
Department of Environmental and Aquatic Animal Health, Virginia Institute of Marine Science, 1208 Greate Rd., College of William and Mary, Gloucester Point, VA 23062, USA.
E-mail: kaattari@vims.edu

Katzenback Barbara A.
Department of Biological Sciences, University of Alberta, Edmonton, Alberta, Canada.

Meseguer J.
Department of Cell Biology, University of Murcia, 30100 Murcia, Spain.
E-mail: meseguer@um.es

Neumann Norman F.
Department of Microbiology and Infectious Diseases, University of Calgary, Calgary, Canada.

Ottinger Christopher A.
USGS, Leetown Science Center, Fish Health Branch, Kearneysville, WV 25430, USA.

Sakai Masahiro
Faculty of Agriculture, University of Miyazaki, Gakuen kibanadai nishi 1-1, Miyazaki, 889-2192, Japan.
E-mail: m.sakai@cc.miyazaki-u.ac.jp

Savan Ram

Faculty of Agriculture, University of Miyazaki, Gakuen kibanadai nishi 1-1, Miyazaki, 889-2192, Japan.

Present address: National Cancer Institute, Frederick, MD 21701, USA.

E-mail: savanr@mail.nih.gov

Scapigliati Giuseppe

Laboratory of Animal Biotechnology, Department of Environmental Sciences, University of Tuscia, Largo dell' Università, I-01100 Viterbo, Italy.

E-mail: scapigg@unitus.it

Secombes C.J.

Scottish Fish Immunology Research Centre, University of Aberdeen, Aberdeen, AB24 2TZ, Scotland.

E-mail: c.secombes@abdn.ac.uk

Smith Valerie J.

Comparative Immunology Group, Gatty Marine Laboratory, School of Biology, University of St Andrews, St Andrews, Fife, KY16 8LB, Scotland, UK.

E-mail: vjs1@st-andrews.ac.uk

Stafford James L.

Department of Biological Sciences, University of Alberta, Edmonton, Alberta, Canada.

Walsh John G.

Department of Genetics, Trinity College, Dublin, Ireland.

Ye J.

Department of Environmental and Aquatic Animal Health, Virginia Institute of Marine Science, 1208 Greate Rd., College of William and Mary, Gloucester Point, VA 23062, USA.

Zou J.

Scottish Fish Immunology Research Centre, University of Aberdeen, Aberdeen, AB24 2TZ, Scotland.

CHAPTER

1

Fish Cytokines: Discovery, Activities and Potential Applications

C.J. Secombes*, J. Zou and S. Bird

INTRODUCTION

Over the last few years, there has been a tremendous increase in our knowledge of the fish cytokine network, largely due to the increasing number of EST sequences in the databases and the availability of sequenced fish genomes. This chapter will highlight some of these advances in terms of genes discovered and, where elucidated, the role of the proteins in the immune system of fish. The potential applications of these molecules in studies to improve fish health in aquaculture will also be discussed.

Authors' address: Scottish Fish Immunology Research Centre, University of Aberdeen, Aberdeen, AB24 2TZ, Scotland.

**Corresponding author:* E-mail: c.secombes@abdn.ac.uk

CYTOKINES INVOLVED IN INNATE IMMUNITY

Cytokines are the key regulators of the immune system. Their role in initiating inflammatory events in response to bacterial exposure is well known in mammals, where a cytokine cascade leads to the attraction of particular leucocyte types and activation of their antimicrobial pathways. Tumor necrosis factor-α (TNF-α) is the first cytokine released in this cascade and leads to the downstream expression of interleukin-1β (IL-1β) and chemokines such as IL-8. Following infection with viruses, cytokines can again activate various cellular pathways but also have direct effects on cells that lead to an antiviral state, as seen with the interferons (IFN). Such molecules represent a crucial component of the innate defences, although cross talk and activation of adaptive (specific) immunity may also be triggered in the medium term.

Pro-inflammatory Cytokines

Discovery

One of the first cytokine genes discovered in fish was IL-1β, found in rainbow trout by homology cloning (Zou *et al.*, 1999a,b) and one of the key pro-inflammatory cytokines known from mammalian studies. This quickly led to the discovery of IL-1β in many other teleost fish species, including carp (Fujiki *et al.*, 2000), sea bass (Scapigliati *et al.*, 2001), sea bream (Pelegrin *et al.*, 2001), goldfish (Bird, 2001), turbot (Low *et al.*, 2003), Japanese flounder (Emmadi *et al.*, 2005), zebrafish (Pressley *et al.*, 2005), crocodile ice fish (Buonocore *et al.*, 2006), Atlantic salmon (Ingerslev *et al.*, 2006), Nile tilapia (Lee *et al.*, 2006), Japanese sea perch (Qiu *et al.*, 2006), channel catfish (Wang *et al.*, 2006) and haddock (Corripio-Miyar *et al.*, 2007). Homology cloning of this molecule in cartilaginous fish was also possible following this initial discovery (Bird *et al.*, 2002a; Inoue *et al.*, 2003a). Further analysis has shown that in some species a second IL-1β gene exists, especially in the salmonids (Pleguezuelos *et al.*, 2000) and cyprinids (Engelsma *et al.*, 2003; Wang *et al.*, 2006), and that even allelic variation is detectable in fish (Wang *et al.*, 2004). Curiously, the exon-intron organization varies between different fish groups, with the classical sevon exon arrangement seen in mammals present in cyprinids (Engelsma *et al.*, 2001), with six exons present in salmonids (Zou *et al.*, 1999b) and five present in more advanced acanthopterygian teleosts (Buonocore *et al.*, 2003a; Lee *et al.*, 2006).

Most conservation of the molecule, when translated into the amino acid sequence, can be seen within the beta sheet regions that form the secondary structure of IL-1β, that is a β-trefoil cytokine (Koussounadis *et al.*, 2004). Modelling of the receptor binding sites across species reveals a high level of variability in terms of the positions involved but indicates conservation of the overall shape of the ligand-receptor complex (Koussounadis *et al.*, 2004). However, analysis of the protein sequence fails to find an obvious processing site for cleavage by interleukin converting enzyme (ICE), required for generation of the bioactive mature peptide in mammals. Nevertheless, evidence to date suggests that fish IL-1β is processed (Hong *et al.*, 2004).

A second major pro-inflammatory cytokine discovered in fish was TNF-α. It was discovered during EST analysis of PMA/ConA stimulated blood leucocytes from Japanese flounder (Hirono *et al.*, 2000), and was the first non-mammalian TNF discovered. Again, this was quickly followed with the cloning of TNF by homology in several teleost fish species, including brook trout (Bobe and Goetz, 2001), rainbow trout (Laing *et al.*, 2001), sea bream (Garcia-Castillo *et al.*, 2002), carp (Saeij *et al.*, 2003), catfish (Zou *et al.*, 2003a), pufferfish and zebrafish (Savan *et al.*, 2005), Atlantic salmon (Ingerslev *et al.*, 2006), turbot (Ordás *et al.*, 2007), ayu (Uenobe *et al.*, 2007) and sea bass (Nascimento *et al.*, 2007a). As with IL-1β, in certain teleost groups multiple isoforms were found to exist (Zou *et al.*, 2002; Savan and Sakai, 2003).

Subsequently, the teleost TNF-α gene was found to be adjacent to another TNF family member in the genomes of pufferfish and zebrafish (Savan *et al.*, 2005), although the exact relationship to other known TNFs, namely TNF-β and lymphotoxin-β (LT-β), was not entirely clear. This allowed the cloning of two isoforms of the equivalent molecule from rainbow trout (Kono *et al.*, 2006), and further analysis revealed that this was the teleost LT-β molecule. Thus, it is apparent that either TNF-β does not exist in fish, as it is not adjacent to the other TNF family members in fish genomes, or it resides in a different part of the genome to mammals.

Chemokines, or chemotactic cytokines, are a family of cytokines that serve to attract leucocytes to particular sites (Laing and Secombes, 2004a). They are subdivided into two main groups, those that are involved in lymphocyte trafficking and immune surveillance as part of homeostatic mechanisms within the immune system, and those that are released during an infection and direct leucocyte migration to an injured or infected site.

Based on the arrangement of the first two cysteines in the molecule they are classified as CXC chemokines (α), CC chemokines (β), C chemokines (γ) or CX_3C chemokines (δ). Only members of these first two families have been found in fish to date.

The first CXC chemokine discovered in fish was IL-8 (CXCL8), initially found in the lamprey (Najakshin *et al.*, 1999) and, subsequently, cloned in Japanese flounder (Lee *et al.*, 2001), rainbow trout (Laing *et al.*, 2002a), carp (Huising *et al.*, 2003—referred to as CXCa by these authors), banded dogfish (Inoue *et al.*, 2003b), silver chimaera (Inoue *et al.*, 2003c), catfish (Chen *et al.*, 2005) and haddock (Corripio-Miyar *et al.*, 2007). An important feature of the IL-8 molecule is the so-called ELR motif immediately upstream of the CXC residues, and responsible for the ability to attract and activate neutrophils. This motif has conservative substitutions in most fish species (e.g., DLR in trout), although the haddock molecule possesses an unsubstituted motif. The gene expression is markedly induced upon stimulation with pro-inflammatory stimuli such as LPS or IL-1β (Fig. 1.1). A second CXC chemokine discovered in fish has clear homology to the gamma interferon-induced chemokines CXCL9, CXCL10 and CXCL11, that all share a common receptor (CXCR3) and are all ELR^- CXC chemokines. This molecule was initially discovered in trout (Laing *et al.*, 2002b; O'Farrell *et al.*, 2002) and, subsequently, discovered in carp (Savan *et al.*, 2003a—referred to as CXCb by Huising *et al.*, 2003) and channel catfish (Baoprasertkul *et al.*, 2004). The trout molecule is more CXCL10-like (based on analyses of individual exon sequences) and is induced by trout recombinant interferon gamma

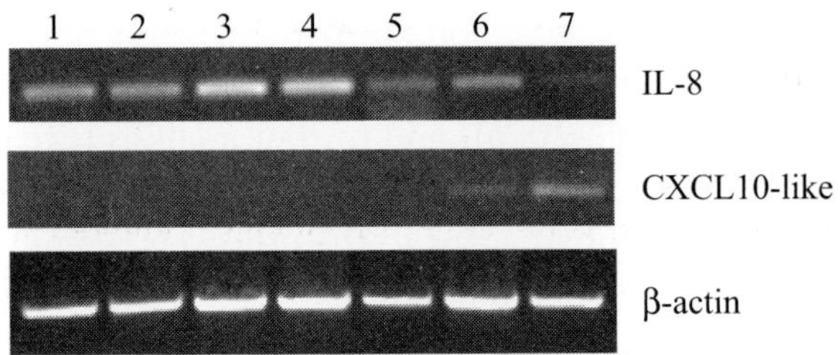

Fig. 1.1 Modulation of chemokine expression by the proinflammatory cytokines IL-1β and IFN-γ. Freshly isolated trout head kidney leucocytes were stimulated for 4 h with trout rIL-1β (lanes 2-4) and rIFN-γ (lanes 5-7) at doses of 1, 10 and 100 ng/ml respectively. Control cells (lane 1) were incubated with an equal volume of elution buffer (20 mM Tris, 100 mM KCl, 5 mM $MgCl_2$, 10 mM mercaptoethanol, 0.1% NP40, 20% glycerol, 250 mM imidazole) used to dissolve the recombinant proteins. Total RNA was extracted for semi-quantitative RT-PCR analysis of gene expression for IL-8 and CXCL10-like chemokines.

(IFN-γ), as expected (Zou *et al.*, 2005a; Fig. 1.1) and is, thus, a useful marker of IFN-γ action. CXC chemokines homologous to CXCL12 and 14 have also been identified in fish (Long *et al.*, 2000; Huising *et al.*, 2004; Baoprasertkul *et al.*, 2005) and are particularly highly conserved, suggesting they have critical roles, potentially in development of the organism (see section on **Activities** below). Two further CXC chemokines have been reported that have less clear homology to known genes, a so-called CXCd gene in trout (Wiens *et al.*, 2006) and a CXCL2-like molecule (17-24% amino acid identity to mammalian molecules) in catfish (Baoprasertkul *et al.*, 2005). CXCL2 is an ELR-containing chemokine in mammals, but like the catfish CXCL8 molecule, the CXCL2-like gene does not possess an ELR motif although, again, a related sequence (PDR) is present. A recent study in zebrafish and pufferfish suggests that molecules with homology to CXCL5, CXCL9 and CXCL11 may also exist in fish (DeVries *et al.*, 2006), although further work is needed to verify this. In some species multiple isoforms of CXC chemokines are present, as seen in carp (Huising *et al.*, 2004), trout (Wiens *et al.*, 2006) and zebrafish (DeVries *et al.*, 2006). This appears to be related to ancient genome duplication events or more recent tandem gene duplications, including the duplication of multiple genes by duplication of chromosomal segments in the case of zebrafish.

Lastly, many CC chemokines have been identified in fish, especially in trout (Dixon *et al.*, 1998; Laing and Secombes, 2004b), catfish (Bao *et al.*, 2006; Peatman *et al.*, 2006) and zebrafish (DeVries *et al.*, 2006; Peatman and Liu, 2006), where numbers can exceed those seen in mammals (e.g., 24 genes known in humans versus 26 in catfish and 46 in zebrafish). A CC chemokine gene has also been reported in a cartilaginous fish (Kuroda *et al.*, 2003). In trout, where 18 unique transcripts (including 6 pairs of closely related genes) were found by interrogation of EST sequences, phylogenetic analysis revealed some genes grouped with known mammalian clades, whilst others had no obvious homology to mammalian genes but were clustered with other fish sequences. This suggested the possibility that fish specific and species-specific gene duplications of CC chemokines have occurred, and that CC chemokines have high divergence rates. Similarly, in catfish, numerous genes were discovered and were mapped to BAC clones. This revealed that the genes were highly clustered in the genome, with 2-8 identified chemokines within individual assembled contigs. Between contigs, the genes were often very similar, suggesting that segmental gene duplication was involved in generating the chemokine gene clusters. Only five of the chemokine genes were present as a single copy, with 5 as two copies, 8 as three copies, 5 as four copies, 2 as five copies and one as six copies, giving some 75 genes in total. Many of the catfish genes shared highest identity to CCL3 or CCL14, perhaps members of the original CC repertoire in early vertebrates. In zebrafish, analysis of ESTs and the available genome allowed a more extensive evaluation of the CC chemokine repertoire in fish (Peatman and Liu, 2006). Forty-six genes were identified, again with evidence of extensive 'en bloc' duplication events. Many of the genes form species-specific clades in phylogenetic analysis, supporting the concept that species-specific duplication events of fish chemokine genes have occurred. A model of the evolutionary history of the chemokine system has been proposed by DeVries *et al.* (2006) based on analyses of fish, amphibian, bird and mammalian genomes.

ACTIVITIES

The recombinant IL-1β protein has been produced in *E. coli* and its bioactivity studied in trout (Hong *et al.*, 2001; Peddie *et al.*, 2001), carp (Yin and Kwang, 2000a; Matthew *et al.*, 2002) and sea bass (Buonocore *et al.*, 2005). With no clear ICE cut site, the recombinant proteins were started at various locations in the cut site region determined by multiple alignment. In trout, Ala^{95} was used as the starting point for the recombinant protein, whereas in carp Thr^{115} was used and in seabass Ala^{86} was used. As outlined below, in all the cases a bioactive protein was produced, although the exact start of the native mature protein remains unknown.

The trout rIL-1β was shown to have a number of effects on trout leucocytes. It increased its own expression in head-kidney leucocytes, with optimal rIL-1β dose and kinetics studied (Hong *et al.*, 2001). It was also able to induce the expression of cyclooxgenase 2 (COX-2), a potent pro-inflammatory gene, and to increase the expression of the MHC class II β chain in a trout macrophage cell line. Lastly, the rIL-1β increased head-kidney leucocyte phagocytic activity for yeast particles (Hong *et al.*, 2001), and induced leucocyte chemotaxis (Peddie *et al.*, 2001). Since the recombinant protein was produced in bacteria, and LPS is known to be a potent stimulant for leucocytes, a number of controls were included in this study, such as the effect of pre-incubation with a specific polyclonal antiserum for 1 h at 4°C, or of heating the rIL-1β to 95°C for 20 min, and in each case the activity of the trout rIL-β was significantly decreased.

Similar results were found with sea bass rIL-1β produced in *E. coli*, which enhanced phagocytosis of head-kidney leucocytes and expression of IL-1β and COX-2 (Buonocore *et al.*, 2005). The seabass rIL-1β was also shown to induce the proliferation of sea bass thymocytes following stimulation with a sub-optimal dose of Con A. Carp rIL-1β was also shown to stimulate *in vitro* the proliferation of carp head-kidney and splenic leucocytes (Mathew *et al.*, 2002).

IL-1β derived peptides have also been produced, based upon the modelling of IL-1β with its receptor, to identify contiguous regions of the protein that are part of the receptor binding domain (Koussounadis *et al.*, 2004). *In vitro* studies showed that one of the peptides (called P3) was particularly potent, and was able to enhance the locomotory (Peddie *et al.*, 2001), phagocytic and bactericidal (Peddie *et al.*, 2002a) capacity of head-kidney leucocytes. The latter was not a result of enhanced oxygen radical production from the cells, which was not increased by P3. Interestingly, when combined with a second peptide (P1)—which alone had limited effects—a synergistic effect was seen on phagocytic and locomotory activity. Similarly, *in vivo* injection of P3 into the peritoneal cavity increased the number of peritoneal leucocytes harvested at days 1 and 3 post-injection, relative to control groups, including fish injected with P1 or an irrelevant peptide, and increased their phagocytic activity (Peddie *et al.*, 2003). One of the most fascinating findings from the *in vivo* studies, however, was that whilst P3 had clear effects on leucocyte activity, it failed to induce the normal activation of the hypothalamic-pituitary-adrenal (HPI) axis that results in the detectable release of cortisol into the blood stream (Holland *et al.*, 2002). Trout rIL-1β, on the other hand, does activate the HPI axis when administered intraperitoneally (ip), with elevated cortisol levels apparent 3h-8h post-injection.

As an alternative approach to producing the recombinant protein *in vitro* for bioactivity studies, in some experiments, plasmid DNA containing the cloned IL-1β gene has been administered. In carp, intramuscular injection of the construct resulted in increased macrophage respiratory burst and phagocytic activity, as well as increased leucocyte proliferation following PHA stimulation of the cells (Kono *et al.*, 2002a). In Japanese flounder, fish injected intramuscularly with a construct containing the full length IL-1β in the pCL-neo expression vector (Emmadi *et al.*, 2005) had kidney tissue isolated at days 1, 3 and 7 post-injection, and the samples used for expression profiling with a Japanese flounder cDNA micro array containing 871 features. In total 93 genes on the array were found to be affected; 64 genes were upregulated, whilst 29 genes were downregulated. Amongst the immune genes induced by injection with this construct were TNF-α, granulocyte colony stimulating factor, MHC class I, β2-microglobulin, IgM, CD3, CD20 receptor and a CC chemokine receptor, with expression levels highest at days 1 and 3 relative to day 7.

Recombinant TNF-α has also been produced and its activity studied in fish. Unlike IL-1β, the processing site for cleavage (by metalloproteinase in this case) to release the mature peptide is relatively well conserved (Laing *et al.*, 2001), and so the start of the mature peptide can be predicted with much confidence. In rainbow trout, where two isoforms of TNF-α were discovered (Zou *et al.*, 2002), both have been produced as recombinant proteins for bioactivity studies (Zou *et al.*, 2003b), and are able to induce the expression of pro-inflammatory genes, such as IL-1β, IL-8, COX-2 and TNF-α itself, in a macrophage cell line (RTS-11 cells). In addition, they increased head-kidney leucocyte migration and phagocytic activity, with no clear differences apparent between the two molecules. In sea bream, ip injection of the recombinant seabream protein resulted in increased respiratory burst activity and mobilization of cells into the peritoneal cavity (Garcia-Castillo *et al.*, 2004). It also increased granulopoiesis in the head-kidney and increased the proliferation of head-kidney cells following *in vitro* stimulation. In ayu, rTNF stimulated respiratory burst activity of cultured ayu kidney leucocytes (Uenobe *et al.*, 2007), whilst in turbot, rTNF-α has been shown to increase NO production by cultured turbot macrophages, especially when combined with LPS stimulation (Ordás *et al.*, 2007).

Rather few studies have looked at the activity of chemokines in fish, and those that have are focussed on the CXC chemokines. Studies to date on CXCL12 and CXCL14 have already confirmed their crucial role in development in fish. Thus, CXCL12 has been shown to mediate the control of primordial cell migration that gives rise to the lateral line, in zebrafish (David *et al.*, 2002; Doitsidou *et al.*, 2002; Sapede *et al.*, 2005). Use of morpholino oligos to knock down CXCL12 expression prevented both primordial cell migration and posterior lateral line formation. CXCL12 knockdown also causes retinal axons to follow aberrant pathways in the retina, demonstrating its importance in guiding retinal ganglion axons to the optic stalk (Li *et al.*, 2005). Lastly, CXCL12 has been shown to have a critical role in fin regeneration, through its affect on epidermal cell proliferation (Dufourcq and Vriz, 2006). CXCL14 appears to play a role in the development of the acoustico-lateralis system in the nervous system (Long *et al.*, 2000), and is highly upregulated in the ovary during oocyte maturation (Bobe *et al.*, 2006), possibly contributing to the inflammatory-type events seen in the ovary at ovulation. The role of CXCL8 in attracting and activating neutrophils has also been determined to some extent. Rainbow trout rCXCL8 has been produced and shown to increase head-kidney leucocyte migration and respiratory burst activity *in vitro* and also to increase peritoneal cell number and increase the percentage of peritoneal neutrophils after ip administration (Harun, 2006). The trout IL-8 gene has also been cloned into an expression vector and, following an injection into the dorsal musculature of trout, shown to induce a massive neutrophil infiltration at this site (Jimenez *et al.*, 2006). The only CC chemokine produced as a recombinant protein to date is rainbow trout CK-1 (Lally *et al.*, 2003). rCK-1 is able to attract and enhance the locomotion of blood leucocytes at concentrations of 1 and 10 μg/ml.

Potential Applications

One potential application of these pro-inflammatory cytokines is as vaccine adjuvants (Secombes and Scheerlinck, 1999), to augment the initial host responsiveness to the vaccine formulation. To date, most studies have concentrated on determining the effect on specific antibody production post-immunization. For example, in carp, addition of rIL-1β to a formalin-killed *Aeromonas hydrophila* vaccine enhanced the serum antibody response, determined three weeks post-injection, relative to appropriate control groups (Yin and Kwang, 2000a). In seabass, rIL-1β added to a *Vibrio anguillarum* vaccine (to give a dose of 500 ng per fish), was shown to enhance the specific antibody response 60 days after ip injection of the vaccine (Buonocore *et al.*, 2003b). In barramundi, addition of carp rIL-1β to a *Vibrio harveyi* vaccine enhanced antibody responses following ip administration and sampling 42 days later (Bridle *et al.*, 2002).

In trout, the first attempt to augment the action of a DNA vaccine by a cytokine has also been studied in terms of the cytokine response elicited (Jimenez *et al.*, 2006). Co-administration of the DNA vaccine for the glycoprotein of the viral pathogen VHSV and an expression construct (pcDNA3.1/V5-His-TOPO) containing the trout IL-8 gene, resulted in up-regulation of a number of cytokines relative to the DNA vaccine alone or with a control plasmid for the IL-8 construct, especially of the pro-inflammatory cytokines IL-1β, TNF-α and IL-11.

These molecules may also have some merit as stimulants of the immune system in their own right, albeit short term in nature. For example, injection of trout with rIL-1β or IL-1β derived peptides is able to increase resistance against bacterial (*Aeromonas salmonicida*) (Hong *et al.*, 2003) and viral (VHSV) diseases (Peddie *et al.*, 2003) at day 2 post-injection, respectively. However, by day 7 post-injection, this effect was lost. Similarly, in carp, injection of plasmid DNA containing the cloned IL-1β gene increased resistance to A. *hydrophila* infection (Kono *et al.*, 2002a). Perhaps a more useful application is as a marker of immunostimulant action, to help confirm the usefulness of potential immunostimulants and optimize their administration. For example, following ip administration of Ergosan to trout expression of IL-1β, TNF-α and IL-8 were all increased in peritoneal leucocytes, with expression of IL-1β and IL-8 showing particularly high increases relative to control fish (Peddie *et al.*, 2002b). Similarly, injection of peptidoglycan significantly elevates IL-1β levels in carp (Kono *et al.*, 2002b) and CpG containing oligodeoxynucleotides have been shown to increase IL-1β expression in trout macrophages (Jørgensen *et al.*, 2001a) and expression of IL-1β, TNF-α and chemokines in carp head-kidney leucocytes (Tassakka *et al.*, 2006). Thus measurement of the expression level of these cytokines may represent a sensitive way to screen as to whether compounds are pro-inflammatory in fish.

Antiviral Cytokines

Discovery

Type I IFNs belong to the α-helix cytokine family, also known as the hematopoietic growth factor family. Well known are their pivotal roles in the host immune defence against viral infection. The first fish type I IFN gene was identified in zebrafish by analysis of the expressed sequence tag database (Altmann *et al.*, 2003). To date, the IFN sequences are available for several fish species including catfish (Long *et al.*, 2006), goldfish (GenBank accession number: AY452069), the Japanese pufferfish (Lutfalla *et al.*, 2003; Zou *et al.*, 2005b), the spotted green pufferfish (Lutfalla *et al.*, 2003), Atlantic salmon (Robertsen *et al.*, 2003) and rainbow trout (GenBank accession numbers: AJ580911, AJ582754, AM235738, AY78889). In general, type I IFN peptides are diverse in nature, with limited homology between different teleost species. Compared to the intron-lacking type I IFN genes in birds and mammals, fish IFN genes contain 5 exons and 4 introns. Multiple copies of the IFN genes are present in all fish species examined to date and evidence derived from genome analysis indicates they are tandemly clustered in the genome (Robertsen *et al.*, 2003; Zou *et al.*, 2005b; Long *et al.*, 2006). Despite having the same genomic organization with the IFN-λ genes so far discovered only in human and mouse, phylogenetic studies demonstrate that fish IFN genes are homologues of type I IFN in birds and mammals and support the view that IFN-λs and type I IFNs diverged from a common progenitor earlier in evolution than the divergence of type I IFN isoforms. Fish IFN genes encode proteins of 175-194 amino acid with a putative signal peptide, suggesting that they are secreted proteins. However, IFN transcripts encoding intracellular proteins without predicted signal peptides have also been described in catfish and rainbow trout (Long *et al.*, 2006; GenBank accession number, AJ580911).

It has been shown that fish IFN genes are widely expressed in many cell types, including T and B cells, macrophages and fibroblasts and are inducible in the immune response to viral infection or double stranded RNA. For example, in ZFL cells (derived from zebrafish liver) the IFN gene was transiently induced 6-12 h after treatment with polyI:C (Altmann *et al.*, 2003). In Atlantic salmon, a similar expression pattern was observed in TO cells (derived from head-kidney) and primary cultures of head-kidney leucocytes (Robertsen *et al.*, 2003). A glycoprotein of viral haemorrhagic septicaemia virus has been shown to increase IFN expression in transfected trout fibroblast cells (Acosta *et al.*, 2006). In catfish, CCO cells (derived from ovary), the IFN gene was up-regulated within 2 h after stimulation with UV-inactivated catfish reovirus (Long *et al.*, 2006). Constitutive expression was also detected in several catfish cell lines, including macrophages, T cells, B cells and fibroblasts.

ACTIVITIES

Antiviral activities of IFNs in fish have been detected indirectly using the culture medium of the cells stimulated with classical IFN inducers such as polyI:C, CpG, viral surface proteins or virus (Trobridge *et al.*, 1997; Jørgensen *et al.*, 2001b, 2003; Jensen and Robertsen, 2002; Acosta *et al.*, 2006; Saint-Jean and Perez-Prieto, 2006). With the availability of gene sequences, direct evidence of antiviral activity came from recent testing of the recombinant IFN proteins. Culture medium of human HEK293 cells transfected with the salmon IFN expression plasmid was able to induce Mx gene expression in CHSE-214 cells (Robertsen *et al.*, 2003). The Mx promoter was also shown to be up-regulated in a reporter gene system when co-transfected into a zebrafish embryo fibroblast cell line ZF4 with a plasmid expressing zebrafish IFN (Altmann *et al.*, 2003). The recombinant fish IFNs produced from COS cells or human HEK 293 cells were also shown to significantly increase resistance of the cells against viral infection (Long *et al.*, 2006).

Potential Applications

As it is only a few years since the first discovery of IFNs in fish, reports of applications of IFNs to enhance fish antiviral defence are scarce. Studies in mammals have used IFNs in clinic trials for many years, as a potential antiviral or anti-proliferating drug. Due to their involvement in antigen presentation, their use as an immunostimulant in order to boost vaccine efficacy is also possible. In addition, the discovery of the fish IFN genes has paved the way to develop potential systems for screening of antiviral drugs and selection of naturally viral resistant fish stocks.

Cytokines Involved in Adaptive Immunity

Due to the many discoveries made in the last few years, it is now possible to speculate that fish have different populations of T-cells that regulate adaptive immunity. This is due to the discovery of a number of key cytokines in several species of teleosts that are either important in the development of, or are secreted by T-helper-1 (Th-1), Th-2 or regulatory T (T_R) cells. However, not all the cytokines known in mammals have been found and it remains to be determined whether the regulation of adaptive immunity in fish is similar to that found in mammals, and if it is equally as complex.

In mammals, two types of $CD4^+$ Th cell populations exist, Th1 and Th2, characterized by their cytokine repertoire and how they regulate B-cell and T-cell responses (Mosmann and Coffman, 1989; Mosmann and Sad, 1996; Mosmann *et al.*, 2005). There is also a third population of T-cells, T_R cells, that are involved in the regulation of the Th responses via the secretion of cytokines, and help to inhibit harmful immunopathological responses directed against self or foreign antigens (Miller and Morahan, 1992; Maloy and Powrie, 2001; Maloy *et al.*, 2003).

The Th1 and Th2 cells have opposing roles; Th1 cells mediate delayed type hypersensitivity responses and provide protection against intracellular pathogens and viruses, whilst Th2 cells provide help to B-cells and eradicate helminthes and other extracellular parasites (Sher and Coffman, 1992; Mosmann *et al.*, 2005). Therefore, the Th response is important in inducing the most appropriate immune response towards a particular pathogen. Th1 and Th2 cells arise from a common precursor and differentiate according to the nature and dose of the antigen (Seder and Paul, 1994; Hosken *et al.*, 1995), co-stimulatory molecules expressed by the antigen presenting cells, and the cytokine environment in which the T-cell activation takes place.

The cytokines involved in the development of Th1 cells and that are expressed by this cell type and T_R cells have now been characterized in bony fish (Bird *et al.*, 2006) and strongly suggest that Th1 cells and specific cell-mediated immunity arose early in vertebrate evolution. Unlike Th1 cells many of the cytokines involved in the development of Th2 cells and that are released by this cell type have yet to be isolated and characterized in fish.

Cytokines that Drive T Cell Differentiation

Discovery

The early presence of IL-4 is the most potent stimulus for Th2 differentiation, whereas intèrleukin-12 (IL-12), IL-18, IL-23 and IL-27 favour Th1 development (O'Garra, 2000; Szabo *et al.*, 2003). To date, homologues for IL-4 have been found within chicken (Avery *et al.*, 2004) and *Xenopus* (Bird *et al.*, 2006) but not in fish and so the question remains as to whether a Th2 subpopulation of T-cells exists in this vertebrate group, whereas cytokines important for Th1 development have been clearly identified.

IL-12, IL-23 and IL-27 belong to the IL-12 family of cytokines each of which have distinct functions (Hunter, 2005). IL-12 exists as a 70-kDa heterodimer (p70) composed of two subunits, α (p35) and β (p40), linked by disulphide bonds that are essential for biological activity (Gubler *et al.*, 1991; Wolf *et al.*, 1991). IL-12 is produced by APCs within a few hours of infection, especially in response to bacteria and intracellular parasites and acts as a pro-inflammatory cytokine. IL-23 consists of a disulphide linked heterodimer of the IL-12 p40 subunit and a p19 protein, closely related to the IL-12 p35 subunit, that has similar but discrete functions from IL-12 (Oppmann *et al.*, 2000; Trinchieri *et al.*, 2003). The newest member is IL-27, which is also a disulphide linked heterodimer composed of two subunits, Epstein-Barr virus-induced gene 3 (EBI-3) and p28 (Pflanz *et al.*, 2002). IL-27 preferentially induces the proliferation of naive but not memory T-cells in combination with TCR cross linking. To date, in fish, the only member of this family to be fully characterized is IL-12. No p19 subunit has been found in non-mammalian vertebrates, and the recent discovery of EBI-3 in bony fish (Bird *et al.*, 2006) allows us only to speculate upon the presence of IL-27. IL-12 was first characterized within the Japanese pufferfish (Yoshiura *et al.*, 2003) and, more recently, in carp (Huising *et al.*, 2006) and seabass (Nascimento *et al.*, 2007b). Comparison of the human and *Fugu* genomes provided some evidence that the *Fugu* p35 and p40 subunits were indeed the homologues of the human subunits, as synteny was relatively well conserved (Yoshiura *et al.*, 2003). The p35 chain gene organization in sea bass and *Fugu* has seven exons and six introns and differs from mammals in containing an additional exon while lacking a second copy of a duplicated exon. Within the missing exon is encoded a cysteine (C^{74}), which in mammals is important in the disulphide bond between the p35 and p40 subunits to form the heterodimeric molecule (Yoon *et al.*, 2000). However, the most critical residues for the formation of the heterodimer in human p35, such as R^{183}, T^{186}, R^{189} and Y^{193}, are all conserved in the fish molecules. All other cysteines that are found within mammals, and are involved in disulphide bond formation, appear to be preserved in fish p35. The p40 chain gene organization in seabass and *Fugu* is the same as mammals having an eight-exon and seven-intron structure. Unlike the p35 molecule, it is clear that the cysteine (C^{177}) in mammals involved in the disulphide bond between the p35 and p40 is present in the fish p40. Protein modelling of carp IL-12 heterodimers has shown the formation of an inter-chain disulphide bridge is plausible (Huising *et al.*, 2006). A unique feature of teleost IL-12, is the discovery of multiple p40 chains (p40a, p40b and p40c) in pufferfish, zebrafish (Nascimento *et al.*, 2007 b) and carp (Huising *et al.*, 2006), although only a single chain has been found in seabass to date (Nascimento *et al.*, 2007 b). Three chains have been identified in carp (p40a, p40b and p40c), whereas only two have been found in other species (p40a and p40b). P40a and p40b are the most similar to each other and appear to have been the result of an early whole genome duplication in the teleost lineage after the divergence from the tetrapods (Jaillon *et al.*, 2004;

Woods *et al.*, 2005). This is evidenced by fact that p40a and p40b are found on different chromosomes but have similar genes surrounding them that share synteny with the p40 region in mammals (Nascimento *et al.*, 2007 b). The p40c does not appear to be as related to the mammalian p40 subunit and it remains to be determined what relationship this molecule has to the other fish p40 chains and its role, if any, in IL-12 function.

The expression of the p35 and the different p40 chains have been investigated in fish, with differences being found between fish species and what is seen in mammals. In mammals, IL-12 is produced in response to both viral and bacterial components. The p35 chain is expressed ubiquitously and constitutively at low levels, and is regulated both transcriptionally and translationally (Watford *et al.*, 2003). Despite the constitutive synthesis of p35 mRNA in unstimulated cells, little or no protein is secreted due to the presence of an inhibitory ATG in the 5′-UTR. However, upon stimulation with LPS the inhibitory region is excluded and the transcription start site changes to allow translation to occur (Babik *et al.*, 1999). The p40 gene is regulated at the level of transcription and is highly inducible by microbial products. In *Fugu*, the expression of the p35 subunit was limited in its tissue expression and was induced in the head-kidney and the spleen after injection with poly I:C, but not after injection with LPS (Yoshiura *et al.*, 2003). In seabass injected with UV-killed *Photobacterium damselae*, p35 was upregulated in the spleen but not head-kidney (Nascimento *et al.*, 2007 b), whereas in carp it was upregulated in head-kidney macrophages in response to LPS (Huising *et al.*, 2006). The expression profile of p40 also shows some inconsistencies. In *Fugu*, p40 showed constitutive expression in a wide variety of tissues, with no increase in response to poly I:C or LPS injection (Yoshiura *et al.*, 2003). In carp, p40a, p40b and p40c showed constitutive expression in a wide variety of tissues, but increased in mRNA expression in head-kidney macrophages after stimulation with LPS (Huising *et al.*, 2006). p40 was also upregulated in the spleen and head-kidney of seabass injected with UV-killed *P. damselae* (Nascimento *et al.*, 2007 b). Differences in expression may be accounted for by the different approaches used, which involved *in vivo* or *in vitro* experiments and the use of either a mixed cell or purified cell populations. Analyses of the promoters of the *Fugu* and sea bass p40 and p35 genes reveal the presence of binding sites for regulatory elements known to be important for controlling IL-12 expression in mammals (Nascimento *et al.*, 2007 b). This finding does not explain the unresponsiveness of *Fugu* p35 to LPS and so the difference in expression of the p35 and p40 chains and the presence of multiple p40 chains in teleosts requires further investigation.

Interleukin-18 (IL-18) is a cofactor for IL-12 induced Th1 cell development, as well as enhancing IFN-γ production from Th1 effector cells (Xu *et al.*, 1998). IL-18 has been characterized in trout and *Fugu* (Zou *et al.*, 2004a). Unlike teleost IL-1β (Bird *et al.*, 2002b), the fish IL-18 molecules contain a putative ICE cleavage site at Asp^{32} in trout and Asp^{31} in *Fugu*. The presence of this cut site within teleost IL-18 is of great interest since it still remains unknown how non-mammalian IL-1 molecules are processed. The gene organization of trout and *Fugu* IL-18 follows the mammalian organization very closely, with the exon sizes being very conserved (Zou *et al.*, 2004a). As found within mammals, the trout IL-18 gene is constitutively expressed in a wide range of tissues including brain, gill, gut, heart, kidney, liver, muscle, skin and spleen. Transcription is not regulated by LPS, polyI:C or trout recombinant IL-1β in primary head-kidney cells or a macrophage cell line (Zou *et al.*, 2004a). Indeed its expression is downregulated in a fibroblast cell line in response to LPS and rIL-1β, whereas an alternatively spliced form of the trout IL-18 molecule has been shown to be produced and its expression is upregulated in fibroblasts by LPS. It has been speculated that the balance of the two IL-18 transcripts may be an important mechanism in controlling IL-18 expression or processing (Zou *et al.*, 2004a).

Cytokines Released from T Cells: T Helper-1 (Th1) Cell Cytokines

Discovery

Upon activation, Th1 cells secrete IL-2, IFN-γ and TNF-β (also known as LT-α), which can activate antimicrobial defences as well as cytokine production in macrophages (Abbas *et al.*, 1996; Romagnani, 1997; O'Garra, 1998).

IL-2 is a central cytokine in the regulation of T-cell responses (Smith, 1980; Swain, 1991). It controls the amplification of naive T cells by initially stimulating growth following antigen activation and later promotes activation-induced cell death (Smith, 1988; Waldmann *et al.*, 2001). The discovery of IL-2 in teleosts was made by exploiting the conservation of gene order (conservation of synteny) between the human and the *Fugu* genome (Bird *et al.*, 2005a). The *Fugu* sequence contains the

IL-2 family signature along with a pair of cysteines that in mammals are important in the formation of an intramolecular disulphide bond. Interestingly, the *Fugu* molecules contain an additional pair of cysteines, which is characteristic of IL-15, and suggests that four cysteines may have been present in the primordial gene before gene duplication and that an ancestral IL-2/IL-15-like gene duplicated before bony fish diverged from other vertebrate groups. No constitutive expression of IL-2 is seen in *Fugu* tissues, in agreement with mammalian studies but stimulation of kidney cells with PHA *in vitro* or exposure to polyI:C or LPS *in vivo* can upregulate IL-2 expression, with many well known transcription factor binding sites clearly present in the 5′ flanking region of this gene (Fig. 1.2), including multiple sites for T-bet that is required for Th1 cell differentiation in mammals (Szabo *et al.*, 2002). The genomic organization of the *Fugu* IL-2 gene has been well conserved through evolution, and has a four exon, three intron structure identical to that in mammals (Bird *et al.*, 2005a). IL-21 was also discovered with IL-2 in *Fugu* where, as in mammals, it is immediately downstream (Bird *et al.*, 2005a). Although not seen as a classical Th1 cytokine, IL-21 has been implicated in the activation of innate immune responses, in Th1 cell responses and in the regulation of immunoglobulin production by B-cells (Strengeil *et al.*, 2002).

IFN-γ is a pleiotropic cytokine, having roles in both the innate and adaptive phases of the immune response. An essential role of IFN-γ is in activating macrophages, leading to an increase in phagocytosis and MHC class I and II expression, and in inducing IL-12, superoxide production and nitric oxide (Boehm *et al.*, 1997). IFN-γ sequences are now available within teleosts such as *Fugu* (Zou *et al.*, 2004b), trout (Zou *et al.*, 2005a), zebrafish (Igawa *et al.*, 2006) and catfish (Milev-Milovanovic *et al.*, 2006). Initially the *Fugu* IFN-γ was discovered by exploiting the synteny that is found between the *Fugu* and the human genomes (Zou *et al.*, 2004b) and was found to have the equivalent genomic structure to other known mammalian and avian IFN-γ molecules (Kaiser *et al.*, 1998). Based on the sequence information of the *Fugu* homologue, the rainbow trout molecule was characterized along with its biological activities (Zou *et al.*, 2005a). The molecule contains the IFN-γ signature motif and is highly expressed in head-kidney leucocytes stimulated with PHA or polyI:C and in the kidney and spleen of fish injected with polyI:C, with transcription factor binding sites present in the IFN-γ promoter consistent with these findings (Fig. 1.2). A nuclear localization sequence motif is conserved within the trout and *Fugu* IFN-γ C terminal region, which has been shown to be

IL-2

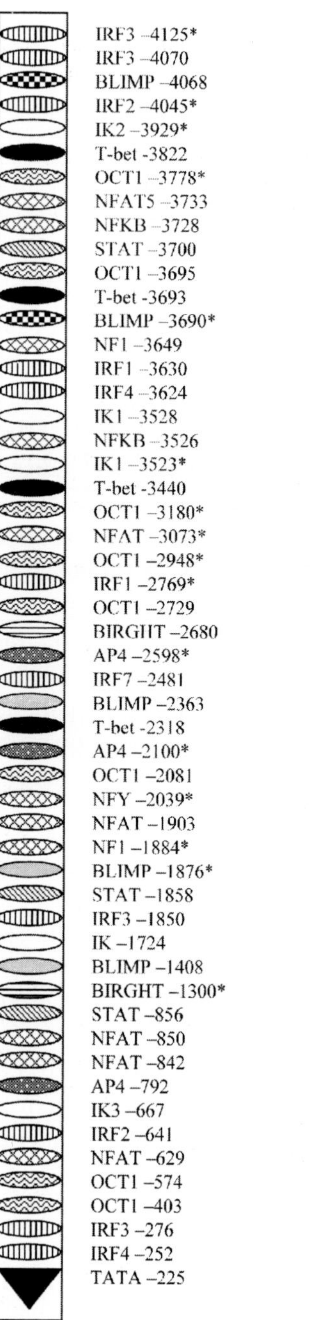

IFN-γ

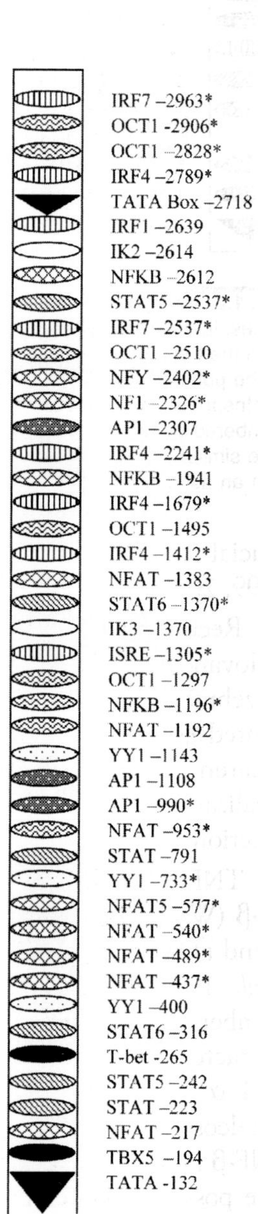

Fig. 1.2 Contd.

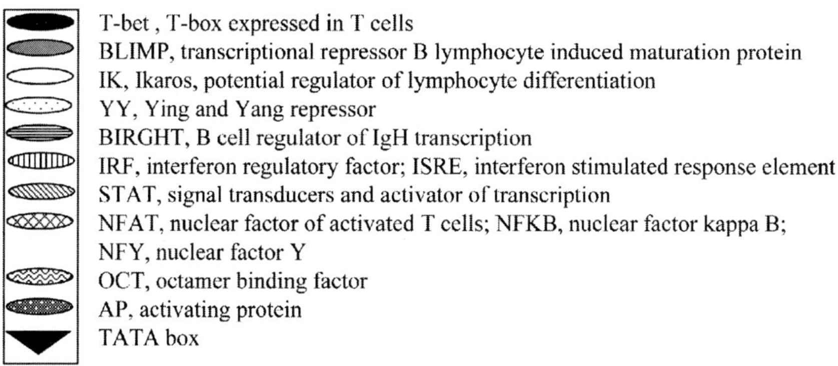

Fig. 1.2 Analysis of key regulatory elements in the promoter region of Fugu Th1 cytokine genes, IL-2 and IFN-γ. The 5′ flanking sequences of the IL-2 and IFN-γ gene were retrieved from the Fugu (*Takifugu rubripes*) genome database (http://www.ensembl.org). Prediction of the putative binding sites for transcription factors was performed using the Genomatix MatInspector programme (http://www.genomatix.de). The translation start codon is numbered as position 0. Selected binding sites for key regulatory transcription factors with core similarity over 90% are presented and those on the complementary strand indicated with an asterisk.

crucial for IFN-γ biological activities in mammals (Subramaniam *et al.*, 1999, 2000).

Recently, in the catfish and zebrafish (Igawa *et al.*, 2006; Milev-Milovanovic *et al.*, 2006), an IFN-γ related gene has been discovered that in zebrafish is closely linked on the same chromosome as IFN-γ. The related gene has a similar gene organization to IFN-γ but shows very different tissue expression patterns as well as lacking the nuclear localization site. More investigation is required to determine the exact function of this gene within the teleost immune response.

TNF-β is a member of the TNF superfamily, along with TNF-α and LT-β (Ware *et al.*, 1998). In both human and mouse, the three genes are found at a single locus, arranged in tandem (Spies *et al.*, 1986; Lawton *et al.*, 1995). As discussed above, TNF-α genes have been cloned in a number of fish species, and more recently LT-β genes have been characterized (Savan *et al.*, 2005; Kono *et al.*, 2006). This suggests that the TNF-α and LT-β genes were present in this locus in an organism ancestral to teleosts. Another gene duplication appears to have taken place to give TNF-β and the classical TNF locus found in amphibians and mammals. The possible absence of TNF-β in fish has a potential impact on LT-β signalling, where in mammals a heterotrimer of TNF-β and LT-β are required for binding to the LT-β receptor, unlike the situation with TNF-α and TNF-β where homotrimers are used (Fig. 1.3).

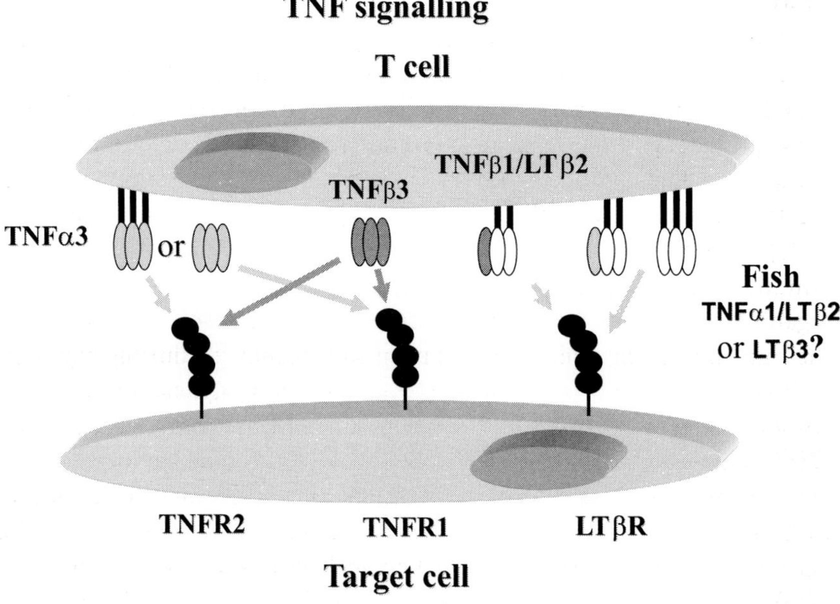

Fig. 1.3 Diagram illustrating the receptor binding of TNF family ligands. Homotrimers of TNF-α, on the cell surface or secreted, can bind to the two TNF receptors. Similarly, homotrimers of TNF-β (LT-α) can bind to the same two receptors. In contrast, heterotrimers of TNF-β and LT-β are required to bind to the LT-β receptor in mammals. This latter ligand-receptor interaction may not be possible in fish, where to date no TNF-β gene has been discovered.

Cytokines Released from T Cells: T Helper-2 (Th2) Cell Cytokines

Discovery

Th2 cells in mammals are associated with humoral immune responses (specific immunity mediated by antibodies), in allergic responses and helminth infections (Sher and Coffman, 1992; Urban *et al.*, 1992; Abbas *et al.*, 1996). Th2 cells produce the cytokines IL-4, IL-5, IL-6, IL-9, IL-10 and IL-13, with IL-4 responsible for the production of immunoglobulin (Ig) E, IL-5 for eosinophilia and the combination of IL-3, IL-4 and IL-10 for mast cell production (Thompsonsnipes *et al.*, 1991). The cytokines

IL-3, IL-4, IL-5, IL-6 and IL-9 are also responsible for antibody isotype switching in B-cells and the production of IgM and noncomplement-activating IgG isotypes as well as IgE (Cocks *et al.*, 1993; Dugas *et al.*, 1993; Petitfrere *et al.*, 1993).

IL-6 is a pleiotropic cytokine that plays a central role in host defence. Its many biological functions include stimulation of Ig synthesis, stimulation of T-cell growth and differentiation and regulation of acute phase protein synthesis from hepatocytes (Kishimoto, 2003). Unlike most of the other Th2 cytokines, IL-6 has been characterized within teleosts, in the Japanese pufferfish (Bird *et al.*, 2005b) and in flounder (Nam *et al.*, 2007). The IL-6 sequence in fish was initially determined by exploiting the synteny found between the human and *Fugu* genomes. The gene organization of *Fugu* and flounder IL-6 and the level of synteny between the human and *Fugu* genomes has been well conserved during evolution with the order and orientation of the genes matching exactly to human chromosome 7. The *Fugu* and flounder IL-6 molecules contain the IL-6/G-CSF/MGF motif, but only contain two of the four cysteines found in the mammalian molecules, important in disulphide bond formation. As found with other cytokine homologues in fish, the amino acid identities of the *Fugu* and flounder IL-6 with other known IL-6 molecules were low (20-29%), although phylogenetic analysis grouped the *Fugu* IL-6 clearly with other IL-6 molecules. The expression of IL-6 within *Fugu* indicated that this molecule was biologically relevant to the bony fish immune response. PHA stimulation of *Fugu* kidney cells *in vitro* resulted in a large increase in the *Fugu* IL-6 transcript whilst *in vivo* LPS (*Fugu*), polyI:C (*Fugu*) and *Edwardsiella tarda* (flounder) injection resulted in a significant increase, especially within spleen cells (Bird *et al.*, 2005b; Nam *et al.*, in press). The apparent absence of other Th2 cytokines in fish and the discovery of only an IL-4 homologue in amphibians (Bird *et al.*, 2006) suggests that the classic Th2 responses seen in mammals may not have evolved prior to the fish tetrapod divergence.

Cytokines Released from T Cells: T Regulatory (T_R) Cell Cytokines

Discovery

T_R cells have been shown to inhibit the activation of other T-cells directly via a cell contact dependent mechanism, or indirectly by down regulating the activities of antigen presenting cells (Maloy and Powrie, 2001;

Maloy *et al.*, 2003). Recent work has shown that T_R cells also inhibit cells of the innate immune system (Maloy *et al.*, 2003). Strong evidence has also been provided for a role of the cytokines IL-10 and TGF-β in the effector function of T_R cells (Read *et al.*, 2000), and both have been characterized in teleosts.

IL-10 potently inhibits production of IL-1α, IL-1β, IL-6, IL-10, IL-12, IL-18, GM-CSF, G-CSF, M-CSF, TNF, LIF, PAF and CC and CXC chemokines by activated monocytes/macrophages (Moore *et al.*, 2001), which are important in activating and sustaining immune and inflammatory responses. IL-10 has been characterized within a number of teleosts, such as *Fugu* (Zou *et al.*, 2003c), *Tetraodon* (Lutfalla *et al.*, 2003), carp (Savan *et al.*, 2003b) and rainbow trout (Inoue *et al.*, 2005). All of these IL-10 peptides contained the IL-10 family signature motif as well as four cysteine residues, found to be important in disulphide bond formation within human IL-10. Similar to the gene structure of human and mouse, the *Fugu*, carp, and trout IL-10 gene contains five exons and four introns, with similar numbers of amino acids encoded by the respective exons across species. In normal fish tissues, the IL-10 expression patterns were quite different for each fish, with IL-10 expressed weakly in trout gill tissue, weakly expressed in *Fugu* kidney and liver and strongly expressed in carp head-kidney and intestine. Carp head-kidney and liver cells show an increase in expression at 1 h post-stimulation, when stimulated with LPS *in vitro* (Savan *et al.*, 2003b). The *Fugu* IL-10 promoter has been characterized but has little identity with the human IL-10 promoter although it contains multiple potential Sp1 binding sites, which play a prominent role in the control of human IL-10 expression (Zou *et al.*, 2003c).

TGF-β belongs to a large group of multifunctional cytokines, called the TGF-β family. In mammals, it consists of three isoforms, TGF-β1, -β2 and -β3 (Graycar *et al.*, 1989). Functional differences have been found using TGF-β knockout mice (Bottinger *et al.*, 1997), and it is known that TGF-β2 and TGF-β3 are important regulators of cellular differentiation and also affect development and embryogenesis, whereas the effects of TGF-β1 are mainly immunological and it is best known as a negative regulator of the immune system (Roberts and Sporn, 1990). TGF-β has been well characterized in a wide variety of lower vertebrates, with all three isoforms of TGF-β found in teleosts. Homologues of TGF-β1 exist in trout (Hardie *et al.*, 1998), carp (Yin and Kwang, 2000b), striped bass

(Harms *et al.*, 2000), plaice (Laing *et al.*, 2000) and sea bream (Tafalla *et al.*, 2003). The trout and seabream gene organization have been characterized. In trout, although the number of exons (seven) is the same as that in human, the introns are in different places, with intron 2 in humans being absent and the trout having an intron within exon 7 of the human gene. In seabream, only five exons are present, with the absence of introns 1, 2 and 4 of the human gene but again the presence of an intron within the human exon 7. Expression analysis of TGF-β1 has also been investigated. In trout and seabream, constitutive expression was detected in blood leucocytes, kidney macrophages, brain, gill, and spleen tissue (Hardie *et al.*, 1998; Tafalla *et al.*, 2003). In seabream, expression was also seen in the liver, muscle, kidney and heart, whereas none could be seen in the trout liver. In carp, TGF-β1 is expressed at low levels in head-kidney, spleen, egg and liver, with its expression increased in head-kidney cells after Con A stimulation (Yin and Kwang, 2000b). Preliminary investigations indicate fish TGF-β1 may have a similar role to that of mammals, where the addition of mammalian TGF-β1 significantly inhibits macrophage respiratory burst activity and the production of macrophage-activating factors by head-kidney leucocytes (Jang *et al.*, 1994, 1995).

ACTIVITIES

To date, most of the cytokines involved in adaptive immunity are yet to be produced as recombinant proteins for bioactivity testing. The exception is IFN-γ, where the effects of the trout recombinant protein produced in *E. coli* correlated well with the mammalian system (Schroder *et al.*, 2004). Trout IFN-γ has been shown to significantly stimulate expression of a trout CXC chemokine found to have homology to CXCL10 (a IFN-γ inducible protein), the MHC class II β-chain and the transcription factor STAT1, as well as enhancing the respiratory burst of trout macrophages. In addition, it has been found to stimulate expression of a newly characterized trout guanylate-binding protein (GBP) (Robertsen *et al.*, 2006). In mammals, GBPs are some of the most abundant proteins accumulating in mammalian cells in response to IFN-γ stimulation. The deletion of the nuclear localization sequence motif within the trout recombinant IFN-γ resulted in a loss of activity with respect to the induction of the CXCL10-like molecule in RTS-11 cells (Zou *et al.*, 2005a). In addition, the activation of protein kinase C (PKC), involved in mediating mammalian IFN-γ signal transduction (Deb *et al.*, 2003), has been shown to be essential for trout IFN-γ functions. IFN-γ induced CXCL10 expression was completely abolished by the PKC inhibitor staurosporine, and partially reduced by U0126, a specific inhibitor for extracellular signal regulated kinases (Zou *et al.*, 2005a).

Potential Applications

It is clear that a complex network exists to regulate the adaptive immune responses of teleost fish from the number of cytokine genes that have been isolated. Once the functional activity of these cytokines has been confirmed, it will be clearer whether classical Th1 responses are present in teleosts, so as to regulate specific cell-mediated immunity. These cytokines may prove useful in directing the immune response towards a Th1 response if added to a vaccine. However, if a classical Th2 response is not found in fish (as discussed above), then the manner in which humoral responses are regulated may need to be re-evaluated. Either way, discovery of the types of Th cells present in fish and the cytokines they produce will help in understanding the balance of such immune responses and aid the effective design of therapeutic strategies to manipulate the immune system towards humoral or cellular immunity in response to specific antigen stimulation.

References

Abbas, A.K., K.M. Murphy and A. Sher. 1996. Functional diversity of helper T lymphocytes. *Nature (London)* 383: 787-793.

Acosta, F., B. Collet, N. Lorenzen and A.E. Ellis. 2006. Expression of the glycoprotein of viral haemorrhagic septicaemia virus (VHSV) on the surface of the fish cell line RTG-P1 induces type 1 interferon expression in neighbouring cells. *Fish and Shellfish Immunology* 21: 272-278.

Altmann, S.M., M.T. Mellon, D.L. Distel and C.H. Kim. 2003. Molecular and functional analysis of an interferon gene from the zebrafish, *Danio rerio*. *Journal of Virology* 77: 1992-2002.

Avery, S., L. Rothwell, W.D.J. Degen, V. Schijns, J. Young, J. Kaufman and P. Kaiser. 2004. Characterization of the first Nonmammalian T2 cytokine gene cluster: The cluster contains functional single-copy genes for IL-3, IL-4, IL-13, and GM-CSF, a gene for IL-5 that appears to be a pseudogene, and a gene encoding another cytokinelike transcript, KK34. *Journal of Interferon Cytokine Research* 24: 600-610.

Babik, J.M., E. Adams, Y. Tone, P.J. Fairchild, M. Tone and H. Waldmann. 1999. Expression of murine IL-12 is regulated by translational control of the p35 subunit. *Journal of Immunology* 162: 4069-4078.

Bao, B.L., E. Peatman, X. Peng, P. Baoprasertkul, G.L. Wang and Z.J. Liu. 2006. Characterisation of 23 CC chemokine genes and analysis of their expression in channel catfish (*Ictalurus punctatus*). *Developmental and Comparative Immunology* 30: 783-796.

Baoprasertkul, P., E. Peatman, L. Chen, C. He, H. Kucuktas, P. Li, M. Simmons and Z. Liu. 2004. Sequence analysis and expression of a CXC chemokine in resistant and susceptible catfish after infection of *Edwardsiella ictaluri*. *Developmental and Comparative Immunology* 28: 769-780.

Baoprasertkul, P., C. He, E. Peatman, S. Zhang, P. Li and Z. Liu. 2005. Constitutive expression of three novel catfish CXC chemokines: homeostatic chemokines in teleost fish. *Molecular Immunology* 42: 1355-1366.

Bird, S. 2001. *Molecular evolution of interleukin-1 beta within vertebrates*. Ph.D. Thesis. University of Aberdeen, UK.

Bird, S., T. Wang, J. Zou, C. Cunningham and C.J. Secombes. 2002a. The first cytokine sequence within cartilaginous fish: IL-1β in the small spotted catshark (*Scyliorhinus canicula*). *Journal of Immunology* 168: 3329-3340.

Bird, S., J. Zou, T.H. Wang, B. Munday, C. Cunningham and C.J. Secombes. 2002b. Evolution of interleukin-1 beta. *Cytokine Growth Factor Reviews* 13: 483-502.

Bird, S., J. Zou, T. Kono, M. Sakai, J.M. Dijkstra and C.J. Secombes. 2005a. Characterisation and expression analysis of interleukin 2 (IL-2) and IL-21 homologues in the Japanese pufferfish, *Fugu rubripes*, following their discovery by synteny. *Immunogenetics* 56: 909-923.

Bird, S., J. Zou, R. Savan, T. Kono, M. Sakai, J. Woo and C.J. Secombes. 2005b. Characterisation and expression analysis of an interleukin 6 homologue in the Japanese pufferfish, *Fugu rubripes*. *Developmental and Comparative Immunology* 29: 775-789.

Bird, S., J. Zou and C.J. Secombes. 2006. Advances in fish cytokine biology give clues to the evolution of a complex network. *Current Pharmaceutical Design* 12: 3051-3069.

Bobe, J. and F.W. Goetz. 2001. Molecular cloning of a TNF receptor and two TNF ligands in the fish ovary. *Comparative Biochemistry and Physiology* B 129: 475-481.

Bobe, J., J. Montfort, T. Nguyen and A. Fostier. 2006. Identification of new participants in the rainbow trout (*Oncorhynchus mykiss*) oocyte maturation and ovulation processes using cDNA microarrays. *Reproductive Biology and Endocrinology* 4: 39.

Boehm, U., T. Klamp, M. Groot and J.C. Howard. 1997. Cellular responses to interferon-gamma. *Annual Review of Immunology* 15: 749-795.

Bottinger, E.P., J.J. Letterio and A.B. Roberts. 1997. Biology of TGF-beta in knockout and transgenic mouse models. *Kidney International* 51: 1355-1360.

Bridle, A.R., P.B.B. Crosbie, R.N. Morrison, J. Kwang and B.F. Nowak. 2002. The immuno-adjuvant effect of carp interleukin-1 beta on the humoral immune response of barramundi, *Lates calcarifer* (Bloch). *Journal of Fish Diseases* 25: 429-432.

Buonocore, F., D. Prugnoli, C. Falasca, C.J. Secombes and G. Scapigliati. 2003a. Peculiar gene organisation and incomplete splicing of sea bass (*Dicentrarchus labrax* L.) interleukin-1β. *Cytokine* 21: 257-264.

Buonocore, F., M. Mazzini, M. Forlenza, E. Randelli, C.J. Secombes, J. Zou and G. Scapigliati. 2003b. Expression in *Escherichia coli* and purification of sea bass (*Dicentrarchus labrax*) interleukin 1β, a possible immunoadjuvant in aquaculture. *Marine Biotechnology* 6: 53-59.

Buonocore, F., M. Forlenza, E. Randelli, S. Benedetti, P. Bossù, S. Meloni, C.J. Secombes, M. Mazzini and G. Scapigliati. 2005. Biological activity of sea bass (*Dicentrarchus labrax* L.) recombinant interleukin-1β. *Marine Biotechnology* 7: 609-617.

Buonocore, F., E. Randelli, F. Paderi, S. Bird, C.J. Secombes, M. Mazzini and G. Scapigliati. 2006. The cytokine IL-1β from the crocodile icefish *Chionodraco hanatus* (Perciformes : Channichthyidae). *Polar Biology* DOI 10.1007/s00300-006-0145-2.

Chen, L., C. He, P. Baoprasertkul, P. Xu, P. Li, J. Serapion, G. Waldbieser, W. Wolters and Z. Liu. 2005. Analysis of a catfish gene resembling interleukin-8: cDNA cloning, gene structure, and expression after infection with *Edwardsiella ictaluri*. *Developmental and Comparative Immunology* 29: 135-142.

Cocks, B.G., R.D. Malefyt, J.P. Galizzi, J.E. Devries and G. Aversa. 1993. IL-13 induces proliferation and differentiation of human B-cells activated by the CD40-ligand. *International Immunology* 5: 657-663.

Corripio-Miyar, Y., S. Bird, K. Tsamopoulos and C.J. Secombes. 2007. Cloning and expression analysis of two pro-inflammatory cytokines, IL-1β and IL-8, in haddock (*Melanogrammus aeglefinus*). *Molecular Immunology* 44: 1361-1373.

David, N.B., D. Sapede, L. Saint-Etienne, C. Thisse, B. Thisse, C. Dambly-Chaudiere, F.M. Rosa and A. Ghysen. 2002. Molecular basis of cell migration in the fish lateral line: role of the chemokine receptor CXCR4 and of its ligand, SDF1. *Proceedings of the National Academy of Sciences of the United States of America* 99: 16297-16302.

Deb, D.K., A. Sassano, F. Lekmine, B. Majchrzak, A. Verma, S. Kambhampati, S. Uddin, A. Rahman, E.N. Fish and L.C. Platanias. 2003. Activation of protein kinase C delta by IFN-gamma. *Journal of Immunology* 171: 267-273.

DeVries, M.E., A.A. Kelvin, L. Xu, L. Ran, J. Robinson and D.J. Kelvin. 2006. Defining the origins and evolution of the chemokine/chemokine receptor system. *Journal of Immunology* 176: 401-415.

Dixon, B., B. Shum, E.J. Adams, K.E. Magor, R.P. Hedrick, D.G. Muir and P. Parham. 1998. CK-1, a putative chemokine of rainbow trout (*Oncorhynchus mykiss*). *Immunological Reviews* 166: 341-348.

Doitsidou, M., M. Reichman-Fried, J. Stebler, M. Koprunner, J. Dorries, D. Meyer, C.V. Esguerra, T. Leung and E. Raz. 2002. Guidance of primordial germ cell migration by the chemokine SDF-1. *Cell* 111: 647-659.

Dufourcq, P. and S. Vriz. 2006. The chemokine SDF-1 regulates blastema formation during zebrafish fin regeneration. *Developmental Genes and Evolution* 216: 635-639.

Dugas, B., J.C. Renauld, J. Pene, J.Y. Bonnefoy, C. Petifrere, P. Braquet, J. Bousquet, J. Vansnick and J.M. Menciahuerta. 1993. Interleukin-9 potentiates the interleukin-4-induced immunoglobulin (IgG, IgM and IgE) production by normal human B-lymphocytes. *European Journal of Immunology* 23: 1687-1692.

Emmadi, D., A. Iwahori, I. Hirono and T. Aoki. 2005. cDNA microarray analysis of interleukin-1β-induced Japanese flounder *Paralichthys olivaceus* kidney cells. *Fisheries Science* 71: 519-530.

Engelsma, M.Y., R.J.M. Stet, H. Schipper and B.M.L. Verburg van Kemenade. 2001. Regulation of interleukin 1 beta RNA expression in the common carp, *Cyprinus carpio*. *Developmental and Comparative Immunology* 25: 195-203.

Engelsma, M.Y., R.J.M. Stet, J.P. Saeij and B.M.L. Verburg van Kemenade. 2003. Differential expression and haplotypic variation of two interleukin-1 beta genes in the common carp (*Cyprinus carpio* L.). *Cytokine* 22: 21-32.

Fujiki, K., D.H. Shin, M. Nakao and T. Yano. 2000. Molecular cloning and expression analysis of carp (*Cyprinus carpio*) interleukin-1 beta, high affinity immunoglobulin E Fc receptor gamma subunit and serum amyloid A. *Fish and Shellfish Immunology* 10: 229-242.

Garcia-Castillo, J., P. Pelegrin, V. Mulero and J. Meseguer. 2002. Molecular cloning and expression analysis of tumor necrosis factor alpha from a marine fish reveal its constitutive expression and ubiquitous nature. *Immunogenetics* 54: 200-207.

Garcia-Castillo, J., E. Chaves-Pozo, P. Olivares, P. Pelegrin, J. Meseguer and V. Mulero. 2004. The tumor necrosis factor a of the bony fish sea bream exhibits the *in vivo* pro-inflammatory and proliferative activities of its mammalian counterparts, yet it functions in a species-specific manner. *Cellular and Molecular Life Sciences* 61: 1331-1340.

Graycar, J.L., D.A. Miller, B.A. Arrick, R.M. Lyons, H.L. Moses and R. Derynck. 1989. Human transforming growth factor-beta-3-recombinant expression, purification, and biological-activities in comparison with transforming growth factor-beta-1 and factor-beta-2. *Molecular Endocrinology* 3: 1977-1986.

Gubler, U., A.O. Chua, D.S. Schoenhaut, C.M. Dwyer, W. McComas, R. Motyka, N. Nabavi, A.G. Wolitzky, P.M. Quinn, P.C. Familletti and M.K. Gately. 1991. Coexpression of 2 distinct genes is required to generate secreted bioactive cytotoxic lymphocyte maturation factor. *Proceedings of the National Academy of Sciences of the United States of America* 88: 4143-4147.

Hardie, L.J., K.J. Laing, G.D. Daniels, P.S. Grabowski, C. Cunningham and C.J. Secombes. 1998. Isolation of the first piscine transforming growth factor beta gene: Analysis reveals tissue specific expression and a potential regulatory sequence in rainbow trout (*Oncorhynchus mykiss*). *Cytokine* 10: 555-563.

Harms, C.A., S. Kennedy-Stoskopf, W.A. Horne, F.J. Fuller and W.A.F. Tompkins. 2000. Cloning and sequencing hybrid striped bass (*Morone saxatilis* × *M. chrysops*) transforming growth factor-beta (TGF-beta), and development of a reverse transcription quantitative competitive polymerase chain reaction (RT-qcPCR) assay to measure TGF-beta mRNA of teleost fish. *Fish and Shellfish Immunology* 10: 61-85.

Harun, N.O. 2006. The role of interleukin-8 in inflammatory responses of fish: Studies in rainbow trout (*Oncorhynchus mykiss*). M.Sc. Thesis. University of Aberdeen, UK.

Hirono, I., B.H. Nam, T. Kurobe and T. Aoki. 2000. Molecular cloning, characterization and expression of TNF cDNA and gene from Japanese flounder *Paralichthys olivaceus*. *Journal of Immunology* 165: 4423-4427.

Holland, J.W., T.G. Pottinger and C.J. Secombes. 2002. Recombinant interleukin-1β activates the hypothalamic-pituitary-interrenal axis in rainbow trout, *Oncorhynchus mykiss*. *Journal of Endocrinology* 175: 261-267.

Hong, S., J. Zou, M. Crampe, S. Peddie, G. Scapigliati, N. Bols, C. Cunningham and C.J. Secombes. 2001. The production and bioactivity of rainbow trout (*Oncorhynchus mykiss*) recombinant IL-1β. *Veterinary Immunology and Immunopathology* 81: 1-14.

Hong, S., S. Peddie, J.J. Campos-Perez, J. Zou and C.J. Secombes. 2003. The effect of intraperitoneally administered recombinant IL-1β on immune parameters and resistance to *Aeromonas salmonicida* in the rainbow trout (*Oncorhynchus mykiss*). *Developmental and Comparative Immunology*. 27: 801-812.

Hong, S., J. Zou, B. Collet, N.C. Bols and C.J. Secombes. 2004. Analysis and characterisation of IL-1β processing in trout, *Oncorhynchus mykiss*. *Fish and Shellfish Immunology* 16: 453-459.

Hosken, N.A., K. Shibuya, A.W. Heath, K.M. Murphy and A. Ogarra. 1995. The effect of antigen dose on CD4(+) T-helper cell phenotype development in a T-cell receptor-alpha-beta-transgenic model. *Journal of Experimental Medicine* 182: 1579-1584.

Huising, M.O., E.H. Stolte, G. Flik, H.F.J. Savelkoul and B.M.L. Verburg-van Kemenade. 2003. CXC chemokines and leukocyte chemotaxis in common carp. *Developmental and Comparative Immunology* 27: 875-888.

Huising, H.O., T. van der Meulen, G. Flik and B.M.L. Verburg-van Kemenade. 2004. Three novel carp chemokines are expressed early in ontogeny and at non-immune sites. *European Journal of Biochemistry* 271: 4094-4106.

Huising, M.O., J.E. van Schijndel, C.P. Kruiswijk, S.B. Nabuurs, H.F.J. Savelkoul, G. Flik and B.M.L. Verburg-van Kemenade. 2006. The presence of multiple and differentially regulated interleukin-12p40 genes in bony fishes signifies an expansion of the vertebrate heterodimeric cytokine family. *Molecular Immunology* 43: 1519-1533.

Hunter, C.A. 2005. New IL-12-family members: IL-23 and IL-27, cytokines with divergent functions. *Nature Reviews Immunology* 5: 521-531.

Igawa, D., M. Sakai and R. Savan. 2006. An unexpected discovery of two interferon gamma-like genes along with interleukin (IL)-22 and -26 from teleost: IL-22 and -26 genes have been described for the first time outside mammals. *Molecular Immunology* 43: 999-1009.

Ingerslev, H.C., C. Cunningham and H.I. Wergeland. 2006. Cloning and expression of TNF-alpha, IL-1 beta and COX-2 in an anadromous and landlocked strain of Atlantic salmon (*Salmo salar* L.) during the smelting period. *Fish and Shellfish Immunology* 20: 450-461.

Inoue, Y., C. Haruta, T. Moritomo and T. Nakanishi. 2003a. Molecular cloning and expression of shark (*Triakis scyllium*) interleukin-1 cDNA. GenBank accession No. AB074142.

Inoue, Y., C. Haruta, K. Usui, T. Moritono and T. Nakanishi. 2003b. Molecular cloning and sequencing of the banded dogfish (*Triakis scyllia*) interleukin-8. *Fish and Shellfish Immunology* 14: 275-281.

Inoue, Y., M. Endo, C. Haruta, T. Taniuchi, T. Moritomo and T. Nakanishi. 2003c. Molecular cloning and sequencing of the silver chimaera (*Chimaera phantasma*) interleukin-8 cDNA. *Fish and Shellfish Immunology* 15: 269-274.

Inoue, Y., S. Kamota, K. Itoa, Y. Yoshiura, M. Ototake, T. Moritomo and T. Nakanishi. 2005. Molecular cloning and expression analysis of rainbow trout (*Oncorhynchus mykiss*) interleukin-10 cDNAs. *Fish and Shellfish Immunology* 18: 335-344.

Jaillon, O., A.M. Aury, F. Brunet, J.L. Petit, N. Stange-Thomann, E. Mauceli, L. Bouneau, C. Fischer, C. Ozouf-Costaz, A. Bernot, S. Nicaud, D. Jaffe, S. Fisher, G. Lutfalla, C. Dossat, B. Segurens, C. Dasilva, M. Salanoubat, M. Levy, N. Boudet, S. Castellano,

R. Anthouard, C. Jubin, V. Castelli, M. Katinka, B. Vacherie, C. Biemont, Z. Skalli, L. Cattolico, J. Poulain, V. de Berardinis, C. Cruaud, S. Duprat, P. Brottier, J.P. Coutanceau, J. Gouzy, G. Parra, G. Lardier, C. Chapple, K.J. McKernan, P. McEwan, S. Bosak, M. Kellis, J.N. Volff, R. Guigo, M.C. Zody, J. Mesirov, K. Lindblad-Toh, B. Birren, C. Nusbaum, D. Kahn, M. Robinson-Rechavi, V. Laudet, V. Schachter, F. Quetier, W. Saurin, C. Scarpelli, P. Wincker, E.S. Lander, J. Weissenbach and H.R. Crollius. 2004. Genome duplication in the teleost fish *Tetraodon nigroviridis* reveals the early vertebrate proto-karyotype. *Nature (London)* 431: 946-957.

Jang, S.I., L.J. Hardie and C.J. Secombes. 1994. Effects of transforming growth-factor beta1 on rainbow-trout *Oncorhynchus mykiss* macrophage respiratory burst activity. *Developmental and Comparative Immunology* 18: 315-323.

Jang, S.I., L.J. Hardie and C.J. Secombes. 1995. Elevation of rainbow trout *Oncorhynchus mykiss* macrophage respiratory burst activity with macrophage-derived supernatants. *Journal of Leukocyte Biology* 57: 943-947.

Jensen, I. and B. Robertsen. 2002. Effect of double-stranded RNA and interferon on the antiviral activity of Atlantic salmon cells against infectious salmon anemia virus and infectious pancreatic necrosis virus. *Fish and Shellfish Immunology* 13: 221-241.

Jimenez, N., J. Coll, F.J. Salguero and C. Tafalla. 2006. Co-injection of interleukin 8 with the glycoprotein gene from viral haemorrhagic septicemia virus (VHSV) modulates the cytokine response in rainbow trout (*Oncorhynchus mykiss*). *Vaccine* 24: 5615-5626.

Jørgensen, J.B., J. Zou, A. Johansen and C.J. Secombes. 2001a. Immunostimulatory CpG oligodeoxynucleotides stimulate expression of IL-1β and interferon-like cytokines in rainbow trout macrophages via a chloroquine-sensitive mechanism. *Fish and Shellfish Immunology* 11: 673-682.

Jørgensen, J.B., A. Johansen, B. Stenersen and A.I. Sommer. 2001b. CpG oligodeoxynucleotides and plasmid DNA stimulate Atlantic salmon (*Salmo salar* L.) leucocytes to produce supernatants with antiviral activity. *Developmental and Comparative Immunology* 25: 313-321.

Jørgensen, J.B., L.H. Johansen, K. Steiro and A. Johansen. 2003. CpG DNA induces protective antiviral immune responses in Atlantic salmon (*Salmo salar* L.). *Journal of Virology* 77: 11471-11479.

Kaiser, P., H.M. Wain and L. Rothwell. 1998. Structure of the chicken interferon-gamma gene, and comparison to mammalian homologues. *Gene* 207: 25-32.

Kishimoto, T. 2003. Interleukin-6. In: *The Cytokine Handbook*, A.N. Thomson and M.T. Lotze (eds.). Academic Press, San Diego, 4th Edition, pp. 281-304.

Kono, T., K. Fujiki, M. Nakao, T. Yano, M. Endo and M. Sakai. 2002a. The immune responses of common carp, *Cyprinus carpio* L., injected with carp interleukin-1 beta gene. *Journal of Interferon and Cytokine Research* 22: 413-419.

Kono, T., H. Watanuki and M. Sakai. 2002b. The activation of interleukin-1 beta in serum of carp, *Cyprinus carpio*, injected with peptidoglycan. *Aquaculture* 212: 1-10.

Kono, T., J. Zou, S. Bird, R. Savan, M. Sakai and C.J. Secombes. 2006. Identification and expression analysis of lymphotoxin-beta like homologues in rainbow trout *Oncorhynchus mykiss*. *Molecular Immunology* 43: 1390-1401.

Koussounadis, A.I., D.W. Ritchie, G.J.L. Kemp and C.J. Secombes. 2004. Analysis of fish IL-1β and derived peptide sequences indicates conserved structures with species-specific IL-1 receptor binding: Implications for pharmacological design. *Current Pharmaceutical Design* 10: 3857-3871.

Kuroda, N., T.S. Uinuk-o, A. Sato, I.E. Samonte, F. Figueroa, W.E. Mayer and J. Klein. 2003. Identification of chemokines and a chemokine receptor in cichlid fish, shark and lamprey. *Immunogenetics* 54: 884-895.

Laing, K.J. and C.J. Secombes. 2004a. Chemokines. *Developmental and Comparative Immunology* 28: 443-460.

Laing, K.J. and C.J. Secombes. 2004b. Trout CC chemokines: comparison of their sequences and expression patterns. *Molecular Immunology* 41: 793-808.

Laing, K.J., C. Cunningham and C.J. Secombes. 2000. Genes for three different isoforms of transforming growth factor-beta are present in plaice (*Pleuronectes platessa*) DNA. *Fish and Shellfish Immunology* 10: 261-271.

Laing, K.J., T. Wang, J. Zou, J. Holland, S. Hong, N. Bols, I. Hirono, T. Aoki and C.J. Secombes. 2001. Cloning and expression analysis of rainbow trout *Oncorhynchus mykiss* tumour necrosis factor-α. *European Journal of Biochemistry* 268: 1315-1322.

Laing, K.J., J.J. Zou, T. Wang, N. Bols, I. Hirono, T. Aoki and C.J. Secombes. 2002a. Identification and analysis of an interleukin 8-like molecule in rainbow trout *Oncorhynchus mykiss*. *Developmental and Comparative Immunology* 26: 433-444.

Laing, K.J., N. Bols and C.J. Secombes. 2002b. A CXC chemokine sequence isolated from the rainbow trout *Oncorhynchus mykiss* resembles the closely related interferon-γ inducible chemokines CXCL9, CXCL10 and CXCL11. *European Cytokine Network* 13: 1-12.

Lally, J., F. Al-Anouti, N. Bols and B. Dixon. 2003. The functional characterisation of CK-1, a putative CC chemokine from rainbow trout (*Oncorhynchus mykiss*). *Fish and Shellfish Immunology* 15: 411-424.

Lawton, P., J. Nelson, R. Tizard and J.L. Browning. 1995. Characterization of the mouse lymphotoxin-beta gene. *Journal of Immunology* 154: 239-246.

Lee, E.Y., H.H. Park, Y.T. Kim and T.J. Choi. 2001. Cloning and sequence analysis of the interleukin-8 gene from flounder (*Paralichthys olivaceous*). *Gene* 274: 237-243.

Lee, D.-S., S.H. Hong, H.-J. Lee, L.J. Jun, J.-K. Chung, K.H. Kim and H.D. Jeong. 2006. Molecular cDNA cloning and analysis of the organization and expression of the IL-1β gene in the Nile tilapia, *Oreochromis niloticus*. *Comparative Biochemistry and Physiology* A143: 307-314.

Li, Q., K. Shirabe, C. Thisse, B. Thisse, H. Okamoto, I. Masai and J.Y. Kuwada. 2005. Chemokine signalling guides axons within the retina in zebrafish. *Journal of Neuroscience* 25: 1711-1717.

Long, Q., E. Quint, S. Lin and M. Ekker. 2000. The zebrafish scyba gene encodes a novel CXC-type chemokine with distinctive expression patterns in the vestibulo-acoustic system during embryogenesis. *Mechanics of Development* 97: 183-186.

Long, S., I. Milev-Milovanovic, M. Wilson, E. Bengten, L.W. Clem, N.W. Miller and V.G. Chinchar. 2006. Identification and expression analysis of cDNAs encoding channel catfish type I interferons. *Fish and Shellfish Immunology* 21: 42-59.

Low, C., S. Wadsworth, C. Burrels and C.J. Secombes. 2003. Expression of immune genes in turbot (*Scophthalmus maximus*) fed a nucleotide supplemented diet. *Aquaculture* 221: 23-40.

Lutfalla, G., H.R. Crollius, N. Stange-thomann, O. Jaillon, K. Mogensen and D. Monneron. 2003. Comparative genomic analysis reveals independent expansion of a lineage-specific gene family in vertebrates: The class II cytokine receptors and their ligands in mammals and fish. *BMC Genomics* 4: 29.

Maloy, K.J. and F. Powrie. 2001. Regulatory T cells in the control of immune pathology. *Nature Immunology* 2: 816-822.

Maloy, K.J., L. Salaun, R. Cahill, G. Dougan, N.J. Saunders and F. Powrie. 2003. CD4(+)CD25(+) T-R cells suppress innate immune pathology through cytokine-dependent mechanisms. *Journal of Experimental Medicine* 197: 111-119.

Mathew, J.A., Y.X. Guo, K.P. Goh, J. Chan, B.M.L. Verburg van Kemnanade and J. Kwang. 2002. Characterisation of a monoclonal antibody to carp IL-1 beta and the development of a sensitive capture ELISA. *Fish and Shellfish Immunology* 13: 85-95.

Milev-Milovanovic, I., S. Long, M. Wilson, E. Bengten, N.W. Miller and V.G. Chinchar. 2006. Identification and expression analysis of interferon gamma genes in channel catfish. *Immunogenetics* 58: 70-80.

Miller, J. and G. Morahan. 1992. Peripheral T-cell tolerance. *Annual Review of Immunology* 10: 51-69.

Moore, K.W., R.D. Malefyt, R.L. Coffman and A. O'Garra. 2001. Interleukin-10 and the interleukin-10 receptor. *Annual Review of Immunology* 19: 683-765.

Mosmann, T.R. and R.L. Coffman. 1989. Th1-cell and Th2-cell - different patterns of lymphokine secretion lead to different functional-properties. *Annual Review of Immunology* 7: 145-173.

Mosmann, T.R. and S. Sad. 1996. The expanding universe of T-cell subsets: Th1, Th2 and more. *Immunology Today* 17: 138-146.

Mosmann, T.R., H. Cherwinski, M.W. Bond, M.A. Giedlin and R.L. Coffman. 2005. Pillars article: Two types of murine helper T cell clone. I. Definition according to profiles of lymphokine activities and secreted proteins. *Journal of Immunology* 175: 5-14.

Najakshin, A.M., L.V. Mechetina, B.Y. Alabyev and A.V. Taranin. 1999. Identification of an IL-8 homolog in lamprey (*Lampetra fluviatilis*): early evolutionary divergence of chemokines. *European Journal of Immunology* 29: 375-382.

Nam, B., J. Byon, Y. Kim, E. Park, Y. Cho and J. Cheong. 2007. Molecular cloning and characterisation of the flounder (*Paralichthys olivaceus*) interleukin-6 gene. *Fish and Shellfish Immunology* 23: 231-236.

Nascimento, D.S., P.J.B. Pereira, M.I.R. Reis, A. do Vale, J. Zou, M.T. Silva, C.J. Secombes and N.M.S. dos Santos. 2007a. Molecular cloning and expression analysis of sea bass (*Dicentrarchus labrax* L.) tumor necrosis factor-α (TNF-α). *Fish and Shellfish Immunology* 23: 701-710.

Nascimento, D.S., A. do Vale, A.M. Tomas, J. Zou, C.J. Secombes and N.M.S. dos Santos. 2007b. Cloning, promoter analysis and expression in response to bacterial exposure of seabass (*Dicentrarchus labrax* L.) interleukin-12 p40 and p35 subunits. *Molecular Immunology* 44: 2277-2291.

O'Farrell, C., N. Vaghefi, M. Cantonnet, B. Buteau, P. Boudinot and A. Benmansour. 2002. Survey of transcript expression in rainbow trout leukocytes reveals a major contribution of interferon-responsive genes in the early response to a rhabdovirus infection. *Journal of Virology* 76: 8040-8049.

O'Garra, A. 1998. Cytokines induce the development of functionally heterogeneous T helper cell subsets. *Immunity* 8: 275-283.

O'Garra, A. 2000. T-cell differentiation: Commitment factors for T helper cells. *Current Biology* 10: R492-R494.

Oppmann, B., R. Lesley, B. Blom, J.C. Timans, Y.M. Xu, B. Hunte, F. Vega, N. Yu, J. Wang, K. Singh, F. Zonin, E. Vaisberg, T. Churakova, M.R. Liu, D. Gorman, J. Wagner, S. Zurawski, Y.J. Liu, J.S. Abrams, K.W. Moore, D. Rennick, R. de Waal-Malefyt, C. Hannum, J.F. Bazan and R.A. Kastelein. 2000. Novel p19 protein engages IL-12p40 to form a cytokine, IL-23, with biological activities similar as well as distinct from IL-12. *Immunity* 13: 715-725.

Ordás, M.C., M.M. Costa, F.J. Roca, G. López-Castejón, V. Mulero, J. Meseguer, A. Figueras and B. Novoa. 2007. Turbot TNFα gene: Molecular characterization and biological activity of the recombinant protein. *Molecular Immunology* 44: 389-400.

Peatman, E. and Z. Liu. 2006. CC chemokines in zebrafish: Evidence for extensive intrachromosomal gene duplications. *Genomics* 88: 381-385.

Peatman, E., B. Bao, X. Peng, P. Baoprasertkul, Y. Brady and Z. Liu. 2006. Catfish CC chemokines: Genomics clustering, duplications, and expression after bacterial infection with *Edwardsiella ictaluri*. *Molecular Genetics and Genomics* 275: 297-309.

Peddie, S., J. Zou, C. Cunningham and C.J. Secombes. 2001. Rainbow trout (*Oncorhynchus mykiss*) recombinant IL-1β and derived peptides induce migration of head-kidney leucocytes *in vitro*. *Fish and Shellfish Immunology* 11: 697-709.

Peddie, S., J. Zou and C.J. Secombes. 2002a. A biologically active IL-1β derived peptide stimulates phagocytosis and bactericidal activity in rainbow trout, *Oncorhynchus mykiss* (Walbaum), head-kidney leucocytes *in vitro*. *Journal of Fish Diseases* 25: 351-360.

Peddie, S., J. Zou and C.J. Secombes. 2002b. Immunostimulation in the rainbow trout (*Oncorhynchus mykiss*) following intraperitoneal administration of Ergosan. *Veterinary Immunology and Immunopathology* 86: 101-113.

Peddie, S., P.E. McLauchlan, A.E. Ellis and C.J. Secombes. 2003. Effect of intraperitoneally administered IL-1β-derived peptides on resistance to viral haemorrhagic septicaemia in rainbow trout *Oncorhynchus mykiss*. *Diseases of Aquatic Organisms* 56: 195-200.

Pelegrin, P., J. Garcia-Castillo, V. Mulero and J. Meseguer. 2001. Interleukin-1β isolated from a marine fish reveals up regulated expression in macrophages following activation with lipopolysaccharide and lymphokines. *Cytokine* 16: 67-72.

Petitfrere, C., B. Dugas, P. Braquet and J.M. Menciahuerta. 1993. Interleukin-9 potentiates the interleukin-4-induced IgE and IgG1 release from murine B-lymphocytes. *Immunology* 79: 146-151.

Pflanz, S., J.C. Timans, J. Cheung, R. Rosales, H. Kanzler, J. Gilbert, L. Hibbert, T. Churakova, M. Travis, E. Vaisberg, W.M. Blumenschein, J.D. Mattson, J.L. Wagner, W. To, S. Zurawski, T.K. McClanahan, D.M. Gorman, J.F. Bazan, R.D. Malefyt,

D. Rennick and R.A. Kastelein. 2002. IL-27, a heterodimeric cytokine composed of EB13 and p28 protein, induces proliferation of naive CD4(+) T cells. *Immunity* 16: 779-790.

Pleguezuelos, O., J. Zou and C.J. Secombes. 2000. Cloning, sequencing and analysis of expression of a second IL-1β gene in rainbow trout (*Oncorhynchus mykiss*). *Immunogenetics* 51: 1002-1011.

Pressley, M.E., P.E. Phelan III, P.E. Witten, M.T. Mellon and C.H. Kim. 2005. Pathogenesis and inflammatory response to *Edwardsiella tarda* infection in the zebrafish. *Developmental and Comparative Immunology* 29: 501-513.

Qiu, L.H., L.S. Song, L.T. Wu, Z.H. Cai and S.G. Jiang. 2006. Molecular cloning and expression analysis of interleukin-1 beta from Japanese sea perch (*Lateolabrax japonicus*). *Acta Oceanologica Sinica* 25: 127-136.

Read, S., V. Malmstrom and F. Powrie. 2000. Cytotoxic T lymphocyte-associated antigen 4 plays an essential role in the function of CD25(+)CD4(+) regulatory cells that control intestinal inflammation. *Journal of Experimental Medicine* 192: 295-302.

Roberts, A.B. and M.B. Sporn. 1990. The transforming growth factor β's. In: *Handbook of Experimental Pharmacology, Peptide Growth Factors and Their Receptors*, A.B. Roberts and M.B. Sporn (eds.). Springer-Verlag, New York, pp. 419-472.

Robertsen, B., V. Bergan, T. Rokenes, R. Larsen and A. Albuquerque. 2003. Atlantic salmon interferon genes: cloning, sequence analysis, expression, and biological activity. *Journal of Interferon and Cytokine Research* 23: 601-612.

Robertsen, B., J. Zou, C. Secombes and J.A. Leong. 2006. Molecular and expression analysis of an interferon-gamma-inducible guanylate-binding protein from rainbow trout (*Oncorhynchus mykiss*). *Developmental and Comparative Immunology* 30: 1023-1033.

Romagnani, S. 1997. The Th1/Th2 paradigm. *Immunology Today* 18: 263-266.

Saeij, J.P., R.J. Stet, B.J. de Vries, W.B. van Muiswinkel and G.F. Wiegertjes. 2003. Molecular and functional characterization of carp TNF: a link between TNF polymorphism and trypanotolerance? *Developmental and Comparative Immunology* 27: 29-41.

Saint-Jean, S.R. and S.I. Perez-Prieto. 2006. Interferon mediated antiviral activity against salmonid fish viruses in BF-2 and other cell lines. *Veterinary Immunology and Immunopathology* 110: 1-10.

Sapede, S., M. Rossel, C. Dambly-Chaudiere and A. Ghysen. 2005. Role of SDF-1 chemokine in the development of lateral line efferent and facial motor neurons. *Proceedings of the National Academy of Sciences of the United States of America* 102: 1714-1718.

Savan, R. and M. Sakai. 2003. Presence of multiple isoforms of TNF alpha in carp (*Cyprinus carpio* L.): genomic and expression analysis. *Fish and Shellfish Immunology* **17**: 87-94.

Savan, R., T. Kono, A. Aman and M. Sakai. 2003a. Isolation and characterization of a novel CXC chemokine in common carp (*Cyprinus carpio* L.). *Molecular Immunology* 39: 829-834.

Savan, R., D. Igawa and M. Sakai. 2003b. Cloning, characterization and expression analysis of interleukin-10 from the common carp, *Cyprinus carpio* L. *European Journal of Biochemistry* 270: 4647-4654.

Savan, R., T. Kono, D. Igawa and M. Sakai. 2005. A novel tumor necrosis factor (TNF) gene present in tandem with the TNF-alpha gene on the same chromosome in teleosts. *Immunogenetics* 57: 140-150.

Scapigliati, G., F. Buonocore, S. Bird, J. Zou, P. Pelegrin, C. Falasca, D. Prugnoli and C.J. Secombes. 2001. Phylogeny of cytokines: molecular cloning and expression analysis of sea bass *Dicentrarchus labrax* interleukin-1β. *Fish and Shellfish Immunology* 11: 711-726.

Schroder, K., P.J. Hertzog, T. Ravasi and D.A. Hume. 2004. Interferon-gamma: an overview of signals, mechanisms and functions. *Journal of Leukocyte Biology* 75: 163-189.

Secombes, C.J. and J.-P.Y. Scheerlinck. 1999. Cytokines as antimicrobials in animals. *Microbiology Australia* 20: 9-10.

Seder, R.A. and W.E. Paul. 1994. Acquisition of lymphokine-producing phenotype by Cd4+ T-cells. *Annual Review of Immunology* 12: 635-673.

Sher, A. and R.L. Coffman. 1992. Regulation of immunity to parasites by T-cells and T-cell derived cytokines. *Annual Review of Immunology* 10: 385-409.

Smith, K.A. 1980. T-cell growth-factor. *Immunological Reviews* 51: 337-357.

Smith, K.A. 1988. Interleukin-2—inception, impact, and implications. *Science* 240: 1169-1176.

Spies, T., C.C. Morton, S.A. Nedospasov, W. Fiers, D. Pious and J.L. Strominger. 1986. Genes for the tumor-necrosis-factor-alpha and factor-beta are linked to the human major histocompatibility complex. *Proceedings of the National Academy of Sciences of the United States of America* 83: 8699-8702.

Strengeil, M., T. Sareneva, D. Foster, I. Julkunen and S. Matikainen. 2002. IL-21 UP-Regulates the expression of genes associated with innate immunity and Th1 response. *Journal of Immunology* 169: 3600-3605.

Subramaniam, P.S., M.G. Mujtaba, M.R. Paddy and H.M. Johnson. 1999. The carboxyl terminus of interferon-gamma contains a functional polybasic nuclear localization sequence. *Journal of Biological Chemistry* 274: 403-407.

Subramaniam, P.S., J. Larkin, M.G. Mujtaba, M.R. Walter and H.M. Johnson. 2000. The COOH-terminal nuclear localization sequence of interferon gamma regulates STAT1 alpha nuclear translocation at an intracellular site. *Journal of Cell Science* 113: 2771-2781.

Swain, S.L. 1991. Lymphokines and the immune-response—the central role of interleukin-2. *Current Opinion in Immunology* 3: 304-310.

Szabo, S.J., B.M. Sullivan, C. Stemmann, A.R. Satoskar, B.P. Sleckman and L.H. Glimcher. 2002. Distinct effects of T-bet in Th1 lineage commitment and IFN-γ production in CD4 and CD8 T cells. *Science* 295: 338-342.

Szabo, S.J., B.M. Sullivan, S.L. Peng and L.H. Glimcher. 2003. Molecular mechanisms regulating Th1 immune responses. *Annual Review of Immunology* 21: 713-758.

Tafalla, C., R. Aranguren, C.J. Secombes, J.L. Castrillo, B. Novoa and A. Figueras. 2003. Molecular characterisation of sea bream (*Sparus aurata*) transforming growth factor beta 1. *Fish and Shellfish Immunology* 14: 405-421.

Tassakka, A.C.M.A.R., R. Savan, H. Watanuki and M. Sakai. 2006. The *in vitro* effects of CpG oligodeoxynucleotides on the expression of cytokine genes in the common carp

(*Cyprinus carpio* L.) head-kidney cells. *Veterinary Immunology and Immunopathology* 110: 79-85.

Thompsonsnipes, L.A., V. Dhar, M.W. Bond, T.R. Mosmann, K.W. Moore and D.M. Rennick. 1991. Interleukin-10—a novel stimulatory factor for mast-cells and their progenitors. *Journal of Experimental Medicine* 173: 507-510.

Trinchieri, G., S. Pflanz and R.A. Kastelein. 2003. The IL-12 family of heterodimeric cytokines: New players in the regulation of T cell responses. *Immunity* 19: 641-644.

Trobridge, G.D., P.P. Chiou, C.H. Kim and J.C. Leong. 1997. Induction of the Mx protein of rainbow trout *Oncorhynchus mykiss in vitro* and *in vivo* with poly I:C dsRNA and infectious hematopoietic necrosis virus. *Diseases of Aquatic Organisms* 30: 91-98.

Uenobe, M., C. Kohchi, N. Yoshioka, A. Yuasa, H. Inagawa, K. Morii, T. Nishizawa, Y. Takahashi and G.-I. Soma. 2007. Cloning and characterization of a TNF-like protein of *Plecoglossus altivelis* (ayu fish). *Molecular Immunology* 44: 1115-1122.

Urban, J.F., K.B. Madden, A. Svetic, A. Cheever, P.P. Trotta, W.C. Gause, I.M. Katona and F.D. Finkelman. 1992. The importance of Th2-cytokines in protective immunity to nematodes. *Immunological Reviews* 127: 205-220.

Waldmann, T.A., S. Dubois and Y. Tagaya. 2001. Contrasting roles of IL-2 and IL-15 in the life and death of lymphocytes: Implications for immunotherapy. *Immunity* 14: 105-110.

Wang, T., N. Johnson, J. Zou, N. Bols and C.J. Secombes. 2004. Sequencing and expression of the second allele of the interleukin-1β1 gene in rainbow trout (*Oncorhynchus mykiss*): identification of a novel SINE in the third intron. *Fish and Shellfish Immunology* 16: 335-358.

Wang, Y., Q. Wang, P. Baoprasertkul, E. Peatman and Z. Liu. 2006. Genomic organization, gene duplication, and expression analysis of interleukin-1β in channel catfish (*Ictalurus punctatus*). *Molecular Immunology* 43: 1653-1664.

Ware, C.F., S. Santee and A. Glass. 1998. Tumor necrosis factor-related ligands and receptors. In: *The Cytokine Handbook*, A.W. Thomson (ed.). Academic Press, San Diego, pp. 549-592. 3rd Edition.

Watford, W.T., M. Moriguchi, A. Morinobu and J.J. O'Shea. 2003. The biology of IL-12: coordinating innate and adaptive immune responses. *Cytokine Growth Factor Reviews* 14: 361-368.

Wiens, G.D., G.W. Glenney, S.E. LaPatra and T.J. Welch. 2006. Identification of novel rainbow trout (*Oncorhynchus mykiss*) chemokines, *CXCd1* and *CXCd2*: mRNA expression after *Yersinia ruckeri* vaccination and challenge. *Immunogenetics* 58: 308-323.

Wolf, S.F., P.A. Temple, M. Kobayashi, D. Young, M. Dicig, L. Lowe, R. Dzialo, L. Fitz, C. Ferenz, R.M. Hewick, K. Kelleher, S.H. Herrmann, S.C. Clark, L. Azzoni, S.H. Chan, G. Trinchieri and B. Perussia. 1991. Cloning of cDNA for natural-killer-cell stimulatory factor, a heterodimeric cytokine with multiple biologic effects on T-cells and natural-killer-cells. *Journal of Immunology* 146: 3074-3081.

Woods, I.G., C. Wilson, B. Friedlander, P. Chang, D.K. Reyes, R. Nix, P.D. Kelly, F. Chu, J.H. Postlethwait and W.S. Talbot. 2005. The zebrafish gene map defines ancestral vertebrate chromosomes. *Genome Research* 15: 1307-1314.

Xu, D.M., W.L. Chan, B.P. Leung, D. Hunter, K. Schulz, R.W. Carter, I.B. McInnes, J.H. Robinson and F.Y. Liew. 1998. Selective expression and functions of interleukin 18 receptor on T helper (Th) type 1 but not Th2 cells. *Journal of Experimental Medicine* 188: 1485-1492.

Yin, Z. and J. Kwang. 2000a. Carp interleukin-1β in the role of an immuno-adjuvant. *Fish and Shellfish Immunology* 10: 375-378.

Yin, Z. and J. Kwang. 2000b. Molecular isolation and characterisation of carp transforming growth factor beta 1 from activated leucocytes. *Fish and Shellfish Immunology* 10: 309-318.

Yoon, C., S.C. Johnston, J. Tang, M. Stahl, J.F. Tobin and W.S. Somers. 2000. Charged residues dominate a unique interlocking topography in the heterodimeric cytokine interleukin-12. *EMBO Journal* 19: 3530-3541.

Yoshiura, Y., I. Kiryu, A. Fujiwara, H. Suetake, Y. Suzuki, T. Nakanishi and M. Ototake. 2003. Identification and characterization of Fugu orthologues of mammalian interleukin-12 subunits. *Immunogenetics* 55: 296-306.

Zou, J., P.S. Grabowski, C. Cunningham and C.J. Secombes. 1999a. Molecular cloning of interleukin 1β from rainbow trout *Oncorhynchus mykiss* reveals no evidence of an ICE cut site. *Cytokine* 11: 552-560.

Zou, J., C. Cunningham and C.J. Secombes. 1999b. The rainbow trout *Oncorhynchus mykiss* interleukin-1β gene has a different organisation to mammals and undergoes incomplete splicing. *European Journal of Biochemistry* 259: 901-908.

Zou, J., T. Wang, I. Hirono, T. Aoki, H. Inagawa, T. Honda, G.-I. Soma, M. Ototake, T. Nakanishi, A.E. Ellis and C.J. Secombes. 2002. Differential expression of two tumor necrosis factor genes in rainbow trout, *Oncorhynchus mykiss*. *Developmental and Comparative Immunology* 26: 161-172.

Zou, J., C.J. Secombes, S. Long, N. Miller, L.W. Clem and V.G. Chinchar. 2003a. Molecular identification and expression analysis of tumor necrosis factor in channel catfish (*Ictalurus punctatus*). *Developmental and Comparative Immunology* 27: 845-858.

Zou, J., S. Peddie, G. Scapigliati, Y. Zhang, N.C. Bols, A.E. Ellis and C.J. Secombes. 2003b. Functional characterisation of the recombinant tumor necrosis factors in rainbow trout, *Oncorhynchus mykiss*. *Developmental and Comparative Immunology* 27: 813-822.

Zou, J., M.S. Clark and C.J. Secombes. 2003c. Characterisation, expression and promoter analysis of an interleukin 10 homologue in the puffer fish, *Fugu rubripes*. *Immunogenetics* 55: 325-335.

Zou, J., S. Bird, J. Truckle, N. Bols, M. Horne and C. Secombes. 2004a. Identification and expression analysis of an IL-18 homologue and its alternatively spliced form in rainbow trout (*Oncorhynchus mykiss*). *European Journal of Biochemistry* 271: 1913-1923.

Zou, J., Y.B. Yasutoshi, J.M. Dijkstra, M. Sakai, M. Ototake and C. Secombes. 2004b. Identification of an interferon gamma homologue in Fugu, *Takifugu rubripes*. *Fish and Shellfish Immunology* 17: 403-409.

Zou, J., A. Carrington, B. Collet, J.M. Dijkstra, N. Bols and C.J. Secombes. 2005a. Identification and bioactivities of interferon gamma in rainbow trout *Oncorhynchus mykiss*: the first Th1 type cytokine characterised functionally in fish. *Journal of Immunology* 175: 2484-2494.

Zou, J., S. Bird and C.J. Secombes. 2005b. Fish cytokine gene discovery and linkage using genomic approaches. *Marine Biotechnology* 6S: 533-539.

CHAPTER

2

Leukocytes and Cytokines Present in Fish Testis: A Review

Alfonsa García-Ayala[*] and Elena Chaves-Pozo

INTRODUCTION

In mammals, most research concerning the testis has concentrated upon germ and Sertoli cells in the seminiferous tubules and the Leydig cells in the interstitial tissue, because these cells serve as essential testis-specific functions of spermatozoa and androgen production, respectively (Hedger, 1997). However, recently, testicular leukocytes, a prominent group of cells located in the interstitial tissue, and the growth factors and cytokines produced by them or by testicular cells, have received considerable attention from reproductive biologists and/or immunologists. Thus, reproductive-immune research has provided substantial insight into interactions between these physiological systems.

In all vertebrates, the testis is considered as an immunologically privileged site as it needs to prevent immune responses against meiotic and

Authors' address: Department of Cell Biology, Faculty of Biology, University of Murcia, 30100 Murcia, Spain.
[*]*Corresponding author:* E-mail: agayala@um.es

haploid germ cells which express 'non-self' antigens, which first appear at the time of puberty, long after the establishment of self-tolerance in the perinatal period. However, defence mechanisms—including both innate and adaptive immunity—are not generally impaired in the testis as they are able to develop inflammatory responses to local and systemic infection (Schuppe and Meinhardt, 2005). Thus, immune responses in the testis are regulated in a manner that provides protection for the developing male germ cells, while permitting qualitatively normal inflammatory responses and protection against infection (Hedger, 2002). There exists a critical balance between health and sickness, when immune-endocrine interactions either drive or repress the reproductive functions in the testis (Hales, 2002). Both 'endocrine hormones' and 'inflammatory mediators', as defined in their original context, play an essential role in the orchestration of spermatogenesis and maintenance of testicular tissue homeostasis (Harris and Bird, 2000). During times of normal reproductive health, it is likely that steroid sex hormones are immunosuppressive, tilting the balance in favour of reproductive functions (i.e., testosterone and gamete production). During times of sickness, infection, inflammation or other forms of biological stress, the functional activities of the endocrine regulators such as androgens and their production are perturbed by the elevated and prolonged expression of the inflammatory mediators (Hales, 2002).

In fish, little is known about reproductive-immune interactions inside the reproductive tissues and most information concerns the modulation of immune responses by circulating hormones, including cortisol, growth hormone (GH), prolactin and reproductive hormones and some proopiomelanocortin-derived peptides (Harris and Bird, 2000; Engelsma *et al.*, 2002). Although the effect of these endocrine mediators depends on the species, in general, these molecules modulate the immune responses by integrating the activities of all the systems to adapt the organism to its environment (Lutton and Callard, 2006). In fact, estrogens, e.g., estradiol (E_2), and androgens, e.g., 11-ketotestosterone (11-KT) and testosterone (T), modulate several immune responses. It has been demonstrated that E_2 and 11-KT stimulate and inhibit lymphocyte proliferation, respectively (Cook, 1994), while T reduces the number of antibody-producing cells and synergizes with cortisol to produce a greater inhibitory effect (Slater and Schreck, 1993). Moreover, intra-peritoneal injection of E_2, progesterone (P) or 11-KT inhibits in a dose-dependent manner phagocytosis and the production of reactive oxygen and nitrogen intermediates (ROIs/RNIs) by the head-kidney macrophages (Watanuki *et al.*, 2002). However, studies *in vitro* with head-kidney macrophages have demonstrated that these hormones inhibit phagocytosis, although only P and 11-KT inhibit RNIs production and none has any effect on the ROI production (Yamaguchi *et al.*, 2001). In salmonids, T can kill leukocytes *in vitro* (Slater and Schreck, 1997) whereas in goldfish the E_2 depresses the immune system and increases its susceptibility to trypanosoma infections (Wang and Belosevic, 1994). In contrast, sex steroids have no immunosuppressive effects on common carp leucocytes and do not induce apoptosis *in vitro* (Saha *et al.*, 2004). In gilthead sea bream, 11-KT enhances but E_2 inhibits ROI production by head-kidney acidophilic granulocytes (Chaves-Pozo *et al.*, 2003). Moreover, levels of reproductive hormones, which vary during the different stages of the reproductive cycle, have been correlated with some immune deficiencies, such as an inability to produce isohaemagglutinins in sexually mature fish (Ridgeway, 1962) and the increased frequency of ectoparasitic infestations, particularly in males (Pickering and Christie, 1980). Related with these observations is the fact that rainbow trout serum shows reduced bactericidal activity during spawning (Iida *et al.*, 1989). Moreover, there is little evidence to support the involvement of the pro-inflammatory cytokines, tumour necrosis factor α (TNFα) and interleukin-1β (IL-1β), in the regulation of goldfish testicular steroid biosynthesis (van der Kraak *et al.*, 1998; Lister and van der Kraak, 2002).

Recently, our research group has focused its attention on the role of leukocytes and cytokines in the reproductive activities of gilthead sea bream males, integrating the views of both immunologists and reproductive biologists. The gilthead sea bream is a protandrous, hermaphrodite, seasonally breeding teleost that develops asynchronous spermatogenesis during the male phase (Fig. 2.1) in which the bisexual gonad has functional testicular and non-functional ovarian areas (Fig. 2.2). The reproductive cycle of the male gilthead sea bream is divided into four gonad stages: gametogenic activity, spawning, post-spawning and resting or involution, this last only when the fish are ready to undergo sex change (Chaves-Pozo *et al.*, 2005a; Liarte *et al.*, unpubl. results). Throughout this cycle, the testis undergoes important morphological changes, passing from being formed by all germinal cell types during spermatogenesis to being formed by spermatogonia with some degenerative cell areas after spawning, as has also been described for other species (Patzner and Seiwald, 1987; Lahnsteiner and Patzner, 1990; Besseau and Faliex, 1994).

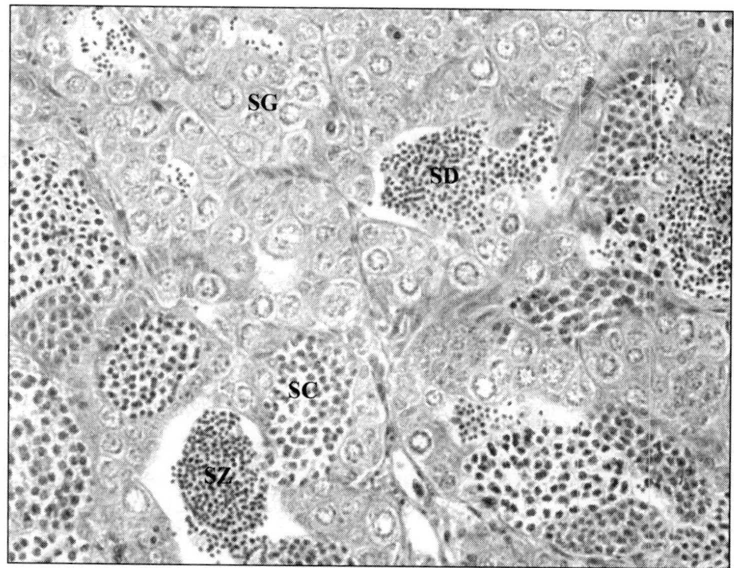

Fig. 2.1 Seminiferous tubules of gilthead sea bream show asynchronous spermatogenesis (Mallory trichromic). SG, spermatogonia; SC, spermatocytes; SD, spermatids; SZ, spermatozoa. ×40.

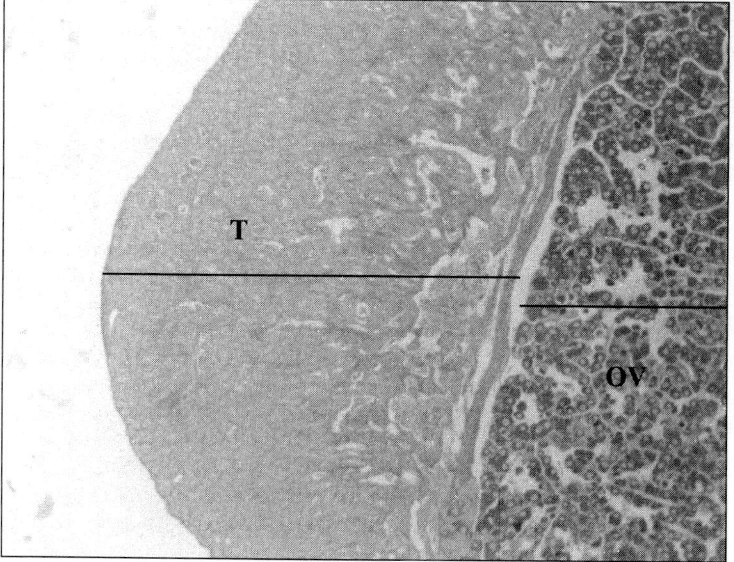

Fig. 2.2 Bisexual gonad of the male phase of gilthead sea bream (Hematoxylin-eosin). T, testis; OV, ovary. ×5.

Testicular Biology

The testis has to fulfil two major functions: the generation of male gametes (spermatogenesis) and the production of sex steroids (steroidogenesis) that predominantly occur in the germinal and interstitial compartments, respectively.

Spermatogenesis, the formation of sperm, is a complex process in which the spermatogonia divide and differentiate into spermatozoa. In fish, spermatogenesis appears to proceed in a similar fashion to that observed in other vertebrates, although with some important differences. Spermatogenesis proceeds in a cystic structure, in which all germ cells (primary spermatogonia, A and B spermatogonia, and spermatocytes) develop synchronously surrounded by a cohort of Sertoli cells (Miura, 1999). Interestingly, fish Sertoli cells proliferate in adult specimens (Fig. 2.3) during the entire reproductive cycle and, depending on the fish species, they divide simultaneously or not with the developing germ cells that evolve (Chaves-Pozo *et al.*, 2005a; Schulz *et al.*, 2005). Moreover,

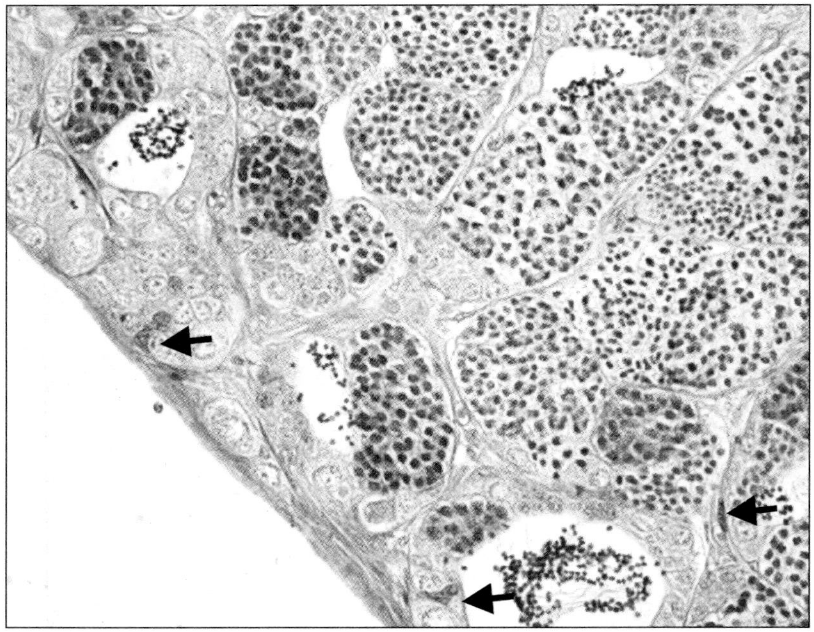

Fig. 2.3 Seminiferous tubules showing proliferative cells (brown) immunostained with the anti-BrdU antibody. Notice the proliferative Sertoli cells (arrows). ×40.

Sertoli cells support the germ cells structurally, nutritionally and with regulatory factors. Initial cysts formed by mitotic proliferation of spermatogonia originate spermatocytes, which after undergoing the first meiotic division give rise to the secondary spermatocytes. These stages complete the second meiotic division and originate spermatids. Upon differentiation (spermiogenesis), the spermatids form the spermatozoa, which are accumulated in the lumina of the tubules before being shed during spawning (for review see Rocha and Rocha, 2006). In seasonal breeding fish, after shedding the spermatozoa, the testes enter into a degenerative process, in which both germinal and interstitial compartments are reorganized and the seminiferous tubules are repopulated with spermatogonia (Besseau and Faliex, 1994; Chaves-Pozo *et al.*, 2005a). Sperm production is a highly conserved process in all vertebrates, although the timing and the number of spermatogonial generations in fish are species-specific (Nagahama, 1983). Although the testicular germinal compartment has been described in detail in several seasonal breeding teleost species (Loir *et al.*, 1995; Pudney, 1995), including the gilthead sea bream (Chaves-Pozo *et al.*, 2005a), little attention has been paid to the interstitial tissue even though it probably plays a pivotal role during the testicular regression process that occurs in both seasonal testicular involution and sex change in seasonal hermaphroditic breeding fish. The interstitial tissue is mainly formed by Leydig cells, the main steroidogenic cell type, fibroblasts, myoid cells or peritubular cells and some types of leukocytes (Grier, 1981; Nagahama, 1983; Loir *et al.*, 1995) and undergoes seasonal changes as the same time as the germinal compartment. Thus, a marked increase in interstitial tissue, the vacuolization of interstitial cells, and the presence of macrophages after the shedding of spermatozoa have been observed by conventional microscopy (Shrestha and Khanna, 1976; Micale *et al.*, 1987; Besseau and Faliex, 1994). Interestingly, although the existence of a blood-testis barrier has been demonstrated in teleost as a barrier between the vascular spaces of the testis and the germ cells (Abraham *et al.*, 1980), macrophages, together with Sertoli cells, have been seen to be involved in germ cell elimination in some teleost species (Billard and Takashima, 1983; Scott and Sumpter, 1989). These morphological data clearly show that both compartments—the germinal and interstitial—interact during the reproductive cycle, the latter being of special interest in the context of Reproductive Immunology as the place in which the leucocytes are located.

Testicular functions are regulated by a well-established hierarchical hormone system. In most of the species studied to date, two gonadotropin hormones (GTH), homologous to tetrapod-stimulating follicle hormone (FSH), and the luteinizing hormone (LH) (Quérat *et al.*, 2000) have been observed, both secreted by the pituitary gland. These glycoproteins share a common a subunit, but differ in their β subunits in a way that confers immunological and biological specificity to each hormone (Pierce and Parsons, 1981). FSH and LH have been isolated and characterized in a number of teleosts (Agulleiro *et al.*, 2006). FSH and LH were found to differ in their pattern of expression at different stages of the reproductive cycle in some species (Nozaki *et al.*, 1990a, b; Schreibman *et al.*, 1990; Naito *et al.*, 1991; Saga *et al.*, 1993; Magliulo-Cepriano *et al.*, 1994; Miranda *et al.*, 2001; García Hernández *et al.*, 2002). In general, FSH gene expression, as well as the synthesis and release of the protein, was higher at the beginning of the reproductive cycle, whereas that of the LH increased in the later stages of the cycle (Prat *et al.*, 1996; Breton *et al.*, 1998; Bon *et al.*, 1999; Gómez *et al.*, 1999; García Ayala *et al.*, 2003), although differences existed between the species. FSH is involved in the control of puberty and gametogenesis, whereas LH mainly regulates final gonadal maturation and spawning (Schulz *et al.*, 2001). The action of the GTHs on the Sertoli cells and Leydig cells induces the secretion of steroids and/or growth factors, which act in the complex network of cellular interactions which control the testicular functions. Thus, LH stimulates gonadal steroidogenesis in testicular Leydig cells, while the role of FSH in the testis is still somewhat unclear, although it appears to have certain functions, such as the stimulation of Sertoli cell proliferation and maintenance of quantitatively normal spermatogenesis by means of indirect effects mediated by Sertoli cells.

Leydig cells have been described as the main steroidogenic site in the teleost testis (Lofts and Bern, 1972; van der Hurk *et al.*, 1978; Kime 1987), and the main source of circulating androgens, 11-KT and T (Borg, 1994), although other sources have also been described (Idler and MacNab, 1967; Kime, 1978). In teleosts, T is converted to 11-KT by the Leydig cells because of the availability of the appropriate converting enzymes (Kime, 1987). Although 11-KT has been considered the main androgen in teleost males (Miura *et al.*, 1991; Borg, 1994; Cavaco *et al.*, 1998), in some sex-changing species, 11-KT levels have been reported to be low or undetectable in both testis and blood (Borg, 1994). In some species (Prat *et al.*, 1990; Chaves-Pozo *et al.*, unpubl. results), and in contrast with the results obtained in most teleost fish (Borg, 1994), T and 11-KT peak at different stages, which suggests that each androgen plays a different role in the reproductive cycle, as has been described in the African catfish, where T seems to be related with gonadotroph cell development of the pituitary, whereas 11-KT is more related with testicular development (Cavaco *et al.*, 2001).

Estrogens have, to date, been considered as female hormones. However, it has recently been demonstrated that estrogens are essential for normal reproductive performance in male vertebrates (O´Donnell *et al.*, 2001; Carreau *et al.*, 2003; Hess, 2003; Sierens *et al.*, 2005). They are synthesized by the enzyme cytochrome P450 aromatasa (P450aro) that, in fish, is expressed in the interstitial tissue of the mature testis (Kobayashi *et al.*, 2004). Two nuclear E_2 receptor (ER) forms have been found expressed in the mammalian testis (O´Donnell *et al.*, 2001) and three in fish (Hawkins *et al.*, 2000; Menuet *et al.*, 2002; Choi and Habibi, 2003; Halm *et al.*, 2004; Pinto *et al.*, 2005). Moreover, a membrane ER-mediated action of E_2 has also been proposed to occur in fish testis (Loomis and Thomas, 2000). The expression of ER in the gonad suggests that E_2 is involved in the testicular physiology (Leger *et al.*, 2000; Andreassen *et al.*, 2003). Although Socorro *et al.* (2000) described the ERa as the only ER expressed in the testis of gilthead sea bream, our data shown that both ERa and b are present in the gonad (Chaves-Pozo *et al.*, unpubl. data). In teleosts, the levels of E_2 change around the reproductive cycle, being higher in spawning and post-spawning (Billard *et al.*, 1982; Lone *et al.*, 2001; Chaves-Pozo *et al.*, unpubl. results). Although the significance of E_2 in the testis is not clear, a role in spermatogonial proliferation and probably in the Sertoli cell physiology has been proposed (Miura *et al.*, 1999; Amer *et al.*, 2001; Miura and Miura, 2001). Interestingly, in spermatogenically active gilthead sea bream males, E_2 blocks spermatogonia stem cell proliferation and triggers the rapid development of all post-meiotic germ cells into spermatozoa whereas pre-meiotic germ cells are induced to undergo apoptosis. However, E_2 is not sufficient to stimulate the shedding of spermatozoa (Chaves-Pozo *et al.*, 2007).

Apart from overall hormonal control, precise regulation of spermatogenesis and steroidogenesis within the testis also depends upon numerous autocrine and paracrine mediators, such as growth factors and cytokines (Schlatt *et al.*, 1997). These signalling molecules, which would be produced by both germ and somatic cells of the testis and also by resident or infiltrated leukocytes, provide the necessary integration and communication pathway between the various different cell types in the testis (Hedger and Meinhardt, 2003). The apparent overlap between the testicular and immune regulatory functions of these cytokines could provide the key to understand the phenomenon of immune privilege and the processes that lead to inflammation-mediated damage in the testis (Schuppe and Meinhardt, 2005).

IMMUNE CELLS

Despite its immunologically privileged status the testis, in mammals, is not isolated from the immune system (Schuppe and Meinhardt, 2005). Thus, immune cells are found in considerable numbers of the normal unaffected testes, including humans (El Demiry *et al.*, 1987; Pöllänen and Niemi, 1987; Hedger, 1997; Schuppe and Meinhardt, 2005). Located in the interstitial compartment, they are involved in Leydig cell development and steroidogenesis as well as in spermatogenesis. Moreover, they are involved in the mechanisms that make the testis an immunologically privileged site, where germ cells are protected from autoimmune attack and foreign tissue grafts may survive for extended periods of time (Schuppe and Meinhardt, 2005). In addition to resident macrophages, which represent the second most abundant cell type after Leydig cells, mast cells are regular components of the interstitial and peritubular tissue (Nistal *et al.*, 1984; Gaytan *et al.*, 1989). The number of lymphocytes in the testis is relatively small, although circulating immune cells have access to the organ and testicular lymphatic vessels allow drainage to regional lymph nodes (Head *et al.*, 1983; Hedger and Meinhardt, 2003). The presence of natural killer cells known to be involved in innate immune responses has been reported in some species. Moreover, dendritic cells as potential professional antigen-presenting cells and the key players during induction of specific immune responses remain to be identified in the normal testis. Under physiological conditions, neither resident nor circulating immune cells are found in seminiferous tubules, while polymorphonuclear cells are completely absent (Schuppe and Meinhard, 2005).

In fish, little attention has been paid to the immune cell population of the testis interstitial tissue. However, our research group has recently begun to focus its attention on the role of leukocytes and cytokines in the reproductive activities of gilthead sea bream males. As no specific markers for fish immune cells were available until 2002, when we developed a monoclonal antibody which is specific to gilthead sea bream acidophilic granulocytes (Sepulcre *et al.*, 2002), the identification of immune cells was only based on microscopic studies carried out on the adult testis, some of them related to different stages of the reproductive cycle. Thus, in the gametogenic and spawning stages, some macrophages are present in the interstitial tissue of the rainbow trout testis (Loir *et al.*, 1995) whereas in the post-spawning stage, a high population of phagocytic cells has been described in several teleost fish (Henderson, 1962; Shrestha and Khanna, 1976; Carrillo and Zanuy, 1977; Billard, 1986; Scott and Sumpter, 1989; Lahnsteiner and Patzner, 1990; Loir *et al.*, 1995). Moreover, macrophages, granulocytes and lymphocytes have been observed in the testis of some sparid fish, although only macrophages have been shown to be phagocytic (Micale *et al.*, 1987; Besseau and Faliex, 1994; Duslé-Sicard and Fourcault, 1997). In general, these cells seem to infiltrate the testis in greater or lesser numbers, depending on the stage of the reproductive cycle, but are not considered to be specific populations of the testis, which responds in a specific manner to specific stimuli, as is the case in mammals. However, our data related to testicular acidophilic granulocytes leads to the conclusion that these cells are a homing tissue subset of the acidophilic granulocyte population with a specific pattern of responses (Chaves-Pozo *et al.*, 2005c). Such reproductive-immune research has only recently been initiated, and so there are few studies that have focused on sexually mature fish and no information exists on this interaction during larval development or in puberty.

Macrophages

Macrophages are ubiquitous cells that play central roles in the innate immune response through secretion of inflammatory cytokines, such as IL-1β and TNFα, the production of cytotoxic ROIs, and the secretion of leukostatic factors and other regulatory molecules. They are also important accessory cells for many other immune responses. In addition, during development, these cells are also thought to have trophic roles enacted through their remodelling capabilities and cytokine production.

In mammals, macrophages are considered as essential accessory cells for normal reproductive functioning (Hunt, 1989; Tachi and Tachi, 1989; Hutson, 1994; Pollard *et al.*, 1997; Cohen *et al.*, 1999), as they abound in the reproductive tract of males. They form a substantial portion of the interstitial cells (~25%) of the testis but none are found in the seminiferous tubule (Niemi *et al.*, 1986). In the immature testis, there are relatively few resident macrophages. However, these cells increase markedly in number around the same time as the appearance of the adult Leydig cell population and beginning of the meiotic development of the spermatogenic cells (Mendis-Handagama *et al.*, 1987; Hardy *et al.*, 1989; Raburn *et al.*, 1993). The number of testicular macrophages continue to increase into adult life and is always related to the number of Leydig cells to which they are physically connected (Hutson, 1992; Hedger, 1997). Regulation of the number of testicular macrophages during pubertal development and in the adult testis principally involves LH, which acts by stimulating the Leydig cells (Raburn *et al.*, 1993; Wang *et al.*, 1994). Moreover, other factors, such as locally produced cytokines, including IL-1 and colony stimulating factor (CSF)-1 (Gérard *et al.*, 1991; Cohen and Pollard, 1994), and other pituitary hormones, particularly GH (Gaytan *et al.*, 1994) and FSH (Yee and Hutson, 1983), also modulate these cell activities. In turn, testicular macrophages influence Leydig cell morphology and the steroidogenic enzymatic content by providing essential growth and differentiation factors for their normal activity (Cohen *et al.*, 1999; Hales, 2002). In mice lacking macrophage-CSF (M-CSF), the reduced numbers of testicular macrophages result in impaired spermatogenesis as a consequence of the dramatically reduced testosterone levels resulting from the abnormal Leydig cells, suggesting that macrophages can contribute to the local regulation of Leydig cell function superimposed upon a larger mechanism that regulates the entire hypothalamic-pituitary-gonadal axis (Cohen *et al.*, 1999). In contrast, when macrophages are activated and produce inflammatory mediators, Leydig cell steroidogenesis is inhibited (Hales, 2002). Thus, activated macrophages produce pro-inflammatory cytokines, such as IL-1 and TNFα, and ROIs such as hydrogen peroxide which appear to act as transcriptional repressors of steroidogenic enzyme gene expression and by perturbing Leydig cell mitochondria, resulting in the inhibition of steroidogenic acute regulatory protein (StAR) protein expression (Hales, 2002), a pivotal enzyme for steroid synthesis.

Interestingly, whereas a similar pattern of functioning has been demonstrated for macrophages resident in different tissues (Laskin *et al.*, 2001; Guillemin and Brew, 2004; Stout and Suttles, 2004), testicular macrophages and their functions are largely determined by the local environment (Hedger, 1997, 2002). Upon inflammatory insult to the tissue, these resident tissue macrophages can contribute to the innate immune response by expressing a variety of inflammatory and effector activities, the pattern of which is differentially regulated by the microenvironment of the different tissues (Stout and Suttles, 2004). Testicular macrophages display numerous immune properties: they can secrete cytokines, present antigens and secrete lysozyme but are somewhat immunosuppressed as compared with other resident macrophage populations (Miller *et al.*, 1983; Wei *et al.*, 1988; Hutson, 1994; Kern *et al.*, 1995; Hales, 1996; Hedger, 1997, 2002; Hales *et al.*, 1999; Jonsson *et al.*, 2000; Meinhardt *et al.*, 2000; Söder *et al.*, 2000; Hedger and Meinhardt, 2003). Apart from their impact on testis-specific functions, macrophages in the testis have to be considered as potential effector cells in the first line of host defence, i.e., activating innate immune responses and, thus, inflammation. Notably, testicular macrophages have been shown to express the major histocompatibility complex class II (MCH II) molecules essential for antigen presentation to CD4+ T cells (Pöllänen and Niemi, 1987; Wang *et al.*, 1994; Hedger, 1997). However, the ability of freshly isolated rat testicular macrophages to release pro-inflammatory cytokines such as IL-1, IL-6 and TNFα is reduced in comparison with macrophages of other origins (Kern *et al.*, 1995; Hayes *et al.*, 1996). Available data suggest that resident macrophages in the normal adult testis mainly exert anti-inflammatory activities (Hedger, 2002). Moreover, the macrophage migration inhibitory factor that is normally produced by activated macrophages (Bernhagen *et al.*, 1993; Calandra *et al.*, 1994) is also produced in the testis by Leydig and also by Sertoli cells, but only when Leydig cells are ablated by drug treatments (Meinhardt *et al.*, 1996).

In fish, only a few morphological studies have described macrophages in the testis and no experimental studies on the possible roles of these cells in the testis exists. In rainbow trout, few macrophages have been observed during spermatogenesis while, after spawning, they are more numerous and appear near the Sertoli cells and phagocytose the non-emitted spermatozoa (Billard and Takashima, 1983; Scott and Sumpter, 1989). In gilthead sea bream, macrophages have been observed in the interstitial tissue of the testis. In addition, expression of the receptors of M-CSF (M-CSFR), a monocyte/macrophage specific marker, further confirms the presence of these cells in the testis of the gilthead sea bream throughout the whole reproductive cycle (Fig. 2.4) (Liarte *et al.*, unpubl. results). During spermatogenesis, these cells appeared close to Leydig cells clusters, suggesting a physiological connection between both, as occurs in mammals

(Chaves-Pozo *et al.*, unpubl. results). Interestingly, real-time PCR experiments show that the highest mRNA levels of M-CSFR occur at spermatogenesis and spawning, and also at the beginning of testicular involution (Chaves-Pozo *et al.* and Liarte *et al.*, unpublished results), suggesting a pivotal role for macrophages in the regulation of spermatogenesis, as occurs in mammals, but also in the involutive process that takes place during sex change.

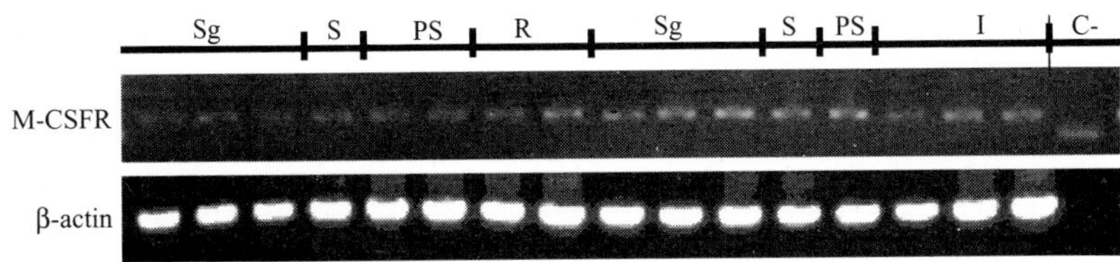

Fig. 2.4 M-CSFR expression assayed throughout two consecutive reproductive cycles by RT-PCR. Sg, spermatogenesis; S, spawning; PS, post-spawning; R, resting and I, involution stages; C-, negative control.

Lymphocytes

T and B cells are the acknowledged cellular pillars of adaptive immunity. T cells are primarily responsible for cell-mediated immunity, while B lymphocytes are responsible for humoral immunity, but they work together and with other types of cells to mediate effective adaptive immunity (Pancer and Cooper, 2006). Approximately 15% of immune cells in the normal adult testis were shown to be lymphocytes (Hedger, 1997). Most of these lymphocytes expressed T cell markers with a predominance of CD8+ T cells, whereas B cells were not detectable (El Demiry *et al.*, 1987; Pöllänen and Niemi, 1987; Hedger, 1997). In spite of the relatively small number of lymphocytes, the testicular immune-privilege may be a localized phenomenon affecting T cell activation and maturation events (Hedger, 1997). T cell inhibition within the testis occurs by specific immunosuppressive cytokines produced by macrophages and/or Sertoli cells, including transforming growth factor β (TGFβ), activin and the cell-surface receptor Fas ligand (FasL) (De Cesaris *et al.*, 1992; Hedger and Clarke, 1993; Pöllänen *et al.*, 1993; Bellgrau *et al.*, 1995). Recently, it has been demonstrated that gonadotropin-releasing hormone (GnRH)-1 is synthesized in lymphocytes of red drum (Mohamed and Khan, 2006), suggesting that testicular lymphocytes, by producing GnRH-1, might contribute to the hormonal axis feedback that regulate testicular functions.

In gilthead seabream, lymphocytes have been observed in the testis and the expression of immunoglobulins (Ig) M and T cell receptor (TCR), the specific markers for lymphocytes B and T respectively, has also been demonstrated (Fig. 2.5) (Liarte *et al.*, unpubl. results). Moreover, real-time PCR analysis show that lymphocytes increased during the testicular regression process that occurs in both seasonal testicular involution and sex change (Chaves-Pozo *et al.* and Liarte *et al.*, unpubl. results).

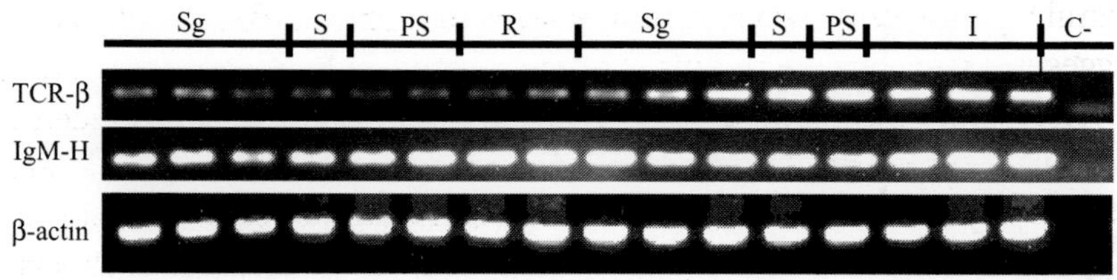

Fig. 2.5 IgM heavy chain and TCR-β chain gene expression assayed throughout two consecutive reproductive cycles by RT-PCR. Sg, spermatogenesis; S, spawning; PS, post-spawning; R, resting and I, involution stages; C-, negative control.

Acidophilic Granulocytes

In gilthead sea bream, acidophilic granulocytes have been characterized by using a monoclonal antibody, which is specific to these cells (Sepulcre *et al.*, 2002; Chaves-Pozo *et al.*, 2005b). The acidophilic granulocytes of the testis showed the ultrastructural characteristics of the head-kidney (Meseguer *et al.*, 1994) and blood (López Ruiz *et al.*, 1992) acidophilic granulocytes. However, during the testicular involution prior to ovarian development, the ultrastructural features of testicular acidophilic granulocytes secretory granules are heavily modified probably due to their involvement in tissue remodelling (Liarte *et al.*, unpubl. results).

The acidophilic granulocytes of gilthead seabream display similar functions to human neutrophils despite their opposite staining pattern. In short, they are the most abundant circulating granulocytes and are recruited rapidly from the head-kidney—the main haematopoietic organ in fish—to the site of inflammation (Chaves-Pozo *et al.*, 2004, 2005c) and are highly specialized to attach, internalize, and kill bacteria by the production of ROIs (Meseguer *et al.*, 1994; Sepulcre *et al.*, 2002; Chaves-Pozo *et al.*, 2004, 2005c).

In contrast with mammalian neutrophils, acidophilic granulocytes of the gilthead seabream are present in the testis in all stages of the reproductive cycle located in the interstitial tissue. However, during testis, remodelling after the shedding of spermatozoa (Chaves-Pozo *et al.*, 2005a) and prior to sex change (Liarte *et al.*, unpubl. results), they sharply increased in number and appeared in the germinal epithelium, closed to spermatogonia, and in the lumen of the seminiferous tubules (Fig. 2.6). In fact, the highest number of testicular acidophilic granulocytes occurred at testicular involution prior to sex change when the remodelling process reached its maximum level. Interestingly, the testicular acidophilic granulocytes do not proliferate in the testis (Chaves-Pozo *et al.*, 2003), suggesting that they are recruited from the head-kidney, depending on the physiological status of the testis. In fact, soluble factors produced by testicular cells are positive recruitment factors for head-kidney acidophilic granulocytes (Chaves-Pozo *et al.*, 2005b). In addition, these data also support the hypothesis that in teleost fish, the testicular leukocytes

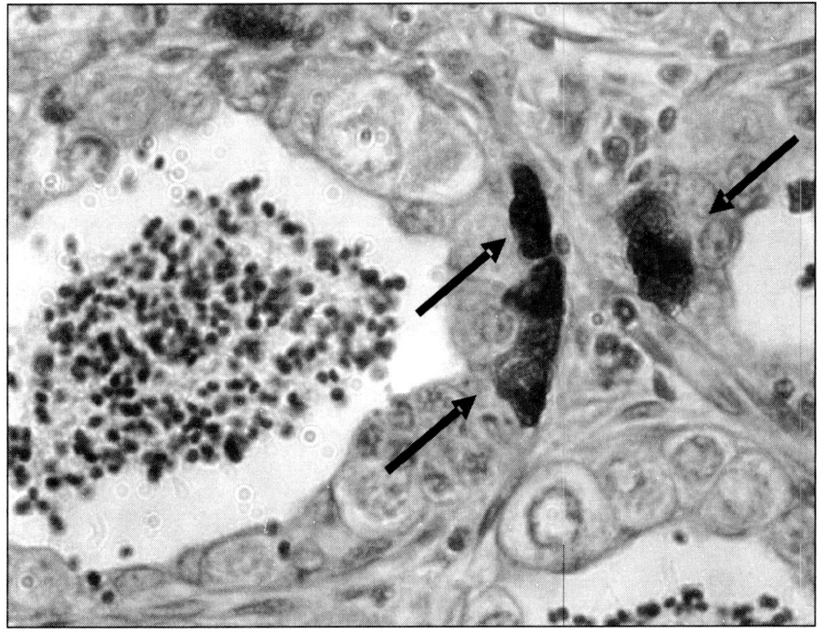

Fig. 2.6 Testicular acidophilic granulocytes (arrows) located in the germinal epithelium immunostained with the monoclonal antibody specific against acidophilic granulocytes from the gilthead sea bream and counterstained with hematoxylin. ×100.

infiltrate the testis at certain stages (Besseau and Faliex, 1994; Bruslé-Sicard and Fourcault, 1997), while in condrictian fish, the testis is a haematopoietic organ (Zapata *et al.*, 1996).

Until now, monocytes/macrophages were believed to be the only innate immune cells able to develop into functional subsets, targeting a tissue and displaying tissue-specific functional pattern presumably under the influence of tissue-specific factors (Stout and Suttles, 2004), whereas neutrophils only infiltrate the tissues upon infection or inflammation (Roit *et al.*, 2001). One exception to this pattern is the female reproductive tract where the recruitment and infiltration of large numbers of circulating eosinophils have been described (Rytomaa, 1960; Gouon-Evans and Pollard, 2001). Interestingly, the acidophilic granulocytes of gilthead sea bream males are able to specifically target the testis, in response to a physiological need and display modified functions. More specifically, testicular acidophilic granulocytes show impaired phagocytic and ROI production activities compared with their head-kidney counterparts, although they are the main type able to produce ROIs in the testis upon phorbol miristate acetate (PMA) stimulation (Chaves-Pozo *et al.*, 2005b). Furthermore, testicular gilthead sea bream acidophilic granulocytes are able to constitutively produce IL-1β (Chaves-Pozo *et al.*, 2003), whereas head-kidney, peripheral blood, and peritoneal exudates in gilthead sea bream only produce IL-1β upon activation (Chaves-Pozo *et al.*, 2004). In fact, testicular soluble factors or cells, pointing to the functional plasticity of acidophilic granulocytes, modulate the phagocytic activities of head-kidney acidophilic granulocytes. It is of note that isolated head-kidney acidophilic granulocytes respond better to testicular conditioned medium than whole head-kidney cell suspensions, suggesting that acidophilic granulocyte activities are also influenced by other immune cells. Notably, the presence of testicular cells dramatically inhibits the production of ROIs by acidophilic granulocytes (Chaves-Pozo *et al.*, 2005b). As regards the functional differences between gilthead sea bream immune and testicular acidophilic granulocytes, it is possible to establish several similarities with the testicular monocyte/macrophage system of mammals (Kern *et al.*, 1995; Meinhardt *et al.*, 1998; Ariyaratne and Mendis-Handagama, 2000; Aubry *et al.*, 2000; Gerdprasert *et al.*, 2002; Hedger, 2002).

In mammals, matrix metalloproteases (MMPs) have been shown to be involved in leukocyte infiltration and tissue remodelling at ovulation, when the changes observed in the ovary resemble a local inflammatory reaction (Espey, 1980, 1994; Gaytan *et al.*, 2003). Furthermore, TNFα induces the rapid release of the MMP9 present in the tertiary granules of human neutrophils (Chakrabarti *et al.*, 2006). It has been suggested that the release of tertiary granules is regulated differently from primary and secondary granules, which may be related with the severe tissue damage than leads to the inappropriate activation of neutrophils in the tissues. Notably, gilthead sea bream testicular acidophilic granulocytes express MMP9- and MMP2-like activities, while their head-kidney counterparts only express MMP2 (Chaves-Pozo *et al.*, unpubl. data). Moreover, gilthead sea bream testis constitutively expresses TNFα (Liarte *et al.*, unpubl.). These results together with the ultrastructure showing by acidophilic granulocytes during involution (see above) suggest that specifically regulated acidophilic granulocytes are involved in the tissue remodelling occurring in the testis of this species.

Taking all these data together, it is tempting to speculate that the activities and secretory granules of acidophilic granulocytes are regulated in the testis to prevent or permit, depending on the stage, the elimination/ damage of germ cells.

CYTOKINES

Cytokines are a broadly defined group of polypeptide mediators involved in the communication network of the immune system (Bellanti *et al.*, 1994). In mammals, an inflammatory insult will result in a cytokine cascade, whereby TNFα is released, followed by IL-1β and the IL-6. Downstream of these cytokines, chemokines are released as potent chemoattractants to induce the migration of neutrophils and macrophages to the site of inflammation (Abbas *et al.*, 1995).

Cytokines have been implicated as novel growth and differentiation factors in the regulation of cells in both compartments of the testis (Schlatt *et al.*, 1997). They are produced within the testis even in the absence of inflammation or immune activation events (Hedger and Meinhardt, 2003) and have direct effects on testicular cell functions, controlling spermatogenic growth and differentiation, although they also have direct, mostly inhibitory, effect on Leydig cell steroidogenesis (Moore and Hutson, 1994). Cytokines are also important for the integration of the neuro-endocrine-immune network that controls testicular function (Rivier and Rivest, 1991; Turnbull and Rivier, 1995; Mayerhofer *et al.*,

1996). In the testis, Sertoli and germ cells produce a number of cytokines, including members of the TGFβ superfamily (e.g., TGFβs, activins, inhibins), platelet-derived growth factor (PDGF), ILs (e.g., IL-1, IL-6, IL-11), tumor necrosis factor (e.g., TNFα, Fas L), interferons (e.g., IFNα, IFNγ), fibroblast growth factor (FGF), nerve growth factor (NFG), and stem cell factor (SCF) (Xia *et al.*, 2005). These cytokines probably mediate cross talk between Sertoli and germ cells to facilitate germ cell movement across the seminiferous epithelium and other cellular events during the epithelial cycle as germ cell differentiation (Xia *et al.*, 2005).

Recently, cytokines, which have been well characterized within mammals, have begun to be cloned and sequenced within non-mammalian vertebrates, including amphibians, birds, bony fish, cartilaginous fish and jawless fish (Scapigliati *et al.*, 2000; Bird *et al.*, 2002). Molecules related to vertebrate cytokine receptors have also been cloned in many vertebrate groups (Bird *et al.*, 2002). This chapter focuses on three of the best characterized fish cytokines—IL-1β, TNFα and TGFβ1—which have recently been cloned in gilthead sea bream and produced as recombinant proteins; their biological activities has been tested *in vitro* and *in vivo* (García-Castillo *et al.*, 2004; Pelegrín *et al.*, 2004; Fernández-Alacid *et al.*, unpubl. results). Although general data on the role of cytokines in fish testis are not available as yet, our data in gilthead sea bream suggest that they are specific factors involved in the physiology of the fish testis.

Interleukin-1β

In mammals, IL-1 occurs as two isoforms, IL-1α and IL-1β. They are quite different structures, but are able to bind to the same receptor and exert similar effects (Dinarello, 1996). Both isoforms are produced in abundance by activated monocytes and macrophages in response to lipopolysacharides (LPS) (Roux-Lombard, 1998), but they can also be induced in other cell types. Other cells of a non-immune origin such as fibroblasts, epithelial cells and keratinocytes have also been demonstrated to produce IL-1. Within the immune response, they have numerous effects most of them pro-inflammatory, or immune stimulatory. However, IL-1 also triggers the acute phase response and the release of glucocorticoids from the adrenal glands (Engelsma *et al.*, 2002).

In the mammalian testis, IL-1 system has been suggested to be involved in cell-cell cross talk (Huleihel and Lunenfeld, 2002). Sertoli cells, Leydig cells and macrophages produce both IL-1α and IL-1β in the testis (Hedger, 1997; Hoek *et al.*, 1997; Hedger and Meinhardt, 2003), although its role in this organ is controversial. Some studies have found that both IL-1α and IL-1β are potent growth factors for spermatogonia and Leydig cells (Pöllänen *et al.*, 1989; Parvinen *et al.*, 1991; Khan *et al.*, 1992) and inhibitors of Leydig cell androgen production (Calkins *et al.*, 1988). However, Cohen and Pollard (1998) have reported that mice lacking a functional type I IL-1 receptor are fertile and have normal testosterone levels.

Fish seem to produce only IL-1β, since no apparent homologues for IL-1α have been identified in the fish genome (Bird *et al.*, 2002). It has been demonstrated that IL-1β is intracellularly accumulated by testicular acidophilic granulocytes of the gilthead sea bream, although we cannot exclude the possibility that other testicular cell types, such as Sertoli cells, Leydig cells or even macrophages, produce this cytokine at lower levels (Chaves-Pozo *et al.*, 2003), as has been described in mammals (Kern *et al.*, 1995; Cudicini *et al.*, 1997; Hedger and Meinhardt, 2003). However, it remains to be elucidated whether this intracellular accumulation of IL-1β by testicular acidophilic granulocytes represents the production of the cytokine by these cells or its uptake (Chaves-Pozo *et al.*, 2003). Interestingly, 11-KT enhanced, while E_2 inhibited, pro-IL-1β intracellular accumulation by LPS/DNA-stimulated head-kidney acidophilic granulocytes in a dose-dependent manner. These effects might suggest a role for these hormones in the regulation of IL-1β intracellular accumulation via acidophilic granulocytes of the testis, although other factors are probably involved in this process since these two hormones on their own failed to promote the intracellular accumulation of IL-1β in resident head-kidney acidophilic granulocytes. Despite the fact that IL-1β is produced by phagocytic cells, it cannot be ruled out that 11-KT and E_2 may also affect the lymphocytes present in the head-kidney cell suspensions, which, in turn, might modulate the accumulation of IL-1β by acidophilic granulocytes (Chaves-Pozo *et al.*, 2003). Finally, although little is known about the biological activity of IL-1 in the fish testis, a heterologous recombinant cytokine, murine IL-1β, has been seen to inhibit basal and human chorionic gonadotrophin (hCG)-stimulated testosterone production in the goldfish testis (Lister and van der Kraak, 2002). IL-1β is expressed during the reproductive cycles of gilthead sea bream (Fig. 2.7).

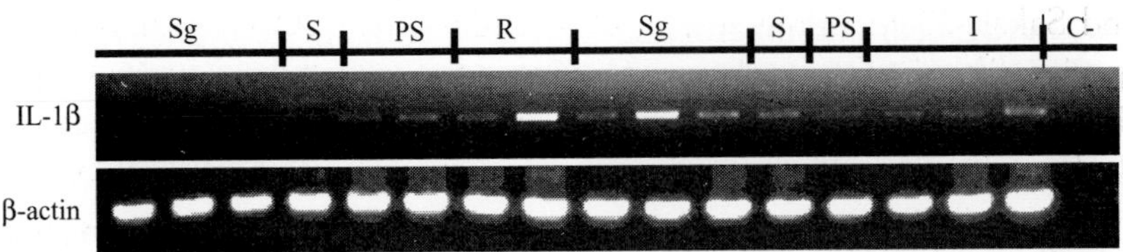

Fig. 2.7 IL-1β gene expression assayed throughout two consecutive reproductive cycles by RT-PCR. Sg, spermatogenesis; S, spawning; PS, post-spawning; R, resting and I, involution stages; C-, negative control.

Tumor Necrosis Factor α

The TNF ligand superfamily includes 19 proteins that share a common structure and biological activities (Aggarwal, 2003). Some of the most studied members of the TNF superfamily are TNFα, TNFβ and Fas L. At least 41 TNF receptors have been described to date. While two TNFs (α and β) are present in mammals, it appears that only one form of TNF is found in fish where it is more similar in structure and genomic organization to mammalian TNFα (Goetz *et al.*, 2004).

TNFα is a pleitropic pro-inflammatory cytokine produced by numerous immune cells during acute inflammation (Wang *et al.*, 2003). TNFα acts as a mediator for different cellular responses, including lymphocyte and leukocyte activation and migration, cell proliferation, differentiation and apoptosis (Wang *et al.*, 2003). However, its influence can be either beneficial or harmful, depending on the amount produced, time course and distribution of the released protein (Manogue *et al.*, 1991; Aggarwal, 2003). Moreover, TNF is a key regulator of inflammation, which is found at the sites of acute and chronic inflammation (Tracey and Cerami, 1994) and is often associated with neutrophil infiltration and activation, resulting in ROI generation and release of granule contents (Chakrabarti *et al.*, 2006). TNFα ligand exerts its different cellular and pathological effects by binding to its receptors, TNFR1 and TNFR2 (MacEwan, 2002).

TNFα of various species of fish has been cloned, including gilthead sea bream (Hirono *et al.*, 2000; Bobe and Goetz, 2001; Laing *et al.*, 2001; García-Castillo *et al.*, 2002; Zou *et al.*, 2002, 2003; Saeij *et al.*, 2003; Savan and Sakai, 2004; Savan *et al.*, 2005; Ordás *et al.*, 2007), although its expression is apparently dependent on the cell type or tissue (Laing *et al.*, 2001; Zou *et al.*, 2003; Ordás *et al.*, 2007) and on the stimulus used (García-Castillo *et al.*, 2002; Mackenzie *et al.*, 2003; Zou *et al.*, 2003; Savan and Sakai, 2004; Bridle *et al.*, 2006; Ordás *et al.*, 2007). Moreover, fish species differ in their inflammatory response as well as in the pathogens that are able to induce the expression of pro-inflammatory cytokines (Van Reth *et al.*, 1999, 2002; Thanawongnuwech *et al.*, 2004). The biological activity of fish TNFα has been studied in the gilthead sea bream (García-Castillo *et al.*, 2004), where TNFα conserves *in vivo* pro-inflammatory activities. Thus, when injected intraperitoneally, gilthead sea bream TNFα (sbTNFα) is biologically active and able to regulate the main activities of innate immune cells at both the local and systemic levels, including the recruitment of phagocytes to the site of injection, the priming of phagocyte respiratory burst and the induction of granulopoiesis. Further, sbTNFα is also able to regulate cellular proliferation *in vitro* and does not show any apoptotic or cytotoxic effect on leukocytes, but rather a strong growth-promoting effect both *in vitro* and *in vivo*. One of the most important findings revealed by these studies is that mammalian and fish TNFα show restricted species specificity. Thus, human TNFα is able to kill murine L929 cells, whereas sea bream TNFα cannot. Conversely, human TNFα is unable to affect the proliferation of sea bream leukocytes, while seabream TNFα is a strong growth-promoting factor for such cells.

In the murine testis, the TNFα has been found to be expressed in pachytene spermatocytes and round spermatids (De *et al.*, 1993). Moreover, TNFa is produced by activated testicular macrophages *in vitro* (Xiong and Hales, 1993). Similar to IL-1, TNF inhibits Leydig cells steroidogenesis (Gómez *et al.*, 1997; Hong *et al.*, 2004). Observations made in the human testis suggest that TNF might play a role in controlling the efficiency of spermatogenesis, inhibiting germ cell apoptosis by regulating the level of FasL (Pentikainen *et al.*, 2001). The Fas system is a potential mechanism for transmission of the apoptotic signal to germ cells during regression in seasonal breeders, although multiple apoptotic pathways probably contribute to testicular regression (Young and Nelson, 2001). TNFα has also been proposed to play a pivotal role in blood-testis barrier dynamics through its effects on the homeostasis of extracellular matrix proteins (Siu and Cheng, 2004a, b).

In gilthead sea bream, the TNFα is expressed during spermatogenesis and post-spawning (Fig. 2.8) (Fernández-Alacid *et al.*, unpubl. results), suggesting a role for this cytokine not only in sperm production but also in tissue remodelling and/or spermatogonia proliferation.

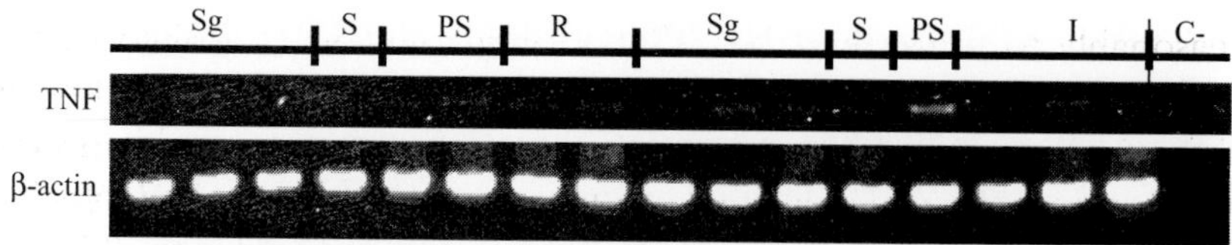

Fig. 2.8 TNFα gene expression assayed throughout two consecutive reproductive cycles by RT-PCR. Sg, spermatogenesis; S, spawning; PS, post-spawning; R, resting and I, involution stages; C-, negative control.

Transforming Growth Factor β

TGFβs are regulatory molecules with pleiotropic effects on cell proliferation, differentiation, migration and survival, affecting multiple biological processes, including development, carcinogenesis, fibrosis, wound healing and immune responses (Blobe *et al.*, 2000). TGFβs belongs to the TGFβ superfamily, with additional members including bone morphogenetic proteins, activins and growth differentiation factors (Chang *et al.*, 2002). There are three homologous TGFβ isoforms in mammals, TGFβ1, TGFβ2, and TGFβ3, which are encoded by different genes (Govinden and Bhoola, 2003). TGFβ1 is the predominant isoform expressed in the immune system, but all three isoforms have similar properties *in vitro*. The TGFβ superfamily mediates its biological functions via binding type I (Tβ-RI), II (Tβ-RII) and III (Tβ-RIII) transmembrane serine/threonine kinase receptors. Recently, genes corresponding to the TGFβ family have been cloned (Hardie *et al.*, 1998; Laing *et al.*, 1999, 2000) including the TGFβ1 in gilthead sea bream (Tafalla *et al.*, 2003).

The pivotal function of TGFβ in the immune system is to maintain tolerance by regulating lymphocyte proliferation, differentiation and survival. In addition, TGFβ controls the initiation and resolution of the inflammatory responses through the regulation of chemotaxis, and the activation and survival of lymphocytes, natural killer cells, dendritic cells, macrophages, mast cells and granulocytes. The regulatory activity of TGFβ is modulated by the cell differentiation state and by the presence of inflammatory cytokines and co-stimulatory molecules. Collectively, TGFβ inhibits the development of immunopathology to self or non-harmful antigens without compromising immune responses to the pathogens (Li *et al.*, 2006).

Given the diverse roles of TGFβ in the regulation of cell differentiation and proliferation in tissue development and repair, it seems reasonable to think that the TGFβ family members contribute to the molecular regulation of reproductive events. Many studies implicate the three isoforms in almost every aspect of the reproductive function, spermatogenesis and steroidogenesis (Ingman and Robertson, 2002). The primary actions of the TGFβ are to enhance formation of the extracellular matrix and to inhibit the proliferation of most cells (Lawrence, 1996), inhibiting progression into late G1 of the cell cycle (Itman *et al.*, 2006). There is a differential production of TGFβ and TGFβ receptors in the testis during development (Ingman and Robertson, 2002; Lui *et al.*, 2003). TGFβ1 is expressed by both somatic cells (Sertoli cells, peritubular myoid cells and macrophages) (Mullaney and Skinner, 1993) and germ cells (Watrin *et al.*, 1991), while TGFβ2 and TGFβ3 are expressed only by somatic cells. TGFβ-RI and II are expressed in greatest abundance in the immature testis (Le Magueresse *et al.*, 1995). Thus, it has been demonstrated that the TGFβ isoforms are differentially expressed in aged rat testis and co-localized in interstitial tissues (Jung *et al.*, 2004). Moreover, they induce pubertal male germ death dose-dependently mediated by the mitochondrial pathway (Konrad *et al.*, 2006). TGFβ1 and TGFβ2 have an age-dependent negative effect on the development of the gonocytes in the rat testis *in vitro* by apoptosis (Olaso *et al.*, 1998), just as TGFβ1 does in various other cell types (Rotello *et al.*, 1991, Bursch *et al.*, 1993), acting directly on these cells (Olaso *et al.*, 1998). The actions of TGFβ in testicular target cells are influenced by endocrine hormones and sex steroids (Ingman and Robertson, 2002).

TGFβ has different functions in the mammal testis, where it stimulates or inhibits Leydig cell steroidogenesis depending on its level in the microenvironment of the testis. It also determines germ cell numbers in the seminiferous epithelium via its effect on germ cell division and apoptosis, activates gene transcription, increases the synthesis of ECM proteins, antagonizes LH action in Leydig cells, attenuates FSH action in Sertoli cells and regulates cell shape and chemotrophic effects on cell migration (Lui *et al.*, 2003).

No information on the expression or biological activity of TGFβ in the testis of fish exists. However, we have found that TGFβ1 is mainly expressed in spermatogenesis and post-spawning in the gilthead sea bream testis (Fig. 2.9), coinciding with the highest expression of TNFα (see above) (Fernández-Alacid *et al.*, unpubl. results). These results suggest an important role for cytokines in the regulation of testicular functions in fish. However, studies aimed at the characterization of the biological activity of these cytokines in the testicular cell functions and the cells responsible for their production are needed to shed more light on the interactions between the immune and reproductive systems of fish.

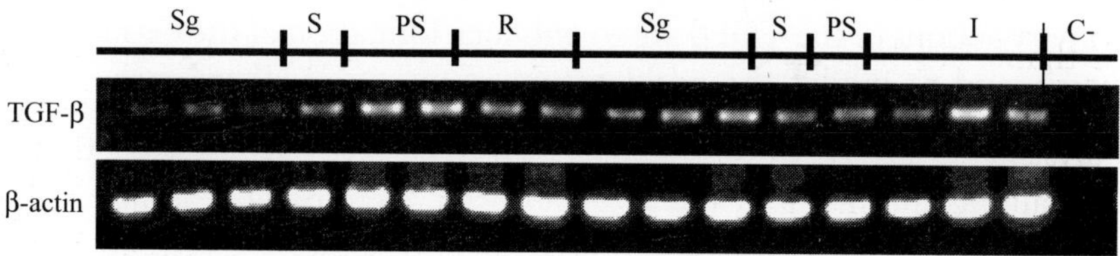

Fig. 2.9 TGFβ1 gene expression assayed throughout two consecutive reproductive cycles by RT-PCR. Sg, spermatogenesis; S, spawning; PS, post-spawning; R, resting and I, involution stages; C-, negative control.

SPECULATION AND CONCLUSIONS

Being a protandric seasonal breeding teleost fish, gilthead sea bream provides a useful model for studying the manner in which immune cells and cytokines influence the testicular involution that occurs after spawning of males and prior to sex change. Testicular acidophilic granulocytes constitute a population whose exact localization and numbers are related with the stage of the reproductive cycle, but whose main activities are immunosuppressed. Moreover, testicular cells and/or soluble factors modulate both their infiltration into the testis and their activities. Other immune cells such as lymphocytes and macrophages have also been observed in the testis of fish but the exact role of these cells is unknown. More studies are necessary to know the possible involvements of these cells in the immunosurveillance of this organ as well as in the regulation of sperm and steroid hormone production. Finally, the expression of several cytokines in the testis of fish during different reproductive stages further suggests an important contribution of the immune system in the physiology of the testis in this group of animals. As spermatogenesis and its regulation are highly conserved in all the vertebrates, we believe that further knowledge of the mechanisms that orchestrate the physiological function of these cells and factors in the fish testis might throw light on the privileged immune status of the testis and even on the development of autoimmune diseases, which could well be of use in clinical applications. Perhaps the best approach will be to use the detailed knowledge that we have of zebrafish genetics together with *in vivo* imaging using fluorescent markers targeting germ and/or somatic cells of the testis of this species, which may contribute to illuminating interactions between the immune and reproductive systems.

Acknowledgments

We thank the 'Servicio de Apoyo a las Ciencias Experimentales' of the University of Murcia for their assistance with electron microscopy, image analysis and cell culture, the Spanish Oceanographic Institute for maintaining the fish and Dr. V. Mulero for his critical reading of the manuscript. We also thank the financial support of the Fundación Séneca, Coordination Centre for Research, CARM (grant 07702/GERM/07 to A. García-Ayala) and University of Murcia (post-doctoral contract to E. Chaves-Pozo).

References

Abbas, A.K., A.H. Lichtman and J.S. Pober. 1995. Citoquinas. In: *Inmunología Celular y Molecular*, A.K. Abbas, A.H. Lichtman and J.S. Pober (eds.). McGraw-Hill-Interamericana de España, Madrid, pp. 267-292.

Abraham, M., E. Rahamim, H. Tibika and E. Golenser. 1980. The blood-testis barrier in *Aphanius dispar* (Teleostei). *Cell and Tissue Research* 211: 207-214.

Aggarwal, B.B. 2003. Signalling pathways of the TNF superfamily: a double-edged sword. *Nature Review of Immunology* 3: 745-756.

Agulleiro, B., M.P. García Hernández and A. García Ayala. 2006. Teleost adenohypophysis: Developmental aspects. In: *Fish Endocrinology*, M Reinecke, G. Zaccone and B.G. Kapoor (eds.). Science Publishers, Enfied, NH, USA, Vol. 1, pp. 289-323.

Amer, M.A., T. Miura, C. Miura and K. Yamauchi. 2001. Involvement of sex steroid hormones in the early stages of spermatogenesis in Japanese huchen (*Hucho perryi*). *Biology of Reproduction* 65: 1057-1066.

Andreassen, T.K., K. Skjoedt, I. Anglade, O. Kah and B. Korsgaard. 2003. Molecular cloning, characterisation, and tissue distribution of oestrogen receptor alpha in eelpout (*Zoarces viviparous*). *General and Comparative Endocrinology* 132: 356-368.

Ariyaratne, H.B. and S.M. Mendis-Handagama. 2000. Changes in the testis interstitium of Sprague-Dawley rats from birth to sexual maturity. *Biology of Reproduction* 62: 680-690.

Aubry, F., C. Habasque, A.P. Sat, B. Jégou and M. Samsom. 2000. Expression and regulation of the CC-chemokine monocyte chemoattractanct protein-1 in rat testicular cells in primary culture. *Biology of Reproduction* 62: 1427-1435.

Bellanti, J.A., J.V. Kadlec and A. Escobar-Gutiérrez. 1994. Cytokines and the immune response. *Pediatrics Clinic of North America* 41: 597-621.

Bellgrau, D., D. Gold, H. Selawry, J. Moore, A. Franzusoff and R.C. Duke. 1995. A role for CD95 ligand in preventing graft rejection. *Nature (London)* 377: 630-632.

Bernhagen, J., T. Calandra, R.A. Mitchell, S.B. Martin, K.J. Tracey, W. Voelter, K.R. Manogue, A. Cerami and R. Bucala. 1993. MIF is a pituitary-derived cytokine that potentiates lethal endotoxaemia. *Nature (London)* 365: 756-759.

Besseau, L. and E. Faliex. 1994. Resorption of unemitted gametes in *Lithognathus mormyrus* (Sparidae, Teleostei): a possible synergic action of somatic immune cells. *Cell and Tissue Research* 276: 123-132.

Billard, R. 1986. Spermatogenesis and spermatology of some teleost fish species. *Reproduction Nutritional Development* 28: 877-920.

Billard, R. and F. Takashima. 1983. Resorption of spermatozoa in the sperm duct of rainbow trout during the post-spawning period. *Bulletin of Japanese Society of Scientific Fisheries* 49: 387-392.

Billard, R., A. Fostier, C. Weil and B. Breton. 1982. Endocrine control of spermatogonia in teleost fish. *Canadian Journal of Fisheries and Aquatic Sciences* 39: 65-79.

Bird, S., J. Zou, T. Wang, B. Munday, C. Cunningham and C.J. Secombes. 2002. Evolution of interleukin-1 beta. *Cytokine Growth Factor Reviews* 13: 483-502.

Blobe, G.C., W.P. Schiemann and H.F. Lodish. 2000. Role of transforming growth factor β in human disease. *The New England Journal of Medicine* 342: 1350-1358.

Bobe, J. and F.W. Goetz. 2001. Molecular cloning of a TNF receptor and two TNF ligands in the fish ovary. *Comparative Biochemistry and Physiology* B 129: 475-481.

Bon, E., B. Breton, M.S. Govoroun and F. Le Menn. 1999. Effects of accelerated photoperiod regimes on the reproductive cycle of the female rainbow trout: seasonal variations of plasma gonadotropins (GTH I and GTH II) levels correlated with ovarian follicle growth and egg size. *Fish Physiology and Biochemistry* 20: 143-154.

Borg, B. 1994. Androgens in teleost fishes. *Comparative Biochemistry and Physiology* C109: 219-245.

Breton, B., M. Govoroun and T. Mikolajczyk. 1998. GTH I and GTH II secretion profiles during the reproductive cycle in female rainbow trout: relationship with pituitary responsiveness to GnRHa stimulation. *General and Comparative Endocrinology* 111: 38-50.

Bridle, A.R., R.N. Morrison and B.F. Nowak. 2006. The expression of immune-regulatory genes in rainbow trout, *Oncorhynchus mykiss*, during amoebic gill disease (AGD). *Fish and Shellfish Immunology* 20: 346-364.

Puslé-Sicard, S. and B. Fourcault. 1997. Recognition of sex-inverting protandric *Sparus aurata*: ultrastructural aspects. *Journal of Fish Biology* 50: 1094-1103.

Bursch, W., F. Oberhammer, R.L. Jirtle, M. Askari, R. Sedivy, B. Grasl-Kraupp, A.F. Purchio and R. Schulte-Hermann. 1993. Transforming growth factor-beta 1 as a signal for induction of cell death by apoptosis. *British Journal of Cancer* 67: 531-536.

Calandra, T., J. Bernhagen, R.A. Mitchell and R. Bucala. 1994. The macrophage is an important and previously unrecognized source of macrophage migration inhibitory factor. *Journal of Experimental Medicine* 179: 1895-1902.

Calkins, J.H., M.M. Sigel, H.R. Nankin and T. Lin. 1988. Interleukin-1 inhibits Leydig cell steroidogenesis in primary culture. *Endocrinology* 123: 1605-1610.

Carreau, S., S. Lambard, C. Delalande, I. Denis-Galeraud, B. Bilinska and S. Bourguiba. 2003. Aromatase expression and role of estrogens in male gonad: A review. *Reproductive Biology and Endocrinology* 1: 35.

Carrillo, M. and S. Zanuy. 1977. Quelques observations sur le testicule chez *Spicara chryselis*. *Investigaciones Pesqueras* 41: 121-146.

Cavaco, J.E.B., C. Vilrokx, V.L. Trudeau, R.W. Schulz and H.J.T. Goos. 1998. Sex steroids and the initiation of puberty in male African catfish (*Clarias gariepinus*). *American Journal of Physiology* 44: R1793-R1802.

Cavaco, J.E.B., J. Bogerd, H.J.T.h. Goos and R.W. Schulz. 2001. Testosterone inhibits 11-ketotestosterone-induced spermatogenesis in African catfish (*Clarias gariepinus*). *Biology of Reproduction* 65: 1807-1812.

Chakrabarti, S., J.M. Zee and K.D. Patel. 2006. Regulation of matrix metalloproteinase-9 (MMP-9) in TNF-stimulated neutrophils: novel pathways for tertiary granule release. *Journal of Leukocyte Biology* 79: 214-222.

Chang, H., C.W. Brown and M.M. Matzuk. 2002. Genetic analysis of the mammalian transforming growth factor-beta superfamily. *Endocrine Reviews* 23: 787-823.

Chaves-Pozo, E., P. Pelegrín, V. Mulero, J. Meseguer and A. García Ayala. 2003. A role for acidophilic granulocytes in the testis of the gilthead sea bream (*Sparus aurata* L., Teleostei). *Journal of Endocrinology* 179: 165-174.

Chaves-Pozo, E., P. Pelegrín, J. García-Castillo, A. García Ayala, V. Mulero and J. Meseguer. 2004. Acidophilic granulocytes of the marine fish gilthead sea bream (*Sparus aurata* L.) produce interleukin-1β following infection with *Vibrio anguillarum*. *Cell and Tissue Research* 316: 189-195.

Chaves-Pozo, E., V. Mulero, J. Meseguer and A. García Ayala 2005a. An overview of cell renewal in the testis throughout the reproductive cycle of a seasonal breeding teleost, the gilthead sea bream (*Sparus aurata* L.). *Biology of Reproduction* 72: 593-601.

Chaves-Pozo, E., V. Mulero, J. Meseguer and A. García Ayala. 2005b. Professional phagocytic granulocytes of the bony fish gilthead sea bream display functional adaptation to testicular microenvironment. *Journal of Leukocyte Biology* 78: 345-351.

Chaves-Pozo, E., P. Muñoz, A. López-Muñoz, P. Pelegrín, A. García Ayala, V. Mulero and J. Meseguer. 2005c. Early innate immune response and redistribution of inflammatory cells in the bony fish gilthead sea bream experimentally infected with *Vibrio anguillarum*. *Cell and Tissue Research* 320: 61-68.

Chaves-Pozo, S., S. Liarte, L. Vargas-Chacoff, A. García-López, V. Mulero, J. Meseguer, J.M. Mancera and A. García Ayala. 2007. 17beta-estradiol triggers postspawning in spermatogenically active gilthead sea bream (*Sparus aurata* L.) males. *Biology of Reproduction* 76: 142-148.

Choi, C.Y. and H.R. Habibi. 2003. Molecular cloning of estrogen receptor a and expression pattern of estrogen receptor subtypes in male and female goldfish. *Molecular and Cellular Endocrinology* 204: 169-177.

Cohen, P.E. and J.W. Pollard. 1994. Use of the osteopetrotic mouse for studying macrophages in the reproductive tract. In: *Immunobiology of Reproduction*, J.S. Hunt (ed.). Springer-Verlag, New York, pp. 104-122.

Cohen, P.E. and J.W. Pollard. 1998. Normal sexual function in male mice lacking a functional type I interleukin-1 (IL-1) receptor. *Endocrinology* 139: 815-818.

Cohen, P.E., K. Nishimura, L. Zhu and J.W. Pollard. 1999. Macrophages: important accessory cells for reproductive function. *Journal of Leukocyte Biology* 66: 765-772.

Cook, J. 1994. *The effects of stress, background colour and steroid hormones on the lymphocytes of rainbow trout (Oncorhynchus mykiss)*. Ph.D. Thesis, University of Sheffield, UK.

Cudicini, C., H. Lejeune, E. Gómez, E. Bosmanas, F. Ballet, J. Sáez and B. Jégou. 1997. Human Leydig cells and Sertoli cells are producers of interleukins-1 and -6. *Journal of Clinical Endocrinology and Metabolism* 82: 1426-1433.

De, S.K., H.L. Chen, J.L. Pace, J.S. Hunt, P.F. Terranova and G.C. Enders. 1993. Expression of tumor necrosis factor-α in mouse spermatogenic cells. *Endocrinology* 133: 389-396.

De Cesaris, P., A. Filippini, C. Cervelli, A. Riccioli, S. Muci, G. Starace, M. Stefanini and E. Ziparo. 1992. Immunosuppressive molecules produced by Sertoli cells cultured in vitro: biological effects on lymphocytes. *Biochemical and Biophysical Research Communications* 186: 1639-1646.

Dinarello, C.A. 1996. Biologic basis for interleukin-1 in disease. *Blood* 87: 2095-2147.

El Demiry, M.I., T.B. Hargreave, A. Busuttil, R. Elton, K. James and G.D. Chisholm. 1987. Immunocompetent cells in human testis in health and disease. *Fertility and Sterility* 48: 470-479.

Engelsma, M.Y., M.O. Huising, W.B. van Muiswinkel, G. Flik, J. Kwang, H.F.J. Savelkoul and L. Verburg-van Kemenade. 2002. Neuroendocrine-immnune interactions in fish: a role for interleukin-1. *Veterinary Immunology and Immunopathology* 87: 467-479.

Espey, L.L. 1980. Ovulation as an inflammatory reaction—a hypothesis. *Biology of Reproduction* 22: 73-106.

Espey, L.L. 1994. Current status of the hypothesis that mammalian ovulation is comparable to an inflammatory reaction. *Biology of Reproduction* 50: 233-238.

García Ayala, A., M. Villaplana, M.P. García Hernández, E. Chaves-Pozo and B. Agulleiro. 2003. FSH-, LH-, and TSH-expressing cells during development of *Sparus aurata* L. (Teleostei). An immunocytochemical study. *General and Comparative Endocrinology* 134: 72-79.

García-Castillo, J., P. Pelegrín, V. Mulero and J. Meseguer. 2002. Molecular cloning and expression analysis of tumor necrosis factor a from a marine fish reveal its constitutive expression and ubiquitous nature. *Immunogenetics* 54: 200-207.

García-Castillo, J., E. Chaves-Pozo, P. Olivares, P. Pelegrín, J. Meseguer and V. Mulero. 2004. The tumor necrosis factor α of the bony fish sea bream exhibits the *in vivo* proinflammatory and proliferative activities of its mammalian counterparts, yet it functions in a species-specific manner. *Cell and Molecular Life Sciences* 61: 1331-1340.

García Hernández, M.P., A. García Ayala, M.A. Zandbergen and B. Agulleiro. 2002. Investigation into the duality of gonadotropic cells of Mediterranean yellowtail (*Seriola dumerili*, Risso 1810): immunocytochemical and ultrastructural studies. *General and Comparative Endocrinology* 128: 25-35.

Gaytan, F., G. Carrera, L. Pinilla, R. Aguilar and C. Bellido. 1989. Mast cells in the testis, epididymis and accesory glands of the rat: effects of neonatal steroid treatment. *Journal of Andrology* 10: 351-358.

Gaytan, F., L. Romero, C. Bellido, C. Morales, C. Reymundo and E. Aguilar. 1994. Effects of growth hormone and prolactin on testicular macrophages in long-term hypophysectomized rats. *Journal of Reproductive Immunology* 27: 73-84.

Gaytan, F., C. Morales, C. Bellido, E. Tarradas and J.E. Sánchez-Criado. 2003. Effects of indomethacin on ovarian leukocytes during the periovulatory period in the rat. *Reproductive Biology and Endocrinology* 1: 26-37.

Gérard, N., V. Syed, C.W. Bardin, N. Genetet and B. Jégou. 1991. Sertoli cells are the site of interleukin-1α synthesis in rat testis. *Molecular and Cellular Endocrinology* 82: R13-R16.

Gerdprasert, O., M.K. O´Bryan, D.J. Nikolic-Paterson, K. Sebire, D.M. de Kretser and M.P. Hedger. 2002. Expression of monocyte chemoattractant protein-1 and macrophage colony-stimulating factor in the normal and inflamed rat testis. *Molecular Human Reproduction* 8: 518-524.

Goetz, F.W., J.V. Planas and S. Mackenzie. 2004. Tumor necrosis factors. *Developmental and Comparative Immunology* 57: 774-777.

Gómez, E., G. Morel, A. Cavalier, M.O. Lienard, F. Haour, J.L. Courtens and B. Jégou. 1997. Type I and type II interleukin-1 receptor expressions in rat, mouse, and human testes. *Biology of Reproduction* 56: 1513-1526.

Gómez, J.M., C. Weil, M. Ollitrault, P.Y. Le Bail, B. Breton and F. Le Gac. 1999. Growth hormone (GH) and gonadotropin subunit gene expression and pituitary and plasma changes during spermatogenesis and oogenesis in rainbow trout (*Oncorhynchus mykiss*). *General and Comparative Endocrinology* 113: 413-428.

Gouon-Evans, V. and J.W. Pollard. 2001. Eotaxin is required for eosinophil homing into the stroma of the pubertal and cycling uterus. *Endocrinology* 142: 4515-4521.

Govinden, R. and K.D. Bhoola. 2003. Genealogy, expression, and cellular function of transforming growth factor-β. *Pharmacological Therapeutics* 98: 257-265.

Grier, H.J. 1981. Cellular organisation of the testis and spermatogenesis in fishes. *American Zoologist* 21: 345-357.

Guillemin, G.J. and B.J. Brew. 2004. Microglia, macrophages, perivascular macrophages, and pericytes: A review of function and identification. *Journal of Leukocyte Biology* 75: 388-397.

Hales, D.B. 1996. Leydig cell-macrophage interactions: an overview. In: *The Leydig Cell*, A.H. Payne, M.P. Hardy and L.D. Russell (eds.). IL: Cahe River Press, Vienna, pp. 451-466.

Hales, D.B. 2002. Testicular macropahge modulation of Leydig cell steroidogenesis. *Journal of Reproductive Immunology* 57: 3-18.

Hales, D.B., T. Diemer and K. Hales. 1999. Role of cytokines in testicular function. *Endocrine* 10: 201-217.

Halm, S., G. Martínez-Rodríguez, L. Rodríguez, F. Prat, C.C. Mylonas, M. Carrillo and S. Zanuy. 2004. Cloning, characterisation, and expression of three oestrogen receptors (ERα, ERβ1 and ERβ2) in the European sea bass, *Dicentrarchus labrax*. *Molecular and Cellular Endocrinology* 223: 63-75.

Hardie, L.J., K.J. Laing, G.D. Daniels, P.S. Grabowski, C. Cunningham and C.J. Secombes. 1998. Isolation of the first piscine transforming growth factor beta gene: analysis reveals tissue specific expression and a potential regulatory sequence in rainbow trout (*Oncorhynchus mykiss*). *Cytokine* 10: 555-563.

Hardy, M.P., B.R. Zirkin and L.L. Ewing. 1989. Kinetic studies on the development of the adult population of Leydig cells in the testes of the pubertal rat. *Endocrinology* 124: 762-770.

Harris, J. and D.J. Bird. 2000. Modulation of the fish immune system by hormones. *Veterinary Immunology and Immunopathology* 77: 163-176.

Hawkins, M.B., J.W. Thornton, D. Crews, J.K. Skipper, A. Dotte and P. Thomas. 2000. Identification of a third distinct estrogen receptor and reclassification of estrogen receptors in teleosts. *Proceedings of the National Academy of Sciences of the United States of America* 97: 10751-10756.

Hayes, R., S.J. Chalmers, D.P. Nikolic-Paterson, R.C. Atkins and M.P. Hedger. 1996. Secretion of bioactive interleukin-1 by rat testicular macrophages *in vitro*. *Journal of Andrology* 17: 41-49.

Head, J.R., W.B. Neaves and R.E. Billingham. 1983. Immune privilege in the testis. I. Basic parameters of allograft survival. *Transplantation* 35: 91-95.

Hedger, M.P. 1997. Testicular leukocytes: What are they doing?. *Reviews of Reproduction* 2: 38-47.

Hedger, M.P. 2002. Macrophages and the immune responsiveness of the testis. *Journal of Reproductive Immunology* 57: 19-34.

Hedger, M.P. and L. Clarke. 1993. Isolation of rat blood lymphocytes using a two-step Percoll density gradient: effect of activin (erythroid differentiation factor) on peripheral T-lymphocyte proliferation *in vitro*. *Journal of Immunological Methods* 163: 133-136.

Hedger, M.P. and A. Meinhardt. 2003. Cytokines and the immune testicular axis. *Journal of Reproductive Immunology* 58: 1-26.

Henderson, N.E. 1962. The annual cycle in the testis of eastern brook trout *Salvelinus fontinalis* (Mitchell). *Canadian Journal of Zoology* 40: 631-641.

Hess, R.A. 2003. Estrogen in the adult male reproductive tract: a review. *Reproductive Biology and Endocrinology* 1: 38-52.

Hirono, I., B.-H. Nam, T. Kurobe and T. Aoki. 2000. Molecular cloning, characterization, and expression of TNF cDNA and gene from Japanese flounder *Paralichthys olivaceus*. *Journal of Immunology* 165: 4423-4427.

Hoek, A., W. Allaerts, P.J. Leenen, J. Schoemaker and H.A. Drexhage. 1997. Dendritic cells and macrophages in the pituitary and the gonads. Evidence for their role in the fine regulation of the reproductive endocrine response. *European Journal of Endocrinology* 136: 8-24.

Hong, C.Y., J.H. Park, R.S. Ahn, S.Y. Im, H.S. Choi, J. Soh, S.H. Mellon and K. Lee. 2004. Molecular mechanism of suppression of testicular steroidogenesis by proinflammatory cytokine tumor necrosis factor alpha. *Molecular and Cellular Biology* 24: 2593-2604.

Huleihel, M. and E. Lunenfeld. 2002. Involvement of intratesticular IL-1 systems in the regulation of Sertoli cell functions. *Molecular and Cellular Endocrinology* 187: 125-132.

Hunt, J.S. 1989. Cytokine networks in the uteroplacental unit: macrophages as pivotal regulatory cells. *Journal of Reproductive Immunology* 16: 1-17.

Hutson, J.C. 1992. Development of cytoplasmic digitations between Leydig cells and testicular macrophages of the rat. *Cell and Tissue Research* 267: 385-389.

Hutson, J.C. 1994. Testicular macrophages. *International Review of Cytology* 149: 99-143.

Idler, D.R. and H.C. MacNab. 1967. The biosynthesis of 11-ketotestosterone and 11β-hydroxy-testosterone by Atlantic salmon tissues *in vitro*. *Canadian Journal of Biochemistry* 45: 581-589.

Iida, T., K. Takahashi and H. Wakabayashi. 1989. Decrease in the bacterial activity of normal serum during the spawning period of rainbow trout. *Bulletin of the Japanese Society of Scientific Fisheries* 55: 463-465.

Ingman, W.V. and S.A. Robertson. 2002. Defining the actions of transforming growth factor beta in reproduction. *BioEssays* 24: 904-914.

Itman, C., S. Mendis, B. Barakat and K.L. Loveland. 2006. Focus on TGF-β signalling. All in the family: TGF-β family action in the testis development. *Reproduction* 1470-1626.

Jonsson, C., B.P. Setchell, T. Sultana, M. Holst, M. Parvinen and O. Soder. 2000. Constitutive and inducible production of proinflammatory cytokines by the rat testis. *Andrologia* 32: 63-74.

Jung, J.C., G.T. Park, K.H. Kim, J.H. Woo, J.M. Ann, K.C. Kim, H.Y. Cheng, Y.S. Bae, J.W. Park, S.S. Kang and Y.S. Lee. 2004. Differential expression of transforming growth factor-β in the interstitial tissue of testis during aging. *Journal of Cell Biochemistry* 92: 92-98.

Kern, S., S.A. Robertson, V.J. Mau and S. Maddocks. 1995. Cytokine secretion by macrophages in the rat testis. *Biology of Reproduction* 53: 1407-1416.

Khan, S.A., S.J. Khan and J.H. Dorrington. 1992. Interleukin-1 stimulates deoxyribonucleic acid synthesis in immature rat Leydig cells *in vitro*. *Endocrinology* 131: 1853-1857.

Kime, D.E. 1978. The hepatic catabolism of cortisol in teleost fish: adrenal origin of 11-oxotestosterone precursors. *General and Comparative Endocrinology* 35: 322-328.

Kime, D.E. 1987. The steroids. In: *Fundamentals of Comparative Vertebrate Endocrinology*, I. Chester-Jones, P. Ingleton and J.G. Phillips (eds.). Plenum Press, New York, pp. 3-56.

Kobayashi, Y., T. Kobayashi, M. Nakamura, T. Sunobe, C.E. Morrey, N. Suzuki and Y. Nagahama. 2004. Characterization of two types of cytochrome P450 aromatase in the serial-sex changing gobiid fish, *Trimma okinawae*. *Zoological Science* 21: 417-425.

Konrad, L., M.M. Keilani, L. Laible, U. Nottelmann and R. Hofmann. 2006. Effects of TGF-betas and a specific antagonist on apoptosis of immature rat male germ cells *in vitro*. *Apoptosis* 11: 739-748.

Lahnsteiner, F. and R.A. Patzner. 1990. The mode of male germ cell renewal and ultrastructure of early spermatogenesis in *Salaria* (=*Blennius*) *pavo* (Teleostei: Blenniidae). *Zoologischer Anzeiger* 224: 129-139.

Laing, K.J., L. Pilstrom, C. Cunningham and C.J. Secombes. 1999. TGF-beta3 exists in bony fish. *Veterinary Immunology and Immunopathology* 72: 45-53.

Laing, K.J., C. Cunningham and C.J. Secombes. 2000. Genes for three different isoforms of transforming growth factor-beta are present in plaice (*Pleuronectes platessa*) DNA. *Fish and Shellfish Immunology* 10: 261-271.

Laing, K.J., T. Wang, J. Zou, J. Holland, S. Hong, N. Bols, I. Hirono, T. Aoki and C.J. Secombes. 2001. Cloning and expression analysis of rainbow trout *Oncorhynchus mykiss* tumour necrosis factor-α. *European Journal of Biochemistry* 268: 1315-1322.

Laskin, D.L., B. Weinberg and J.D. Laskin. 2001. Functional heterogeneity in liver and lung macrophages. *Journal of Leukocyte Biology* 70: 163-170.

Lawrence, D.A. 1996. Transforming growth factor-beta: A general review. *European Cytokine Network* 7: 363-374.

Leger, J., J.L.M. Broekhof, A. Brouwer, P.H. Lanser, A.J. Murk, P.T. Van der Saag, A.D. Vethaak, P. Wester, D. Zivkovic and B. van der Burg. 2000. A novel *in vivo* bioassay for (xenooestrogens using transgenic zebrafish. *Environmental Science and Technology* 34: 4439-4444.

Le Magueresse-Battistoni, B., A.M. Morera, I. Goddard and M. Benahmed. 1995. Expression of mRNAs for transforming growth factor-β receptors in the rat testis. *Endocrinology* 136: 2788-2791.

Li, M.O., Y.Y. Wan, S. Sanjabi, A-K.L. Robertson and R.A. Flavell. 2006. Transforming growth factor-β regulation of immune responses. *Annual Review of Immunology* 24: 99-146.

Lister, A. and G. van der Kraak. 2002. Modulation of goldfish testicular testosterone production *in vitro* by tumor necrosis factor alpha, interleukin-1 beta, and macrophage conditioned media. *Journal of Experimental Zoology* 292: 477-486.

Lofts, B. and H.A. Bern. 1972. The functional morphology of steroidogenic tissues. In: *Steroids in Non-mammalian Vertebrates*, D.R. Idler (ed.). Academic Press, New York, pp. 37-126.

Loir, M., P. Sourdaine, S.M. Mendis-Handagama and B. Jégou. 1995. Cell-cell interactions in the testis of teleosts and elasmobranchs. *Microscopic Research and Technique* 32: 533-552.

Lone, K.P., S. Al-Ablani and A. Al-Yaqout. 2001. Steroid hormone profiles and correlative gonadal histological changes during natural sex reversal of sobaity kept in tanks and sea-cages. *Journal of Fish Biology* 58: 305-324.

Loomis, A.K. and P. Thomas. 2000. Effects of estrogens and xenoestrogens on androgen production by atlantic croaker testes *in vitro*: evidence for a nongenomic action mediated by an estrogen membrane receptor. *Biology of Reproduction* 62: 995-1004.

López-Ruiz, A., M.A. Esteban and J. Meseguer. 1992. Blood cells of the gilthead seabream (*Sparus aurata* L.): light and electron microscopic studies. *Anatomical Record* 234: 161-171.

Lui, W-Y., W.M. Lee and C.Y. Cheng. 2003. TGF-βs: their role in testicular function and Sertoli cell tight junction dynamics. *International Journal of Andrology* 26: 147-160.

Lutton, B. and I. Callard. 2006. Evolution of reproductive-immune interactions. *International Comparative Biology* 46: 1060-1071.

MacEwan, D.J. 2002. TNF receptor subtype signalling: differences and cellular consequences. *Cellular Signaling* 14: 477-492.

Mackenzie, S., J.V. Planas and F.W. Goetz. 2003. LPS-stimulate expression of a tumor necrosis factor-alpha mRNA in primary trout monocytes and *in vitro* differentiated macrophages. *Developmental and Comparative Immunology* 27: 393-400.

Magliulo-Cepriano, L., M.P. Schreibman and V. Blüm. 1994. Distribution of variant forms of immunoreactive gonadotropin-releasing hormone and β-gonadotropins I and II in the platyfish, *Xiphophorus maculatus*, from birth to sexual maturity. *General and Comparative Endocrinology* 94: 135-150.

Manogue, K.R., S.J.H. van Deventer and A. Cerami. 1991. Tumour necrosis factor alpha or cachectin. In: *The Cytokine Handbook*, A. Thomson (ed.). Academic Press, New York, pp. 241-256.

Mayerhofer, A., G. Lahr, K. Seidl, B. Eusterschulte, A. Christoph and M. Gratzl. 1996. The neural cell adhesion molecule (NCAM) provides clues to the development of testicular Leydig cells. *Journal of Andrology* 17: 223-230.

Meinhardt, A., M. Bacher, J.R. McFarlane, C.N. Metz, J. Seitz, M.P. Hedger, D.M. de Kretser and R. Bucala. 1996. Macrophage migration inhibitory factor production by Leydig cells: Evidence for a role in the regulation of testicular function. *Endocrinology* 137: 5090-5095.

Meinhardt, A., M. Bacher, C. Metz, R. Bucala, N. Wreford, H. Lan, R. Atkins and M. Hedger. 1998. Local regulation of macrophages subsets in the adult rat testis: examination of the roles of the seminiferous tubules, testosterone, and macrophage migration inhibitory factor. *Biology of Reproduction* 59: 371-378.

Meinhardt, A., J.R. McFarlane, J. Seitz and D.M. de Kretser. 2000. Activin maintains the condensed type of mitochondria in germ cells. *Journal of Cell Science* 112: 1337-1344.

Mendis-Handagama, S.M.L.C., G.P. Risbridger and D.M. de Kretser. 1987. Morphometric analysis of the components of the neonatal and the adult rat testis interstitium. *International Journal of Andrology* 10: 525-534.

Menuet, A., E. Pellegrini, I. Anglade, O. Blaise, V. Laudet, O. Kah and F. Pakdel. 2002. Molecular characterization of three estrogen receptor forms in zebrafish: binding characteristics, transactivation properties, and tissue distributions. *Biology of Reproduction* 66: 1881-1892.

Meseguer, J., A. López-Ruiz and M.A. Esteban. 1994. Cytochemical characterization of leukocytes from the seawater teleost, gilthead sea bream (*Sparus aurata* L.). *Histochemistry* 102: 37-44.

Micale, V., F. Perdichizzi and G. Sanatagelo. 1987. The gonadal cycle of captive white bream, *Diplodus sargus* (L.). *Journal of Fish Biology* 31: 435-440.

Miller, S.C., B.M. Bowman and H.G. Rowland. 1983. Structure, cytochemistry, endocytic activity, and immunoglobulin (Fc) receptors of rat testicular interstitial-tissue macrophages. *American Journal of Anatomy* 168: 1-13.

Miranda, L.A., C.A. Strüssmann and G.M. Somoza. 2001. Immunocytochemical identification of GTH1 and GTH2 cells during the temperature-sensitive period for sex determination in pejerrey, *Odontesthes bonariensis*. *General and Comparative Endocrinology* 124: 45-52.

Miura, T. 1999. Spermatogenetic cycle in fish. In: *Encyclopedia of Reproduction*, E. Knobil and J.D. Neil (eds.). Academic Press, New York, pp. 571-578.

Miura, T. and C. Miura. 2001. Japanese eel: a model for analysis of spermatogenesis. *Zoological Science* 18: 1055-1063.

Miura, T., K. Yamauchi, H. Takahashi and Y. Nagahama. 1991. Hormonal induction of all stages of spermatogenesis *in vitro* in the male Japanese eel (*Anguilla japonica*). *Proceedings of the National Academy of Sciences* 88: 5774-5778.

Miura, T., C. Miura, T. Ohta, M.R. Nader, T. Todo and K. Yamauchi. 1999. Estradiol-17β stimulates the renewal of spermatogonial stem cells in males. *Biochemical and Biophysical Research Communications* 264: 230-234.

Mohamed, J.S. and I.A. Khan. 2006. Molecular cloning and differential expression of three GnRH mRNAs in discrete brain areas and lymphocytes in red drum. *Journal of Endocrinology* 188: 407-416.

Moore, C. and J.C. Hutson. 1994. Physiological relevance of tumor necrosis factor in mediating macrophage-Leydig cell interactions. *Endocrinology* 134: 63-69.

Mullaney, B.P. and M.K. Skinner. 1993. Transforming growth factor-beta (beta 1, beta 2, and beta 3) gene expression and action during pubertal development of the seminiferous tubule: Potential role at the onset of spermatogenesis. *Molecular Endocrinology* 7: 67-76.

Nagahama, Y. 1983. The functional morphology of teleost gonads. In: *Fish Physiology*, W.S. Hoar, D.J. Randall and E.M. Donaldson (eds.). Academic Press, New York, Vol. 9A, pp. 223-264.

Naito, N., S. Hyodo, N. Okumoto, A. Urano and Y. Nakai. 1991. Differential production and regulation of gonadotropins (GTH I and GTH II) in the pituitary gland of rainbow trout, *Oncorhynchus mykiss*, during ovarian development. *Cell and Tissue Research* 266: 457-467.

Niemi, M., R.M. Sharpe and W.R. Brown. 1986. Macrophages in the interstitial tissue of the rat testis. *Cell and Tissue Research* 243: 337-344.

Nistal, M., L. Santamaría and R. Paniagua. 1984. Masts cells in the human testis and epididymis from birth to adulthood. *Acta Anatomica* 119: 155-160.

Nozaki, M., N. Naito, P. Swanson, K. Miyata, Y. Nakai, Y. Oota, K. Suzuki and H. Kawauchi. 1990a. Salmonid pituitary gonadotrophs. I. Distinct cellular distributions of two gonadotropins, GTH I and GTH II. *General and Comparative Endocrinology* 77: 348-357.

Nozaki, M., N. Naito, P. Swanson, W.W. Dickhoff, Y. Nakai, K. Suzuki and H. Kawauchi. 1990b. Salmonid pituitary gonadotrophs. II. Ontogeny of GTH I and GTH II cells in the rainbow trout (*Salmo gairdneri irideus*). *General and Comparative Endocrinology* 77: 358-367.

O'Donnell, L., K.M. Robertson, M.E. Jones and E.R. Simpson. 2001. Estrogen and spermatogenesis. *Endocrinology Reviews* 22: 289-318.

Olaso, R., C. Pairault, B. Boulogne, P. Durand and R. Habert. 1998. Transforming growth factor β1 and β2 reduce the number of gonocytes by increasing apoptosis. *Endocrinology* 139: 733-740.

Ordás, M.C., M.M. Costa, F.J. Roca, G. López-Castejón, V. Mulero, J. Meseguer, A. Figueras and B. Novoa. 2007. Turbot TNFα gene: molecular characterization and biological activity of the recombinant protein. *Molecular Immunology* 44: 389-400.

Pancer, Z. and M.D. Cooper. 2006. The evolution of adaptive immunity. *Annual Review of Immunology* 24: 497-518.

Parvinen, M., O. Söder, P. Mali, B. Froysa and E.M. Ritzen. 1991. *In vitro* stimulation of stage-specific deoxyribonucleic acid synthesis in rat seminiferous tubule segments by interleukin-1 alpha. *Endocrinology* 129: 1614-1620.

Patzner, R.A. and M. Seiwald. 1987. The reproduction of *Blennius pavo* (Teleostei, Blenniidae). VI. Testicular cycle. *Zoologischer Anzeiger* 219: 265-273.

Pelegrín, P., E. Chaves-Pozo, V. Mulero and J. Meseguer. 2004. Production and mechanism of secretion of interleukin-1 beta from the marine fish gilthead seabream. *Developmental and Comparative Immunology* 28: 229-237.

Pentikainen, V., K. Erkkila, L. Suomalainen, M. Otala, M.O. Pentikainen, M. Parvinen and L. Dunkel. 2001. TNFalpha down-regulates the Fas ligand and inhibits germ cell apoptosis in the human testis. *Journal of Clinical Endocrinology and Metabolism* 86: 4480-4488.

Pickering, A.D. and P. Christie. 1980. Sexual differences in the incidence and severity of ectoparasitic infestation of the brown trout, *Salmo trutta* L. *Journal of Fish Biology* 16: 669-683.

Pierce, J.G. and T.F. Parsons. 1981. Glycoprotein hormones: structure and function. *Annual Review of Biochemistry* 50: 465-495.

Pinto, P., A.L. Passos, D.M. Power and A.V. Canario. 2005. Sea bream (*Sparus auratus*) estrogen receptors: phylogeny and tissue distribution. *Annals of the New York Academy of Sciences* 1040: 436-438.

Pöllänen, P. and M. Niemi. 1987. Immunohistochemical identification of macrophages, lymphoid cells and HLA antigens in the human testis. *International Journal of Andrology* 10: 37-42.

Pöllänen, P., O. Söder and M. Parvinen. 1989. Interleukin-1 alpha stimulation of spermatogonial proliferation *in vivo*. *Reproduction, Fertility and Development* 1: 85-87.

Pöllänen, P., M. von Euler, K. Jahnukainen, T. Saari, M. Parvinen, S. Sainio-Pöllänen and O. Söder. 1993. Role of transforming growth factor β in testicular immunosuppression. *Journal of Reproductive Immunology* 24: 123-137.

Pollard, J.W., M.G. Dominguez, S. Mocci, P.E. Cohen and E.R. Stanley. 1997. The effect of the colony-stimulating factor-1 null mutation, osteopetrotic ($csfm^{op}$), on the distribution of macrophages in the male mouse reproductive tract. *Biology of Reproduction* 56: 1290-1300.

Prat, F., S. Zanuy, M. Carrillo, A. Mones and A. Fostier. 1990. Seasonal changes in plasma levels of gonadal steroids of sea bass, *Dicentrarchus labrax* L. *General and Comparative Endocrinology* 78: 361-373.

Prat, F., J.P. Sumpter and C.R. Tyler. 1996. Validation of radioimmunoassay for two salmon gonadotropins (GTH-I and GTH-II) and their plasma concentrations throughout the reproductive cycle in male and female rainbow trout (*Oncorhynchus mykiss*). *Biology of Reproduction* 54: 1328-1375.

Pudney, J. 1995. Spermatogenesis in non-mammalian vertebrates. *Microscopic Research and Technique* 32: 459-497.

Quérat, B., A. Sellouk and C. Salmon. 2000. Phylogenetic analysis of the vertebrate glycoprotein hormone family including new sequences of sturgeon (*Acipenser baeri*) β subunits of the two gonadotropins and the thyroid-stimulating hormone. *Biology of Reproduction* 63: 222-228.

Raburn, D.J., A.J. Reinhart, A. Coquelin and J.C. Hutson. 1993. Regulation of the macrophage population in postnatal rat testis. *Journal of Reproductive Immunology* 24: 139-151.

Ridgeway, G.J. 1962. Demonstration of blood types in rainbow trout and salmon by isoimmunisation. *Annals of the New York Academy of Sciences* 97: 111-118.

Rivier, C. and S. Rivest. 1991. Effect of stress on the activity of the hypothalamic-pituitary-gonadal axis: peripheral and central mechanisms. *Biology of Reproduction* 45: 523-532.

Rocha, M.J. and E. Rocha. 2006. Morphofunctional aspects of reproduction from synchronous to asynchronous fishes—An overview. In: *Fish Endocrinology*, M. Reinecke, G. Zaccone and B.G. Kapoor (eds.). Science Publishers, Enfield, NH, USA, Vol. 2, pp. 571-624.

Roit, I., J. Brostoff and D. Male. 2001. *Immunology*. C.V. Mosby, London. 6th Edition.

Rotello, R.J., R.C. Lieberman, A.F. Purchio and L.E. Gerschenson. 1991. Coordinated regulation of apoptosis and cell proliferation by transforming growth factor beta 1 in cultured uterine epithelial cells. *Proceedings of the National Academy of Sciences of the United States of America* 88: 3412-3415.

Roux-Lombard, P. 1998. The interleukin-1 family. *European Cytokine Network* 9: 565-567.

Rytomaa, T. 1960. Organ distribution and histochemical properties of eosinphil granulocytes in rat. *Acta Pathologica, Microbiologica et Immunologica Scandinavica* 50: 1-118.

Saeij, J.P.J., R.J.M. Stet, B.J. de Vries, W.B. van Muiswinkel and G.F. Wiegertjes. 2003. Molecular and functional characterization of carp TNF: a link between TNF polymorphism and trypanotolerance? *Developmental and Comparative Immunology* 27: 29-41.

Saga, T., Y. Oota, M. Nozaki and P. Swanson. 1993. Salmonid pituitary gonadotrophs. III. Chronological appearance of GTH I and other adenohypophysial hormones in the pituitary of the developing rainbow trout (*Oncorhynchus mykiss irideus*). *General and Comparative Endocrinology* 92: 233-241.

Saha, N.R., T. Usami and Y. Suzuki. 2004. *In vitro* effects of steroid hormones on IgM-secreting cells and IgM secretion in common carp (*Cyprinus carpio*). *Fish and Shellfish Immunology* 17: 149-158.

Savan, R. and M. Sakai. 2004. Presence of multiple isoforms of TNF alpha in carp (*Cyprinus carpio* L.): genomic and expression analysis. *Fish and Shellfish Immunology* 17: 87-94.

Savan, R., T. Kono, D. Igawa and M. Sakai. 2005. A novel tumor necrosis factor (TNF) gene present in tandem with the TNF-α gene on the same chromosome in teleosts. *Immunogenetics* 57: 140-150.

Scapigliati, G., S. Bird and C.J. Secombes. 2000. Invertebrate and fish cytokines. *European Cytokine Network* 11: 354-361.

Schlatt, S., A. Meinhardt and E. Nieschlag. 1997. Paracrine regulation of cellular interactions in the testis: factors in search of a function. *European Journal of Endocrinology* 137: 107-117.

Schreibman, M.P., S. Holtzman and L. Cepriano. 1990. The life cycle of the brain–pituitary–gonad-axis in teleosts. In: *Progress in Comparative Endocrinology*, A. Epple, C.G. Scanes and M.H. Stetson (eds.). Wiley-Liss, New York, pp. 399-408.

Schulz, R.W., H.F. Vischer, J.E.B. Cavaco, E.M. Santos, C.R. Tyler, H.J.Th. Goos and J. Bogerd. 2001. Gonadotropins, their receptors, and the regulation of testicular functions in fish. *Comparative Biochemistry and Physiology* B129: 407-417.

Schulz, R.W., S. Menting, J. Bogerd, L.R. Franca, D.A. Vilela and H.P. Godinho. 2005. Sertoli cell proliferation in the adult testis-evidence from two fish species belonging to different orders. *Biology of Reproduction* 73: 891-898.

Schuppe, H.C. and A. Meinhardt. 2005. Immune privilege and inflammation. In: *Immunology of Gametes and Embryo Implantation*, U.R. Markert (ed.). *Chemistry of Immunological Allergy*. S. Karger, Basel, Vol. 88, pp. 1-14.

Scott, A.P. and J.P. Sumpter. 1989. Seasonal variations in testicular germ cell stages and in plasma concentration of sex steroids in male rainbow trout (*Salmo gairdneri*) maturing at 2 years old. *General and Comparative Endocrinology* 73: 46-58.

Sepulcre, M.P., P. Pelegrín, V. Mulero and J. Meseguer. 2002. Characterisation of gilthead sea bream acidophilic granulocytes by a monoclonal antibody unequivocally points to their involvement in fish phagocytic response. *Cell and Tissue Research* 308: 97-102.

Shrestha, T.K. and S.S. Khanna. 1976. Histology and seasonal changes in the testes of a hill-stream fish, *Schizothorax plagiostomus*. *Zeitschrift für Mikroskopisch-Anatomische Forschung* 90: 749-761.

Sierens, J.E., S.F. Sneddon, F. Collin, M.R. Millar and P.T.K. Saunders. 2005. Estrogens in testis biology. *Annals of the New York Academy of Sciences* 1061: 65-76.

Siu, M.K. and C.Y. Cheng. 2004a. Dynamic cross-talk between cells and the extracellular matrix in the testis. *Bioessays* 26: 978-992.

Siu, M.K. and C.Y. Cheng. 2004b. Extracellular matrix: recent advances on its role in junction dynamics in the seminiferous epithelium during spermatogenesis. *Biology of Reproduction* 71: 375-391.

Slater, C.H. and C.B. Schreck. 1993. Testosterone alters the immune response of chinook salmon, *Oncorhynchus tshawytscha*. *General and Comparative Endocrinology* 89: 291-298.

Slater, C.H. and C.B. Schreck. 1997. Physiological levels of testosterone kill salmonid leukocytes *in vitro*. *General and Comparative Endocrinology* 106: 113-119.

Socorro, S., D.M. Power, P.E. Olsson and A.V.M. Canario. 2000. Two estrogen receptors expressed in the teleost fish, *Sparus aurata*: cDNA cloning, characterization and tissue distribution. *Journal of Endocrinology* 166: 293-306.

Söder, O., T. Sultana, C. Jonsson, A. Wahlgren, C. Petersen and M. Holst. 2000. The interleukin-1 system in the testis. *Andrologia* 32: 52-55.

Stout, R.D. and J. Suttles. 2004. Functional plasticity of macrophages: reversible adaptation to changing microenvironments. *Journal of Leukocyte Biology* 76: 509-513.

Tachi, C. and S. Tachi. 1989. Role of macrophages in the maternal recognition of pregnancy. *Journal of Reproductive Fertility* 37: 63-68.

Tafalla, C., R. Aranguren, C.J. Secombes, J.L. Castrillo, B. Novoa and A. Figueras. 2003. Molecular characterisation of sea bream (*Sparus aurata*) transforming growth factor beta1. *Fish and Shellfish Immunology* 14: 405-421.

Thanawongnuwech, R., B. Thacker, P. Halbur and E. Thacker. 2004. Increased production of proinflammatory cytokines following infections with porcine reproductive and respiratory syndrome virus and *Mycoplasma hyopneumoniae*. *Clinical and Diagnostic Laboratory Immunology* 11: 901-908.

Tracey, K.J. and A. Cerami. 1994. Tumor necrosis factor: a pleiotropic cytokine and therapeutic target. *Annual Review of Medicine* 45: 491-503.

Turnbull, A.V. and C. Rivier. 1995. Regulation of the HPA axis by cytokines. *Brain Behavior and Immunity* 9: 253-275.

Van der Hurk, R., J. Peute and J.A.J. Vermeij. 1978. Morphological and enzyme cytochemical aspects of the testis and vas deferens of the rainbow trout, *Salmo gairdneri*. *Cell and Tissue Research* 186: 309-325.

Van der Kraak, G.J., J. Chang and D. Janz. 1998. Reproduction. In: *The Physiology of Fishes*, D.H. Evans (ed.). 2nd Edition. CRC Press, Boca Raton, pp. 465-488.

Van Reeth, K., G. Labarque, H. Nauwynck and M. Pensaert. 1999. Differential production of proinflammatory cytokines in the pig lung during different respiratory virus infections: correlations with pathogenicity. *Research in Veterinary Science* 67: 47-52.

Van Reeth, K., S. Van Gucht and M. Pensaert. 2002. *In vivo* studies on cytokine involvement during acute viral respiratory disease of swine: troublesome but rewarding. *Veterinary Immunology and Immunopathology* 87: 161-168.

Wang, R. and M. Belosevic. 1994. Estradiol increases susceptibility of goldfish to *Trypanosoma danilewskyi*. *Developmental and Comparative Immunology* 18: 377-387.

Wang, J., N.G.M. Wreford, H.Y. Lan, R. Atkins and M.P. Hedger. 1994. Leukocyte populations of the adult rat testis following removal of Leydig cells by treatment with ethane dimethane sulfonate and subcutaneous testosterone implants. *Biology of Reproduction* 51: 551-561.

Wang, H., C.J. Czura and K.J. Tracey. 2003. Tumor necrosis factor. In: *The Cytokine Handbook*, A.W. Thomson and M.T. Lotze (eds.). Academic Press, New York, pp. 837-860. 4th Edition.

Watanuki, H., T. Yamaguchi and M. Sakai. 2002. Suppression in function of phagocytic cells in common carp *Cyprinus carpio* L. injected with estradiol, progesterone or 11-ketotestosterone. *Comparative Biochemistry and Physiology* C132: 407-413.

Watrin, F., L. Scotto, R.K. Assoian and D.J. Wolgemuth. 1991. Cell lineage specificity of expression of the murine transforming growth factor beta 3 and transforming growth factor beta 1 genes. *Cell Growth Differentiation* 2: 77-83.

Wei, R.Q., J.B. Yee, D.C. Straus and J.C. Hutson. 1988. Bactericidal activity of testicular macrophages. *Biology of Reproduction* 38: 830-835.

Xia, W., D.D. Mruk, W.M. Lee and C. Yan Cheng. 2005. Cytokines and junction restructuring during spermatogenesis—a lesson to learn from the testis. *Cytokine Growth Factors Reviews* 16: 469-493.

Xiong, Y. and D.B. Hales. 1993. Expression, regulation, and production of tumor necrosis factor-alpha in mouse testicular interstitial macrophages *in vitro*. *Endocrinology* 133: 2568-2573.

Yamaguchi, T., H. Watanuki and M. Sakai. 2001. Effects of estradiol, progesterone and testosterone on the function of carp, *Cyprinus carpio*, phagocytes *in vitro*. *Comparative Biochemistry and Physiology* C 129: 49-55.

Yee, J.B. and J.C. Hutson. 1983. Testicular macrophages: isolation, characterization and hormonal responsiveness. *Biology of Reproduction* 29: 1319-1326.

Young, K.A. and R.J. Nelson. 2001. Mediation of seasonal testicular regression by apoptosis. *Reproduction* 122: 677-685.

Zapata, A.G., A. Chibá and A. Varas. 1996. Cells and tissues of the immune system of fish. In: *The Fish Immune System. Organism, Pathogen, and Environment*, G.K. Iwama and T. Nakanishi (eds.). Academic Press, San Diego, pp. 1-62.

Zou, J., T. Wang, I. Hirono, T. Aoki, H. Inagawa, T. Honda, G-I. Soma, M. Ototake, T. Nakanishi, A.E. Ellis and C.J. Secombes. 2002. Differential expression of two tumor necrosis factor genes in rainbow trout, *Oncorhynchus mykiss*. *Developmental and Comparative Immunology* 26: 161-172.

Zou, J., C.J. Secombes, S. Long, N. Miller, L.W. Clem and V.G. Chinchar. 2003. Molecular identification and expression analysis of tumor necrosis factor in channel catfish (*Ictalurus punctatus*). *Developmental and Comparative Immunology* 27: 845-858.

CHAPTER

3

The Cellular and Developmental Biology of the Teleost Antibody Response

S. Kaattari*, G. Brown, I. Kaattari, J. Ye, A. Haines and E. Bromage

INTRODUCTION

The purpose of this chapter will be to provide a vista point from which the reader may view the unfamiliar and, at times, the seemingly exotic landscape of the teleost antibody response. We hope to provide a guide of what we believe to be some of the more unique and significant landmarks, annotated with some, hopefully, provocative hypotheses that may encourage greater participation and deeper exploration into this exciting and relatively unmapped region of the immunological world.

This chapter begins with the molecular organization of the antibody genes, followed by the structural and functional features of the antibody molecule. The second half of the chapter deals with the development of

Authors' address: Department of Environmental and Aquatic Animal Health, Virginia Institute of Marine Science, 1208 Greate Rd., College of William and Mary, Gloucester Point VA 23062, USA.

**Corresponding author:* E-mail: kaattari@vims.edu

the antibody producing cell and the organization and function of the system that leads to this antibody response. In a final summary, key features are re-emphasized and brought into a context that is contra to numerous other portrayals of the ectothermic immune system (i.e., a dynamic and uniquely evolving immune system rather than an evolutionary relic).

TELEOST IMMUNOGLOBULIN LOCI

Most teleost serum Ig is a tetrameric molecule, termed IgM, although serum IgM in mammals and in the elasmobranchs is pentameric (Fig. 3.1;

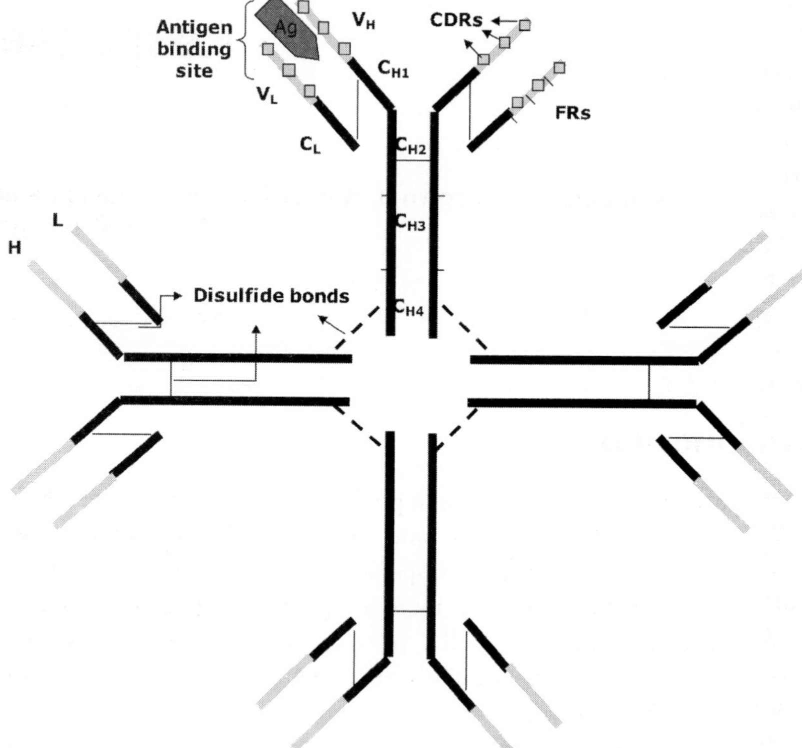

Fig. 3.1 Schematic representation of teleost IgM. The teleost IgM molecule is a tetramer composed of eight identical heavy (H) and eight light (L) chains which are linked by disulfide bridges, contributing eight antigen binding sites. Each H and L chain is composed of a V_H and V_L (variable regions), respectively, as well as a C_{H1-4} and C_L. H and L chain complementarity determining regions, CDRs (squares) which are separated by framework regions (FRs), interact with antigens (Ag) in the antigen binding site. Certain interchain disulfide bonds are uniformly present (solid lines), while other intersubunit disulfides are not (dashed lines).

also see Wilson and Warr, 1992, for an early review). The monomeric building block—comparable to IgG in other vertebrates—consists of equimolar amounts of heavy chains and light chains encoded by separate loci within the genome (Warr, 1995). Until recently, IgM was believed to be the only teleost immunoglobulin isotype. However, with advances in molecular biology and genome sequencing, the true complexity engendered in teleost immunoglobulins is becoming apparent.

The Heavy Chain Locus

The Variable Heavy Chain (V_H) Region

The teleost heavy chain variable region, as with other jawed vertebrates, consists of four framework regions bracketing three hypervariable or complementarity determining regions (CDR) (Kabat *et al.*, 1979). The three heavy chain CDRs represent three of the six antigen contact points that confer antigenic specificity, the other three being encoded by the variable light (V_L) chain gene. Each FR1-CDR1-FR2-CDR2-FR3 is encoded primarily by a variable (V_H) gene, whereas the CDR3 is encoded by the 3′ end of V_H, a D_H gene and the 5′ end of a joining (J_H) gene, which are randomly fused through the process of somatic recombination (Tonegawa, 1983). Recombination follows the 12 (one helix turn)/23 (two helix turn) rule, requiring a spacer region of either 12 or 23 bp separating a conserved palindromic heptamer or a conserved AT-rich nonomer (Amemiya and Litman, 1990; Hayman and Lobb, 2000). Sequences containing the 12 bp spacer can combine only with sequences containing the 23 bp spacer and vice versa. The fourth framework (FR4) is encoded entirely by the remaining portion of a J_H gene.

The arrangement of these genes within the genome is similar to the translocon arrangement found in the mammalian heavy and light chain loci (Ghaffari and Lobb, 1989b; Amemiya and Litman, 1990). Multiple V_H genes are arranged in tandem, upstream of separate groups encoding the D_H and J_H genes followed by different constant (C_H) genes (Fig. 3.2). Therefore, the number of V_H genes and V_H gene families, as well as the combinatorial diversity produced through somatic rearrangement with D_H and J_H genes are likely the primary factors generating antibody diversity (Table 3.1). In general, multiple V_H genes (>40) are available for recombination with the various D_H (3-10) and J_H (2-11) genes (see Table 3.1), therefore engendering the capability of generating

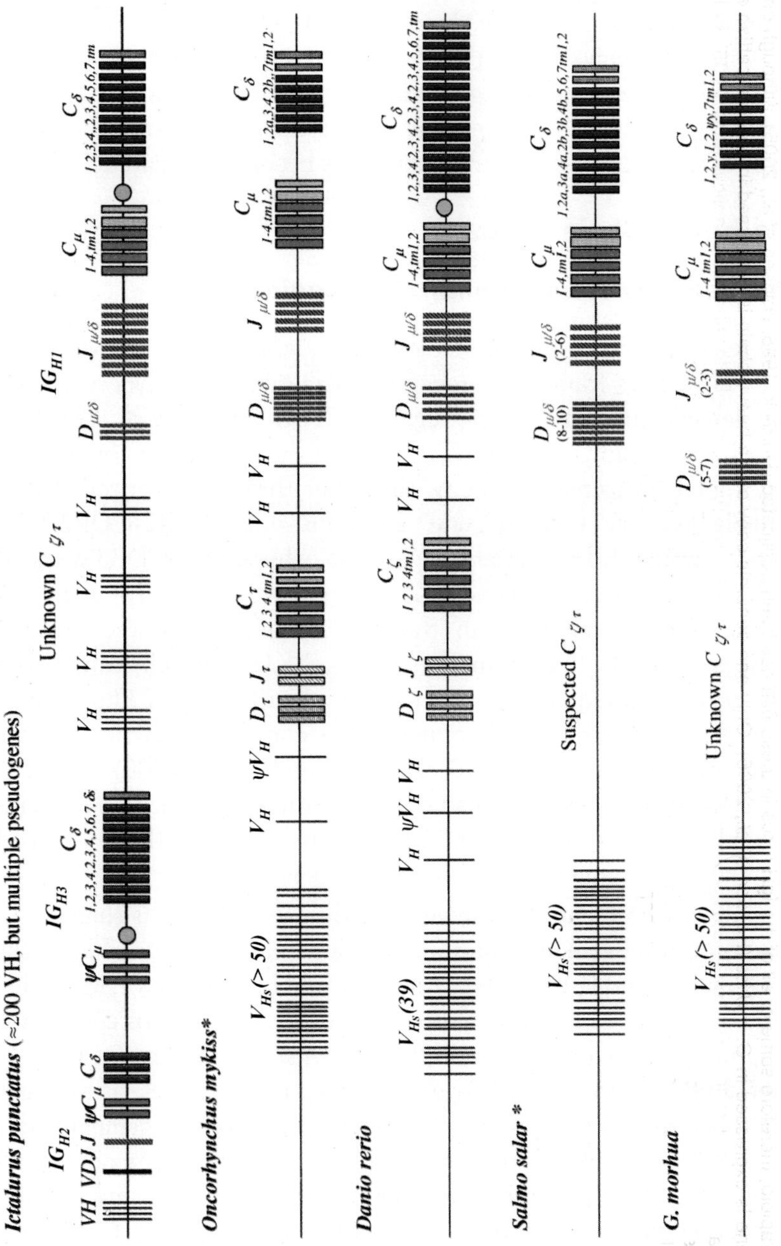

Ictalurus punctatus (≈200 VH, but multiple pseudogenes)

IG_{H2} VH VDJ J ψC_μ C_δ — IG_{H3} ψC_μ C_δ 1,2,3,4,2,3,4,5,6,7,δs — Unknown $C_{\zeta/\tau}$ V_H V_H V_H V_H — IG_{H1} $D_{\mu/\delta}$ $J_{\mu/\delta}$ C_μ 1-4,tm1,2 C_δ 1,2,3,4,,2,3,4,5,6,7,tm

Oncorhynchus mykiss*

V_{Hs}(> 50) V_H ψV_H D_τ J_τ C_τ 1 2 3 4 tm1,2 V_H V_H $D_{\mu/\delta}$ $J_{\mu/\delta}$ C_μ 1-4,tm1,2 C_δ 1,2a,3,4,2b,,7tm1,2

Danio rerio

V_{Hs}(39) V_H ψV_H V_H D_ζ J_ζ C_ζ 1 2 3 4tm1,2 V_H V_H $D_{\mu/\delta}$ $J_{\mu/\delta}$ C_μ 1-4,tm1,2 C_δ 1,2,3,4,2,3,4,2,3,4,2,3,4,5,6,7,tm

***Salmo salar* ***

V_{Hs}(> 50) Suspected $C_{\zeta/\tau}$ $D_{\mu/\delta}$ (8-10) $J_{\mu/\delta}$ (2-6) C_μ 1-4,tm1,2 C_δ 1,2a,3a,4a,2b,3b,4b,5,6,7tm1,2

G. morhua

V_{Hs}(> 50) Unknown $C_{\zeta/\tau}$ $D_{\mu/\delta}$ (5-7) $J_{\mu/\delta}$ (2-3) C_μ 1-4 tm1,2 C_δ 1,2,y,1,2,ψy,7tm1,2

Fig. 3.2 The IgH loci of teleosts. A variety of different H chain loci found in teleost fish are depicted. **Oncorhynchus mykiss* and *Salmo salar* are tetraploid, therefore some or all of the IgH locus in these fish has been duplicated (Hordvik, 1998; Hansen *et al.*, 2005), although only one IgM gene is expressed in *O. mykiss* (Hansen *et al.*, 2005). Only one genomic copy of the heavy chain is expressed in channel catfish as well (Ghaffari and Lobb, 1989b) and there is no evidence of more than one copy of the heavy chain locus in zebrafish (Danilova *et al.*, 2005). Enhancer regions (blue dot) are identified. Drawings not to scale. Reference sources for species: ***I. punctatus***—(Ghaffari and Lobb, 1989a, 1989b, 1992, 1999; Hayman *et al.*, 1993; Andersson and Matsunaga, 1995; Wilson *et al.*, 1997; Hayman and Lobb, 2000; Ventura-Holman and Lobb, 2001; Bengtén *et al.*, 2002, 2006; Yang *et al.*, 2003). ***S. salar/S. alpinus***—(Hordvik *et al.*, 1997, 1998, 1999, 2002; Andersson and Matsunaga, 1993, 1995, 1998; Solem *et al.*, 2001). ***O. mykiss***—(Matsunaga *et al.*, 1990; Roman and Charlemagne, 1994; Roman *et al.*, 1995, 1996; Hansen *et al.*, 2005; Brown *et al.*, 2006). ***D. rerio***—(Danilova *et al.*, 2000, 2005; Ellestad and Magor, 2005). ***G. morhua***—(Bengtén *et al.*, 1994; Stenvik and Jorgensen, 2000; Stenvik *et al.*, 2000; Solem and Stenvik, 2006).

Table 3.1 *V, D, J* gene estimates for various teleost species[1].

Species[2]	V (Families)	$D_{\zeta/\tau}$	$J_{\zeta/\tau}$	$D_{\mu/\delta}$	$J_{\mu/\delta}$	VDJ[3]	IgH Loci
I. punctatus	200[4] (13)	ND[5]	ND	3	11	1	1[6]
S. salar/S. alpinus	>50 (8-13)	?[7]	?	8-10	2-6	ND	2
O. mykiss	>50 (13)	3	2	6	5	ND	2
D. rerio	36-39 (14)	2	2	5	5	ND	1
G. morhua	>50 (4)	?	?	5-7	2-3	ND	1

[1]Projected numbers were obtained from Southern blots, cDNA and/or germline sequences.
[2]References included in Fig. 3.2 legend
[3]Combined in the germline
[4]Half may be pseudogenes
[5]None Discovered
[6]Three different IgH regions have been identified in tandem; however, only one contains a functional *Igμ*
[7]Not determined

thousands of different *V(D)J* combinations. Fourteen V_H gene families have been identified in the zebrafish (*Danio rerio*) (Danilova *et al.*, 2005), 13 in the channel catfish (*Ictalurus punctatus*) (Yang *et al.*, 2003), and 13 in the rainbow trout (*Oncorhynchus mykiss*) (Brown *et al.*, 2006). However, some fish apparently utilize fewer V_H genes and/or families. Only four families have been found in Atlantic cod (*Gadus morhua*), and those utilized are predominantly from family III (Stenvik *et al.*, 2001). In such cases, antibody variability can still be generated through somatic mutation (Yang *et al.*, 2006) and *V(D)J* junctional diversity. Pseudogenes are also common (Yang *et al.*, 2003) and can contribute to antibody diversity through gene conversion. This is the primary process for generating Ig diversity in chicken where only one functional V_H gene exists. Here, the process of hyperconversion utilizes a donor pool of pseudogenes to generate significant V_H diversity from the single functional gene (Reynaud *et al.*, 1989).

In addition to the potential junctional diversity generated simply by the number and variety of V_H, D_H, and J_H genes, the actual process of recombination adds to the diversity in the *CDR3* region through the deletion of nucleotides, the addition of non-templated nucleotides (N-region additions), and the addition of nucleotides palindromic to the 3′ end of a coding region (P-additions). Deletion of nucleotides from both the 5′ and 3′ ends of the D_H genes, as well as N and P-additions, have been observed in channel catfish (Hayman and Lobb, 2000). The addition or deletion of a nucleotide upstream or at the 5′ end of the *D* region can significantly alter the amino acid sequence of the CDR3. It is common for

D_H regions to have more than one open reading frame (ORF) (Hayman and Lobb, 2000). In the channel catfish, Hayman and Lobb (2000) detected utilization of alternate D_H reading frames where one ORF encoded for hydrophobic residues while a second encoded for glycine and polar/hydrophilic residues; therefore, depending on the number of additions or deletions upstream of the D_H gene, either a hydrophobic or a hydrophilic CDR3 region could be produced. The presence of similar observations in rainbow trout (Roman *et al.*, 1995), Atlantic salmon (*Salmo salar*) (Solem *et al.*, 2001) and Atlantic cod (Solem and Stenvik, 2006) would suggest that flexible reading frame usage, N and P additions, and D_H segment nucleotide deletions are utilized by most teleosts to generate diversity of the CDR3.

The Constant Heavy Chain (C_H) Region

Following somatic heavy chain gene rearrangement, transcription, and finally through RNA processing, the leader sequence and the rearranged *V(D)J* segment are spliced to the exon of a constant region creating the mRNA transcript responsible for encoding the complete immunoglobulin heavy chain. Three to five different heavy chain isotypes have been identified in teleost fish, depending on the species: IgM_1, IgM_2 (Hordvik *et al.*, 1997, 2002), IgD (Wilson *et al.*, 1997), IgZ (Danilova *et al.*, 2005) and IgT (Hansen *et al.*, 2005), determined by the constant region (C_μ, C_τ, C_ζ) that is joined to the *V(D)J*. Unlike tetrapods, however, there is no evidence of class switch recombination. Class switch recombination may not be necessary in teleosts due to the structure of the immunoglobulin heavy chain locus (Danilova *et al.*, 2005), or due to the lack of evolved switch regions (Barreto *et al.*, 2005). Interestingly, the gene encoding for activation-induced cytidine deaminase (AID)—which is required for class switching as well as for somatic hypermutation—was recently identified from zebrafish (Zhao *et al.*, 2005). Although class switching has not been observed in teleosts, teleost AID can catalyze class switching in mammalian cell lines (Barreto *et al.*, 2005).

The C_μ region (IgM)

IgM is the most prevalent Ig in teleosts and, prior to the 1990's, it was the only isotype identified. In the serum of teleosts, it exists predominantly as a tetramer [4 monomeric units, $(H_2L_2)_4$, for a total of 8 heavy chains and

8 light chains (Fig. 3.1)] with eight antigen binding sites (Acton *et al.*, 1971; Clem, 1971), but is also found bound to the membrane as a monomer (Clem and McLean, 1975; Warr *et al.*, 1976; Wilson *et al.*, 1992). The transcript for the secreted form contains all four $C\mu$ exons; however, alternative splicing patterns join the $C_{\mu}3$ exon to the two transmembrane domains to generate the membrane form. This alternative RNA splicing pattern, first discovered in the channel catfish (Wilson *et al.*, 1992) and apparently common to all teleosts examined thus far (Bengtén *et al.*, 1991; Hordvik *et al.*, 1992; Lee *et al.*, 1993), is also present in more primitive holostean fish (Wilson *et al.*, 1995). Alternative RNA splicing overcomes the lack of a cryptic donor splice site, which is present in other vertebrate $C_{\mu}4$s, but not present in teleost $C_{\mu}4$ (Ross *et al.*, 1998).

Transcription of μ occurs early in development and gradually increases. The recombination activating gene, *rag1*, necessary for somatic *V(D)J* recombination was detected by day 4 post-hatch in the zebrafish, which coincided with the first detection of *V(D)J* rearrangements (Danilova and Steiner, 2002). Transcription of membrane μ was then detected by day 7 and secreted μ by day 13. *In situ* hybridization of μ transcripts were detected in 10-day-old zebrafish (Danilova *et al.*, 2005) and day 22 trout embryos (Hansen *et al.*, 2005).

The C_{δ} region (IgD)

In the late 1990's, an additional Ig heavy chain constant region gene was discovered in the channel catfish that: (1) bore some sequence similarity to human δ; (2) was similarly located immediately downstream of μ; (3) possessed separate exons for secretory and membrane forms; and (4) due to a lack of switch mechanisms could be co-expressed with μ utilizing the same *V(D)J* rearrangement (Wilson *et al.*, 1997). Since then, *IgD* has also been identified in Atlantic salmon (Hordvik *et al.*, 1999), Atlantic cod (Stenvik and Jorgensen, 2000), Atlantic halibut *(Hippoglossus hippoglossus)* (Hordvik, 2002), Japanese flounder (*Paralichthys olivaceus*) (Hirono *et al.*, 2003), fugu (*Takifugu rubripes*) (Saha *et al.*, 2004), rainbow trout (Hansen *et al.*, 2005), and zebrafish (Danilova *et al.*, 2005). Similar to teleost IgM, IgD also utilizes an alternative pathway of RNA processing. Somatically, the same pod cf *V(D)J* genes utilized with μ for the production of *IgM* are recombined and spliced to the $C_{\mu}1$ exon followed by splicing to the C_{δ} exons, producing a chimeric Ig molecule with both μ and δ genes. It has been suggested that $C_{\mu}1$ is necessary for heavy/light chain disulfide bonding since $C_{\delta}1$ does not code for the necessary cysteine (Wilson *et al.*, 1997). In addition, teleost IgD also differs structurally from the IgD of other organismal groups, in the sense that it lacks a hinge region (Wilson *et al.*, 1997).

The number and usage of C_{δ} exons is not consistent among teleosts. As seen in Figure 3.2, δ 2-4 are duplicated in the genomes of several fish species, for example, Atlantic salmon (Hordvik *et al.*, 1999), channel catfish (Bengten *et al.*, 2002), Atlantic halibut (Hordvik, 2002) and zebrafish (Danilova *et al.*, 2005). In the Atlantic salmon, the duplicated δ exons were maintained in the transcripts and apparently not spliced out. In fugu, δ1-6 are duplicated in a tandem repeat (Saha *et al.*, 2004), whereas in the Atlantic cod domains δ3-6 have been deleted and only δ1 and δ2 are duplicated (Stenvik and Jorgensen, 2000). The Japanese flounder is one of the few species that has a simple δ1-7 exon arrangement without deletion or duplication of δ domains (Srisapoome *et al.*, 2004). In another interesting variant, the channel catfish contains separate exons for secretory and membrane bound I_gD, each associated with a δ exon pattern of δ1,2,3,4,2,3,4,5,6,7, but located between different regions of the heavy chain (Bengten *et al.*, 2002). A similar arrangement has not been observed in any other fish and could either indicate that other fish may only express the membrane form of IgD (Srisapoome *et al.*, 2004), or simply that the secretory genes have not been located. Additional δ domain diversity exists in zebrafish, which have only one membrane exon; however, the resulting amino acid sequence is similar to the sequences encoded by two membrane exons (Danilova *et al.*, 2005).

Although the function of IgD is still unknown, the dominant form appears to be membrane bound, suggesting a role as a B cell receptor (Stenvik *et al.*, 2001). In the Japanese flounder, expression of the δ gene was found predominantly in the peripheral blood lymphocytes. However, expression levels were much lower than μ (Hirono *et al.*, 2003). To date, there has only been one report of secreted IgD, which was in the channel catfish (Miller *et al.*, 1998).

The $C_{\tau/\zeta}$ region (I_gT and I_gz)

IgT and *IgZ*, which may be orthologous, could also be representatives of the same isotype (Danilova *et al.*, 2005; Hansen *et al.*, 2005). There does not appear to be a previously described Ig equivalent to τ or ζ in other vertebrates. Due to the similarity between the isotypes, i.e., their location, size, exclusive DJ usage, and restriction to fish, for the purposes of this chapter they will be addressed together.

Genes similar to τ and ζ have been identified from deposited sequences of fugu, carp (*Cyprinus carpio*) and Atlantic salmon (Danilova *et al.*, 2005), although not in channel catfish (Bengtén *et al.*, 2006). In both the zebrafish and in the rainbow trout, 4 $C_{\tau/\zeta}$ exons are found upstream of regions encoding the $D_{\mu/\delta}$, $J_{\mu/\delta}$ C_{μ} and C_{δ} domains (Fig. 3.2). Both possess a secretory tail and a transmembrane domain. IgT and IgZ, unlike IgM, retain all four $C_{\zeta/\tau}$ domains in membrane bound form (Hansen *et al.*, 2005). Interestingly, additional *D* and *J* genes were identified upstream of the C_{ζ} and C_{τ} exons, which are not utilized in IgM or IgD. Following examination of cDNAs encoding IgT and IgZ, these *D* and *J* genes were utilized exclusively with τ and ζ transcripts, further supporting the identification of a new isotype. The different D_{τ} and J_{τ} genes utilized in IgT produced Ig with longer CDR3 regions (5-10 aa) when compared to IgM CDR3 (4-5 aa) (Hansen *et al.*, 2005).

Transcription of τ and ζ appears to occur very early, being detected in 22-day-old rainbow trout and in 6-day-old zebrafish (Danilova *et al.*, 2005; Hansen *et al.*, 2005). In zebrafish, the ζ transcripts were more prominent than μ at two weeks. However, τ transcription was never shown to be greater than μ transcription in the rainbow trout. Interestingly, in zebrafish, the *V(D)J* arrangement, known to take place by day 4, may be responsible for $V(D_{\zeta})J_{\zeta}$ arrangements and not $V(D_{\mu})J_{\mu}$ as previously thought (Danilova *et al.*, 2005). Apparently, both membrane-bound and secreted forms of IgT are expressed in rainbow trout, as the two bands are detected by Northern blots, presumably one for secreted τ and one for membrane τ (Hansen *et al.*, 2005). In addition, as with the δ genes, τ genes are also duplicated in the rainbow trout. Transcription of ζ appears to be restricted more to the thymus, mid-kidney and anterior kidney when compared to τ transcription, which was also detected in the spleen, intestine, heart, and weakly in the mid-kidney. Although this difference may imply a functional difference between the two isotypes, other similarities (i.e., phylogenetic analysis, DJ usage, and location) would suggest a closer relationship. The expression of additional τ/ζ genes from other species needs to be examined before a conclusion can be made.

The degree of polymerization of IgZ or of IgT is also unknown. $C_{\tau/\zeta}$ possesses a cysteine at position 13 or 14 (rainbow trout or zebrafish, respectively), which typically forms a bond with a light chain cysteine in other immunoglobulins, however, there is only one other cysteine near the

C terminus of secreted IgZ or IgT, that is available to form a covalent bond with another monomeric unit (Danilova *et al.*, 2005; Hansen *et al.*, 2005).

An additional, novel chimeric immunoglobulin has recently been observed in carp where the deduced amino acid sequence of the first constant region was 95% similar to carp $C_{\mu}1$, but the second constant region (C_H2) showed similarity (52.6%) to the predicted amino acid sequence for C_H4 of IgZ. The carp immunoglobulin is unique as it contains only two constant domains (Savan *et al.*, 2005a). Fugu also appears to have an *IgH* gene containing two constant regions. However, the second constant region encodes a hinge region and neither constant region is similar to any other fish *IgH* genes (Savan *et al.*, 2005b).

Light Chain Loci

L Chain Interaction with the H Chain

Each monomeric subunit of an immunoglobulin molecule is composed of two H chains and two L chains. The individual L chains are usually associated with the H chains through a disulfide bridge between the single C domain of the light chain and the C1 domain of the H chain (Bengtén *et al.*, 2000b). The antigen-binding region of the entire Ig molecule is stabilized by hydrophobic interactions between the V_L and V_H domains (Secher *et al.*, 1977). Although the importance of the V_L in antigen binding is thought to be limited (Pilström *et al.*, 1998; Pilström, 2002), the presence and variability of V_L domains in most immunoglobulins suggests V_L's contribution to the antigen-binding site can be significant.

In contrast to C_H regions, C_L regions do not appear to have special effector functions. However, as Pilström (2002) emphasizes, most H chains of immunoglobulins cannot be expressed (and later secreted) without L chains. Thus, an important function of the L chain is to facilitate expression of the complete B cell membrane-bound and secreted immunoglobulins. If the initial rearrangement of the teleost L chain locus fails to produce a functional antibody, another version of the L chain may be produced to combine with the original H chain. There are also rare instances where immunoglobulin molecules are only composed of heavy chains (Hamers-Casterman *et al.*, 1993; Greenberg *et al.*, 1995). These immunoglobulins function as well as conventional immunoglobulins (Nguyen *et al.*, 2000).

L Chain Isotypes

Most vertebrates, with the exception of birds, express two or more L chain isotypes (i.e., κ and λ), which are expressed in a translocon manner. The mammalian κ complex possesses multiple V_L genes, several J_L genes, and a single C_L gene (Kirschbaum *et al.*, 1996). The λ locus possesses multiple V_L genes upstream of several J_L-C_L clusters (Bauer and Blomberg, 1991; Frippiat *et al.*, 1995). In humans and mice, if the initial $V_\kappa J_\kappa$ rearrangement is unsuitable (i.e., incompatibility with the H chain, being out-of-frame, or generating an autoreactive receptor), secondary rearrangements occur between other upstream V_κ genes and any available downstream J_κ genes. After exhausting the potential κ rearrangements, the κ locus can be inactivated by deletion of its C_κ exon, and the second L chain locus, λ, is subsequently utilized (Gorman and Alt, 1998; Klein *et al.*, 2005). Thus, the λ locus thus can be visualized as a reserve in the event of failure at the κ locus to produce an appropriate B cell receptor (BCR) (Wardemann *et al.*, 2004).

In both cartilaginous and teleost fish, the L chain genes are not arranged in a translocon manner, but are in multiple 'clusters'. The *IgL* loci of the teleosts, however, differ from those of the elasmobranchs in at least two ways: the clusters are more proximal to each other, and the V_L genes have an opposite transcriptional direction in relation to the J_L and C_L genes. Both of these factors may facilitate rearrangements between clusters and thus increase variability. Another unique feature of teleost L chains is the abundance of sterile transcripts and non-rearranged transcripts in cDNA libraries screened via nonbiased selection procedures (Daggfeldt *et al.*, 1993; Partula *et al.*, 1996; Haire *et al.*, 2000; Timmusk *et al.*, 2000).

Although V_H regions are encoded by the recombination of V_H, D_H and J_H genes, D_L genes do not exist, only V_L and J_L genes (Shamblott and Litman, 1989a, b). Despite this difference, light chains also employ a process of random somatic recombination between the V_L and J_L genes thus contributing to the large repertoire of antigen-binding sites. If any of the antigen-binding sites are auto-reactive, their specificity is subsequently altered by replacement of the original L chain gene via secondary rearrangements, otherwise known as receptor editing (Gay *et al.*, 1993; Radic *et al.*, 1993; Tiegs *et al.*, 1993).

Teleost IgL chains cannot be classified as either κ or λ, although the V_L genes of Atlantic cod, channel catfish, and rainbow trout have somewhat higher sequence identity to κ rather than λ and have recombination signal sequences (RSS) that are of κ type (Pilström *et al.*, 1998). Initially, different teleost L chain isotypes were distinguished antigenically, by the use of monoclonal antibodies for light chains (Lobb *et al.*, 1984; Sanchez and Dominguez, 1991), or by anion exchange chromatography (Havarstein *et al.*, 1988). Sanchez and Dominguez produced two mAbs, 2H9 and 2A1, which identified two different rainbow trout light chains (26 and 24 kDa, respectively). These mAbs only reacted with 20% (2H9) and 11% (2A1) of Ig from a pool of sera (Sanchez and Dominguez, 1991), implying the existence of at least another light chain isotype. This was corroborated by subsequent FACS analyses, which also demonstrated distinct recognition of two mutually exclusive lymphocyte populations (Sanchez *et al.*, 1995) totaling less than 100% of those cells stained by an anti-H chain antibody.

A similar line of research, SDS-PAGE analysis of channel catfish serum antibody by Lobb and coworkers (1984), revealed light chains of approximately 26, 24, and 22 kDa. Two mAbs, 3F12 and 1G7, were found to detect different populations of channel catfish Ig. MAb 3F12 reacted with a subpopulation of channel catfish Ig that contained two of these L chain variants, ~24 and ~22 kDa. A second mAb, 1G7, reacted only with the third light chain variant, ~26 kDa. Peptide mapping demonstrated that these L chains were structurally different. Additional validation for these channel catfish L chain isotypes came with discovery that different haptens induced varying, yet additive, percentages of reactivity.

More recently, numerous studies have utilized genetic analyses to compare putative L chain isotypes. Hsu and Criscitiello (2006) utilized the genome database of the zebrafish to further characterize the three previously described types of zebrafish *L* chain genes (Haire *et al.*, 2000) and to specify their transcriptional polarity. Haire and coworkers (2000) had originally classified the zebrafish L chain isotypes by their C gene identity, typically having ~30% homology between isotypes. A similar degree of homology (35-37%) exists between mammalian κ and λ C regions.

The first light chain locus of zebrafish (type 1, or *L1*) possesses both small *V-J-C* clusters and an 'expanded' cluster, which is also found in type 2 (*L2*) and type 3 (*L3*) loci. The expanded cluster spans 516 kb, with four clusters of *V-J-C*, separated by intervals of 3.7-418 kb. Several other teleosts, including channel catfish, Atlantic cod, fugu, Atlantic salmon, and rainbow trout also have non-extended, *L1* chain isotypes. The V_{L1} genes are the least diverse among the three isotypes. The *V* genes are primarily in the opposite transcriptional orientation as the *J* and C genes.

The zebrafish *L2* locus is the most complex, containing 12 *V* genes, 4 *J* genes, and 2 C genes. This arrangement is termed an extended cluster. The two C genes are 59% identical at the nucleotide level and 45% similar at the derived amino acid sequence. *L2 C* genes share 22-31% identity with *L1/L3* sequences. Based on the fact that type 2 light chains are found in a broad diversity of teleosts, including carp, fugu, Atlantic salmon, rainbow trout, and zebrafish, and the *V-J-C* genes are in the same, closely clustered transcriptional orientation, Hsu and Criscitiello (2006) hypothesized a common, early existence for L2 L chains in the Teleostei.

The *L3* L chain locus has 8 *V* gene sequences (zebrafish), one *J* gene segment, and one C gene. These *V* genes are positioned on both sides of the *J* and C genes, and seven are in opposite transcriptional polarity to the *J* and C. At this point, the only teleosts known to have L3 L chains are carp, channel catfish, and zebrafish (Hsu and Criscitiello, 2006). Phylogenetic analyses and shared transcriptional characteristics strongly indicate a common derivation of L1 and L3 L chains (Hsu and Criscitiello, 2006).

Studies on the regulation of light chain expression in teleost fish have only recently begun. No functional data are yet available for L chain promoters of ectothermic vertebrates. In mammals, a TATA-box and an octamer motif are considered necessary and sufficient for promoter function. Putative promoter regions identified in the *L1* and *L3* loci of rainbow trout, *L1* of Atlantic cod, and *L2* of zebrafish have either consensus or very similar TATA and octamer motifs (Timmusk *et al.*, 2002). However, in the *L2* locus of rainbow trout, Timmusk and co-workers (2002) have been unable to detect any octamer motif. There is also some question as to whether every *V-J-C* cluster of the light chain has its own enhancer. In mice, there are two enhancers in each locus: one *J-C* intronic and one 3′ of the C gene in the κ locus and two 3′ enhancers in the λ locus. Research by Bengtén *et al.* (2000a) shows that Atlantic cod *IgL* clusters do not always have their own enhancer, but at times have to share some control elements that occur 3′ of some C genes.

THE TELEOST ANTIBODY MOLECULE

Antibody Expression and Structure

As described above, recent molecular studies demonstrate that multiple isotypes are present in teleosts. Specifically, although μ exists for all teleost species, two μ heavy chain genes have been found in Atlantic salmon and brown trout (*Salmo trutta*) (Hordvik *et al.*, 1992, 1997, 2002), δ genes with various organizations have been identified in channel catfish (Wilson *et al.*, 1997), Atlantic salmon (Hordvik *et al.*, 1999), Atlantic cod (Stenvik and Jorgensen, 2000), Atlantic halibut (Hordvik, 2002), fugu (Saha *et al.*, 2004), Japanese flounder (Hirono *et al.*, 2003), zebrafish (Danilova *et al.*, 2005) and rainbow trout (Hansen *et al.*, 2005), τ was recently characterized in rainbow trout (Hansen *et al.*, 2005), as well as ζ in zebrafish (Danilova *et al.*, 2005), and unknown *IgHs* were found in carp and fugu (Savan *et al.*, 2005a, b). To date, studies on the expression of isotypes other than μ have been limited to relatively few transcriptional analyses (Bengtén *et al.*, 2002; Hirono *et al.*, 2003; Saha *et al.*, 2004; Danilova *et al.*, 2005; Hansen *et al.*, 2005). These analyses have demonstrated that δ is preferentially transcribed in peripheral blood leukocytes (PBLs), but also found in the spleen and kidney. The copy number of δ mRNA is much less than that of μ (Hirono *et al.*, 2003). Thus far, only a monomeric IgD has been detected in channel catfish serum (Miller *et al.*, 1998; Bengtén *et al.*, 2002). The molecular weight of this molecule is approximately 160-180 kDa (Miller *et al.*, 1998; Bengtén *et al.*, 2002). Recent studies suggest that IgD's primary location is bound to the surface of phagocytic cells, which may indicate a role as an opsonin (Stafford *et al.*, 2006). Studies on τ and ζ suggest that they may be differentially transcribed in specific tissues: i.e., in the adult zebrafish, ζ is limited to the primary lymphoid tissues, whereas τ is expressed in a variety of rainbow trout tissues (Danilova *et al.*, 2005; Hansen *et al.*, 2005).

Regardless of the molecular evidence for teleost isotypic diversity, tetrameric molecule appears to be the predominant immunoglobulin (Ig) found in the circulation in all species (Bengtén *et al.*, 2000b, 2006; Solem and Stenvik, 2006). Additionally, this molecule has also been identified in the mucus (Lobb and Clem, 1981b; Lobb, 1987; Rombout *et al.*, 1993; Bromage *et al.*, 2006), egg (Fuda *et al.*, 1992; Hayman and Lobb, 2000; Bromage *et al.*, 2006) and ovarian fluid (Bromage *et al.*, 2006; Nakamura *et al.*, 2006). Despite being tetrameric, teleost IgM appears structurally analogous to IgM (pentamer) of mammals in that they: (1) are both polymeric, (2) have similar overall domain size, (3) have similar carboxyl-terminal regions, (4) lack an associated hinge region and (5) possess similar masses and carbohydrate contents (Marchalonis and Cone, 1973; Ghaffari and Lobb, 1989b). The heterodimer subunit of teleost IgM is composed of H chains of 60-75 kDa and L chains of 22-27 kDa (Fig. 3.1). Size-exclusion chromatography or denaturing, non-reducing acrylamide gel electrophoresis of the native molecule yields apparent molecular weights between 650 and 850 kDa, depending upon the species (Acton *et al.*, 1971; Lobb and Clem, 1981a; Shelby *et al.*, 2002; Bromage *et al.*, 2004b, 2006; Solem and Stenvik 2006).

Antibody Affinity and Avidity

In addition to the structural similarities described above, both mammalian and teleost IgM possess the functional similarity of binding sites with a low intrinsic affinity for antigen, but a rather high avidity (multivalency). Generally, the range of teleost IgM affinities for a variety of species is between 10^4 M^{-1} and 10^6 M^{-1} (Fiebig and Ambrosius, 1977; Fiebig *et al.*, 1977; Voss *et al.*, 1978; Lobb, 1985; Killie *et al.*, 1991; Kaattari *et al.*, 2002). The affinity of carp anti-DNP antibodies increases from about 10^4 M^{-1} in the early immune response and up to 10^5 M^{-1} in the late primary response (Fiebig and Ambrosius, 1977). The average affinity of coho salmon (*Oncorhynchus kisutch*) anti-fluorescyl antibodies is about 10^5 M^{-1} at 2 months post-primary immunization (Voss *et al.*, 1978), similar to that of Atlantic salmon anti-NIP antibodies at 4 months post-primary immunization (Killie *et al.*, 1991). Channel catfish anti-DNP and rainbow trout anti-TNP antibodies possessed a higher affinity of approximately 10^6 M^{-1} found at different time points post immunization (catfish at 7 months post-primary, and rainbow trout at 3 months post-primary) (Lobb, 1985; Kaattari *et al.*, 2002).

Past methodologies such as equilibrium dialysis have simply permitted an average affinity assessment. Alternatively, the use of a partition-based immunoassay (see following section on **Affinity Maturation**) has resolved a wide range of antibody affinity subpopulations (10^4 to 10^7 M^{-1}) within individual rainbow trout anti-TNP sera (Kaattari *et al.*, 2002). Although this latter study demonstrated that rainbow trout are capable of generating higher affinity antibody subpopulations ($Ka = 10^7$ M^{-1}) relatively late in the antibody response, the average affinity is still low (i.e., 10^6 M^{-1}).

It has been proposed that the binding activity of this low-intrinsic affinity of teleost antibody may be compensated by the octavalent nature of the tetrameric Ig, leading to a much higher avidity (Voss *et al.*, 1978). For example, carp anti-DNP IgM has high avidity of 10^{10} M^{-1} late in the primary response, and 10^{12} M^{-1} in the secondary response, although the affinity remains low (10^5 M^{-1}) throughout the response (Fiebig and Ambrosius, 1977). Another study confirmed this hypothesis by demonstrating that mammalian IgM antibodies exhibit 10,000-fold higher avidity than IgG antibodies, although the affinities of both antibodies were comparable (Kitov and Bundle, 2003). Such high avidities would only occur when the antigens with which these antibodies are reacting possess multiple repeating epitopes, such as found with polysaccharides, bacteria, or viruses.

Affinity Maturation

The affinity Maturation of teleost and other lower vertebrate responses appears fairly limited (Du Pasquier, 1982). Specifically in teleosts Voss and co-workers (1978) demonstrated that the average affinity constant for the coho salmon anti-fluorescyl antibodies increased slightly from 4.3 to 4.7×10^5 M^{-1} between 3 and 8 weeks post-primary immunization. Similarly, only slight increases in the intrinsic affinity of channel catfish anti-DNP antibodies have been demonstrated. Lobb (1985) observed that the intrinsic affinity of the anti-DNP antibodies increased 10-fold, but did so over a prolonged period. Initially, at one month post-primary, the affinity achieved a value of 1.2×10^6 M^{-1} which rose to 5.6×10^6 M^{-1} at 3 months, 8.6×10^6 M^{-1} at 7 months and finally to a maximum of 11×10^6 M^{-1} at 15 months. Killie and co-workers (1991) reported a 2- to 3-fold increase in affinity of Atlantic salmon anti-NP antibodies from a K_0 value of 3.4×10^5 M^{-1} at 2 months to 8.4×10^5 M^{-1} at 4 months. These studies demonstrated a significantly lower degree of affinity maturation in the teleost response, which is also typical of other ectothermic vertebrates (Wilson *et al.*, 1992; Diaz *et al.*, 1999), as compared to mammalian IgG responses (Eisen and Siskind, 1964; Hirose *et al.*, 1993; Rajewsky, 1996) (Table 3.2).

The opinion that affinity maturation is restricted in teleost fish (Du Pasquier, 1982) may have primarily been due to the fact that the increase in intrinsic affinity had been difficult to analyze (Fiebig and Ambrosius,

Table 3.2 Affinity Maturation in Various Vertebrate Species

Species	*Initial* $K_o^{(7)}$	*Late* $K_o^{(7)}$	*Increase in* $K_o^{(7)}$
Rainbow Trout IgM[1]	$1\text{-}5 \times 10^4\ M^{-1}$	$0.5\text{-}5.0 \times 10^6\ M^{-1}$	10-100
Channel Catfish IgM[2]	$1.2 \times 10^6\ M^{-1}$	$11.1 \times 10^6\ M^{-1}$	<10
Carp IgM[3]	$2.0\text{-}10 \times 10^4\ M^{-1}$	$4.9\text{-}11.9 \times 10^5\ M^{-1}$	10
Frog IgM[4]	$3.0\text{-}4.0 \times 10^4\ M^{-1}$	$1.0\text{-}2.0 \times 10^5\ M^{-1}$	<10
Mouse IgM[5]	$1.0\text{-}10 \times 10^4\ M^{-1}$	$1.0\text{-}10 \times 10^5\ M^{-1}$	10
Mouse IgG[6]	$1.0\text{-}10 \times 10^5\ M^{-1}$	$1.0\text{-}10 \times 10^8\ M^{-1}$	$10^3\text{-}10^4$

1. Arkoosh and Kaattari 1991; Kaattari *et al.* 2002
2. Lobb 1985
3. Fiebig and Ambrosius 1977
4. Wabl and Du Pasquier 1976; Wilson *et al.* 1992
5. Fiebig *et al.* 1979
6. Nieto *et al.* 1984
7. Value equal to aK for Rainbow trout IgM and Mouse IgG

1977; Voss *et al.*, 1978; Lobb 1985; Killie *et al.*, 1991). Further, this lack of a significant degree of affinity maturation has suggested to some researchers that somatic mutation may not occur in these species (Clem and Small, 1970). Studies employing the BIAcore biosensor (Cain *et al.*, 2002), a more sensitive method of analysis than equilibrium dialysis or fluorescence quenching, demonstrated a 2- to 3-fold increase in antibody affinity during the anti-FITC response in rainbow trout. Unfortunately, all of the above techniques are limited to providing only an average estimate of antibody affinity for a complex mixture of antibodies (e.g., antisera). An alternative means of assessing serum antibody affinities, which affords resolution of affinity subpopulations via partitioning, was developed by Nieto *et al.* (1984). We originally employed this technique in the analysis of rainbow trout serum antibodies, as we believed that shifts in affinity, in lieu of significant somatic mutation, may be restricted solely to antigen-driven selection of the original germline repertoire (Khor, 1996; Kaattari *et al.*, 2002). Such shifts would not have been discernible by earlier techniques, which do not resolve the emergence of affinity subpopulations. Partition affinity ELISAs however were found to resolve the emergence and eventual clonal dominance of high affinity antibody subpopulations that were not expressed earlier in the response, thus suggesting the possibility of somatic mutations (Lewis, 2000; Kaattari *et al.*, 2002).

Lewis (2000) further demonstrated that somatic mutations were temporally observed over the course of anti-TNP immune response in rainbow trout. She found that the affinity maturation of anti-TNP rainbow trout antibodies and the emergence of new, higher affinity subpopulations correlated with the accumulation of unique somatic variants, at least within the *CDR2* region of Ig V_H genes (possible mutational events within the *CDR3* have yet to be ascertained). The probability of somatic mutation occurring would appear to be likely, as AID has been identified within piscine genomic databases (Saunders and Magor, 2004). Not only have these investigators identified AID sequences in fugu, zebrafish, and channel catfish, but their expression in channel catfish has been observed via RT-PCR in tissues co-expressing Ig heavy chains. Recently, Yang and co-workers (2006) demonstrated that somatic mutation occurs in channel catfish V_H and J_H genes. However, whether these mutations become clonally dominant, as would be anticipated through antigen-driven selection, has yet to be determined through the analysis of antigen-specific lymphocytes.

Redox Structure

A unique and intriguing difference between teleost and mammalian IgM is the disulfide-based structural heterogeneity that is universally observed in teleost antibodies (Fig. 3.3). This structural diversity is created by the non-uniform disulfide cross-linking of the halfmer or monomeric constituents of the native tetramer. This structural diversity has been observed by the use of denaturing, non-reducing PAGE analysis of immunoglobulins from channel catfish (Lobb and Clem, 1983; Ghaffari and Lobb, 1989a), rainbow trout (Kaattari *et al.*, 1998; Bromage *et al.*, 2004b), sheepshead (*Archosargus probatocephalus*) (Lobb and Clem, 1981a), toadfish (*Spheroides glaber*) (Warr, 1983), chum salmon (*Oncorhynchus keta*) (Kobayashi *et al.*, 1982), carp (Romalde *et al.*, 1996), Atlantic salmon, striped bass (*Morone saxatilis*), barramundi (*Lates calcarifer*), Mosambique tilapia (*Oreochromis mossambicus*), Nile tilapia (*O. niloticus*) (Bromage *et al.*, 2004b), Atlantic cod, haddock (*Melanogrammus aeglefinus*), pollock (*Pollachius pollachius*), and cusk (*Brosme brosme*) (Kofod *et al.*, 1994). We have termed these differentially disulfide-bonded isomers, redox forms (Evans *et al.*, 1998; Kaattari *et al.*, 1998), referring to the variably reduced or oxidized state of the inter-subunit disulfides. These forms represent a post-translational modification of the Ig structure and not an isotypic difference, as a single C_μ gene sequence yields all redox forms (Ledford *et al.*, 1993). As these differences are solely dependent

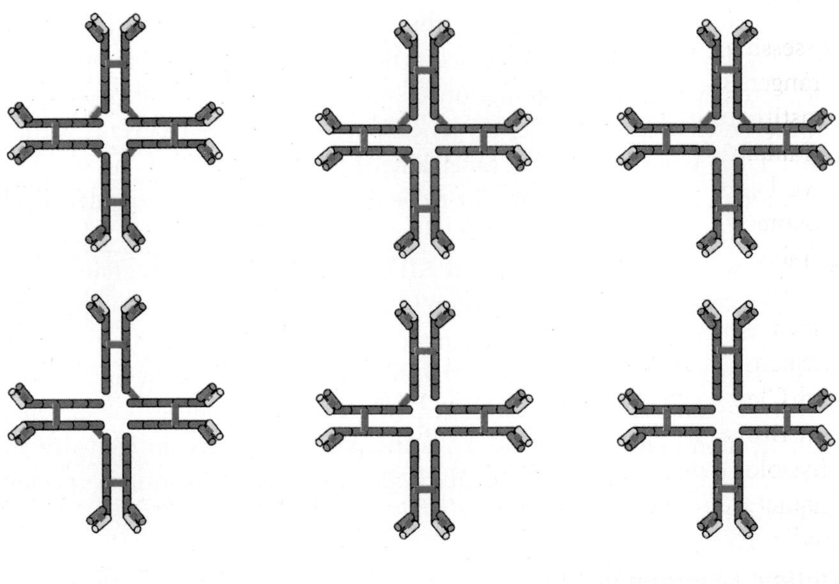

Fig. 3.3 Schematic representation of rainbow trout IgM redox structure. Each tetrameric antibody is composed of four individual monomeric subunits of approximately 200 kDa. These tetrameric antibodies can be of one of six different redox forms, depending upon the number of intermonomeric disulfide bonds.

upon disulfide polymerization, the native redox forms have not yet been physically isolated from one another. Therefore, analysis of this structural diversity is limited to denaturing, non-reducing chromatography or electrophoresis and, as such, the proportion of the different, native redox structures can only be deduced via determination of the relative molar ratios of the constituent forms after denaturation (Evans *et al.*, 1998). Differential disulfide cross-linking has not been reported for mammalian IgM. However, comparable structures have been observed with rat IgA (Chintalacharuvu *et al.*, 1993) and human IgG_4 (Schuurman *et al.*, 1999). Although a wide variety of teleosts routinely exhibit this redox structural diversity (all reported species to date), the degree and form of polymerization of the Ig subunits varies between the species. For example, denaturing, non-reducing electrophoretic analysis of channel catfish serum Ig reveals that the native tetramer yields eight different covalent constituents (Lobb and Clem, 1983), the smallest being a halfmer (HL), the largest being a fully cross-linked tetramer $(H_2L_2)_4$, with six forms possessing incremental halfmeric increases in size. In contrast to this arrangement, most studies on rainbow trout serum antibodies reveal four constituents under similar conditions; monomers, dimers, trimers and tetramers (Kaattari *et al.*, 1998). Other possible species-specific differences have been observed; for example, toadfish and carp serum Ig contain monomeric, dimeric subunits and tetramers, but no trimers (Warr, 1983; Rombout *et al.*, 1993). Four species of Gadidae fish exhibit monomeric constituents, dimers and trimers, but possess no fully cross-linked tetramers (Kofod *et al.*, 1994). Sheepshead Ig constituents resolve to covalent dimers and tetramers, without monomer or trimer subunits (Lobb and Clem, 1981a). It is not clear whether the absence of all possible constituent forms reflects the outcome of a stochastic or perhaps physiologically regulated process, rather than being a species-specific characteristic. A physiologically regulated process is suggested by our own studies (Bromage *et al.*, 2006), wherein serum Igs from rainbow trout could be resolved to collections of monomers, dimers, trimers, and tetramers; whereas mucus uniquely possessed a high proportion of halfmers. This, again, may suggest that production and distribution of these forms may reflect an, as yet, unresolved regulatory mechanism of the assembly process and not a rigid programmatic species-specific function.

Aside from the apparent regional or tissue-dependent differences described above (Bromage *et al.*, 2006), very little functional significance has been attributed to this structural diversity. It has, however, been observed that in teleost species with protein A-reactive antibodies, the lower order forms (less polymerized) have higher affinity for protein A than do higher order forms (more polymerized) (Bromage *et al.*, 2004b).

One hypothesis as to the function of redox diversity posed by Kaattari and co-workers (1999) was that the greater degree of disulfide polymerization between these subunits might result in greater rigidity. Thus, such redox forms may be less able to flexibly accommodate multiple epitopes (such as within a polysaccharide, bacterium, or virus), as might a less polymerized antibody. A similar concept has been postulated by Feinstein and co-workers who demonstrated that a loss in murine IgM rigidity in binding multivalent antigens might result in an inability to bind C1q and activate complement (Feinstein *et al.*, 1986). Thus, not only may less disulfide polymerization lead to greater flexibility and ability to bind multivalently to more substituted carriers but, in turn, it may exert regulatory control over complement activation. Potentially, this could apply to other Fc-mediated effector functions, such as opsonization or antibody-dependent cellular cytotoxicity. Thus, it is possible that teleost fish may have evolved a mode of regulating isotype-like control over Fc-mediated effector functions by post-translational modification processes rather than through employing alternate isotypes.

There are rare cases in mammals where similar redox structural heterogeneity has been observed. Chintalacharuvu *et al.* (1993) demonstrated that variability in interchain disulfide cross-linking of rat dimeric IgA is due to molecular instability, presumably due to the influence of nearby free sulfhydryl groups, and that non-covalent forces are critical for stabilizing the dimeric IgA complex. Similarly Schuurman and co-workers (1999) reported that human IgG_4 antibodies in plasma, unlike other IgG subclasses, possess redox diversity leading to instability of inter-monomeric disulfides and halfmeric exchange (Schuurman *et al.*, 2001). This halfmeric exchange leads to bi-functional (hybrid) antibodies. Such a phenomenon may explain the observation of grouper (*Epinephelus itaira*) antibodies possessing half-high and half-low affinity binding sites within a single molecule (Clem and Small, 1970) and prompts examination of cysteine placement within grouper heavy chains.

B CELL DEVELOPMENT, DIFFERENTIATION, AND FUNCTION

Hematopoiesis/Lymphopoiesis

In contrast to tetrapods, teleosts do not possess hematopoietic tissues or marrow in their bones (Harder, 1975) but rather within soft tissues. In all the vertebrates, the ventral mesoderm gives rise to hematopoietic tissues (Hansen and Zapata, 1998). Depending on the species being examined, these mesodermally derived hematopoietic tissues migrate to different sites. Teleosts employ a diversity of sites for early hematopoiesis during development, including the yolk sac, intermediate cells mass (ICM), and aorta-gonad-mid-kidney (AGM). The relative contribution of each site to early hematopoiesis it still unknown (Zapata *et al.*, 2006). Eventually, the aglomerular, anterior kidney becomes the primary hematopoietic organ (Fänge, 1986). This organ is located immediately posterior of the braincase and lies just ventral to the spinal cord along the body wall. The size and shape of the anterior kidney and its relationship with the glomerular posterior kidney can differ significantly between species.

At first glance, the basic histology of the teleost anterior kidney is somewhat unremarkable. The tissue appears fairly uniform, composed of primarily basophilic and some acidophilic cells punctuated by occasional groups of melanomacrophages and blood vessels (Fig. 3.4). In fact, the anterior kidney is a highly unusual and complex organ that houses structurally and functionally distinct tissues, including components of the neuroendocrine, reticuloendothelial, and hematopoietic systems (Zapata and Cooper, 1990). While our primary focus will be the hematopoietic role of the anterior kidney, the interactions between these tissues are also discussed .

The hypothalamus-pituitary-interrenal (HPI)-axis, which is the teleost equivalent of the hypothalamus-pituitary-adrenal (HPA)-axis, is so called due to the endocrine (interrenal) cells found in the anterior kidney (Engelsma *et al.*, 2002). These include cortisol-producing cells and

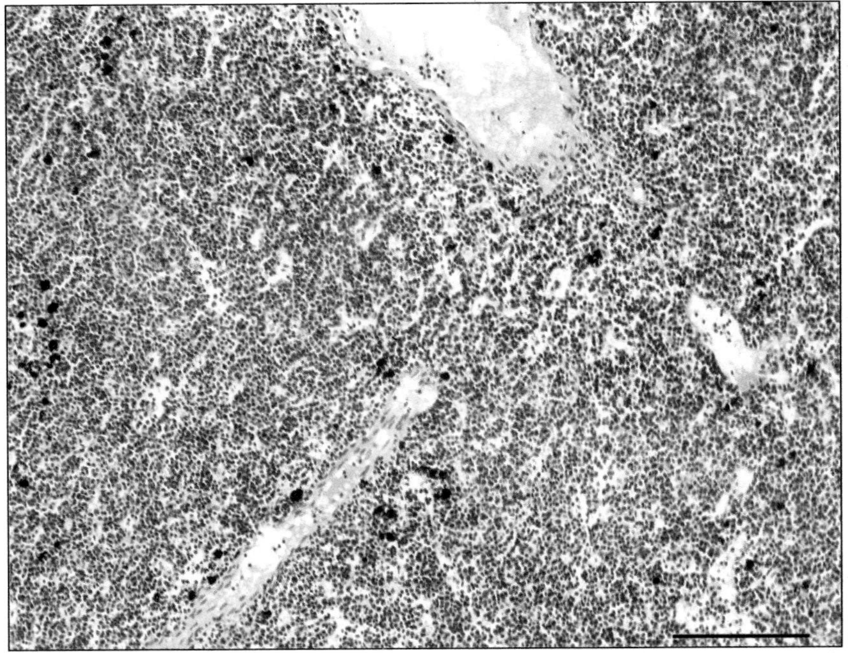

Fig. 3.4 Histology of the rainbow trout anterior kidney (hematoxylin and eosin). The anterior kidney is a fairly homogeneous tissue composed primarily of basophilic cells amongst stromal elements, punctuated by groups of melanomacrophages (dark brown) and blood vessels (pink). Scale bar = 100 µm.

chromaffin cells similar to those found in the mammalian adrenal gland (Grassi Milano *et al.*, 1997; Wendelaar Bonga, 1997). Cortisol is widely recognized as a potent regulator of immune function in both mammals and fish. For example, of particular interest in this context is the finding that cortisol induces apoptosis in circulating B cells (Weyts *et al.*, 1998). Notably, though, B cells from anterior kidney are less affected by cortisol (as its effects seem to be dependent upon the differentiation and activation state of the cell (Krammer *et al.*, 1994)), suggesting one way in which these tissues may co-exist (Verburg van Kemenade *et al.*, 1999). Given that the anterior kidney combines the production of stress hormones and other endocrine mediators with lymphopoiesis and antibody production, the potential for paracrine modulation of immune responses by stress hormones seems likely (Engelsma *et al.*, 2002).

The anterior kidney also includes a reticulo-endothelial stroma made up of endothelial cells lining the sinusoids, adventitial cells covering the abluminal surface of the endothelial cells and reticular cells which are either macrophage-like or fibroblast-like (Meseguer *et al.*, 1995). These cell types parallel those found in the bone marrow of mammals and suggest the potential for analogous functions (Meseguer *et al.*, 1995). For example, the stromal cells may form part of the hematopoietic microenvironment, providing cytokines, adhesion factors and the physical niche that supports and regulates stem cell development (Zapata, 1979; Razquin *et al.*, 1990; Weiss and Geduldig, 1991; Alvarez *et al.*, 1996). This concept is further supported by work with the rainbow trout spleen derived cell line RTS34st which 'provided a hematopoietic inductive microenvironment' for differentiation *in vitro* of precursor cells from the anterior kidney (Ganassin and Bols, 1999).

In addition to physically supporting anterior kidney hematopoiesis, the reticulo-endothelial stroma is capable of secondary lymphoid tissue functions. Several of the stromal cell types have phagocytic capacity—including sinusoidal macrophages and endothelial cells—which participate in the trapping of particles and substances from the bloodstream (MacArthur *et al.*, 1983; Dannevig *et al.*, 1994; Meseguer *et al.*, 1995; Brattgjerd and Evensen, 1996; Press and Evensen, 1999). For example, when radiolabeled-bacteria are injected into rainbow trout, more than 70% of the radioactivity lodges in the kidney (Ferguson *et al.*, 1982). Melanomacrophages are also prominent in the anterior kidney and have been shown to retain antigen for long periods (Lamers and De Haas, 1985;

Herraez and Zapata, 1986; Tsujii and Seno, 1990; Brattgjerd and Evensen, 1996; Grove *et al.*, 2003a). This function may facilitate immunological memory (Press *et al.*, 1996) by providing a persistent source of antigen for prolonged stimulation. Induction of *in vitro* antibody responses with antigen indicates that anterior kidney leukocytes can process and present antigen, thus functioning as a secondary lymphoid tissue (Arkoosh and Kaattari, 1991).

Within this stromal cell environment, lymphoid and myeloid cells can be found in numerous states of differentiation. B cells are observed to be scattered as single cells or, less frequently, small relatively compact clusters often associated with blood vessels (Meseguer *et al.*, 1995; Grove *et al.*, 2006). A loose association between IgM$^+$ cells and melanomacrophages centers was observed in Atlantic halibut but was not found to be particularly striking (Grove *et al.*, 2006). Beyond these data, no systematic survey of B cell distribution across the different regions of the anterior kidney has been conducted. Such studies may prove valuable in identifying spatial separation between the primary hematopoietic function and the secondary immune functions of this organ.

B Cell Ontogeny

Previous morphological studies have shown that in fish, the thymus is the first lymphoid organ to contain lymphocytes during ontogeny (Ellis *et al.*, 1977; Grace and Manning, 1980; Josefsson and Tatner, 1993; Abelli *et al.*, 1996; Breuil *et al.*, 1997; Zapata *et al.*, 1997). However, in at least some marine teleosts, the kidney is the first tissue to become lymphoid, followed by the spleen and thymus. In the rainbow trout, the kidney is well developed prior to hatching, producing erythrocytes and granulocytes. RAG$^+$ cells have been detected by *in situ* hybridization (Hansen and Zapata, 1998) and the first mIgM$^+$ cells, by immunohistochemistry (IHC), appear at 4 days post-hatching (dph) (Razquin *et al.*, 1990). However, RAG expression has been reported as early as 10 days post-fertilization (dpf) in rainbow trout embryos, cytoplasmic IgM$^+$ cells (cIgM$^+$) as early as 12-14 dpf (Castillo *et al.*, 1993), and L chain expression was demonstrated by ELISA in 8-day pre-hatched embryos (Sanchez *et al.*, 1995).

B Cell Development

B Cell Differentiation Factors

Expression of many early B cell differentiation markers supports the idea that the anterior kidney is the major lymphopoietic organ. This had long been proposed on the basis of histological studies and is now supported by molecular data in several species. In the early stages of teleost B cell differentiation recombination activating gene (RAG-1, -2) and terminal deoxynucleotidyl transferase (TdT) initiate immunoglobulin gene rearrangement. In zebrafish, rainbow trout and fugu, these genes are expressed in the thymus and kidney, suggesting that lymphopoiesis can occur in those organs (Hansen and Kaattari, 1995, 1996; Hansen, 1997; Willett *et al.*, 1997; Peixoto *et al.*, 2000). In addition, expression of several transcription factors associated with B cell early differentiation have been reported in bony and/or cartilaginous fish. These include PAX-5, Bcl-6, EBF-1, E2A, Ikaros and PU.1 (Hansen *et al.*, 1997; Anderson *et al.*, 2004; Hikima *et al.*, 2005; Park *et al.*, 2005; Zwollo *et al.*, 2005; Ohtani *et al.*, 2006a, c).

Within the B-cell lineage, paired box region 5 (PAX-5) is expressed in the pre- and mature B-cell stages and activates transcription of the genes encoding CD19, CD79a and AID (Ohtani *et al.*, 2006b). In mammals, AID is required for both class switching of immunoglobulin genes and somatic hypermutation (Muramatsu *et al.*, 2000; Arakawa *et al.*, 2002). While there is evidence of the existence of AID and experimental class switching in mammalian B-cells by fish AID, questions remain about its ability to induce hypermutation (and class switching) in fish (Barreto *et al.*, 2005; Zhao *et al.*, 2005; Ohtani *et al.*, 2006b). In contrast to these activating effects, PAX-5 represses the expression of XBP-1, indicating that PAX-5 also has an important role in late B cell development and activation in mammals (Kozmik *et al.*, 1992; Neurath *et al.*, 1994; Reimold *et al.*, 1996; Gonda *et al.*, 2003). In keeping with this pattern of expression, studies of PAX-5 in rainbow trout indicate that functional PAX-5 is expressed during all B developmental stages except for the plasma cell stage (Zwollo *et al.*, 2005).

B cell lymphoma-6 (Bcl-6) is a transcriptional repressor that regulates lymphocyte differentiation primarily by repressing B lymphocyte-induced maturation protein-1 (*Blimp-1*) (Reljic *et al.*, 2000; Shaffer *et al.*, 2000;

Tunyaplin *et al.*, 2004). As such, it is downregulated in plasma cells (Cattoretti *et al.*, 1995; Ye *et al.*, 1997; Angelin-Duclos *et al.*, 2000; Tunyaplin *et al.*, 2004). Although *Bcl-6* has been cloned in fugu and is highly expressed in the anterior kidney, its role in B cell development remains to be determined (Ohtani *et al.*, 2006b, c).

The B cell developmental cascade also involves the transcription factors E2A and EBF-1 (Early B cell Factor-1). In mammals, E2A transcription factors regulate the transcription of many B-lineage genes including *EBF-1*, *TdT*, *RAG-1* (Schlissel *et al.*, 1991; Choi *et al.*, 1996; Bain *et al.*, 1997; Kee and Murre 1998), and also influence processes such as Ig gene rearrangement and the expression of AID (Romanow *et al.*, 2000; Goebel *et al.*, 2001; Sayegh *et al.*, 2003). In teleosts, the role of E2A is significantly less clear. A single study in channel catfish describes the molecular cloning and expression studies of teleost *E2A* and indicates some differences in the expression and ability to drive transcription of the *IgH* locus as compared to what is described in mammals (Hikima *et al.*, 2005).

While E2A family members are found in multiple cell types, EBF-1 is highly restricted to B cells. It is a DNA-binding protein that directs progenitor cells to undergo B lymphopoiesis and activates transcription of B cell-specific genes in the absence of upstream regulators. The manner in which EBF mediates these effects in mammals remains unclear (Hagman and Lukin, 2005). A partial sequence for *EBF* was reported from a study of Atlantic halibut expressed sequence tags (ESTs) and it has been described in an elasmobranch (*Raja* sp.) (Anderson *et al.*, 2001; Park *et al.*, 2005). However, the full-length sequence of the gene, as well as expression studies, are needed to assess whether it is involved in transcriptional control of B cell development.

Ikaros was one of the first identified members of a small family of DNA-binding proteins required for lymphocyte development. The genes targeted by Ikaros, however, have not been conclusively identified in fish, although a few potential targets have been suggested (Cobb and Smale, 2005). *Ikaros* transcription in rainbow trout and zebrafish thymocytes has been reported (Hansen *et al.*, 1997; Hansen and Zapata, 1998). In rainbow trout, *Ikaros* is transcribed at 6 dpf in the yolk sac and embryo, presumably at sites of early hematopoiesis such as the yolk sac blood islands and the developing ICM. Early expression of this gene has led to the suggestion that in teleosts, Ikaros may be involved in early primitive/definitive hematopoiesis rather than solely lymphocyte differentiation (Trede and Zon, 1998).

The PU.1 transcription factor family is a divergent subclass of the Ets transcription factor family identified only in vertebrates (Anderson *et al.*, 2001). PU.1 (Spi-1) is required for normal development of T cells, B cells, macrophages, and granulocytes in mammals (Scott *et al.*, 1994; McKercher *et al.*, 1996). Its critical role in lymphocyte development suggests that at least orthologs should be present in all animals with lymphocytes (Anderson *et al.*, 2001). As such, PU.1 has been reported in zebrafish and in the skate (*Raja eglanteria*) (Anderson *et al.*, 2001; Rhodes *et al.*, 2005) and its expression pattern is consistent with a role in hematopoiesis in these species.

B Cells Distribution and Function

mIgM$^+$ B Cells

The distribution of mIgM$^+$ B cells generally follows the same pattern in all species of fish, with large numbers in the peripheral blood, spleen and kidney (Rodrigues *et al.*, 1995; Milston *et al.*, 2003) (Fig. 3.5). Typically, the peripheral blood and spleen contain the highest ratio of mIgM$^+$ to total leukocytes (>50%). However, the large blood volume compared to the capacity of the spleen make the peripheral blood the dominant tissue for harboring these cells (Hansen *et al.*, 2005; Bengtén *et al.*, 2006). Comparatively, the posterior and anterior kidneys harbor a much lower proportion (<25%). Recently, this distribution was quantified following experimental parasitic infection in turbot (*Scophthalmus maximus*) (Bermudez *et al.*, 2006). During infection, the number of mature B cells increased in the intestine, the organ targeted by the parasite, while their numbers decreased in both the spleen and kidney.

Antibody Secreting Cell (ASC)

The ASC is ultimately responsible for maintaining antibody production or humoral immunity. Until recently, comparative immunologists had presumed that only one type of ASC existed in fish, the plasma cell (Boesen *et al.*, 1997; Davidson *et al.*, 1997; Meloni and Scapigliati, 2000; Dos Santos *et al.*, 2001). However, recent evidence suggests that at least two types of ASCs exist in fish: the plasmablast and the plasma cell (Bromage *et al.*, 2004a; Zwollo *et al.*, 2005). These cells are found in

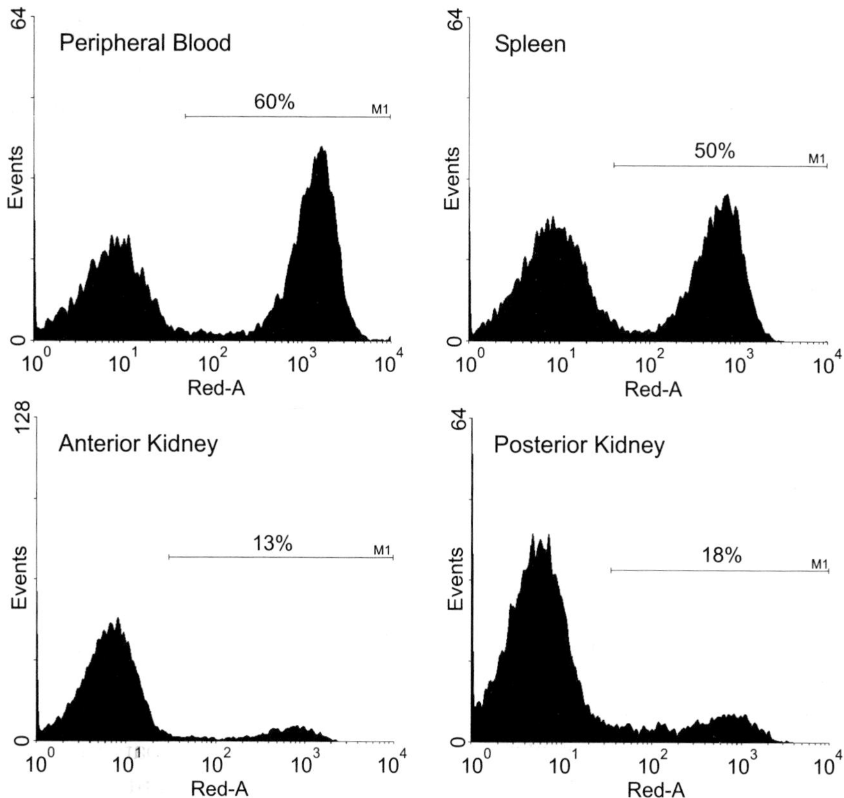

Fig. 3.5 The distribution of mIgM⁺ B-cells in rainbow trout immune tissues. Histograms showing the relative percentage of mIgM⁺ in the peripheral blood, spleen, anterior and posterior kidney of rainbow trout, as determined by flow cytometry using a pan-specific anti-heavy chain monoclonal antibody 1-14 directly labeled with Alexa-647.

varying numbers in the different immune tissues and their functions are strikingly different.

Plasmablasts are defined as proliferating ASCs that retain some expression of mIgM, and appear to be homologous to those described in mammals (Martin and Kearney, 2000; Sze *et al.*, 2000; Wehrli *et al.*, 2001). The identification of plasmablasts in a teleost has only occurred recently (Bromage *et al.*, 2004a). However, previous studies have described cells displaying plasmablast characteristics. For example, *in vitro* LPS stimulation of chinook salmon (*Oncorhynchus tshawytscha*) lymphocytes isolated from peripheral blood and spleen demonstrated the generation of blasting lymphocytes that possessed significant levels of mIgM, 4 and

7 days after induction (Milston *et al.*, 2003). Also, Miller and co-workers have described a channel catfish B cell line that possesses an 18-hour cell cycle, retains mIgM and has a low antibody secretion rate (Miller *et al.*, 1994).

Lymphocytes isolated from all teleost immune tissues can give rise to ASCs *in vitro* (Davidson *et al.*, 1992, 1997; Dos Santos *et al.*, 2001; Milston *et al.*, 2003; Bromage *et al.*, 2004a; Zwollo *et al.*, 2005). However, peripheral blood lymphocytes appear to be capable of plasmablast generation only (Bromage *et al.*, 2004a). The functions of plasmablasts in the teleost immune response have yet to be fully elucidated. Plasmablasts may be an essential stage in the production of terminally differentiated plasma cells, by providing the opportunity for somatic hypermutation during rounds of proliferation. Plasmablast predominance in peripheral blood lymphocyte cultures may also reflect the lack of a suitable microenvironment (e.g., stromal cells) for plasma cell maintenance. Our recent data suggests that early in the immune response, plasmablasts are the predominant ASCs found in all immune tissues (unpublished data). Thus, plasmablasts must serve an important role in the initial phase of humoral immunity and their specific function is the focus of ongoing investigations.

Plasma cells represent the end cells of B cell differentiation with their primary function to produce large quantities of specific antibody (Shapiro-Shelef and Calame, 2005). Studies on antibody secretion rates from plasmablasts and plasma cells in rainbow trout have revealed that plasma cells secrete approximately twice the amount of antibody than do plasmablasts over the same period of time (Zwollo *et al.*, 2005). Most research has focused on the generation of ASCs without distinguishing between plasmablasts and plasma cells (Davidson *et al.*, 1992, 1997; Dos Santos *et al.*, 2001; Shaffer *et al.*, 2004). Our research indicates that between 30-50% of the ASCs generated via *in vitro* LPS stimulation in the spleen and anterior kidney are plasma cells, as determined by their resistance to the cell-cycle inhibitor, hydroxyurea (Bromage *et al.*, 2004a). The spleen and anterior kidney also possess a number of 'spontaneous' *ex vivo* ASCs (Davidson *et al.*, 1997; Bromage *et al.*, 2004a), which are likely the consequence of previous exposure to an antigen.

Plasma Cell Differentiation

Studies on the latter stages of teleost B cell differentiation to a plasma cell are beginning to emerge. BLIMP-1, a zinc-finger protein transcriptional repressor has recently been described in fugu (Ohtani *et al.*, 2006a). This gene plays a key role in mammalian B cell terminal differentiation and represses the expression of many genes encoding other transcription factors, such as PAX-5, BCL-6, c-myc, CIITA, SpiB and Id3 (Lin *et al.*, 1997, 2002; Piskurich *et al.*, 2000; Shaffer *et al.*, 2000, 2002; Tunyaplin *et al.*, 2004), while permitting the expression of Xbox binding protein (XBP-1; Shaffer *et al.*, 2002). XBP-1 is a positive-activating transcription factor in the CREB/ATF family that plays a role in protein transport and folding (Shaffer *et al.*, 2004) and is necessary for the production and secretion of immunoglobulin by plasma cells (Reimold *et al.*, 1996; Iwakoshi *et al.*, 2003). Late B cell differentiation is also associated with repression of *c-myc* as well as the expression of BLIMP-1 and XBP-1 (Lin *et al.*, 1997; Shaffer *et al.*, 2002; Lee *et al.*, 2003). It is thought that *c-myc* is one of the most important genes in controlling B cell activation and proliferation (Roy *et al.*, 1993). Its expression is subsequently downregulated by BLIMP-1 prior to terminal differentiation (Lin *et al.*, 1997). Zhang *et al.* (1995) and Futami *et al.* (2001) have reported the molecular cloning and expression pattern of two *c-myc* genes from carp. No additional functional data is available for teleost *c-myc*. Despite the limited data currently available, we speculate that B cell differentiation in teleosts is likely to parallel that seen in mammalian bone marrow. This is not unreasonable in the light of the morphological data on the anterior kidney and the emerging molecular data describing common cell markers and transcription factors.

While the plasma cell is a terminally differentiated end cell, disruption of any of the signaling molecules such as BLIMP or XBP-1 will result in the loss of plasma cell function (Shaffer *et al.*, 2004). Considering the disruption of the cellular matrix that occurs during the harvest of teleost lymphocytes, it may be profitable to consider how this may impact the requisite cellular interactions that regulate transcription factor expression *ex vivo*. The high initial ('spontaneous') ASC numbers observed in the spleen and anterior kidney (Dos Santos *et al.*, 2001; Bromage *et al.*, 2004a) and their rapid reduction *in vitro* may be due to this cell-matrix disruption. Thus, it would be intriguing to examine the *ex vivo* antibody response when lymphocytes are cultured either with a kidney or spleen stromal cell line, or with conditioned cultured media.

The anterior kidney appears to be the final site of plasma cell migration and persistence late in the teleost immune response (Dos Santos *et al.*, 2001; Bromage *et al.*, 2004a). A number of investigators have found that teleosts can generate persistent antibody responses upon a single exposure to antigen (Thuvander *et al.*, 1987; Bricknell *et al.*, 1999; Bowden *et al.*, 2003). Our research indicates that the longevity of the antibody response correlates with the development of stable numbers of plasma cells in the anterior kidney for at least 9 months (Bromage *et al.*, 2004a). Further, in an ongoing study, we have observed significant numbers of antigen-specific ASCs in the anterior kidney of rainbow trout up to 2 years after vaccination (unpublished data). A large number of these cells were found to be capable of secreting antibody in the presence of hydroxyurea for up to 15 days *in vitro* without the need of antigen. This indicated that they were not cycling and were long lived. Thus, if persistent antibody responses can be correlated with ASC activity in the anterior kidney, the induction and maintenance of these cells will be pivotal to the development of successful vaccines. The factor(s) critical to the migration and survival of these cells in the anterior kidney are yet to be determined. Mammalian models are no further along in this regard, but it is hypothesized that the expression of specific receptors as well as the availability of vacant niches may determine the plasma cell fate (Shapiro-Shelef and Calame, 2005). The rainbow trout model may offer an advantage for studying B cell differentiation and plasma cell generation as compared to the mouse. As with the mouse models, syngeneic strains of rainbow trout exist but far greater numbers of histocompatible lymphocytes can be procured, allowing greater possibilities for cell-based analyses. Further, any discoveries from this research can be translated into productive vaccine strategies to enhance aquaculture.

MODELS AND FUTURE STUDIES

Niches, Trafficking and Organization

The existing data on anterior kidney morphology and gene expression have led our group to describe a working model for B cell development in teleosts (Fig. 3.6). B cells develop in the anterior kidney and then migrate from the kidney to sites of antigen presentation in the periphery. We hypothesize that the anterior kidney must contain an antigen-privileged site, thus serving as a primary immune organ where B cell progenitors mature into antigen-responsive B cells. It is not clear if developing B cells change their location in the anterior kidney during maturation, as is the case with developing B cells in the bone marrow of mammals (Tang *et al.*,

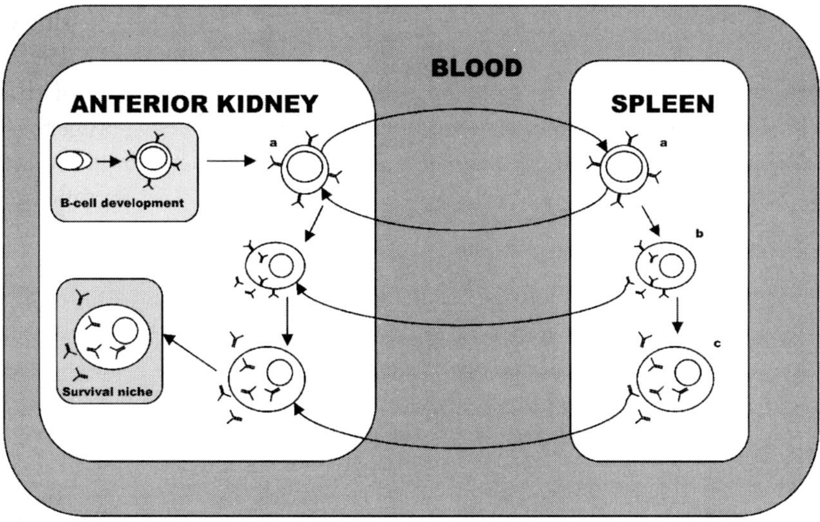

Fig. 3.6 Proposed model for plasmablast/plasma cell distribution within the anterior kidney, spleen and blood. The anterior kidney is posed as the major site of B lymphocyte development. Mature, naïve B cells (a) arise within the anterior kidney and are distributed to peripheral tissue via the blood. These B cells encounter antigen and mature into plasmablast (b) or plasma cells (c) in either the anterior kidney or periphery. Plasmablast or plasma cell differentiating within the periphery home to the anterior kidney, wherein plasma cells from all tissues, compete for long-term maintenance within the survival niches of the anterior kidney (Bromage *et al.*, 2004a). (Reproduced by permission).

1993). Typically, maturing B cells migrate to microenvironments within the bone marrow important to that life stage. Failure at any step of the maturation process will result in apoptosis (Defrance *et al.*, 2002). One of the most important steps in this process is the elimination or tolerization of B cells that recognize self during maturation (Nossal, 1994; Rathmell *et al.*, 1996; Weintraub and Goodnow, 1998). This process poses a unique challenge in teleosts; typically, B cell maturation in mammals occurs in only in the absence of non-self antigens (Nossal, 1994; Nagasawa, 2006). However, the anterior kidney of teleosts cannot be classified as antigen-free (Lamers and De Haas, 1985; Herraez and Zapata, 1986; Tsujii and Seno, 1990; Brattgjerd and Evensen, 1996; Bader *et al.*, 2003; Grove *et al.*, 2003b). Indeed, a recent study on the trafficking of antigen-coated microspheres in channel catfish demonstrated that a significant number of the microspheres was found in both the anterior and posterior kidney post-challenge (Glenney and Petrie-Hanson, 2006). Thus, if foreign antigens are present in the sites of B cell development and maturation, how can teleosts undergo self-tolerization while developing a non-self repertoire? Might there exist the structural organization within the anterior kidney that would enable the development of 'antigen-privileged' zones even during infection? Alternatively, might teleosts utilize endocrine or some other physiological control mechanism to downregulate the production of immature B cells during the period of infection or extensive antigen exposure? Obviously, there are considerable gaps in the knowledge of B cell development in teleosts, and detailed immunohistological studies coupled with the development of immunological or molecular tools targeting progenitor, pro- and pre-B cell markers will aid in elucidating how and where B cells develop within the anterior kidney.

The proportionately large population of $mIgM^+$ cells in the peripheral blood prompts speculation as to the site of antigen encounter, processing and presentation. It is thought that most teleost pathogens rapidly enter their hosts through permeable membranes such as the gills, nares, skin or intestine (Romalde *et al.*, 1996; Bromage and Owens 2002; Bader *et al.*, 2003). Therefore, it may be beneficial for teleosts to have a large number of $mIgM^+$ cells in circulation, so that they are rapidly disseminated to sites of antigen infiltration and inflammation. The recent description of partially activated B cells in rainbow trout peripheral blood (Zwollo *et al.*, 2005) may reflect the transit of antigen-sensitized B cells to secondary immune tissues for continued differentiation. This scenario, of early activated B cells transiting in the peripheral blood, is further supported by the lack of both proliferation and ASCs (Bromage *et al.*, 2004a).

Once a B cell is activated, the current paradigm suggests that it can follow a number of different pathways, leading to plasmablasts, short-lived plasma cells, long-lived plasma cells or memory cells (Kaattari *et al.*, 2005). In the mammalian system, B cell proliferation, affinity maturation and isotype switching occur in the germinal centers of the spleen that develop following antigenic challenge (Thorbecke *et al.*, 1994). There is a notable absence of clearly defined germinal centers in teleosts (Zapata, 1980). However, this does not necessarily result in a restricted or limited splenic response. The transcription factor PAX-5 maintains B cell identity prior to terminal differentiation, but it is also required to activate target genes such as AID and *BCL-6* in the germinal centers of mice (Shapiro-Shelef and Calame 2005). The expressions of PAX-5, AID, and BCL-6 have been observed in the teleost spleen (Saunders and Magor, 2004; Zwollo *et al.*, 2005; Ohtani *et al.*, 2006c). There is also evidence for somatic hypermutation (Saunders and Magor, 2004), which may suggest that the teleost spleen facilitates a similar function. Teleosts also utilize alternative approaches for maintaining humoral immunity such as antibody redox diversity (Kaattari *et al.*, 1998), predominant plasmablast production and the generation of long-lived plasma cells (Bromage *et al.*, 2004a).

A potential model for B cell trafficking in teleosts is that $mIgM^+$ B cells encounter antigen in the periphery and begin internalization and processing of that particular antigen (activation and blasting). Subsequently, they traffic to the secondary immune tissues—such as the spleen and posterior kidney—where they present their antigen to T cells and receive secondary signaling from cytokines to begin the process of proliferation and antibody secretion. Indeed, secondary signaling has been found to be critical in the proliferation and differentiation of mammalian B cells (Bartlett *et al.*, 1989; Horikawa and Takatsu, 2006). The spleen and possibly the posterior kidney are strictly secondary immune tissues involved in B cell activation and subsequent differentiation into plasmablasts. Subsets of plasmablasts from these tissues home back to the anterior kidney where they may become long-lived or short-lived plasma cells. In this model, the blood serves as a reservoir to store mature B cells and perhaps memory cells, while providing a means of transporting plasmablasts and/or plasma cells to their various homing sites, as suggested by other groups (Davidson *et al.*, 1992, 1997; Shaffer *et al.*, 2002).

Model validation and development requires an expansion of the antibody repertoire against stage-specific markers that are capable of distinguishing and isolating B cell subpopulations. Attempts to search for subpopulation markers that correspond to mammalian CD family proteins have failed thus far (Ohtani *et al.*, 2006b).

CONCLUSION

Phylogenetic Chauvinism and Comparative Immunology

Historically, the field of comparative immunology has been plagued with the perception that ectothermic vertebrate immune systems embody an ancestral form of the mammalian system. The past depiction of isotypic diversity in many standard immunology texts (Coleman *et al.*, 1989; Horton and Ratcliffe, 1993) illustrates this biased concept of assigning a linear evolutionary relationship to living, yet phylogenetically distant species. Often, the phylogenetic relationship is depicted as a 'flowering' of isotypic diversity, where fish possess only a single IgM-like isotype, increasing to three isotypes with amphibians and finally to eight or nine with mammals. Although such tree-like analogies fit well into the view that mammals are an epitome of evolutionary focus, this concept of evolution is without foundation. It can confound or obscure essential evolutionary developments that achieve comparable and, perhaps, even more efficient function in distantly related vertebrate species. Alternatively, the supposed linear increase in isotypic complexity from fish to mammals often reflects our past inability to recognize complexity in species phylogenetically distant from ourselves.

With the advent of molecular analyses, the genetic basis for antibody diversity in non-mammalian species is proving to be far greater than previously believed. Most startling was the initial discovery that elasmobranchs possess hundreds of constant region (C) genes. In addition, molecular analyses revealed a completely different organization, with mini-clusters of C genes scattered throughout the genome and not within a single complex. Also, as discussed above (Section on **The Constant Heavy Chain (C_H) Region**), the increased isotypic complexity and organization has been revealed in teleosts, with as many as three to five isotypes identified as of 2006. In addition, studies have revealed that structural diversity of the Ig molecule need not be simply relegated to C gene diversity, as it is in mammals. Structural diversity can be conferred via unique post-translational processes, such as differential disulfide cross-linking of the halfmer or monomeric subunits of the tetrameric Ig, which by and large, is not shared with mammalian species (see Section on Redox Structure).

Another area of supposed evolutionary primitiveness has been the rather restricted level of affinity maturation with the antibody response of teleosts (see **Section on Affinity Maturation**). It seems doubtful that if the generation of higher affinity antibodies were important for optimal immune function, then such capabilities would not have evolved in the teleost. However, as mammals also possess a marked degree of restriction among their multimeric immunoglobulins (Table 3.2), perhaps high-affinity multimeric antibodies are not important in any taxa (i.e., high avidity is compensatory in some fashion). Further, in the few cases where these more 'primitive' species have expressed monomeric Igs (similar to IgG), these antibodies also demonstrate higher, or mammalian-like levels of affinity maturation (Voss and Sigel, 1972; Dooley and Flajnik, 2005). Thus, it would seem that restricted levels of affinity maturation might have more to do with the efficient function of multimeric antibodies than some innate persistence of primitiveness or inability to undergo affinity maturation within these extant species.

This view of supposed complexity being a reflection of evolutionary advancement is not restricted to the perceived molecular sophistication of the antibody molecule, but also with respect to the organization of the immune system itself. As recently as 2001 (Du Pasquier, 2001), the evolution of immune systems has been depicted as a growing diversity of immune organs/tissues within extant species. In this particular example, teleosts were depicted as possessing only gut-associated lymphoid tissue, thymus, and spleen with numbers of immune tissues increasing in amphibians, reptiles, birds and finally mammals. However, missing from the teleost portrayal was the complex array of differentiated immune tissues within the kidney (including the lymphopoietic tissue in the anterior kidney progressing to the more mature cellularity of lymphocytes within the posterior kidney). However, even if there were less tissue complexity in the teleost immune system, this should not imply that fish should be considered more evolutionarily primitive. Perhaps, these taxa have evolved more adaptable organ systems. A more revealing approach would be to investigate the role of such differences as the 100-fold higher lymphocellularity in teleost blood (see **Section on Plasma Cell Differentiation**) or the presence of a distinct B cell subpopulation that possesses phagocytic capabilities (Li *et al.*, 2006). Rather than considering the teleost an evolutionary holdback with a primitive immune system, they should be considered to be as sophisticated as mammals.

Rarely has the perspective been suggested that these vertebrates have been evolving continuously from the time of our common ancestor. Thus, teleost immune systems are best not considered to be prototypical, but rather lateral developments, possessing not only ancestral similarities but also the potential for as much diversity and sophistication as found in mammals. Confusing phylogenetic distance with evolutionary advancement has been elegantly addressed by the late Stephen J. Gould (1994), 'Moreover, when we consider that for each mode of life involving greater complexity, there probably exists an equally advantageous style based on greater simplicity of form…, then preferential evolution toward complexity seems unlikely a priori. Our impression that life evolves toward greater complexity is probably only a bias inspired by parochial focus on ourselves, and consequent over attention to complexifying creatures, while we ignore just as many lineages adapting equally well by becoming simpler in form.'

The presumption that evolution is directed towards a mammalian form of complexity can obscure recognition of the sophistication and complexity that has evolved in all extant species. Comparative approaches to the study of immunology will be of the greatest value when alternate modes of analogous function are understood in the context of their own unique selective advantages.

Acknowledgements

This project was supported by the National Research Initiative of the USDA Cooperative State Research, Education and Extension Service, grant numbers 2002-35204-11685, 2004-35205-14199, 2005-35204-16271. This is VIMS contribution number 2819. The authors wish to acknowledge the technical assistance of M. Vogelbein, C. Felts, and K. Dowless.

References

Abelli, L., S. Picchietti, N. Romano, L. Mastrolia and G. Scapigliati. 1996. Immunocytochemical detection of thymocyte antigenic determinants in developing lymphoid organs of sea bass *Dicentrarchus labrax* (L.). *Fish and Shellfish Immunology* 6: 493-505.

Acton, R.T., P.F. Weinheimer, S.J. Hall, W. Niedermeier, E. Shelton and J.C. Bennett. 1971. Tetrameric immune macroglobulins in three orders of bony fishes. *Proceedings of the National Academy of Sciences of the United States of America* A 68: 107-111.

Alvarez, F., E. Flano, A. Castillo, P. Lopez-Fierro, B. Razquin and A. Villena. 1996. Tissue distribution and structure of barrier cells in the hematopoietic and lymphoid organs of salmonids. *Anatomical Record* 245: 17-24.

Amemiya, C.T. and G.W. Litman. 1990. Complete nucleotide sequence of an immunoglobulin heavy-chain gene and analysis of immunoglobulin gene organization in a primitive teleost species. *Proceedings of the National Academy of Sciences of the United States of America* 87: 811-815.

Andersson, E. and T. Matsunaga. 1993. Complete cDNA sequence of a rainbow trout IgM gene and evolution of vertebrate constant domains. *Immunogenetics* 38: 243-250.

Andersson, E. and T. Matsunaga. 1995. Evolution of immunoglobulin heavy chain variable region genes: a *VH* family can last for 150-200 million years or longer. *Immunogenetics* 41: 18-28.

Andersson, E. and T. Matsunaga. 1998. Evolutionary stability of the immunoglobulin heavy chain variable region gene families in teleost. *Immunogenetics* 47: 272-277.

Anderson, M.K., X. Sun, A.L. Miracle, G.W. Litman and E.V. Rothenberg. 2001. Evolution of hematopoiesis: Three members of the PU.1 transcription factor family in a cartilaginous fish, *Raja eglanteria*. *Proceedings of the National Academy of Sciences of the United States of America* 98: 553-558.

Anderson, M.K., R. Pant, A.L. Miracle, X. Sun, C.A. Luer, C.J. Walsh, J.C. Telfer, G.W. Litman and E.V. Rothenberg. 2004. Evolutionary origins of lymphocytes: Ensembles of T cell and B cell transcriptional regulators in a cartilaginous fish. *Journal of Immunology* 172: 5851-5860.

Angelin-Duclos, C., G. Cattoretti, K.I. Lin and K. Calame. 2000. Commitment of B lymphocytes to a plasma cell fate is associated with Blimp-1 expression *in vivo*. *Journal of Immunology* 165: 5462-5471.

Arakawa, H., J. Hauschild and J.M. Buerstedde. 2002. Requirement of the activation-induced deaminase (AID) gene for immunoglobulin gene conversion. *Science* 295: 1301-1306.

Arkoosh, M.R. and S.L. Kaattari. 1991. Development of immunological memory in rainbow trout (*Oncorhynchus mykiss*). I. An immunochemical and cellular analysis of the B cell response. *Developmental and Comparative Immunology* 15: 279-293.

Bader, J.A., K.E. Nusbaum and C.A. Shoemaker. 2003. Comparative challenge model of *Flavobacterium columnare* using abraded and unabraded channel catfish, *Ictalurus punctatus* (Rafinesque). *Journal of Fish Diseases* 26: 461-467.

Bain, G., E.C. Robanus Maandag, H.P. te Riele, A.J. Feeney, A. Sheehy, M. Schlissel, S.A. Shinton, R.R. Hardy and C. Murre. 1997. Both E12 and E47 allow commitment to the B cell lineage. *Immunity* 6: 145-154.

Barreto, V.M., Q. Pan-Hammarstrom, Y. Zhao, L. Hammarstrom, Z. Misulovin and M.C. Nussenzweig. 2005. AID from bony fish catalyzes class switch recombination. *Journal of Experimental Medicine* 202: 733-738.

Bartlett, W.C., A. Michael, J. McCann, D. Yuan, E. Claassen and R.J. Noelle. 1989. Cognate interactions between helper T cells and B cells. II. Dissection of cognate help by using a class II-restricted, antigen-specific, IL-2-dependent helper T cell clone. *Journal of Immunology* 143: 1745-1754.

Bauer, T.R., Jr. and B. Blomberg. 1991. The human lambda L chain Ig locus. Recharacterization of JC lambda 6 and identification of a functional JC lambda 7. *Journal of Immunology* 146: 2813-2820.

Bengtén, E., T. Leanderson and L. Pilström. 1991. Immunoglobulin heavy chain cDNA from the teleost Atlantic cod (*Gadus morhua* L.): Nucleotide sequences of secretory and membrane form show an unusual splicing pattern. *European Journal of Immunology* 21: 3027-3033.

Bengtén, E., S. Strömberg and L. Pilström. 1994. Immunoglobulin V_H regions in Atlantic cod (*Gadus morhua* L.): Their diversity and relationship to V_H families from other species. *Developmental and Comparative Immunology* 18: 109-122.

Bengtén, E., S. Strömberg, A. Daggfeldt, B.G. Magor and L. Pilström. 2000a. Transcriptional enhancers of immunoglobulin light chain genes in Atlantic cod (*Gadus morhua*). *Immunogenetics* 51: 647-658.

Bengtén, E., M. Wilson, N. Miller, L.W. Clem, L. Pilstrom and G.W. Warr. 2000b. Immunoglobulin isotypes: structure, function, and genetics. *Current Topics in Microbiological Immunology* 248: 189-219.

Bengtén, E., S.M. Quiniou, T.B. Stuge, T. Katagiri, N.W. Miller, L.W. Clem, G.W. Warr and M. Wilson. 2002. The IgH locus of the channel catfish, *Ictalurus punctatus*, contains multiple constant region gene sequences: Different genes encode heavy chains of membrane and secreted IgD. *Journal of Immunology* 169: 2488-2497.

Bengtén, E., L.W. Clem, N.W. Miller, G.W. Warr and M. Wilson. 2006. Channel catfish immunoglobulins: repertoire and expression. *Developmental and Comparative Immunology* 30: 77-92.

Bermudez, R., F. Vigliano, A. Marcaccini, A. Sitja-Bobadilla, M. Quiroga and J. Nieto. 2006. Response of Ig-positive cells to *Enteromyxum scophthalmi* (Myxozoa) experimental infection in turbot, *Scophthalmus maximus* (L.): A histopathological and immunohistochemical study. *Fish and Shellfish Immunology* 21: 501-512.

Boesen, H.T., K. Pedersen, C. Koch and J.L. Larsen. 1997. Immune response of rainbow trout (*Oncorhynchus mykiss*) to antigenic preparations from *Vibrio anguillarum* serogroup O1. *Fish and Shellfish Immunology* 7: 543-553.

Bowden, T.J., K. Adamson, P. MacLachlan, C.C. Pert and I.R. Bricknell. 2003. Long-term study of antibody response and injection-site effects of oil adjuvants in Atlantic halibut (*Hippoglossus hippoglossus* L.). *Fish and Shellfish Immunology* 14: 363-369.

Brattgjerd, S. and O. Evensen. 1996. A sequential light microscopic and ultrastructural study on the uptake and handling of *Vibrio salmonicida* in phagocytes of the head kidney in experimentally infected Atlantic salmon (*Salmo salar* L.). *Veterinary Pathology* 33: 55-65.

Breuil, G., B. Vassiloglou, J.F. Pepin and B. Romestand. 1997. Ontogeny of IgM-bearing cells and changes in the immunoglobulin M-like protein level (IgM) during larval stages in sea bass (*Dicentrarchus labrax*). *Fish and Shellfish Immunology* 7: 29-43.

Bricknell, I.R., J.A. King, T.J. Bowden and A.E. Ellis. 1999. Duration of protective antibodies, and the correlation with protection in Atlantic salmon (*Salmo salar* L.), following vaccination with an *Aeromonas salmonicida* vaccine containing iron-regulated outer membrane proteins and secretory polysaccharide. *Fish and Shellfish Immunology* 9: 139-151.

Bromage, E.S. and L. Owens. 2002. Infection of barramundi *Lates calcarifer* with *Streptococcus iniae*: Effects of different routes of exposure. *Diseases of Aquatic Organisms* 52: 199-205.

Bromage, E.S., J. Ye and S.L. Kaattari. 2006. Antibody structural variation in rainbow trout fluids. *Comparative Biochemistry and Physiology* B143: 61-69.

Bromage, E.S., I.M. Kaattari, P. Zwollo and S.L. Kaattari. 2004a. Plasmablast and plasma cell production and distribution in trout immune tissues. *Journal of Immunology* 173: 7317-7323.

Bromage, E.S., J. Ye; L. Owens, I.M. Kaattari and S.L. Kaattari. 2004b. Use of staphylococcal protein A in the analysis of teleost immunoglobulin structural diversity. *Developmental and Comparative Immunology* 28: 803-814.

Brown, G.D., I.M. Kaattari and S.L. Kaattari. 2006. Two new Ig VH gene families in *Oncorhynchus mykiss*. *Immunogenetics* 58: 933-936.

Cain, K.D., D.R. Jones and R.L. Raison. 2002. Antibody-antigen kinetics following immunization of rainbow trout (*Oncorhynchus mykiss*) with a T-cell dependent antigen. *Developmental and Comparative Immunology* 26: 181-190.

Castillo, A., C. Sanchez, J. Dominguez, S.L. Kaattari and A.J. Villena. 1993. Ontogeny of IgM and IgM-bearing cells in rainbow trout. *Developmental and Comparative Immunology* 17: 419-424.

Cattoretti, G., C.C. Chang, K. Cechova, J. Zhang, B.H. Ye, B. Falini, D.C. Louie, K. Offit, R.S. Chaganti and R. Dalla-Favera. 1995. BCL-6 protein is expressed in germinal-center B cells. *Blood* 86: 45-53.

Chintalacharuvu, K.R., M.E. Lamm and C.S. Kaetzel. 1993. Unstable inter-H chain disulfide bonding and non-covalently associated J chain in rat dimeric IgA. *Molecular Immunology* 30: 19-26.

Choi, J.K., C.P. Shen, H.S. Radomska, L.A. Eckhardt and T. Kadesch. 1996. E47 activates the Ig-heavy chain and TdT loci in non-B cells. *EMBO Journal* 15: 5014-5021.

Clem, L.W. 1971. Phylogeny of immunoglobulin structure and function. IV. Immunoglobulins of the giant grouper, *Epinephelus itaira*. *Journal of Biological Chemistry* 246: 9-15.

Clem, L.W. and P.A. Small, Jr. 1970. Phylogeny of immunoglobulin structure and function. V. Valences and association constants of teleost antibodies to a haptenic determinant. *Journal of Experimental Medicine* 132: 385-400.

Clem, L.W. and W.E. McLean. 1975. Phylogeny of immunoglobulin structure and function. VII. Monomeric and tetrameric immunoglobulins of the margate, a marine teleost fish. *Immunology* 29: 791-799.

Cobb, B.S. and S.T. Smale. 2005. Ikaros-family proteins: in search of molecular functions during lymphocyte development. *Current Topics in Microbiological Immunology* 290: 29-47.

Coleman, R.M., M.F. Lombard, R.E. Sicard and N.J. Rencricca. 1989. Comparative immunobiology. In: *Fundamental Immunology*. William C. Brown Company Publishers, Dubuque, IA.

Daggfeldt, A., E. Bengten and L. Pilström. 1993. A cluster type organization of the loci of the immunoglobulin light chain in Atlantic cod (*Gadus morhua* L.) and rainbow trout (*Oncorhynchus mykiss* Walbaum) indicated by nucleotide sequences of cDNAs and hybridization analysis. *Immunogenetics* 38: 199-209.

Danilova, N. and L.A. Steiner. 2002. B cells develop in the zebrafish pancreas. *Proceedings of the National Academy of Sciences of the United States of America* 99: 13711-13716.

Danilova, N., V.S. Hohman, E.H. Kim and L.A. Steiner. 2000. Immunoglobulin variable-region diversity in the zebrafish. *Immunogenetics* 52: 81-91.

Danilova, N., J. Bussmann, K. Jekosch and L.A. Steiner. 2005. The immunoglobulin heavy-chain locus in zebrafish: Identification and expression of a previously unknown isotype, immunoglobulin Z. *Nature Immunology* 6: 295-302.

Dannevig, B.H., A. Lauve, C.M. Press and T. Landsverk. 1994. Receptor-mediated endocytosis and phagocytosis by rainbow trout head kidney sinusoidal cells. *Fish and Shellfish Immunology* 4: 3-18.

Davidson, G.A., A.E. Ellis and C.J. Secombes. 1992. An ELISPOT assay for the quantification of specific antibody secreting cells to *Aeromonas salmonicida* in rainbow trout, *Oncorhynchus mykiss* (Walbaum). *Journal of Fish Diseases* 15: 85-89.

Davidson, G.A., S.H. Lin, C.J. Secombes and A.E. Ellis. 1997. Detection of specific and 'constitutive' antibody secreting cells in the gills, head kidney and peripheral blood leucocytes of dab (*Limanda limanda*). *Veterinary Immunology and Immunopathology* 58: 363-374.

Defrance, T., M. Casamayor-Palleja and P.H. Krammer. 2002. The life and death of a B cell. *Advances in Cancer Research* 86: 195-225.

Diaz, M., J. Velez, M. Singh, J. Cerny and M.F. Flajnik. 1999. Mutational pattern of the nurse shark antigen receptor gene (NAR) is similar to that of mammalian Ig genes and to spontaneous mutations in evolution: the translesion synthesis model of somatic hypermutation. *International Immunology* 11: 825-833.

Dooley, H. and M.F. Flajnik. 2005. Shark immunity bites back: affinity maturation and memory response in the nurse shark, *Ginglymostoma cirratum*. *Journal of Immunology* 35: 935-945.

Dos Santos, N.M., J.J. Taverne-Thiele, A.C. Barnes, A.E. Ellis and J.H.M. Rombout. 2001. Kinetics of juvenile sea bass (*Dicentrarchus labrax* L.) systemic and mucosal antibody secreting cell response to different antigens (*Photobacterium damselae* spp. *piscicida*, *Vibrio anguillarum* and DNP). *Fish and Shellfish Immunology* 11: 317-331.

Du Pasquier, L. 1982. Antibody diversity in lower vertebrates—why is it so restricted? *Nature (London)* 296: 311-313.

Du Pasquier, L. 2001. The immune system of invertebrates and vertebrates. *Comparative Biochemistry and Physiology* B129: 1-15.

Eisen, H.N. and G.W. Siskind. 1964. Variations in affinities of antibodies during the immune response. *Biochemistry* 3: 996-1008.

Ellestad, K.K. and B.G. Magor. 2005. Evolution of transcriptional enhancers in the immunoglobulin heavy-chain gene: functional characteristics of the zebrafish Eμ3'. *Immunogenetics* 57: 129-139.

Ellis, A.E., J.B. Solomon and J.D. Horton. 1977. Ontogeny of the immune response in *Salmo salar*. Histogenesis of the lymphoid organs and appearance of membrane immunoglobulin and mixed leukocyte reactivity. *Developmental Immunobiology*. Anonymous. Elsevier/North Holland, Amsterdam.

Engelsma, M.Y., M.O. Huising, W.B. van Muiswinkel, G. Flik, J. Kwang, H.F. Savelkoul and B.M. Verburg-van Kemenade. 2002. Neuroendocrine-immune interactions in fish: A role for interleukin-1. *Veterinary Immunology and Immunopathology* 87: 467-479.

Evans, D.A., J. Klemer and S.L. Kaattari. 1998. Heuristic models of the intermonomeric disulfide bonding process. *Journal of Theoretical Biology* 195: 505-524.

Fänge, R. 1986. Lymphoid organs in sturgeons (Acipenseridae). *Veterinary Immunology and Immunopathology* 12: 153-161.

Feinstein, A., N. Richardson and M.I. Taussig. 1986. Immunoglobulin flexibility in complement activation. *Immunology Today* 7: 169-174.

Ferguson, H.W., M.J. Claxton, R.D. Moccia and E.J. Wilkie. 1982. The quantitative clearance of bacteria from the bloodstream of rainbow trout (*Salmo gairdneri*). *Veterinary Pathology* 19: 687-699.

Fiebig, H. and H. Ambrosius. 1977. Regulation of the IgM immune response. I. Changes in the affinity of carp anti-DNP-antibodies during immunization depending on the antigen dose. *Acta Biologie et Medicine Germanicas* 36: 79-86.

Fiebig, H., R. Gruhn and H. Ambrosius. 1977. Studies on the control of IgM antibody synthesis—III. Preferential formation of anti-DNP-antibodies of high functional affinity in the course of the immune response in carp. *Immunochemistry* 14: 721-726.

Fiebig, H., S. Hornig, I. Scherbaum and H. Ambrosius. 1979. Studies on the regulation of IgM immune response. VII. Changes in the affinity of antibodies and cell receptors after immunization with the T-cell independent antigen DNP-Ficoll. *Acta Biologie et Medicine Germanicas* 38: 1627-1637.

Frippiat, J.P., S.C. Williams, I.M. Tomlinson, G.P. Cook, D. Cherif, D. Le Paslier, J.E. Collins, I. Dunham, G. Winter and M.P. Lefranc. 1995. Organization of the human immunoglobulin lambda light-chain locus on chromosome 22q11.2. *Human Molecular Genetics* 4: 983-991.

Fuda, H., A. Hara, F. Yamazaki and K. Kobayashi. 1992. A peculiar immunoglobulin M (IgM) identified in eggs of chum salmon (*Oncorhynchus keta*). *Developmental and Comparative Immunology* 16: 415-423.

Futami, K., T. Komiya, H. Zhang and N. Okamoto. 2001. Differential expression of max and two types of *c-myc* genes in a tetraploid fish, the common carp (*Cyprinus carpio*). *Gene* 269: 113-119.

Ganassin, R.C. and N.C. Bols. 1999. A stromal cell line from rainbow trout spleen, RTS34ST, that supports the growth of rainbow trout macrophages and produces conditioned medium with mitogenic effects on leukocytes. *In Vitro Cellular and Developmental Biology: Animal* 35: 80-86.

Gay, D., T. Saunders, S. Camper and M. Weigert. 1993. Receptor editing: an approach by autoreactive B cells to escape tolerance. *Journal of Experimental Medicine* 177: 999-1008.

Ghaffari, S.H. and C.J. Lobb. 1989a. Cloning and sequence analysis of channel catfish heavy chain cDNA indicate phylogenetic diversity within the IgM immunoglobulin family. *Journal of Immunology* 142: 1356-1365.

Ghaffari, S.H. and C.J. Lobb. 1989b. Nucleotide sequence of channel catfish heavy chain cDNA and genomic blot analyses. Implications for the phylogeny of Ig heavy chains. *Journal of Immunology* 143: 2730-2739.

Ghaffari, S.H. and C.J. Lobb. 1992. Organization of immunoglobulin heavy chain constant and joining region genes in the channel catfish. *Molecular Immunology* 29: 151-159.

Ghaffari, S.H. and C.J. Lobb. 1999. Structure and genomic organization of a second cluster of immunoglobulin heavy chain gene segments in the channel catfish. *Journal of Immunology* 162: 1519-1529.

Glenney, G.W. and L. Petrie-Hanson. 2006. Fate of fluorescent microspheres in developing *Ictalurus punctatus* following prolonged immersion. *Fish and Shellfish Immunology* 20: 758-768.

Goebel, P., N. Janney, J.R. Valenzuela, W.J. Romanow, C. Murre and A.J. Feeney. 2001. Localized gene-specific induction of accessibility to V(D)J recombination induced by E2A and early B cell factor in nonlymphoid cells. *Journal of Experimental Medicine* 194: 645-656.

Gonda, H., M. Sugai, Y. Nambu, T. Katakai, Y. Agata, K.J. Mori, Y. Yokota and A. Shimizu. 2003. The balance between Pax5 and Id2 activities is the key to AID gene expression. *Journal of Experimental Medicine* 198: 1427-1437.

Gorman, J.R. and F.W. Alt. 1998. Regulation of immunoglobulin light chain isotype expression. *Advances in Immunology* 69: 113-181.

Gould, S.J. 1994. The Evolution of Life on Earth. *Scientific American* 271: 85-91.

Grace, M.F. and M.J. Manning. 1980. Histogenesis of the lymphoid organs in rainbow trout, *Salmo gairdneri* Rich. 1836. *Developmental and Comparative Immunology* 4: 255-264.

Grassi Milano, E., F. Basari and C. Chimenti. 1997. Adrenocortical and adrenomedullary homologs in eight species of adult and developing teleosts: morphology, histology, and immunohistochemistry. *General and Comparative Endocrinology* 108: 483-496.

Greenberg, A.S., D. Avila, M. Hughes, A. Hughes, E.C. McKinney and M.F. Flajnik. 1995. A new antigen receptor gene family that undergoes rearrangement and extensive somatic diversification in sharks. *Nature (London)* 374: 168-173.

Grove, S., S. Hoie and O. Evensen. 2003a. Distribution and retention of antigens of *Aeromonas salmonicida* in Atlantic salmon (*Salmo salar* L.) vaccinated with a Δ *aro*A mutant or formalin-inactivated bacteria in oil-adjuvant. *Fish and Shellfish Immunology* 15: 349-358.

Grove, S., R. Johansen, B.H. Dannevig, L.J. Reitan and T. Ranheim. 2003b. Experimental infection of Atlantic halibut *Hippoglossus hippoglossus* with nodavirus: tissue distribution and immune response. *Diseases in Aquatic Organisms* 53: 211-221.

Grove, S., R. Johansen, L.J. Reitan and C.M. Press. 2006. Immune and enzyme histochemical characterisation of leukocyte populations within lymphoid and mucosal tissues of Atlantic halibut (*Hippoglossus hippoglossus*). *Fish and Shellfish Immunology* 20: 693-708.

Hagman, J. and K. Lukin. 2005. Early B-cell factor 'pioneers' the way for B-cell development. *Trends in Immunology* 26: 455-461.

Haire, R.N., J.P. Rast, R.T. Litman and G.W. Litman. 2000. Characterization of three isotypes of immunoglobulin light chains and T-cell antigen receptor alpha in zebrafish. *Immunogenetics* 51: 915-923.

Hamers-Casterman, C., T. Atarhouch, S. Muyldermans, G. Robinson, C. Hamers, E.B. Songa, N. Bendahman and R. Hamers. 1993. Naturally occurring antibodies devoid of light chains. *Nature (London)* 363: 446-448.

Hansen, J.D. 1997. Characterization of rainbow trout terminal deoxynucleotidyl transferase structure and expression. TdT and RAG1 co-expression define the trout primary lymphoid tissues. *Immunogenetics* 46: 367-375.

Hansen, J.D. and S.L. Kaattari. 1995. The recombination activation gene 1 (RAG1) of rainbow trout (*Oncorhynchus mykiss*): Cloning, expression, and phylogenetic analysis. *Immunogenetics* 42: 188-195.

Hansen, J.D. and S.L. Kaattari. 1996. The recombination activating gene 2 (RAG2) of the rainbow trout *Oncorhynchus mykiss*. *Immunogenetics* 44: 203-211.

Hansen, J.D. and A.G. Zapata. 1998. Lymphocyte development in fish and amphibians. *Immunological Reviews* 166: 199-220.

Hansen, J.D., E.D. Landis and R.B. Phillips. 2005. Discovery of a unique Ig heavy-chain isotype (IgT) in rainbow trout: Implications for a distinctive B cell developmental pathway in teleost fish. *Proceedings of the National Academy of Sciences of the United States of America* 102: 6919-6924.

Hansen, J.D., P. Strassburger and L. Du Pasquier. 1997. Conservation of a master hematopoietic switch gene during vertebrate evolution: Isolation and characterization of *Ikaros* from teleost and amphibian species. *European Journal of Immunology* 27: 3049-3058.

Harder, W. 1975. *Anatomy of Fishes*. E. Schweizerbart'sche Verlagsbuchhandlung, Stuttgart.

Havarstein, L.S., P.M. Aasjord, S. Ness and C. Endresen. 1988. Purification and partial characterization of an IgM-like serum immunoglobulin from Atlantic salmon (*Salmo salar*). *Developmental and Comparative Immunology* 12: 773-785.

Hayman, J.R. and C.J. Lobb. 2000. Heavy chain diversity region segments of the channel catfish: Structure, organization, expression and phylogenetic implications. *Journal of Immunology* 164: 1916-1924.

Hayman, J.R., S.H. Ghaffari and C.J. Lobb. 1993. Heavy chain joining region segments of the channel catfish: Genomic organization and phylogenetic implications. *Journal of Immunology* 151: 3587-3596.

Herraez, M.P. and A.G. Zapata. 1986. Structure and function of the melano-macrophage centres of the goldfish *Carassius auratus*. *Veterinary Immunology and Immunopathology* 12: 117-126.

Hikima, J., D.L. Middleton, M.R. Wilson, N.W. Miller, L.W. Clem and G.W. Warr. 2005. Regulation of immunoglobulin gene transcription in a teleost fish: Identification, expression and functional properties of E2A in the channel catfish. *Immunogenetics* 57: 273-282.

Hirono, I., B.H. Nam, J. Enomoto, K. Uchino and T. Aoki. 2003. Cloning and characterization of a cDNA encoding Japanese flounder *Paralichthys olivaceus* IgD. *Fish and Shellfish Immunology* 15: 63-70.

Hirose, S., M. Wakiya, Y. Kawano-Nishi, J. Yi, R. Sanokawa, S. Taki, T. Shimamura, T. Kishimoto, H. Tsurui, H. Nishimura and T. Shirai. 1993. Somatic diversification and affinity maturation of IgM and IgG anti-DNA antibodies in murine lupus. *European Journal of Immunology* 23: 2813-2820.

Hordvik, I. 1998. The impact of ancestral tetraploidy on antibody heterogeneity in salmonid fishes. *Immunological Reviews* 166: 153-157.

Hordvik, I. 2002. Identification of a novel immunoglobulin delta transcript and comparative analysis of the genes encoding IgD in Atlantic salmon and Atlantic halibut. *Molecular Immunology* 39: 85-91.

Hordvik, I., F.S. Berven, S.T. Solem, F. Hatten and C. Endresen. 2002. Analysis of two IgM isotypes in Atlantic salmon and brown trout. *Molecular Immunology* 39: 313-321.

Hordvik, I., J. Thevarajan, I. Samdal, N. Bastani and B. Krossoy. 1999a. Molecular cloning and phylogenetic analysis of the Atlantic salmon immunoglobulin D gene. *Scandinavian Journal of Immunology* 50: 202-210.

Hordvik, I., A.M. Voie, J. Glette, R. Male and C. Endresen. 1992. Cloning and sequence analysis of two isotypic IgM heavy chain genes from Atlantic salmon, *Salmo salar* L. *European Journal of Immunology* 22: 2957-2962.

Hordvik, I., C. De Vries Lindstrom, A.M. Voie, A. Lilybert, J. Jacob and C. Endresen. 1997. Structure and organization of the immunoglobulin M heavy chain genes in Atlantic salmon, *Salmo salar*. *Molecular Immunology* 34: 631-639.

Horikawa, K. and K. Takatsu. 2006. Interleukin-5 regulates genes involved in B-cell terminal maturation. *Immunology* 118: 497-508.

Horton, J. and N. Ratcliffe. 1993. Evolution of immunity. In: *Immunology*, I. Roitt, J. Brostoff and D. Male (eds.). C.V. Mosby, London.

Hsu, E. and M.F. Criscitiello. 2006. Diverse immunoglobulin light chain organizations in fish retain potential to revise B cell receptor specificities. *Journal of Immunology* 177: 2452-2462.

Iwakoshi, N.N., A.H. Lee and L.H. Glimcher. 2003. The X-box binding protein-1 transcription factor is required for plasma cell differentiation and the unfolded protein response. *Immunological Reviews* 194: 29-38.

Josefsson, S. and M.F. Tatner. 1993. Histogenesis of the lymphoid organs in sea bream, *Sparus auratus* L. *Fish and Shellfish Immunology* 3: 35-49.

Kaattari, S., D. Evans and J. Klemer. 1998. Varied redox forms of teleost IgM: an alternative to isotypic diversity? *Immunological Reviews* 166: 133-142.

Kaattari, S.L., J.V. Klemer and D.A. Evans. 1999. Teleost antibody structure: simple prototype or elegant alternative? *Bulletin of the European Association of Fish Pathologists* 19: 245-249.

Kaattari, S.L., H.L. Zhang, I.W. Khor, I.M. Kaattari and D.A. Shapiro. 2002. Affinity maturation in trout: Clonal dominance of high affinity antibodies late in the immune response. *Developmental and Comparative Immunology* 26: 191-200.

Kaattari, S., E. Bromage and I. Kaattari. 2005. Analysis of long-lived plasma cell production and regulation: Implications for vaccine design for aquaculture. *Aquaculture* 246: 1-9.

Kabat, E.A., T.T. Wu and H. Bilofsky. 1979. Sequence of immunoglobulin chains. *National Institute of Health Publication* No. 80-2008: 1-107.

Kee, B.L. and C. Murre. 1998. Induction of early B cell factor (EBF) and multiple B lineage genes by the basic helix-loop-helix transcription factor E12. *Journal of Experimental Medicine* 188: 699-713.

Khor, I. 1996. The development of affinity maturation in rainbow trout (*Oncorhynchus mykiss*). M.Sc. Thesis, Virginia Institute of Marine Science, The College of William and Mary, Gloucester Point, VA, USA.

Killie, J.E., S. Espelid and T.O. Jorgensen. 1991. The humoral immune response in Atlantic salmon (*Salmo salar* L.) against the hapten carrier antigen NIP-LPH; the effect of determinant (NIP) density and the isotype profile of anti-NIP antibodies. *Fish and Shellfish Immunology* 1: 33-46.

Kirschbaum, T., R. Jaenichen and H.G. Zachau. 1996. The mouse immunoglobulin kappa locus contains about 140 variable gene segments. *European Journal of Immunology* 26: 1613-1620.

Kitov, P.I. and D.R. Bundle. 2003. On the nature of the multivalency effect: a thermodynamic model. *Journal of American Chemical Society* 125: 16271-16284.

Klein, F., N. Feldhahn, J.L. Mooster, M. Sprangers, W.K. Hofmann, P. Wernet, M. Wartenberg and M. Muschen. 2005. Tracing the pre-B to immature B cell transition in human leukemia cells reveals a coordinated sequence of primary and secondary IGK gene rearrangement, IGK deletion, and IGL gene rearrangement. *Journal of Immunology* 174: 367-375.

Kobayashi, K., A. Hara, K. Takano and H. Hirai. 1982. Studies on subunit components of immunoglobulin M from a bony fish, the chum salmon (*Oncorhynchus keta*). *Molecular Immunology* 19: 95-103.

Kofod, H., K. Pedersen, J.L. Larsen and K. Buchmann. 1994. Purification and characterization of IgM-like immunoglobulin from turbot (*Scophthalmus maximus* L.). *Acta Veterinaria Scandinavica* 35: 1-10.

Kozmik, Z., S. Wang, P. Dorfler, B. Adams and M. Busslinger. 1992. The promoter of the CD19 gene is a target for the B-cell-specific transcription factor BSAP. *Molecular and Cell Biology* 12: 2662-2672.

Krammer, P.H., I. Behrmann, P. Daniel, J. Dhein and K.M. Debatin. 1994. Regulation of apoptosis in the immune system. *Current Opinion in Immunology* 6: 279-289.

Lamers, C.H. and M.J. De Haas. 1985. Antigen localization in the lymphoid organs of carp (*Cyprinus carpio*). *Cell and Tissue Research* 242: 491-498.

Ledford, B.E., B.G. Magor, D.L. Middleton, R.L. Miller, M.R. Wilson, N.W. Miller, L.W. Clem and G.W. Warr. 1993. Expression of a mouse-channel catfish chimeric IgM molecule in a mouse myeloma cell. *Molecular Immunology* 30: 1405-1417.

Lee, M.A., E. Bengtén, A. Daggfeldt, A.S. Rytting and L. Pilström. 1993. Characterisation of rainbow trout cDNAs encoding a secreted and membrane-bound Ig heavy chain and the genomic intron upstream of the first constant exon. *Molecular Immunology* 30: 641-648.

Lee, S.C., A. Bottaro and R.A. Insel. 2003. Activation of terminal B cell differentiation by inhibition of histone deacetylation. *Molecular Immunology* 39: 923-932.

Lewis, T.D. (2000). The specific immune response in rainbow trout: Somatic hyperimmunization and V_H gene utilization. Ph.D. Thesis, Virginia Institute of Marine Science, College of William and Mary, Gloucester Point, VA, USA.

Li, J., D.R. Barreda, Y.A. Zhang, H. Boshra, A.E. Gelman, S. LaPatra, L. Tort and J.O. Sunyer. 2006. B lymphocytes from early vertebrates have potent phagocytic and microbicidal abilities. *Natural Immunology* 7: 1116-1124.

Lin, K.I., C. Angelin-Duclos, T.C. Kuo and K. Calame. 2002. Blimp-1-dependent repression of Pax-5 is required for differentiation of B cells to immunoglobulin M-secreting plasma cells. *Molecular and Cell Biology* 22: 4771-4780.

Lin, Y., K. Wong and K. Calame. 1997. Repression of *c-myc* transcription by Blimp-1, an inducer of terminal B cell differentiation. *Science* 276: 596-599.

Lobb, C.J. 1985. Covalent structure and affinity of channel catfish anti-dinitrophenyl antibodies. *Molecular Immunology* 22: 993-999.

Lobb, C.J. 1987. Secretory immunity induced in catfish, *Ictalurus punctatus*, following bath immunization. *Developmental and Comparative Immunology* 11: 727-738.

Lobb, C.J. and L.W. Clem. 1981a. Phylogeny of immunoglobulin in structure and function. X. Humoral immunoglobulins of the sheepshead, *Archosargus probatocephalus*. *Developmental and Comparative Immunology* 5: 271-282.

Lobb, C.J. and L.W. Clem. 1981b. Phylogeny of immunoglobulin structure and function. XI. Secretory immunoglobulins in the cutaneous mucus of the sheepshead, *Archosargus probatocephalus*. *Developmental and Comparative Immunology* 5: 587-596.

Lobb, C.J. and L.W. Clem. 1983. Distinctive subpopulations of catfish serum antibody and immunoglobulin. *Molecular Immunology* 20: 811-818.

Lobb, C.J., M.O. Olson and L.W. Clem. 1984. Immunoglobulin light chain classes in a teleost fish. *Journal of Immunology* 132: 1917-1923.

MacArthur, J.I., T.C. Fletcher and A.W. Thomson. 1983. Distribution of radiolabeled erythrocytes and the effect of temperature on clearance in the plaice (*Pleuronectes platessa* L.). *Journal of Reticuloendothelial Society* 34: 13-21.

Marchalonis, J.J. and R.E. Cone. 1973. The phylogenetic emergence of vertebrate immunity. *Australian Journal of Experimental Biology and Medical Science* 51: 461-488.

Martin, F. and J.F. Kearney. 2000. B-cell subsets and the mature preimmune repertoire. Marginal zone and B1 B cells as part of a 'natural immune memory'. *Immunological Reviews* 175: 70-79.

Matsunaga, T., T. Chen and V. Tormanen. 1990. Characterization of a complete immunoglobulin heavy-chain variable region germ-line gene of rainbow trout. *Proceedings of the National Academy of Sciences of the United States of America* 87: 7767-7771.

McKercher, S.R., B.E. Torbett, K.L. Anderson, G.W. Henkel, D.J. Vestal, H. Baribault, M. Klemsz, A.J. Feeney, G.E. Wu, C.J. Paige and R.A. Maki. 1996. Targeted disruption of the PU.1 gene results in multiple hematopoietic abnormalities. *EMBO Journal* 15: 5647-5658.

Meloni, S. and G. Scapigliati. 2000. Evaluation of immunoglobulins produced *in vitro* by head-kidney leucocytes of sea bass *Dicentrarchus labrax* by immunoenzymatic assay. *Fish and Shellfish Immunology* 10: 95-99.

Meseguer, J., A. Lopez-Ruiz and A. Garcia-Ayala. 1995. Reticulo-endothelial stroma of the head-kidney from the seawater teleost gilthead sea bream (*Sparus aurata* L.): An ultrastructural and cytochemical study. *Anatomical Record* 241: 303-309.

Miller, N., M. Wilson, E. Bengtén, T. Stuge, G. Warr and W. Clem. 1998. Functional and molecular characterization of teleost leukocytes. *Immunological Reviews* 166: 187-197.

Miller, N.W., M.A. Rycyzyn, M.R. Wilson, G.W. Warr, J.P. Naftel and L.W. Clem. 1994. Development and characterization of channel catfish long term B cell lines. *Journal of Immunology* 152: 2180-2189.

Milston, R.H., A.T. Vella, T.L. Crippen, M.S. Fitzpatrick, J.A. Leong and C.B. Schreck. 2003. *In vitro* detection of functional humoral immunocompetence in juvenile chinook salmon (*Oncorhynchus tshawytscha*) using flow cytometry. *Fish and Shellfish Immunology* 15: 145-158.

Muramatsu, M., K. Kinoshita, S. Fagarasan, S. Yamada, Y. Shinkai and T. Honjo. 2000. Class switch recombination and hypermutation require activation-induced cytidine deaminase (AID), a potential RNA editing enzyme. *Cell* 102: 553-563.

Nagasawa, T. 2006. Microenvironmental niches in the bone marrow required for B-cell development. *Nature Reviews Immunology* 6: 107-116.

Nakamura, O., R. Kudo, H. Aoki and T. Watanabe. 2006. IgM secretion and absorption in the materno-fetal interface of a viviparous teleost, *Neoditrema ransonneti* (Perciformes; Embiotocidae). *Developmental and Comparative Immunology* 30: 493-502.

Neurath, M.F., W. Strober and Y. Wakatsuki. 1994. The murine Ig 3' alpha enhancer is a target site with repressor function for the B cell lineage-specific transcription factor BSAP (NF-HB, S alpha-BP). *Journal of Immunology* 153: 730-742.

Nguyen, V.K., R. Hamers, L. Wyns and S. Muyldermans. 2000. Camel heavy-chain antibodies: diverse germline V(H)H and specific mechanisms enlarge the antigen-binding repertoire. *EMBO Journal* 19: 921-930.

Nieto, A., A. Gaya, M. Jansa, C. Moreno and J. Vives. 1984. Direct measurement of antibody affinity distribution by hapten-inhibition enzyme immunoassay. *Molecular Immunology* 21: 537-543.

Nossal, G.J. 1994. Negative selection of lymphocytes. *Cell* 76: 229-239.

Ohtani, M., T. Miyadai and S. Hiroishi. 2006a. B-lymphocyte-induced maturation protein-1 (Blimp-1) gene of torafugu (*Takifugu rubripes*). *Fish and Shellfish Immunology* 20: 409-413.

Ohtani, M., T. Miyadai and S. Hiroishi. 2006b. Identification of genes encoding critical factors regulating B-cell terminal differentiation in torafugu (*Takifugu rubripes*). *Comparative Biochemistry and Physiology* D Genomics and Proteomics 1: 109-114.

Ohtani, M., T. Miyadai and S. Hiroishi. 2006c. Molecular cloning of the BCL-6 gene, a transcriptional repressor for B-cell differentiation, in torafugu (*Takifugu rubripes*). *Molecular Immunology* 43: 1047-1053.

Park, C.I., I. Hirono, J.Y. Hwang and T. Aoki. 2005. Characterization and expression of a CD40 homolog gene in Japanese flounder *Paralichthys olivaceus*. *Immunogenetics* 57: 682-689.

Partula, S., J. Schwager, S. Timmusk, L. Pilström and J. Charlemagne. 1996. A second immunoglobulin light chain isotype in the rainbow trout. *Immunogenetics* 45: 44-51.

Peixoto, B.R., Y. Mikawa and S. Brenner. 2000. Characterization of the recombinase activating gene-1 and 2 locus in the Japanese pufferfish, *Fugu rubripes*. *Gene* 246: 275-283.

Pilström, L. 2002. The mysterious immunoglobulin light chain. *Developmental and Comparative Immunology* 26: 207-215.

Pilström, L., M.L. Lundqvist and N.E. Wermenstam. 1998. The immunoglobulin light chain in poikilothermic vertebrates. *Immunological Reviews* 166: 123-132.

Piskurich, J.F., K.I. Lin, Y. Lin, Y. Wang, J.P. Ting and K. Calame. 2000. BLIMP-I mediates extinction of major histocompatibility class II transactivator expression in plasma cells. *Nature Immunology* 1: 526-532.

Press, C.M. and O. Evensen. 1999. The morphology of the immune system in teleost fishes. *Fish and Shellfish Immunology* 9: 309-318.

Press, C.M., O. Evensen, L.J. Reitan and T. Landsverk. 1996. Retention of furunculosis vaccine components in Atlantic salmon, *Salmo salar* L., following different routes of vaccine administration. *Journal of Fish Diseases* 19: 215-224.

Radic, M.Z., J. Erikson, S. Litwin and M. Weigert. 1993. B lymphocytes may escape tolerance by revising their antigen receptors. *Journal of Experimental Medicine* 177: 1165-1173.

Rajewsky, K. 1996. Clonal selection and learning in the antibody system. *Nature (London)* 381: 751-758.

Rathmell, J.C., S.E. Townsend, J.C. Xu, R.A. Flavell and C.C. Goodnow. 1996. Expansion or elimination of B cells in vivo: dual roles for CD40- and Fas (CD95)-ligands modulated by the B cell antigen receptor. *Cell* 87: 319-329.

Razquin, B.E., A. Castillo, P. Lopez-Fierro, F. Alvarez, A. Zapata and A. Villena. 1990. Ontogeny of IgM-producing cells in the lymphoid organs of rainbow trout, *Salmo gairdneri* Richardson: an immunological and enzyme-histochemical study. *Journal of Fish Biology* 36: 159-173.

Reimold, A.M., P.D. Ponath, Y.S. Li, R.R. Hardy, C.S. David, J.L. Strominger and L.H. Glimcher. 1996. Transcription factor B cell lineage-specific activator protein regulates the gene for human X-box binding protein 1. *Journal of Experimental Medicine* 183: 393-401.

Reljic, R., S.D. Wagner, L.J. Peakman and D.T. Fearon. 2000. Suppression of signal transducer and activator of transcription 3-dependent B lymphocyte terminal differentiation by BCL-6. *Journal of Experimental Medicine* 192: 1841-1848.

Reynaud, C.A., A. Dahan, V. Anquez and J.C. Weill. 1989. Somatic hyperconversion diversifies the single Vh gene of the chicken with a high incidence in the D region. *Cell* 59: 171-183.

Rhodes, J., A. Hagen, K. Hsu, M. Deng, T.X. Liu, A.T. Look and J.P. Kanki. 2005. Interplay of PU.1 and GATA1 determines myelo-erythroid progenitor cell fate in zebrafish. *Developmental Cell* 8: 97-108.

Rodrigues, P.N., T.T. Hermsen, J.H. Rombout, E. Egberts and R.J. Stet. 1995. Detection of MHC class II transcripts in lymphoid tissues of the common carp (*Cyprinus carpio* L.). *Developmental and Comparative Immunology* 19: 483-496.

Romalde, J.L., B. Magarinos, S. Nunez, J.L. Barja and A.E. Toranzo. 1996. Host range susceptibility of *Enterococcus* sp. strains isolated from diseased turbot: possible routes of infection. *Applied Environmental Microbiology* 62: 607-611.

Roman, T. and J. Charlemagne. 1994. The immunoglobulin repertoire of the rainbow trout (*Oncorhynchus mykiss*): Definition of nine *Igh-V* families. *Immunogenetics* 40: 210-216.

Roman, T., A. De Guerra and J. Charlemagne. 1995. Evolution of specific antigen recognition: size reduction and restricted length distribution of the CDRH3 regions in the rainbow trout. *European Journal of Immunology* 25: 269-273.

Roman, T., E. Andersson, E. Bengten, J.D. Hansen, S. Kaattari, L. Pilström, J. Charlemagne and T. Matsunaga. 1996. Unified nomenclature of Ig VH genes in rainbow trout (*Oncorhynchus mykiss*): Definition of eleven *VH* families. *Immunogenetics* 43: 325-326.

Romanow, W.J., A.W. Langerak, P. Goebel, I.L. Wolvers-Tettero, J.J. van Dongen, A.J. Feeney and C. Murre. 2000. E2A and EBF act in synergy with the V(D)J recombinase to generate a diverse immunoglobulin repertoire in nonlymphoid cells. *Molecular Cell* 5: 343-353.

Rombout, J.H., N. Taverne, M. van de Kamp and A.J. Taverne-Thiele. 1993. Differences in mucus and serum immunoglobulin of carp (*Cyprinus carpio* L.). *Developmental and Comparative Immunology* 17: 309-317.

Ross, D.A., M.R. Wilson, N.W. Miller, L.W. Clem and G.W. Warr. 1998. Evolutionary variation of immunoglobulin μ heavy chain RNA processing pathways: Origins, effects, and implications. *Immunological Reviews* 166: 143-151.

Roy, A.L., C. Carruthers, T. Gutjahr and R.G. Roeder. 1993. Direct role for Myc in transcription initiation mediated by interactions with TFII-I. *Nature (London)* 365: 359-361.

Saha, N.R., H. Suetake, K. Kikuchi and Y. Suzuki. 2004. Fugu immunoglobulin D: A highly unusual gene with unprecedented duplications in its constant region. *Immunogenetics* 56: 438-447.

Sanchez, C. and J. Dominguez. 1991. Trout immunoglobulin populations differing in light chains revealed by monoclonal antibodies. *Molecular Immunology* 28: 1271-1277.

Sanchez, C., A. Alvarez, A. Castillo, A. Zapata, A. Villena and J. Dominguez. 1995. Two different subpopulations of Ig-bearing cells in lymphoid organs of rainbow trout. *Developmental and Comparative Immunology* 19: 79-86.

Saunders, H.L. and B.G. Magor. 2004. Cloning and expression of the AID gene in the channel catfish. *Developmental and Comparative Immunology* 28: 657-663.

Savan, R., A. Aman, M. Nakao, H. Watanuki and M. Sakai. 2005a. Discovery of a novel immunoglobulin heavy chain gene chimera from common carp (*Cyprinus carpio* L.). *Immunogenetics* 57: 458-463.

Savan, R., A. Aman, K. Sato, R. Yamaguchi and M. Sakai. 2005b. Discovery of a new class of immunoglobulin heavy chain from fugu. *European Journal of Immunology* 35: 3320-3331.

Sayegh, C.E., M.W. Quong, Y. Agata and C. Murre. 2003. E-proteins directly regulate expression of activation-induced deaminase in mature B cells. *Nature Immunology* 4: 586-593.

Schlissel, M., A. Voronova and D. Baltimore. 1991. Helix-loop-helix transcription factor E47 activates germ-line immunoglobulin heavy-chain gene transcription and rearrangement in a pre-T-cell line. *Genes Development* 5: 1367-1376.

Schuurman, J., G.J. Perdok, A.D. Gorter and R.C. Aalberse. 2001. The inter-heavy chain disulfide bonds of IgG4 are in equilibrium with intra-chain disulfide bonds. *Molecular Immunology* 38: 1-8.

Schuurman, J., R. Van Ree, G.J. Perdok, H.R. Van Doorn, K.Y. Tan and R.C. Aalberse. 1999. Normal human immunoglobulin G4 is bispecific: It has two different antigen-combining sites. *Immunology* 97: 693-698.

Scott, E.W., M.C. Simon, J. Anastasi and H. Singh. 1994. Requirement of transcription factor PU.1 in the development of multiple hematopoietic lineages. *Science* 265: 1573-1577.

Secher, D.S., C. Milstein and D.K. Adetugbo. 1977. Somatic mutants and antibody diversity. *Immunological Reviews* 36: 51-72.

Shaffer, A.L., X. Yu, Y. He, J. Boldrick, E.P. Chan and L.M. Staudt. 2000. BCL-6 represses genes that function in lymphocyte differentiation, inflammation, and cell cycle control. *Immunity* 13: 199-212.

Shaffer, A.L., K.I. Lin, T.C. Kuo, X. Yu, E.M. Hurt, A. Rosenwald, J.M. Giltnane, L. Yang, H. Zhao, K. Calame and L.M. Staudt. 2002. Blimp-1 orchestrates plasma cell differentiation by extinguishing the mature B cell gene expression program. *Immunity* 17: 51-62.

Shaffer, A.L., M. Shapiro-Shelef, N.N. Iwakoshi, A.H. Lee, S.B. Qian, H. Zhao, X. Yu, L. Yang, B.K. Tan, A. Rosenwald, E.M. Hurt, E. Petroulakis, N. Sonenberg, J.W. Yewdell, K. Calame, L.H. Glimcher and L.M. Staudt. 2004. XBP1, downstream of Blimp-1, expands the secretory apparatus and other organelles, and increases protein synthesis in plasma cell differentiation. *Immunity* 21: 81-93.

Shamblott, M.J. and G.W. Litman. 1989a. Complete nucleotide sequence of primitive vertebrate immunoglobulin light chain genes. *Proceedings of the National Academy of Sciences of the United States of America* 86: 4684-4688.

Shamblott, M.J. and G.W. Litman. 1989b. Genomic organization and sequences of immunoglobulin light chain genes in a primitive vertebrate suggest coevolution of immunoglobulin gene organization. *EMBO Journal* 8: 3733-3739.

Shapiro-Shelef, M. and K. Calame. 2005. Regulation of plasma-cell development. *Nature Reviews Immunology* 5: 230-242.

Shelby, R.A., J.J. Evans and P.H. Klesius. 2002. Isolation, purification, and molecular weight determination of serum immunoglobulin from gulf menhaden: Development of an enzyme-linked immunosorbent assay to assess serum immunoglobulin concentrations from Atlantic menhaden. *Journal of Aquatic Animal Health* 14: 254-262.

Solem, S.T. and J. Stenvik. 2006. Antibody repertoire development in teleosts—a review with emphasis on salmonids and *Gadus morhua* L. *Developmental and Comparative Immunology* 30: 57-76.

Solem, S.T., I. Hordvik, J.A. Killie, G.W. Warr and T.O. Jorgensen. 2001. Diversity of the immunoglobulin heavy chain in the Atlantic salmon (*Salmo salar* L.) is contributed by genes from two parallel IgH isoloci. *Developmental and Comparative Immunology* 25: 403-417.

Srisapoome, P., T. Ohira, I. Hirono and T. Aoki. 2004. Genes of the constant regions of functional immunoglobulin heavy chain of Japanese flounder, *Paralichthys olivaceus*. *Immunogenetics* 56: 292-300.

Stafford, J.L., E. Bengtén, G.W. Warr, N.W. Miller and M. Wilson. 2006. Channel catfish *Ictalurus punctatus* granulocytes bind IgD. 10th Congress of ISDCI, Charleston, SC.

Stenvik, J. and T.O. Jorgensen. 2000. Immunoglobulin D (IgD) of Atlantic cod has a unique structure. *Immunogenetics* 51: 452-461.

Stenvik, J., T.O. Jorgensen and L. Pilström. 2000. Variable region diversity of the Atlantic cod (*Gadus morhua* L.) immunoglobulin heavy chain. *Immunogenetics* 51: 670-680.

Stenvik, J., M.B. Schroder, K. Olsen, A. Zapata and T.O. Jorgensen. 2001. Expression of immunoglobulin heavy chain transcripts (VH-families, IgM, and IgD) in head kidney and spleen of the Atlantic cod (*Gadus morhua* L.). *Developmental and Comparative Immunology* 25: 291-302.

Sze, D.M.-Y., K.-M. Toellner, C.G. de Vinuesa, D.R. Taylor and I.C.M. MacLennan. 2000. Intrinsic constraint on plasmablast growth and extrinsic limits of plasma cell survival. *Journal of Experimental Medicine* 192: 813-822.

Tang, J., G. Scott and D.H. Ryan. 1993. Subpopulations of bone marrow fibroblasts support VLA-4-mediated migration of B-cell precursors. *Blood* 82: 3415-3423.

Thorbecke, G.J., A.R. Amin and V.K. Tsiagbe. 1994. Biology of germinal centers in lymphoid tissue. *Federation of American Societies of Experimental Biology* 8: 832-840.

Thuvander, A., T. Hongslo, E. Jansson and B. Sundquist. 1987. Duration of protective immunity and antibody titres measured by ELISA after vaccination of rainbow trout, *Salmo gairdneri* Richardson, against vibriosis. *Journal of Fish Diseases* 10: 479-486.

Tiegs, S.L., D.M. Russell and D. Nemazee. 1993. Receptor editing in self-reactive bone marrow B cells. *Journal of Experimental Medicine* 177: 1009-1020.

Timmusk, S., S. Partula and L. Pilström. 2000. Different genomic organization and expression of immunoglobulin light-chain isotypes in the rainbow trout. *Immunogenetics* 51: 905-914.

Timmusk, S., S. Strömberg and L. Pilström. 2002. Light chain promoter regions in rainbow trout *Oncorhynchus mykiss*: Function of a classical and an atypical Ig promoter. *Developmental and Comparative Immunology* 26: 785-796.

Tonegawa, S. 1983. Somatic generation of antibody diversity. *Nature (London)* 302: 575-581.

Trede, N.S. and L.I. Zon. 1998. Development of T-cells during fish embryogenesis. *Developmental and Comparative Immunology* 22: 253-263.

Tsujii, T. and S. Seno. 1990. Melano-macrophage centers in the aglomerular kidney of the sea horse (teleosts): Morphologic studies on its formation and possible function. *Anatomical Record* 226: 460-470.

Tunyaplin, C., A.L. Shaffer, C.D. Angelin-Duclos, X. Yu, L.M. Staudt and K.L. Calame. 2004. Direct repression of prdm1 by Bcl-6 inhibits plasmacytic differentiation. *Journal of Immunology* 173: 1158-1165.

Ventura-Holman, T. and C.J. Lobb. 2001. Structural organization of the immunoglobulin heavy chain locus in the channel catfish: the IgH locus represents composite of two gene clusters. *Molecular Immunology* 38: 557-564.

Verburg-van Kemenade, B.M., B. Nowak, M.Y. Engelsma and F.A. Weyts. 1999. Differential effects of cortisol on apoptosis and proliferation of carp B-lymphocytes from head kidney, spleen and blood. *Fish and Shellfish Immunology* 9: 405-415.

Voss, E.W., Jr., W.J. Groberg, Jr. and J.L. Fryer. 1978. Binding affinity of tetrameric coho salmon Ig anti-hapten antibodies. *Immunochemistry* 15: 459-464.

Voss, E.W. and M.M. Sigel. 1972. Valence and temporal changes in affinity of purified 7S and 18S nurse shark anti-2,4-dinitrophenyl antibodies. *Journal of Immunology* 109: 665-673.

Wabl, M.R. and L. Du Pasquier. 1976. Antibody patterns in genetically identical frogs. *Nature (London)* 264: 642-644.

Wardemann, H., J. Hammersen and M.C. Nussenzweig. 2004. Human autoantibody silencing by immunoglobulin light chains. *Journal of Experimental Medicine* 200: 191-199.

Warr, G.W. 1983. Immunoglobulin of the toadfish, *Spheroides glaber*. *Comparative Biochemistry and Physiology* B 76: 507-514.

Warr, G.W. 1995. The immunoglobulin genes of fish. *Developmental and Comparative Immunology* 19: 1-12.

Warr, G.W., D. DeLuca and J.J. Marchalonis. 1976. Phylogenetic origins of immune recognition: Lymphocyte surface immunoglobulins in the goldfish, *Carassius auratus*. *Proceedings of the National Academy of Sciences of the United States of America* 73: 2476-2480.

Wehrli, N., D.F. Legler, D. Finke, K.M. Toellner, P. Loetscher, M. Baggiolini, I.C. MacLennan and H. Acha-Orbea. 2001. Changing responsiveness to chemokines allows medullary plasmablasts to leave lymph nodes. *European Journal of Immunology* 31: 609-616.

Weintraub, B.C. and C.C. Goodnow. 1998. Immune Responses: costimulatory receptors have their say. *Current Biology* 8: 575-577.

Weiss, L. and U. Geduldig. 1991. Barrier cells: stromal regulation of hematopoiesis and blood cell release in normal and stressed murine bone marrow. *Blood* 78: 975-990.

Wendelaar Bonga, S.E. 1997. The stress response in fish. *Physiological Reviews* 77: 591-625.

Weyts, F.A., G. Flik, J.H. Rombout and B.M. Verburg-van Kemenade. 1998. Cortisol induces apoptosis in activated B cells, not in other lymphoid cells of the common carp, *Cyprinus carpio* L. *Developmental and Comparative Immunology* 22: 551-562.

Willett, C.E., J.J. Cherry and L.A. Steiner. 1997. Characterization and expression of the recombination activating genes (*rag1* and *rag2*) of zebrafish. *Immunogenetics* 45: 394-404.

Wilson, M. and G. Warr. 1992. Fish immunoglobulins and the genes that encode them. *Annual Review of Fish Diseases* 2: 201-221.

Wilson, M., E. Hsu, A. Marcuz, M. Courtet, L. Du Pasquier and C. Steinberg. 1992. What limits affinity maturation of antibodies in *Xenopus*—the rate of somatic mutation or the ability to select mutants? *EMBO Journal* 11: 4337-4347.

Wilson, M., E. Bengtén, N.W. Miller, L.W. Clem, L. Du Pasquier and G.W. Warr. 1997. A novel chimeric Ig heavy chain from a teleost fish shares similarities to IgD. *Proceedings of the National Academy of Sciences of the United States of America* 94: 4593-4597.

Wilson, M.R., E. van Ravenstein, N.W. Miller, L.W. Clem, D.L. Middleton and G.W. Warr. 1995. cDNA sequences and organization of IgM heavy chain genes in two holostean fish. *Developmental and Comparative Immunology* 19: 153-164.

Yang, F., G.C. Waldbieser and C.J. Lobb. 2006. The nucleotide targets of somatic mutation and the role of selection in immunoglobulin heavy chains of a teleost fish. *Journal of Immunology* 176: 1655-1667.

Yang, F., T. Ventura-Holman, G.C. Waldbieser and C.J. Lobb. 2003. Structure, genomic organization, and phylogenetic implications of six new VH families in the channel catfish. *Molecular Immunology* 40: 247-260.

Ye, B.H., G. Cattoretti, Q. Shen, J. Zhang, N. Hawe, R. de Waard, C. Leung, M. Nouri-Shirazi, A. Orazi, R.S. Chaganti, P. Rothman, A.M. Stall, P.P. Pandolfi and R. Dalla-Favera. 1997. The BCL-6 proto-oncogene controls germinal-centre formation and Th2-type inflammation. *Nature Genetics* 16: 161-170.

Zapata, A. 1979. Ultrastructural study of the teleost fish kidney. *Developmental and Comparative Immunology* 3: 55-65.

Zapata, A. 1980. Ultrastructure of elasmobranch lymphoid tissue. 1. Thymus and spleen. *Developmental and Comparative Immunology* 4: 459-571.

Zapata, A.G. and E.L. Cooper. 1990. *The Immune System: Comparative Histopathology.* Wiley, Chichester, U.K.

Zapata, A.G., M. Torroba, A. Varas and A.V. Jimenez. 1997. Immunity in fish larvae. *Developments in Biological Standardization* 90: 23-32.

Zapata, A., B. Diez, T. Cejalvo, C. Gutierrez-de Frias and A. Cortes. 2006. Ontogeny of the immune system of fish. *Fish and Shellfish Immunology* 20: 126-136.

Zhang, H., N. Okamoto and Y. Ikeda. 1995. Two *c-myc* genes from a tetraploid fish, the common carp (*Cyprinus carpio*). *Gene* 153: 231-236.

Zhao, Y., Q. Pan-Hammarstrom, Z. Zhao and L. Hammarstrom. 2005. Identification of the activation-induced cytidine deaminase gene from zebrafish: an evolutionary analysis. *Developmental and Comparative Immunology* 29: 61-71.

Zwollo, P., S. Cole, E. Bromage and S. Kaattari. 2005. B cell heterogeneity in the teleost kidney: evidence for a maturation gradient from anterior to posterior kidney. *Journal of Immunology* 174: 6608-6616.

CHAPTER

4

Use of CpG ODNs in Aquaculture: A Review

M.A. Esteban, A. Cuesta and J. Meseguer*

INTRODUCTION

Disease control is of prime concern to the aquaculture industry, where it is generally accepted that the prevention of disease during the culture of commercially important species is a vital goal. The fish innate immune system, which includes both humoral and cellular defence mechanisms such as the complement system and the activities carried out by phagocytes and non-specific cytotoxic cells, is extremely important as a defence against pathogens. To improve resistance in aquaculture, several substances which enhance fish innate immune defences, such as immunostimulants, have been test as an alternative way of disease prevention (Logambal and Michael, 2000; Esteban *et al.*, 2001, 2005; Ortuño *et al.*, 2002).

Immunostimulants come from a variety of sources (Bricknell and Dalmo, 2005) such as microbial polymers, which include cell wall

Authors' address: Department of Cell Biology, University of Murcia, 30100 Murcia, Spain.
**Corresponding author:* E-mail: meseguer@um.es

polysaccharides, double-stranded viral RNA and bacterial DNA. These components, although not specific to a given microorganism, are associated with a particular class of pathogen (for example, lipopolysaccharide is a component of Gram-negative bacteria), for which reason they (presently named pathogen-associated molecular patterns, PAMPs) allow the immune system to make the distinction between self and non-self (Häcker *et al.*, 2002), in other words, to recognize some molecules as 'danger' signal and to start an immune response (Hemmi *et al.*, 2000; Krieg *et al.*, 2000). These PAMPs have been conserved throughout the evolution of microorganisms, but are not normal constituents of higher organisms (Robertsen, 1999), a fact which suggests that innate immune responses have evolved towards the recognition of conserved pathogen structures. These molecules trigger several mechanisms in vertebrates and it is very difficult to identify which are the most important for pathogen defence.

An interesting observation was made some years ago by a Japanese group working on the way in which vaccination with bacilli of Calmette-Guérin (BCG, the low-virulence strain of *Mycobacterium bovis*) reduced tumour growth. Initially, the mechanisms of action were traced to the stimulation of the immune system by BCG (Tokunaga *et al.*, 1999), but in a second step, this group was able to attribute a major stimulatory effect to mycobacterial DNA and, finally, they extended these observations to a more general description of bacterial DNA as a non-specific immune stimulatory agent in mammals (Yamamoto *et al.*, 1992a). Numerous investigations since then have demonstrated the antigen-like capacity of prokaryotic DNA.

Furthermore, it has been demonstrated that DNA, a molecule present in all living organisms, could also fit the definition of PAMP ligands, which may bind to the pattern recognition receptors (PRR) defined within the Toll receptor (TLR) family (Modlin, 2000). This stimulatory property of DNA could also be attributed to single-stranded oligodeoxyribonucleotides (ODNs) containing some specific sequences (Yamamoto *et al.*, 1992b) that lower and higher vertebrates recognize as a 'danger' signal (Mutwiri *et al.*, 2003).

The structural requirement for immunostimulation of these ODN was defined as a central cytosine-phosphodiester-guanosine (CG) core, unmethylated and flanked by other less important base sequences. Using sequence-specific synthetic oligonucleotides to trigger B proliferation, unmethylated CpG motifs displaying 5′Pu-Pu-CpG-Pyr-Pyr3′ nucleotide sequences were seen to be the most biologically active (Yamamoto *et al.*, 1992b; Krieg *et al.*, 1995), although other combinations of the flanking sequences are also immunostimulatory. Such investigation also helped show why the mammalian immune system can discriminate between bacterial and host DNA. Mammalian genomic DNA contains very scarce CpG-dinucleotide motifs (CpG suppression) (Bird, 1987) and, when present, they are normally methylated (5′-methylcytosine). However, bacterial genomic DNA presents the above-mentioned structural requirements for recognition and stimulation and, accordingly, it stimulates the mammalian immune system. Similar observations were extended to DNA from yeast and insects (Sun *et al.*, 1996; Wagner, 1999).

Krieg and co-workers developed extensive studies to define the best sequences to produce the optimal stimulation of immune function with the aim of optimizing synthetic ODN. The most immunostimulatory ODN were about 20 nucleotides in length, with a central core of CG (Krieg *et al.*, 1995): the regions adjacent to the CpG, which also affect stimulation, differed from mouse to human (Van Uden and Raz, 2000). However, it has been demonstrated that the immunostimulatory effects of CpG motifs are sequence specific, and that a base change from CpG to GpC in the core motif abolishes the activity (Krieg *et al.*, 1995; Jorgensen *et al.*, 2001a). Furthermore, Kanellos and coworkers (1999a-c) found that CpG motifs are phylum specific.

The immunostimulatory effects of CpG-DNA can be attributed to stimulation of certain immune cells of the innate immune system. In mammal models, these cells are dendritic cells, macrophages, natural killer (NK) cells and also B and T lymphocytes (Wagner *et al.*, 1999) and they express or secrete cytokines (like interleukin (IL)-1, IL-6, IL-12, tumour necrosis factor (TNF)-α and interferon (IFN) (Klinman *et al.*, 1996; Stacey *et al.*, 1996; Lipford *et al.*, 1997; Sparwasser *et al.*, 1998). Under *in vivo* conditions, it was demonstrated that CpG-DNA is an excellent adjuvant and enhances the B and T cell response to an antigen. Usually, in mammals, the response is biased towards the generation of a T helper 1-dependent immunity that produces immunoglobulins (Ig) of class G2a and a Th1-dominated cytokine profile (Roman *et al.*, 1997; Weiner *et al.*, 1997; Cho *et al.*, 2000; Juffermans *et al.*, 2002).

Two suggestions have been proposed to understand how mammalian immune cells can recognize CpG-DNA. The first was that the cell might be able to take up the DNA, which it would hybridize to genomic DNA in the cell nucleus, and thereby regulate the transcription of specific genes. However, there is still no evidence supporting this hypothesis. The second proposal was that a cellular receptor might exist which would specifically bind CpG-DNA and initiate the signal transduction that ultimately results in activation of the transcription of nuclear factor (NF)-κB and AP-1 (Fig. 4.1). Several bodies of evidence support this idea (Hacker *et al.*, 1998; Yi *et al.*, 1998a, b) and it has been proposed that microbial CpG pattern

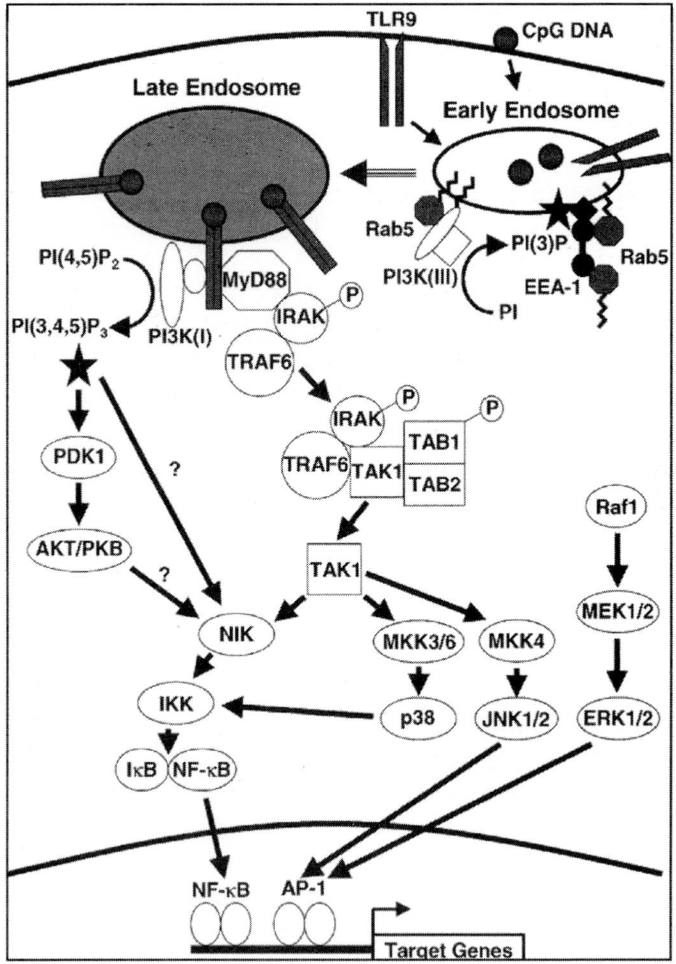

Fig. 4.1 Scheme of mammalian CpG DNA/TLR9-mediated cellular signaling (from Takeshita *et al.*, 2004). In fish, the cellular signaling might be very similar. So far, molecular and functional data have probed the existence and function of TLR9, My88, IRAK, TRAF6, NF-κB, p38 and AP-1 in different fish species.

recognition is an evolved innate defence mechanism against intracellular pathogens that may function through a Toll-like receptor (TLR) for CpG motifs (Hemmi *et al.*, 2000; Krieg *et al.*, 2000), the TLR9, because TLR-deficient mice were unresponsive to CpG-stimulation (Hemmi *et al.*, 2000).

While it is a well-known fact that CpG-DNA or synthetic ODNs can activate immune cells in mammals (mice, primates and humans) (Krieg *et al.*, 1995; Hartmann and Krieg, 2000; Verthelyi *et al.*, 2001), to date there has been limited information concerning the biological effects of CpG-ODN in other species and vertebrate groups, including cold-blooded vertebrates, especially fish. The aim of this chapter is to present the available results on this interesting topic.

In vitro Effects of CpG ODNs on Fish Immune Cells

Different synthetic ODNs have demonstrated an *in vitro* effect on several innate immune cell activities carried out by phagocytes, cytotoxic cells and lymphocytes. The ODNs containing multiple CpGs generally exhibited a greater stimulatory capacity, although CpGs located at the terminus of an ODN were infective. Incubation of common and grass carp (*Cyprinus carpio* L. and *Ctenopharyngodon idellus*, respectively) head-kidney phagocytes with these substances enhances their phagocytic and bactericidal activities, superoxide anion and hydrogen peroxide production, as well as the acid phosphatase content, in a time- and dose-dependent manner (Meng *et al.*, 2003; Tassakka and Sakai, 2003, 2004).

Highly purified non-specific cytotoxic cells (NCC) from catfish (*Ictalurus punctatus*) have been demonstrated to specifically bind to synthetic ODNs and become activated, achieving their highest activity after 24 hours of incubation (Oumouna *et al.*, 2002). This study also demonstrated that the same sequence could have different effects, depending on the vertebrate species and the immune activity studied. For example, the 5′-GACGTT-3′ which stimulate optimum IL-6 and IL-12 release from mammalian B cells (Ye *et al.*, 1996) had a minimal effect in stimulating NCC cytotoxicity.

Furthermore, and similarly to that described in both canine and feline spleen and lymph node cells (Wernette *et al.*, 2002), ODNs also enhance common carp head-kidney lymphocyte proliferation, although higher concentrations of CpG-ODN were needed to achieve that particular effect. As regards phagocytic activity, it has also been demonstrated that the presence of multiple CpGs generally results in greater stimulatory capacity (Malina *et al.*, 2003; Tassakka and Sakai, 2003).

Regarding cytokine production, both plasmid DNA and synthetic ODNs containing CpG motifs induced the production of antiviral (interferon-like, IFN) cytokines in Atlantic salmon (*Salmo salar* L.) and rainbow trout (*Oncorhynchus mykiss*) leukocytes (Jørgensen *et al.*, 2001a,b, 2003). Recently, a synergy between CpG ODNs (at concentrations that are ineffective alone) and cationic histone proteins in the stimulation of type I IFN activity and Mx transcripts has been demonstrated (Pedersen *et al.*, 2006). This more intense antiviral response reflects an increased resistance to infectious pancreatic virus (IPNV) infection (Jørgensen *et al.*, 2003).

Mechanism of Action of CpG DNA on Fish Leucocytes

In spite of all these interesting results concerning the effects of CpG DNA on fish leucocytes, the action mechanism of these molecules still largely remains unknown. It has been demonstrated that synthetic ODNs bind to TLR9 and induce the production of antiviral IFN-α in higher vertebrates (Hemmi *et al.*, 2003). However, the earliest demonstration of this fact in fish concluded that ODNs were inefficient at increasing antibody production towards co-administered β-Gal in goldfish (*Carassius auratus*) (Kanellos *et al.*, 1999c). Later, it was demonstrated that Atlantic salmon (*S. salar*) leucocytes produced antiviral activities and IL-1 as a consequence of their incubation with CpG ODNs, as indicated above (Jørgensen *et al.*, 2001a,b, 2003).

Although the subcellular localization of TLR9 has not yet been determined in mammals, the DNA taken up by leucocytes appears in the endosomes. In this sense, a role in endosomal acidification and/or maturation in the signalling pathways triggered by CpG DNA is suggested by the fact that some compounds that interfere with endosomal processing (such as chloroquine) specifically block all of their stimulatory effects, although they do not block other pathways of leucocyte activation (Hacker *et al.*, 1998; Yi and Krieg, 1998a, b). Based on these findings, a few years later, it was demonstrated that endosomal maturation is essential for CpG signalling in rainbow trout macrophages, as chloroquine inhibits cytokine expression in those cells, although the mechanism(s) by means of which these inhibitors of maturation block the signalling pathways is still unclear. Nevertheless, the ability of LPS to induce cytokine expression was unaffected by chloroquine, demonstrating the specificity of this inhibition (Jørgensen *et al.*, 2001b).

In vivo Effects of CpG ODNs on Fish Immune Response

To date, limited information is available on the *in vivo* effects in fish of CpG ODNs and on their *in vivo* ability to enhance resistance to disease, since most studies in this respect have been carried out in mice. Recently, the immunostimulatory effect of CpG motifs (the same sequences which activated the head-kidney leucocytes after *in vitro* incubation) on the innate immune response of common carp (*Cyprinus carpio*) was demonstrated after being intraperitoneally injected (Tassakka and Sakai, 2003). In that study, carp specimens were intraperitoneally injected daily for 3 days with ODNs containing CpG, as a result of which the carp innate immune response was stimulated. More specifically, serum lysozyme activity and the phagocytic activity and reactive oxygen species production by head-kidney leucocytes were increased, the leucocytes remaining active for at least 7 days post-injection (Tassakka and Sakai, 2002).

Synthetic ODNs have also been injected into olive flounder (*Paralichthys olivaceus*), where they primed the respiratory burst activity of the head-kidney phagocytes 3, 5 and 7 days post-injection. These results correlated with those obtained in common carp but the authors also demonstrated that such increments in phagocytosis correlated with increased protection against the bacterial pathogen *Edwardsiella tarda* after provoking a lethal infection in the specimens (Lee *et al.*, 2003). Similarly, ODN intraperitoneal injection enhanced the resistance of Atlantic salmon to amoebic gill disease (Bridle *et al.*, 2003).

More recently, the cloning of immune-related genes in fish has facilitated analysis of the effects of different treatments (such as peptidoglycan and lipopolysaccharides) at the molecular level (Pelegrín *et al.*, 2001; Savan and Sakai, 2002; Kono *et al.*, 2004; Sakai *et al.*, 2005). There is evidence that intraperitoneal injection of CpG ODNs stimulates the expression of immune genes in the head-kidney of common carp, more specifically, the expression of several pro-inflammatory cytokines (IL-1 beta, CXC and CC-chemokines and TNF-alpha) 1, 5 and 7 days post-injection and the lysozyme-C gene expression after 7 days post-injection (Tassakka and Sakai, 2004, 2005). IL-1 beta is a key mediator in response to microbial invasion and tissue injury and can stimulate immune responses by activating lymphocytes or by inducing the release of other cytokines that activate macrophages (Low *et al.*, 2003). CXC and CC-chemokines are inducible and secrete proteins that cause the migration of leucocytes towards injury or infection sites (Dixon *et al.*, 1998). Lysozymes are potent molecules capable of digesting the peptidoglycan of bacteria. The interest of these findings lies in the fact that CpG ODNs regulate the expression of different cytokines which act as signalling molecules within the immune system, where they play important roles in initiating and regulating the inflammatory process, which is one of the important defenses in innate immunity. Furthermore, these results agree with others obtained in mammals, in so far as, while the CpG ODN are potent activators of the immune system, their biological activity is only transient, subsequently limiting their therapeutic application (Mutwiri *et al.*, 2004). However, different studies are in progress in order to further improve the use of ODNs.

Adjuvant Effect of ODN in Fish

All the above studies mention the possible use of CpG ODNs as a disease control treatment; indeed, immunostimulatory CpG motifs are one of the main features of DNA vaccines, and have been regarded as promising technology, as immunotherapeutic agents, with practical and immunological advantages over traditional antigen vaccines in aquaculture. Furthermore, their properties make these compounds a potentially valuable tool, whether used alone or as adjuvants in vaccines (Pontarollo *et al.*, 2002). Recently, CpG motifs have been defined as novel, non-toxic adjuvants because they can induce a stronger immune response with no apparent adverse effects compared to many conventional adjuvants such as Freunds' complete and incomplete adjuvants. Furthermore, many of the adjuvants used in fish vaccines, particularly oil-based adjuvants, contribute to good protection, but have serious side effects (Midtlyng *et al.*, 1996). CpG DNA used as an adjuvant is reported to induce stronger immune responses with less toxicity than other adjuvants when tested in murine models (Weeratna *et al.*, 2000).

All these abilities mentioned for ODNs make them a potential adjuvants for use in viral vaccines for fish, because DNA vaccination studies with fish have indicated a role for type I IFN in the protection provided by these vaccines (Boudinot *et al.*, 1998; Kim *et al.*, 2000; Lorenzen *et al.*, 2002). And there is some evidence that the type I IFN response may provide a signal for the initiation of adaptive responses.

One of the challenges in developing CpG ODNs as adjuvants is the identification of suitable delivery systems. The ideal delivery system would potentiate the effect of CpG ODNs by promoting their uptake and delivery to antigen-presenting cells. Cationic delivery systems, which function as bridges between the negative charges of DNA fragments and cell membranes, and protect from nuclease digestion, are an effective delivery system for CpG DNA in mammals (Singh *et al.*, 2001).

Different *in vivo* studies have been carried out to ascertain the effects of various DNA vaccines. Plasmid vectors composed of different constructs as well as empty vector controls resulted in increased antibody responses and generally heightened cytokine-like activities in injected fish (Russell *et al.*, 1998; Kanellos *et al.*, 1999a, c; Heppell and Davis, 2000; Tucker *et al.*, 2001). However, the studies carried out to ascertain the immunostimulatory nucleotide motifs within the plasmid DNA that produce ligand-specific activation are inconclusive. For example, in goldfish (*Carassius auratus* L.), it was demonstrated that plasmids containing CpG-ODNs motifs possess adjuvant effects after being injected with a protein subunit vaccine. It was found that the presence of the plasmids potentates antibody responses to the β-galactosidase protein and the motif consisting of -AACGTT- might be the stimulatory sequence, while the -GACGTT- motif did not stimulate the antibody production (Kanellos *et al.*, 1999b). However, leucocytes from Atlantic salmon were activated by -GACGTT- whereas the inverse motif (-GAGCTT-) was not stimulatory (Jørgensen *et al.*, 2001a).

Future Prospects

The major challenge faced by disease control is the full understanding of the interactions between the various defence mechanisms of the host and the host-pathogen interactions. The recognition of different PRRs that trigger different signalling pathways of the immune system is providing the opportunity to drive the immune response to the desired direction. CpG not only activates the innate immune response, but also shifts the specific response, so it must be considered as one of the most useful immunomodulators. More studies are needed to understand the effect of CpG ODNs on the stimulation of innate immunity as also on the adaptive immune response in fish. Concomitantly, different strategies for enhancing the immunostimulatory effect of CpG ODNs are in progress (Mutwiri *et al.*, 2004). Such studies aim to increase the half-life of CpG ODN, enhance its activity, localize ODN in tissue, increase cellular uptake and binding, enhance delivery to intracellular compartments or to find new delivery systems: in short, the studies are trying to ensure that CpG goes to the appropriate cell types and in the correct concentration so as to have the optimal effect. It seems, therefore, that much work on new formulations and delivery systems is needed.

References

Bird, A.P. 1987. CpG islands as gene markers in the vertebrate nucleus. *Trends in Genetics* 3: 342-347.

Boudinot, P., M. Blanco, P. de Kinkelin and A. Benmansour. 1998. Combined DNA immunization with the glycoprotein gene of viral hemorrhagic septicaemia virus and infectious hematopoietic necrosis virus induces double-specific protective immunity and nonspecific response in rainbow trout. *Virology* 249: 297-306.

Bricknell, I. and R.A. Dalmo. 2005. The use of immunostimulants in fish larval aquaculture. *Fish and Shellfish Immunology* 19: 457-472.

Bridle, A.R., R. Butler and B.F. Nowak. 2003. Immunostimulatory CpG oligodeoxynucleotides increase resistance against amoebic gill disease in Atlantic salmon, *Salmo salar* L. *Journal of Fish Diseases* 26: 367-371.

Carrington, A.C., B. Collet, J.W. Holland and C.J. Secombes. 2004. CpG oligodeoxynucleotides stimulate immune cell proliferation but not specific antibody production in rainbow trout (*Oncorhynchus mykiss*). *Veterinary Immunology and Immunopathology* 101: 211-222.

Cho, H.J., K. Takabayashi, P.M. Cheng, M.D. Nguyen, M. Corr, S. Tuck and E. Raz. 2000. Immunostimulatory DNA-based vaccines induce cytotoxic lymphocyte activity by a T-helper cell-independent mechanism. *Natural Biotechnology* 18: 509-514.

Dixon, B., B. Shum, E.J. Adams, K.E. Magor, R.P. Hedrick, D.G. Muir and P. Parham. 1998. CK-1, a putative chemokine of rainbow trout (*Oncorhynchus mykiss*). *Immunological Reviews* 166: 341-348.

Esteban, M.A., A. Cuesta, J. Ortuño and J. Meseguer. 2001. Immunomodulatory effects of dietary intake of chitin on gilthead seabream (*Sparus aurata* L.) innate immune system. *Fish and Shellfish Immunology* 11: 303-315.

Esteban, M.A., Rodríguez, A., Cuesta, A. and J. Meseguer. 2005. Effects of lactoferrin on non-specific immune responses of gilthead sea bream (*Sparus auratus* L.). *Fish and Shellfish Immunology* 18: 109-124.

Häcker, H., V. Redecke and H. Häcker. 2002. Activation of the immune system by bacterial CpG-DNA. *Immunology* 105: 245-251.

Häcker, H., H. Mischak, T. Miethke, S. Liptay, R. Schmid, T. Sparwasser, K. Heeg, G.B. Lipford and H. Wagner. 1998. CpG-DNA-specific activation of antigen-presenting cells requires stress kinase activity and is preceded by non-specific endocytosis and endosomal maturation. *EMBO Journal* 17: 6230-6240.

Hartmann, G. and A.M. Krieg. 2000. Mechanism and function of a newly identified CpG DNA motif in human primary B cells. *Journal of Immunology* 164: 944-953.

Hemmi, H., T. Kaisho, K. Takeda and S. Akira. 2003. The roles of toll-like receptor 9, MyD88, and DNA-PKcs in the effects of two distinct CpG DNAs on dendritic cell subsets. *Journal of Immunology* 170: 3059-3064.

Hemmi, H., O. Takeuchi, T. Kawai, T. Kaisho, S. Sato, H. Sanjo, M. Matsumoto, K. Hoshino, H. Wagner, K. Takeda and S. Akira. 2000. A toll-like receptor recognizes bacterial DNA. *Nature (London)* 408: 740-745.

Heppell, J. and H.L. Davies. 2000. Application of DNA vaccine technology to aquaculture. *Advanced Drug Delivery Reviews* 43: 29-43.

Jørgensen, J.B., J. Zou, A. Johansen and C.J. Secombes. 2001a. Immunostimulatory CpG oligodeoxydenucleotides stimulate expression of IL-1beta and interferon-like cytokines in rainbow trout macrophages via a chloroquine-sensitive mechanism. *Fish and Shellfish Immunology* 11: 673-682.

Jørgensen, J.B., A. Johansen, B. Steersen and A.I. Sommer. 2001b. CpG oligodeoxydenucleotides and plasmid DNA stimulate Atlantic salmon (*Salmo salar* L.) leucocytes to produce supernatants with antiviral activity. *Developmental and Comparative Immunology* 25: 313-321.

Jørgensen, J.B., L.H. Johansen, K. Steiro and A. Johansen. 2003. CpG DNA induces protective antiviral immune responses in Atlantic salmon (*Salmo salar* L.). *Journal of Virology* 77: 11471-11479.

Juffermans, N.P., J.C. Leemans, S. Florquin, A. Verbon, A.H. Kolk, P. Speelman, S.J.H. van Deventer and T. van der Poll. 2002. CpG oligodeoxydenucleotides enhance host defence during murine tuberculosis. *Infection and Immunity* 70: 147-152.

Kanellos, T., I.D. Sylvester, A.G. Ambali, C.R. Howard and P.H. Russell. 1999a. The safety and longevity of DNA vaccines for fish. *Immunology* 96: 307-313.

Kanellos, T.S., I.D. Sylvester, V.L. Butler, A.G. Ambali, C.D. Partidos, A.S. Hamblin and P.H. Russel. 1999b. Mammalian granulocyte-macrophage colony-stimulating factor and some CpG motifs have an effect on the immunogenicity of DNA and subunit vaccines in fish. *Immunology* 96: 507-510.

Kanellos, T., I.D. Sylvester, C.R. Howard and P.H. Russell. 1999c. DNA is as effective as protein to inducing antibody in fish. *Vaccine* 17: 965-972.

Kim, C.H., M.C. Johnson, J.D. Drennan, B.E. Simon, E. Thoman and J.A. Leong. 2000. DNA vaccines encoding viral glycoproteins induce non-specific immunity and Mx protein synthesis in fish. *Journal of Virology* 74: 7048-7054.

Klinman, D., A.K. Yi, S.L. Beaucage, J. Conover and A.M. Krieg. 1996. CpG motifs expressed by bacterial DNA rapidly induce lymphocytes to secrete IL-6, IL-12 and IFN. *Proceedings of the National Academy of Sciences of the United States of America* 93: 2879-2883.

Kono, T., R. Savan, H. Kuragasaki, A.C.M.A.R. Tassakka and M. Sakai. 2004. Expression analysis of immune-related genes in carp *Cyprinus carpio* L. *Journal of Marine Biotechnology* 6: S123-S127.

Krieg, A.M., G. Hartmann and A.K. Yi. 2000. Mechanism of action of CpG DNA. *Current Topics in Microbiology and Immunology* 247: 1-21.

Krieg, A.M., A.K. Yi, S. Matson, T.J. Waldschmidt, G.A. Bishop, R. Teasdale, G.A. Koretzky and D.M. Klinman. 1995. CpG motifs in bacterial DNA trigger direct B-cell activation. *Nature (London)* 374: 546-549.

Lee, C.H., H. Do Jeong, J.K. Chung, H.H. Lee and K.H. Kim. 2003. CpG motif in synthetic ODN primes respiratory burst of live flounder *Paralichthys olivaceus* phagocytes and enhances protection against *Edwardsiella tarda*. *Diseases of Aquatic Organisms* 56: 43-48.

Lipford, G.B., M. Bauer, C. Blank, R. Reiter, H. Wagner and K. Heeg. 1997. CpG-containing synthetic oligonucleotides promote B and cytotoxic cells responses to protein antigen: a new class of vaccine adjuvants. *European Journal of Immunology* 27: 2340-2344.

Logambal, S.M. and R.D. Michael. 2000. Immunostimulatory effect of azadirachtin in *Oreochromis mossambicus* (Peters). *Indian Journal of Experimental Biology* 38: 1092-1096.

Lorenzen, N., E. Lorenzen, K. Einer-Jensen and S.E. LaPatra. 2002. DNA vaccines as a tool for analysing the protective immune response against rhabdoviruses in rainbow trout. *Fish and Shellfish Immunology* 12: 439-453.

Low, C., S. Wadsworth, C. Burrells and C.J. Secombes. 2003. Expression of immune genes in turbot (*Scophthalmus maximus*) fed a nucleotide-supplemented diet. *Aquaculture* 221: 23-40.

Malina, A.C., A.R. Tassakka and M. Sakai. 2003. The in vitro effect of CpG-ODNs on the innate immune response of common carp, *Cyprinus carpio* L. *Aquaculture* 220: 27-36.

Meng, Z., J.Z. Shao and L.X. Xiang. 2003. CpG oligodeoxynucleotides activate grass carp (*Ctenopharyngodon idellus*) macrophages. *Developmental and Comparative Immunology* 27: 313-321.

Midtlyng, P.J., L.J. Reitan and L. Speilbers. 1996. Efficacy and side-effects of injectable furunculosis vaccines in Atlantic salmon (*Salmo salar* L.). *Fish and Shellfish Immunology* 6: 335-350.

Modlin, R.L. 2000. A tool for DNA vaccines. *Nature (London)* 408: 659-660.

Mutwiri, G.K., A.K. Nichani, S. Babiuk and L.A. Babiuk. 2004. Strategies for enhancing the immunostimulatory effects of CpG oligodeoxynucleotides. *Journal of Controlled Release* 97: 1-17.

Mutwiri, G., R. Pontarollo, S. Babiuk, P. Gribel, S. van Drunen Little-van den Hurk, A. Mena, C. Tsang, V. Alcon, A. Nichani, X. Ioannou, S. Gomis, H. Townsend, R. Hecker, A. Potter and L.A. Babiuk. 2003. Biological activity of immunostimulatory CpG DNA motifs in domestic animals. *Veterinary Immunology and Immunopathology* 91: 89-103.

Ortuño, J., A. Cuesta, A. Rodríguez, M.A. Esteban and J. Meseguer. 2002. Oral administration of yeast, *Saccharomyces cerevisiae*, enhances the cellular immune response of gilthead sea bream (*Sparus aurata* L.). *Veterinary Immunology and Immunopathology* 85: 41-50.

Oumouna, M., L. Jaso-Friedmann and D.L. Evans. 2002. Activation of non-specific cytotoxic cells (NCC) with synthetic oligodeoxynucleotides and bacterial genomic DNA: Binding, specificity and identification of unique immunostimulatory motifs. *Developmental and Comparative Immunology* 26: 257-269.

Pedersen, G.M., A. Johansen, R.L. Olsen and J.B. Jørgensen. 2006. Stimulation of type I IFN activity in Atlantic salmon (*Salmo salar* L.) leukocytes: Synergistic effects of cationic proteins and CpG ODN. *Fish and Shellfish Immunology*. (In Press).

Pelegrín, P., J. García Castillo, V. Mulero and J. Meseguer. 2001. Interleukin-1 beta isolated from a marine fish reveals up-regulated expression in macrophages following activation with lipopolysaccharide and lymphokines. *Cytokine* 21: 67-72.

Pontarollo, R.A., R. Rankin, L.A. Babiuk, D.L. Godson, P.J. Griebel, R. Hecker, A.M. Krieg and S. van Drunen Little-van den Hurk. 2002. Monocytes are required for optimum in vitro stimulation of bovine peripheral blood mononuclear cells by non-methylated CpG motifs. *Veterinary Immunology and Immunopathology* 84: 43-49.

Robertsen, B. 1999. Modulation of the non-specific defence of fish by structurally conserved microbial polymers. *Fish and Shellfish Immunology* 9: 269-304.

Roman, M., E. Martin-Orozco, J.S. Goodman, M.D. Nguyen, Y. Sato, A. Ronaghy, R.S. Kornbluth, D.D. Richman, D.A. Carson and E. Raz. 1997. Immunostimulatory DNA sequences function as T helper-1-promoting adjuvants. *Natural Medicine* 3: 849-854.

Russell, P.H., T. Kanellos, I.D. Sylvester, K.C. Chang and C.R. Howard. 1998. Nucleic acid immunization with a reporter gene results in antibody production in goldfish (*Carassius auratus* L.). *Fish and Shellfish Immunology* 8: 121-128.

Sakai, M., T. Kono and R. Savan. 2005. Identification of expressed genes in carp (*Cyprinus carpio*) head kidney cells after in-vitro treatment with immunostimulants. In: *Fish Vaccinology*, P. Midtlying (ed.). S. Karger, Basel, Vol. 121, pp. 45-51.

Savan, R. and M. Sakai. 2002. Analysis of expressed sequence tags (EST) obtained from common carp, *Cyprinus carpio* L., head-kidney cells after stimulation by two mitogens, lipopolysaccharide and concavalin-A. *Comparative Biochemistry and Physiology* B131: 71-82.

Singh, M., G. Ott, J. Kazzaz, M. Ugozzoli, M. Briones, J. Donnelly and D.T. O'Hagan. 2001 Cationic microparticles are an effective delivery system for immune stimulatory cpG DNA. *Pharmaceutical Research* 18: 1476-1479.

Sparwasser, T., E.S. Koch, M.V. Ramunas, K. Heeg, G.B. Lipford, J.W. Ellwart and H. Wagner. 1998. Bacterial DNA and immunostimulatory CpG trigger maturation and activation of murine dendritic cells. *European Journal of Immunology* 28: 2045-2054.

Stacey, K.J., M.J. Sweet and D.A. Hume. 1996. Macrophages ingest and are activated by bacterial DNA. *Journal of Immunology* 157: 2116-2122.

Sun, S., Z. Cai, P. Langlade-Demoyen, H. Kosada, A. Brunmark, M.R. Jackson, P.A. Peterson and J. Sprent. 1996. Dual function of *Drosophila* cells as APCs for naive $CD8^+$ T cells. Implications for tumour immunotherapy. *Immunity*. 4: 555-564.

Takeshita, F., I. Gursel, K.J. Ishii, K. Suzuki, M. Gursel and D.M. Klinman. 2004. Signal transduction pathways mediated by the interaction of CpG DNA with Toll-like receptor 9. *Seminar in Immunology* 16: 17-22.

Tassakka, A.C.M.A.R. and M. Sakai. 2002. CpG oligodeoxydenucleotides enhance the non-specific immune responses on carp, *Cyprinus carpio*. *Aquaculture* 209: 1-10.

Tassakka, A.C.M.A.R. and M. Sakai. 2003. The *in vitro* effect of CpG-ODNs on the innate immune response of common carp, *Cyprinus Carpio* L. *Aquaculture* 220: 27-36.

Tassakka, A.C.M.A.R. and M. Sakai. 2004. Expression of immune related genes in the common carp (*Cyprinus carpio* L.) after stimulation by CpG oligodeoxynucleotides. *Aquaculture* 242: 1-12.

Tassakka, A.C.M.A.R., R. Savan, H. Watanuki and M. Sakai. 2006. The in vitro effects of CpG oligodeoxydenucleotides on the expression of cytokine genes in the common carp (*Cyprinus carpio* L.) head-kidney cells. *Veterinary Immunology and Immunopathology* 110: 79-85.

Tokunaga, T., T. Yamamoto and S. Yamamoto. 1999. How BCG led to the discovery of immunostimulatory DNA. *Japanese Journal of Infectious Diseases* 52: 1-11.

Tucker, C., M. Endo, I. Hirono and T. Aoki. 2001. Assessment of DNA vaccine potential for juvenile Japanese flounder *Paralichthys olivaceus*, through the introduction of reporter genes by particle bombardment and histopathology. *Vaccine* 19: 801-809.

Van Uden, J. and E. Raz. 2000. Introduction to immunostimulatory DNA sequences. *Springer Seminar in Immunopathology* 22: 1-9.

Verthelyi, D., K.J. Ishii, M. Gursel, F. Takeshita and D.M. Klinman. 2001. Human peripheral blood cells differentially recognize and respond to two distinct CpG motifs. *Journal of Immunology* 166: 2372-2377.

Wagner, H. 1999. Bacterial CpG-DNA activates immune cells to signal infectious danger. *Advances in Immunology* 73: 329-368.

Weeratna, R.D., M.J. McCluskie, Y. Xu and H.L. Davis. 2000. CpG DNA induces stronger immune responses with less toxicity than other adjuvants. *Vaccine* 18: 1755-1762.

Weiner, G.J., H.M. Liu, J.E. Wooldridge, C.E. Dahle and A.M. Krieg. 1997. Immunostimulatory oligodeoxynucleotides containing the CpG motif are effective as immune adjuvants in tumor antigen immunization. *Proceedings of the National Academy of Sciences of the United States of America* 94: 10833-10837.

Wernette, C.M., B.F. Smith, Z.L. Barksdale, R. Hecker and H.J. Baker. 2002. CpG oligodeoxydenucleotides stimulate canine and feline immune cell proliferation. *Veterinary Immunology and Immunopathology* 84: 223-236.

Yamamoto, S.O., T. Yamamoto, S. Shimada, E. Kuramoto, O. Yano, T. Kataoka and T. Tokunaga. 1992a. DNA from bacteria, but not from vertebrates, induces interferons, activates natural killer cells and inhibits tumour growth. *Microbiol Immunology* 36: 983-987.

Yamamoto, S., T. Yamamoto, T. Kataoka, E. Kuramoto, O. Yano and T. Tokunaga. 1992b. Unique palindromic sequences in synthetic oligonucleotides are required to induce IFN and augment IFN-mediated natural killer activity. *Journal of Immunology* 148: 4072-4076.

Yi, A.K. and A.M. Krieg. 1998. Rapid induction of mitogen-activated protein kinases by immune stimulatory CpG DNA. *Journal of Immunology* 161: 4493-4497.

Ye, A.K., D.M. Klinman, T.L. Martin, S. Matson and A.M. Krieg. 1996. Rapid immune activation by CpG motifs in bacterial DNA. Systemic induction of IL-6 transcription through an antioxidant-sensitive pathway. *Journal of Immunology* 157: 5394-5402.

Yi, A.K., M. Chang, D.W. Peckman, A.M. Krieg and R.F. Ashman. 1998. CpG oligodeoxiribonucleotides rescue mature spleen B cells from spontaneous apoptosis and promote cell cycle entry. *Journal of Immunology* 160: 5898-5906.

CHAPTER

5

Innate Immunity of Fish: Antimicrobial Responses of Fish Macrophages

Miodrag Belosevic[3,*], George Haddad[3], John G. Walsh[1], Leon Grayfer[3], Barbara A. Katzenback[3], Patrick C. Hanington[3], Norman F. Neumann[2] and James L. Stafford[3]

INTRODUCTION

The fundamental role of the immune system is to recognize the self from the non-self; discriminating the finite structure of foreign molecules from the diverse array of molecular patterns and complexities intrinsic to the host. Non-self recognition mechanisms appear to be an inherent prerequisite for the survival of any living organism, and are present even within the simplest life forms—mycoplasms, as restriction enzymes;

Authors' addresses: [1]Department of Genetics, Trinity College, Dublin, Ireland.
[2]Department of Microbiology and Infectious Diseases, University of Calgary, Calgary, Canada.
[3]Department of Biological Sciences, University of Alberta, Edmonton, Alberta, Canada.
**Corresponding author*: E-mail: mike.belosevic@ualberta.ca

proteins evolved to recognize and degrade invasive foreign genetic material (i.e., bacteriophages) (Gumulak-Smith *et al.*, 2001).

Phagocytosis is the primordial defense mechanism of all metazoan organisms. In fact, it has even been suggested that phagocytosis represents a rudimentary innate defense mechanism in the most primitive eukaryotic kingdom, the Protozoa (Nickel *et al.*, 1999). Haeckel first described phagocytosis in 1862, using blood cells from the gastropod *Tethys* spp. (Vaughan, 1965), but its importance as a defense mechanism was not realized until 1882, by Elie Metchnikoff (Vaughan, 1965).

The most important phagocytic cell in the vertebrate body is the macrophage, and cells functionally reminiscent of the vertebrate macrophage are present in virtually all metazoan organisms, attesting to the importance of phagocytosis in host defense in all multi-cellular organisms. Macrophages play a pivotal role in the detection of pathogenic microorganisms and in the ensuing effector phases in eliminating the infectious agent. Pathogen-associated molecular patterns (PAMPs) on the surface of pathogenic microorganisms are recognized by pattern recognition receptors (PRR) on macrophages, facilitating the phagocytic process and initiating subsequent killing mechanisms (Medzhitov and Janeway, 1977a, b, 2000; Medzhitov, 2001).

The ubiquitous distribution of macrophages within the vertebrate body ensures the continuous surveillance of host tissues for foreign invaders. In many cases, it is the macrophage (or lineage-related cells such as dendritic cells) that provides the first line of cell-mediated host defense against pathogens. Foreign microorganisms are rapidly phagocytosed by macrophages, and are destroyed by lysosomal enzymes released into the phagosome, or by toxic reactive intermediates formed by activated macrophages. In addition, macrophages possess a number of nutrient deprivation mechanisms that are employed to *starve* phagocytosed pathogens of essential micronutrients. The fundamental roles of macrophages in host defense are to limit the initial dissemination and/or growth of infectious organisms and to modulate ensuing immunological reactions.

This chapter focuses on the cellular processes involved in macrophage-mediated innate host defense of bony fishes; from the initial host-pathogen interactions that mediate recognition, to the mechanisms employed by macrophages to destroy and eliminate the pathogens.

RECOGNITION OF INFECTIOUS AGENTS BY MACROPHAGES

How macrophages sense a foreign molecule within the context of a molecularly complex host environment is extraordinary. The possible number of foreign molecular configurations on a diverse assemblage of pathogenic microorganisms is immense. Although vertebrates have evolved adaptive immune responses to recognize large numbers of foreign molecules (i.e., T-cell receptors and antibodies), these same mechanisms appear to be absent from invertebrates, and as such, greater than 99% of all multi-cellular life forms rely on non-self innate immune mechanism to recognize foreign molecules. The extensive diversity of foreign molecular structures, accompanied by limited genome sizes, necessitated the evolution of non-self discriminatory mechanisms that recognized unique molecular configurations on groups of microorganisms. These unique foreign molecular configurations are also known as pathogen-associated molecular patterns (PAMPs) (Medzhitov and Janeway, 2000; Medzhitov, 2001). All PAMPs have the common features of being unique to a group of pathogens, are essential for survival, and are relatively invariant in their basic structure. Specific examples of PAMPs include lipopolysaccharide (LPS; gram-negative bacteria), lipoarbinomannan (mycobacteria), lipoteichoic acids (LTA; gram-positive bacteria), mannans (yeast), and double-stranded RNA (viruses).

PAMPs are recognized by pattern recognition receptors (PRRs) found on the surface of immune cells including macrophages (Medzhitov and Janeway, 1997; Medzhitov, 2001). These innate immune receptors are a germ-line encoded group of receptors with a genetically pre-determined specificity that are highly conserved among different organisms. Pattern recognition receptors have distinctive ligand-binding properties for specific classes of PAMPs, but are less discriminatory towards the subtle differences in a fine molecular structure within that class of PAMPs. For example, CD14, a PRR for bacterial LPS, can bind to LPS from a diverse range of microorganisms, even though the fine structure of LPS can vary among different bacterial species (Landmann *et al.*, 2000). In addition, both Toll and Toll-like receptors (TLRs) are believed to not only bind foreign material, but are also capable of binding endogenous proteins (i.e., recognition of cleaved Spatzle by *Drosophila* Toll and recognition of fibrinogen and fibronectin by TLR-4) (Lemaitre *et al.*, 1996; Okamura *et al.* 2001; Smiley *et al.*, 2001).

Examples of some of the major proteins involved in microbial pattern recognition include mannose-binding proteins, mannose receptor, scavenger receptor, CD14, and TLRs. Functionally, PRRs can be divided into three classes: (1) secreted PRRs, which usually function as opsonins or activators of complement, (2) endocytic PRRs, which function in pathogen binding and phagocytosis and (3) signaling PRRs that activate gene transcriptional mechanisms that lead to cellular activation. This portion of the chapter focuses on the PRR, which are found on the surface of macrophages and we will discuss their roles in pathogen recognition/binding.

Lectins

Lectins play an important role in the recognition of pathogens and serve multiple functions in the immune system, including cell adhesion, recruitment, differentiation and activation (Drickamer and Taylor 1993; Linehan *et al.*, 2000). Many of the lectins involved in the recognition and neutralization of pathogens are members of the C-type or calcium-dependent animal lectin families (Weiss *et al.*, 1998). Within this family are two major groups of PRRs that are important components of the innate immune response; the collectins and the mannose receptor. Collectins are large soluble proteins that mediate pathogen neutralization through the complement pathway and include pulmonary surfactant proteins (SP-A and SP-D) and the serum mannose binding protein (MBP) (Drickamer and Taylor, 1993). These molecules have a collagen tail and a carboxy-terminal lectin domain. The collagen tail is a ligand for the collectin receptor, which is found on a variety of mammalian cells including monocytes, endothelial cells and fibroblasts. The lectin domain recognizes carbohydrate (CHO) moieties on the surface of viruses, bacteria, fungi, and protozoa. The other member of the C-type lectin family is the mannose receptor and is found on antigen presenting cells (i.e., macrophages and dendritic cells) (Fraser *et al.*, 1998). This cell-surface protein directly binds CHO moieties found on the surface of pathogens, leading to phagocytosis of microorganisms (Ross, 1989; Ezekowitz *et al.*, 1990).

The recognition of CHO moieties by members of the C-type lectin super family is mediated by structurally related calcium-dependent CHO-recognition domains (C-type CRDs) (Drickamer and Taylor, 1993). These domains recognize the equatorial orientation of the C_3 and C_4 hydroxyl groups in hexoses, N-acetylglucosamine, glucose, fructose and mannose (Weis *et al.*, 1998). These carbohydrate structures are common components of cell membrane and cell wall structures in a variety of pathogens, and include molecules like LPS, LTA, and mannans. In addition, the specific arrangement of CRDs on the host cell surface favor the spatial orientation of the CHO moieties that decorate the cell walls of microorganisms, providing they span the correct distance between the CRDs (ligands have to span the distance of 45 angstroms) (Ezerkowitz *et al.*, 1990; Weis *et al.*, 1998). These conditions mediate high-affinity binding of ligand to the receptor. The configuration of the hydroxyl groups in the sugars that decorate mammalian glycoproteins (i.e., galactose and sialic acid), and the repetitive CHO nature of many microbial PAMPs, are not accommodated by the CRD of C-type lectins. Thus, the broad selectivity of the monosaccharide-binding site, combined with the geometry of multiple CRDs in the intact lectins, allows these PRRs to mediate discrimination between the self and the non-self.

Complement Receptor (CR3)

The complement system is a proteolytic cascade that is activated either directly or indirectly by non-self recognition (i.e., microorganisms) and plays a major role in mammalian innate immunity (Ross, 1990). The activation of complement by microorganisms can have three major biological effects: (1) fixation of the terminal complement components resulting in complement-mediated lysis through the formation of a membrane attack complex (MAC), (2) fixation of the third complement component (C3) leading to opsonization and phagocytosis by macrophages and (3) elaboration of the complement anaphylotoxins C3a and C5a leading to recruitment of immune cells and initiation of an inflammatory reaction.

Macrophages contain complement receptors on their surface that aid in the receptor-mediated phagocytosis of opsonized microorganisms (Wright *et al.*, 1983). Many pathogens are recognized by CR3 and this receptor has been implicated in their internalization. For example, the yeast stage of *Histoplasma capsulatum* uses the CR3/LFA-1/p150, 95 family of adhesion molecules to enter the human macrophages, employing an opsonin-independent mechanism (Bullock and Wright, 1987). *Legionella pneumophila* uses CR1 and CR3 following opsonization with C3b and C3bi, respectively (Payne and Horwitz, 1987). *Mycobacterium leprae* and both the promastigote and amastigote stages of *Leishmania* spp. require C3 opsonization for uptake by CR3 (Schlesinger and Horwitz, 1991). Some pathogens such as *Leishmania* spp. can even exploit the fixation of the complement onto their surfaces in order to increase their uptake into host macrophages through the CR3 receptor (Brittingham *et al.*, 1995).

Scavanger Receptors (SRs)

Scavanger receptors are a unique type of receptor that bind to both host- and pathogen-derived ligands, using pattern recognition that does not induce macrophage activation (Pearson, 1996; Haworth *et al.*, 1997). These receptors were originally identified by their ability to bind modified low-density lipoproteins (LDLs), such as oxidized LDL (OxLDL) and acetylated LDL (AcLDL), and initially studied for their role in angiogenesis and LDL-cholesterol accumulation by macrophages in atherosclerotic lesions (Brown and Goldstein, 1983; Krieger and Herz, 1994).

Two classes, SR-AI and SR-AII, were the first macrophage SRs to be identified and were generated from alternative splicing of mRNA transcribed from the same gene (Krieger and Herz, 1994). SR-A is found on monocytes, B-lymphocytes, capillary endothelial cells, platelets, and adipocytes. Both classes exhibit nearly identical ligand-binding properties, specifically binding an array of polyionic ligands with high affinity (Krieger and Herz, 1994). SR-A recognizes polyionic molecules via its collagen-like domains, and recognition may be determined by the spatial characteristics of the repeating charged units found on host-derived, synthetic, and microbial origin. While SRs can bind and internalize many different ligands, they do not appear to play a role in the activation of immune cells (Haworth *et al.*, 1997). Therefore, SRs may participate in host defense by clearing foreign products such as LTA, LPS, or intact bacteria from tissues and the circulatory system during bacterial sepsis.

A third class of SRs, termed MARCO or SR-BI, has been identified and shown to bind both AcLDL and bacteria (Kraal *et al.*, 2000). Class B SRs are composed of members of the CD36 family (Endemann *et al.*, 1993) and include SR-BI, the lysosomal protein Limp II (Vega *et al.*, 1991), and *Drosophila* emp (epithelial membrane protein) (Hart and Wilcox, 1993). This class of SRs binds a variety of ligands that must contain negatively charged moieties in order to be recognized (Rigotti *et al.*, 1995). SR-BIs (i.e., MARCO), are primarily lipid-binding proteins that are expressed by cells and tissues involved in host defense and/or lipid metabolism (Acton *et al.*, 1996). LPS associated with high-density lipoproteins and anionic environmental particulates appear to bind to SR-BI. Therefore, this class of SRc may facilitate LPS-clearance by the liver and the removal of particulates by alveolar macrophages in the lung (Greenwalt *et al.*, 1992).

CD14

The recognition of bacterial endotoxin (i.e., LPS) is an important function of the innate immune system, and failure to contain bacterial infections can result in septic shock as a result of the release of LPS from bacteria as they grow, multiply, or die (i.e., after antibiotic treatment). Therefore, eukaryotes have developed sensitive immunological surveillance mechanisms that can detect minute amounts of bacterial LPS (Ulevitch and Tobias, 1995). The PRR responsible for recognition of this bacterial PAMP is CD14 (reviewed by Wright, 1995).

Two forms of the soluble LPS receptor (CD14) are constitutively generated; a 55 kDa form is liberated by escape from GPI-anchoring, and a 49 kDa form that is derived from the cell membrane by proteolytic cleavage with a serine protease (Bazil *et al.*, 1989). sCD14 mediates the binding of LPS to endothelial cells that do not contain a surface LPS receptor (mCD14) (Tobias *et al.*, 1992). This has been shown to potentiate LPS-responsiveness in endothelial cells resulting in the expression of adhesion molecules and cytokines (Pugin *et al.*, 1993). Activation of macrophages following recognition of bacterial LPS leads to the production of tumor necrosis factor-alpha (TNFα) (Wright, 1995) and other pro-inflammatory cytokines including IL-1, IL-6, and IL-8 (Dicks *et al.*, 2001).

Interestingly, the GPI-anchored mCD14 is unable to transduce a signal in response after ligation of LPS and requires an accessory signal in order to initiate a response (i.e., increased gene transcription). The mCD14 found on macrophages is co-expressed and forms a complex with another receptor known as the Toll-like receptor-4 (TLR4). The association of mCD14 with TLR4 has recently provided the investigators with a link between LPS-binding and the transcriptional responses of macrophages to this bacterial PAMP.

Toll-like Receptors (TLRs)

TLRs have received considerable attention as innate PRRs that are important not only in the recognition of PAMPs but are important in the initiation and transduction of the intracellular signals that induce innate immune mechanisms in macrophages (Medzhitov, 2001). The Toll receptor was originally identified as a key mediator of development in *Drosophila* (Anderson *et al.*, 1985), and only later was it found to also initiate innate anti-fungal responses in the fly (Lemaire *et al.*, 1996). Toll-receptors are characterized by an N-terminal extracellular domain containing several leucine-rich repeats (LRRs), and an intracellular C-terminal Toll/Interleukin-1 receptor (TIR) domain, named for its homology to the signaling domain of the IL-1 receptor. The TIR domain interacts with a heterotrimeric complex of death-domain containing adapter proteins (Sun *et al.*, 2002), ultimately leading to the activation of the transcription factor NF-κβ or its homologues, Dif or Relish, in *Drosophila*.

It is apparent that TLRs have an important role in the recognition of endogenous proteins such as necrotic cells and extracellular breakdown products (reviewed by Akira, 2003). A variety of endogenous ligands have also been implicated as potential activators of TLRs, including fibrinogen (Smiley *et al.*, 2001), fibronectin (Okamura *et al.*, 2001) and heat-shock proteins (Zhu and Pisetsky, 2001).

Several EST projects have identified sequence fragments with significant similarity to regions of mammalian TLRs (Zebrafish: GenBanks accessions BM185313, BG304206, BF158452; Japanese flounder AB076709, AU091257; Rainbow trout: AF281346). However, none of these sequences contains both the TIR and LRR domains, which are hallmarks of the TLR family. The *F. ruburipes* genome database (Aparicio *et al.*, 2002), and analysis for fish homologues of the TLR family revealed that the Toll family is shared by fish and humans. The predicted pufferfish TLR-2, -TLR-3, TLR-5, TLR-7, TLR-8, and TLR-9 corresponded structurally to the respective mammalian TLRs and one pufferfish TLR showed equal amino acid similarities to human TLR-1, TLR-6, and TLR-10 (Oshiumi *et al.*, 2003). Interestingly, two of the pufferfish genes were found to be unique to fish and were named TLR-21 and TLR-22. The pufferfish genome provides evidence for the presence of several TLR genes in fish, and we have verified that at least one of these receptors is expressed on goldfish macrophages (Stafford *et al.*, 2003). This goldfish TLR does not have a distinguishable sequence homology with any single mammalian TLR that has been described but contains both a LRR and a TIR domain. Further studies are required to elucidate the specific functional characteristics of this goldfish macrophage TLR.

ANTIMICROBIAL MECHANISMS OF MACROPHAGES

Macrophages possess a repertoire of potent pre-formed antimicrobial molecules stored within their granules and lysosomes. These organelles contain a salvo of degradative enzymes and antimicrobial peptides that are released into the phagolysosome upon ingestion of a foreign organism. In most cases, degradative enzymes such as proteases, nucleases, phosphatases, esterases, lipases, and highly basic antimicrobial peptides actively destroy the phagocytosed organism.

The available evidence points to the conservation of macrophage antimicrobial mechanisms in lower vertebrates that are similar to those reported for mammals. The following section structurally and functionally highlights the role of these macrophage-mediated killing mechanisms in host defense against infectious agents.

Nutrient Deprivation Mechanisms

Recruitment and Mobilization of Iron-binding Proteins

Commonly associated with infection and/or neoplasia is a condition known as anemia of infection and chronic disorder (Lee, 1983; Kent *et al.*, 1994). This condition was originally thought to be a pathological state induced by infection, but is now known to be a physiological response to infection, the desired effect of which is to decrease circulating iron, preventing or limiting the access of this critical metabolic element to pathogens (Kent *et al.*, 1994). Experimental evidence supporting the concept of induced iron deprivation as an antimicrobial mechanism has been summarized by Weinberg (1984): (1) hosts mobilize iron-binding proteins at sites of infection, (2) hosts recruit iron-withholding mechanisms in response to microbial infection, (3) increased iron withholding decreases incidence and intensity of infection, and (4) pathogenic microorganisms attempt to acquire iron from host tissues and fluids.

Iron sequestered in macrophages represents a significant portion of the total metabolically available iron content in mammals (Jacobs, 1977). Iron transported systemically by serum transferrin, enters cells through CD71 receptor-mediated endocytosis, dissociates from its receptor within the acidified endolysosome, and is subsequently transported inside the cell via a transporter system (Nunez *et al.*, 1990). Internalized iron accumulates in labile iron pools and is compounded to low molecular weight proteins (Jacobs, 1977). This intracellular pool acts as a readily available source of iron for metabolic processes or use by pathogens (Nunez *et al.*, 1990; Reif and Simmons, 1990; Byrd and Horwitz, 1991). Excess iron in the cell is transferred to ferritin for storage (Jacobs, 1977). Ferritin is one of many iron-containing proteins that are susceptible to inhibition by NO, which may represent a way of limiting the availability of intracellular iron to a developing pathogen (Reif and Simmons, 1990). Expression of ferritin is downregulated in activated macrophages, possibly through the effects of nitric oxide on iron response factors. These iron response factors are NO sensitive enzymes whose function is to regulate the status of iron in the cell (Reif and Simmons, 1990; Drapier *et al.*, 1993; Weiss *et al.*, 1994).

Transferrin is a serum protein primarily involved in the transport of iron throughout the body and contributes to the deprivation of this essential element. When activated with cytokines, murine macrophages increase production of transferrin (Weiss *et al.*, 1997). Transferrin produced by activated macrophages binds to intracellular iron, which limits the availability of intracellular iron for certain intracellular organisms (Byrd and Horwitz, 1991). Furthermore, activated macrophages have decreased numbers of transferrin receptors on their surface, limiting the influx of extracellular iron (Hamilton *et al.*, 1984a; Byrd and Horwitz, 1989; Byrd and Horwitz, 2000). Interestingly, blood monocytes do not possess transferrin receptors on their surface, but acquire these receptors upon maturation or in response to inflammatory signals (Hamilton *et al.*, 1984b).

Iron metabolism and immune activation of macrophages are intimately linked to other antimicrobial mechanisms such as NO production and tryptophan degradation (see section below). Treatment of murine macrophage with iron (Fe^{3+}) reduces activity of iNOS, and chelation of iron, by addition of desferrioxamine, increases activity of iNOS (Weiss *et al.*, 1993, 1994) and induces increased tryptophan degradation in human macrophages (Weiss *et al.*, 1999).

Natural-resistance-associated Macrophage Proteins (NRAMPs)

NRAMPs are members of the solute carrier family of proteins, and act as divalent metal/proton co-transporter proteins (Forbes and Gros, 2001). These proteins are highly conserved across an extremely diverse range of taxa, and homologues of this protein have been characterized in mammals, birds, fish, insects, yeast and bacteria (Malo *et al.*, 1994; Forbes and Gros, 2001). Two forms of NRAMP exist in mammals, NRAMP-1 and NRAMP-2, and share approximately 63% amino acid sequence homology (Gruenheid *et al.*, 1995). NRAMP-1 is 110 kD integral membrane phosphoglycoprotein, primarily expressed in macrophages and granulocytes (Vidal *et al.*, 1996). NRAMP-1 appears to be exclusively expressed in the lysosomal membrane of phagocytic cells (Gruenheid *et al.*, 1999), and message expression of the NRAMP-1 gene is induced in response to IFNγ and LPS (Govoni *et al.*, 1995). It is generally believed that NRAMP-1 acts as an efflux pump, removing divalent cations from the lumen of the phagolysosome and transporting it into the cytoplasm (Forbes and Gros, 2001). The co-transport activity of NRAMP-1 also leads to increased H^+ concentrations in the phagolysosome, aiding in the acidification process of the intra-phagolysosomal milieu and, subsequently, leading to activation of a number of lysosomal degradative enzymes (Hackam *et al.*, 1998).

The expression for NRAMP-2 is regulated by alternative RNA splicing mechanisms that produce two different transcripts, one that contains an iron response element (isoform I) and another that does not (isoform II) (Lee *et al.*, 1998). NRAMP-2 is localized predominantly to the plasma membrane, and mediates iron uptake from acidified endosomes formed during transferrin receptor-mediated endocystosis (Gruenheid *et al.*, 1999). Iron transported by NRAMP-2 subsequently associates with the cytoplasmic labile iron pool (Picard *et al.*, 2000). Other divalent cations transported by NRAMP-2 include, Zn^{2+}, Cd^{2+}, Mn^{2+}, Cu^{2+} and Co^{2+} (Forbes and Gros, 2001).

Treatment of macrophages with bacterial LPS induces a seven-fold increase in message expression for NRAMP-2 (Wardrop and Richardson, 2000). The sensitivity in expression of NRAMP-2 to stimulation with the non-self molecules suggests a role of NRAMP-2 in host defense, even though NRAMP-2 has a ubiquitous tissue distribution (Gruenheid *et al.*, 1995). Its potential role in macrophage-mediated innate immunity may be most influential during the initial stages of phagocytosis, in which divalent cations can be extruded from the phagosome prior to fusion of the phagosome with endosomes/lysosomes containing NRAMP-1.

The deprivation of iron and other divalent cations from the phagolysosome may induce cascading effects on pathogen survival. Divalent cations are co-factors for many enzymatic reactions, including those enzymes involved in oxidative phosphorylation, mitochondrial respiration and DNA replication. Furthermore, many detoxifying enzymes produced by pathogens, such as superoxide dismutase, are dependent on divalent cations for functionality (Forbes and Gros, 2001). Limiting the availability of divalent cations may not only prevent growth and replication but also increase susceptibility of pathogens to the reactive intermediates generated by macrophages (Forbes and Gros, 2001).

In fish, unlike mammals, it appears that the NRAMP-2 gene is duplicated. In rainbow trout, two NRAMP proteins have been identified (designated as α and β) and both cluster with the mammalian NRAMP-2. The expression of NRAMPα in rainbow trout was limited to the head-kidney and ovary while expression of NRAMPβ was ubiquitous (Dorshner and Philips, 1999). In pufferfish, two NRAMP-2 proteins have also been found—one of them localizing to the late endosomes/lysosomes—consistent with a divergence towards an NRAMP-1-like molecule (Blackwell *et al.*, 2001). Three NRAMP transcripts have been identified in catfish due to alternative splicing in the 3′ UTR and alternative polyadenylation resulting in a single functional protein. Injection of catfish with LPS increased transcription of NRAMP-2 in the kidney, spleen, and a monocyte/macrophage cell line (Overath *et al.*, 1999; Chen *et al.*, 2002). In carp, NRAMP-2 expression has also been observed and was found to be modulated by infections with *Trypanoplasma borreli* (Saeij and Wiegertjes, 1999).

Tryptophan Degradation

Tryptophan is an essential amino acid, and is the least available amino acid for metabolism (Ozaki *et al.*, 1987). Plants and select microorganisms synthesise tryptophan de novo. Indoleamine 2,3-dioxygenase (IDO) is an inducible protein found in virtually all tissues of the body and is involved in the catabolism of tryptophan (Taylor and Feng, 1991). This enzyme is a 42 kD protein that uses superoxide anion to oxidize the pyrrole ring of tryptophan in association with dihydroflavin mononucleotide and tetrahydrobiopterin as cofactors (Hayaishi *et al.*, 1977).

IDO induction in macrophages deprives intracellular pathogens of available tryptophan. Tryptophan degradation as an antimicrobial response was first suggested by Pfefferkorn (1984), who demonstrated that the protozoan *Toxoplasma gondii* required tryptophan for intracellular growth and survival in human fibroblasts. This inhibition of parasite growth was reversed by the addition of exogenous tryptophan to infected fibroblast cultures. The effect was not a result of decreased uptake of tryptophan, but due to an enhanced degradation rate.

Indoleamine 2,3-dioxygenase is an interferon inducible protein (Taylor and Feng, 1991), and cytokine-activated human macrophages have several fold higher IDO activity than do monocytes (Carlin *et al.*, 1989). IFNγ is the most potent stimulator of IDO activity and tryptophan degradation (Carlin *et al.*, 1989), and in some cases can induce an almost complete depletion of intracellular tryptophan stores (MacKenzie *et al.*, 1999).

Metabolic products of tryptophan degradation can induce other antimicrobial functions of activated macrophages. Picolinic acid, a metabolite of tryptophan degradation, augments NO responses in IFNγ-treated murine macrophages, and synergizes with IFNγ for induction of macrophage tumoricidal activity (Melillo *et al.*, 1993). Picolinic acid exhibits its NO-inducing effects through a hypoxia-responsive element located 5′ to the iNOS gene (Melillo *et al.*, 1995). Picolinic acid also induces a rapid increase in production of the inflammatory chemokines MIP-1α and 1β (macrophage inflammatory protein; Bosco *et al.*, 2000) by macrophages; proteins involved in the recruitment of T cells to the site of inflammation (Taub *et al.*, 1993).

The Respiratory Burst

Phagocytosis is often accompanied by a dramatic increase in the consumption of oxygen by phagocytic cells. This burst in oxygen consumption is not solely required for the increased metabolic demands necessary for phagocytosis, since metabolic inhibitors such as cyanide do not significantly affect oxygen consumption by phagocytes (Iyer *et al.*, 1961). A correlation between this oxidative burst and the formation of reactive oxygen intermediates has been clearly demonstrated (Iyer *et al.*, 1961). The respiratory burst is a potent antimicrobial response of phagocytic cells such as macrophages and neutrophils (Nakagawara *et al.*, 1982; Dinauer, 1993; Segal, 1996).

The respiratory burst response has received considerable attention in teleosts. However, little is known about the biochemical structure of the enzymes involved in the respiratory burst of fish phagocytes. Secombes *et al.* (1992) demonstrated the presence of a low potential b-type cytochrome that localized to the plasma membrane of rainbow trout macrophages; a phenomenon similar to cytochrome b_{558} (gp91*phox* and p21*phox*) in mammalian phagocytes. A polyclonal antibody raised against a carboxy terminal sequence of human cytochrome b_{558} also recognized a 90 kD protein in eel neutrophils (Itou *et al.*, 1998).

Significantly more work has been done regarding the regulation of respiratory burst activity in fish phagocytes stimulated by soluble mediators (i.e., cytokines). In early studies, the regulation of these responses in teleost macrophages has relied on the use of crude cytokine-like preparations. These crude cytokine preparations are obtained by stimulating kidney leukocytes with mitogens or macrophages with LPS and collecting the supernatants from stimulated cells. These crude cytokine preparations contain soluble mediators that have been shown to activate fish macrophages and neutrophils (Graham and Secombes, 1988, 1990a, b; Chen and Ainsworth, 1991; Secombes *et al.*, 1992; Jang *et al.*, 1995a; Neumann *et al.*, 1995, 1998, 2000a; Neumann and Belosevic, 1996; Novoa *et al.*, 1996; Stafford *et al.*, 1999; Tafalla *et al.*, 1999; Qin *et al.*, 2001). These preparations have also been shown to contain factors that deactivate antimicrobial responses of fish macrophages (Chung and Secombes, 1987; Jang *et al.*, 1994; Neumann and Belosevic, 1996; Stafford *et al.*, 1999; Tafalla and Novoa, 2000) indicating that the control of fish macrophage antimicrobial responses is mediated by a variety of endogenously derived factors that exhibit 'cytokine-like' activities.

The respiratory burst of rainbow trout macrophages can be primed *in vitro* by stimulating resident kidney macrophages with culture supernatants obtained from mitogen-stimulated kidney leukocytes (Graham and Secombes, 1988, 1990a, b). Priming the respiratory burst response of rainbow trout resident kidney macrophage requires extensive cultivation periods [i.e., 48 h] with these supernatants; an effect shared with their mammalian counterparts (Novoa *et al.*, 1996). The macrophage activating factor (MAF) responsible for priming respiratory burst activity appears to be a product of fish T-cells (Graham and Secombes, 1990a), and is both heat and acid labile (Graham and Secombes, 1990b). Production of a MAF that primes trout macrophage respiratory burst activity can also be induced by antigen-specific stimulation of lymphocyte cultures *in vitro* (Francis and Ellis, 1994; Yin *et al.*, 1997). Crude MAF preparations also induce spreading and adherence of macrophages in culture (Secombes *et al.*, 1987).

Interestingly, different macrophage sub-populations in fish appear to display distinct priming kinetics of respiratory burst activity when stimulated with these crude-cytokine preparations. Our laboratory has developed a culture system for obtaining high yields of macrophages from the kidneys of goldfish (Wang *et al.*, 1995b; Neumann, 1999; Barreda *et al.*, 2000; Neumann *et al.*, 2000b). In this culture system, macrophages are generated by incubating kidney leukocytes in the presence of cell-conditioned medium (CCM) containing macrophage growth factor(s) (MGFs). We previously demonstrated that both kidney leukocytes and a goldfish macrophage cell line secrete an endogenous growth factor(s) that selectively induces the proliferation and differentiation of cells in the macrophage lineage. These *in vitro*-derived kidney macrophage (IVDKM) cultures appear to contain three distinct macrophage morphotypes, represented by macrophage progenitor cells, monocytes, and mature macrophages. Characterization of these different macrophage sub-populations was performed using: (1) flow cytometric, (2) function (phagocytosis, respiratory burst, nitric oxide production), (3) cytochemical profiles (non-specific esterase, acid phosphatase, myeloperoxidase), (4) morphology and (5) *in vitro* proliferation and differentiation pathways (Neumann, 1999; Barreda *et al.*, 2000; Neumann *et al.*, 2000b).

The monocyte-like cells found in IVDKM cultures have a significantly greater basal respiratory burst response than do the more mature macrophage sub-population (Stafford *et al.*, 2001). The monocytes present in goldfish IVDKM cultures can be rapidly primed for respiratory burst activity using crude MAF preparations, displaying enhanced respiratory burst responses after only 6-24 h of stimulation with crude MAF preparations. Interestingly, after 24 h of stimulation with MAF, these cells gradually lose their primed respiratory burst potential (Neumann, 1999; Barreda *et al.*, 2000; Neumann *et al.*, 2000b).

The mature macrophage-like cells within IVDKM cultures display a different pattern of priming kinetics compared to the monocyte-like cells.

The longer the macrophages are stimulated with MAF, the greater their respiratory burst response (Neumann, 1999; Barreda *et al.*, 2000; Neumann *et al.*, 2000a). These data are consistent with those observed by others using resident kidney macrophages isolated from various fish species (Chung and Secombes, 1987; Secombes *et al.*, 1987; Graham and Secombes, 1990b; Marsden *et al.*, 1994; Marsden and Secombes, 1997; Mulero and Meseguer, 1998). In these studies, respiratory burst capacity was measured ≥ 48h of stimulation with MAF. Recently, differentiation-mediated alterations in antimicrobial functions have been observed in rainbow trout at the molecular level. MacKenzie *et al.* (2003) reported that following stimulation with LPS, monocyte-like cells appeared to differentiate into more mature macrophage-like cells, which exhibited increased phagocytic capacities and expression of inflammatory genes.

The two functional sub-populations identified in IVDKM (i.e., monocytes and mature macrophages) display distinct priming kinetics that are similar to those described for mammalian phagocytes. In mammals, macrophages need extensive stimulation with IFNγ (48-72 h) for induction of maximal respiratory burst activity (Nathan *et al.*, 1983). Mammalian neutrophils, on the other hand, can be primed with IFNγ after only 6 h of stimulation, a consequence of protein upregulation and expression, and not an increased affinity change in the oxidase (Cassatella *et al.*, 1988). Thus, although the machinery required for respiratory burst activity may be similar in different cell types, functional regulation of this response may be specific to individual cells populations.

Although the native molecule(s) responsible for priming respiratory burst activity in fish have not been identified, others and we have attempted to purify these molecules from crude-cytokine preparations. Crude cytokine supernatants contain two distinct MAF activities that modulate macrophage respiratory burst activity (Chen and Ainsworth, 1991; Neumann *et al.*, 2000a). One activity (corresponding to a protein of 50 kD) induces a rapid but transient priming effect on the respiratory burst capacity of IVDKM. Stimulation of IVDKM with this molecule for only 6 h results in a greatly enhanced respiratory burst response compared to controls. However, 48 h after stimulation the respiratory burst capacity of stimulated macrophages is significantly reduced compared to those macrophages stimulated for only 6 h (Chen and Ainsworth, 1991; Neumann and Belosevic, 1999). We have also demonstrated the presence of a 30 kD factor present in crude cytokine preparations that also modulates respiratory burst activity in goldfish macrophages. This molecule may be similar to one characterized by Graham and Secombes (1990b), who fractionated a respiratory burst enhancing molecule with similar molecular weight from rainbow trout. This molecule induces unique priming effects on IVDKM and cells stimulated with the 30 kD MAF continue to increase their priming potential the longer they are stimulated with this molecule (Neumann *et al.*, 2000a). Interestingly, when IVDKM are co-stimulated with the 50 kD and 30 kD MAF, the respiratory burst potential of IVDKM is greater than that induced by either factor alone (Neumann *et al.*, 2000a). However, the effect is transient, and IVDKM co-stimulated for 48 h with these factors have significantly lower respiratory burst potential compared to those stimulated for 24 h. The 50 kD MAF has also been shown to induce potent NO induction in goldfish macrophages (Neumann *et al.*, 2000a).

The respiratory burst of fish macrophages can also be primed by bacterial LPS (Waterstrat *et al.*, 1991; Solem *et al.*, 1995; Taylor and Hoole, 1995; Neumann and Belosevic, 1996; Campos-Perez *et al.*, 1997), β-glucans from yeast cell walls (Brattgjerd *et al.*, 1994; Jorgensen and Robertsen, 1995; Dalmo *et al.*, 1996; Tahir and Secombes, 1996; Robertsen, 1999), and bacterial proteins (Francis and Ellis, 1994). Macrophage respiratory burst activity can also be primed *in vivo* by administration of killed or attenuated bacterial pathogens (Chung and Secombes, 1987; Enane *et al.*, 1993; Marsden *et al.*, 1994; Yin *et al.*, 1997), and neutrophils collected after injection of irritants such as casein or heat killed bacteria or different dietary regimens also display elevated respiratory burst responses (Itou *et al.*, 1996; Nikoskelainen *et al.*, 2006; Watanuki *et al.*, 2006).

There is evidence to suggest that a TNFα-like molecule may also be responsible for the respiratory burst-inducing activity exhibited by crude cytokine supernatants (Hardie *et al.*, 1994; Jang *et al.*, 1995b; Novoa *et al.*, 1996; Campos-Perez *et al.*, 1997; Hirono *et al.*, 2000; Qin *et al.*, 2001). Studies have shown that human recombinant TNFα synergizes with crude-cytokine preparations to enhance priming of respiratory burst activity in rainbow trout macrophages (Hardie *et al.*, 1994; Tahir *et al.*, 1996; Campos-Perez *et al.*, 1997). Moreover, priming of the respiratory burst can be partially inhibited using anti-human TNFα receptor 1 monoclonal antibodies, suggesting a certain degree of conservation in both the TNF molecule and its receptor between mammals and teleosts (Jang *et al.*, 1995b). Supernatants derived from rainbow trout were also highly toxic to murine L929 cells, which are highly sensitive to mammalian

TNFα, further suggesting that a functional TNFα-like molecule was present in fish (Qin *et al.*, 2001). The presence of this cytokine has recently been confirmed by the cloning of the TNFα gene from a variety of fish species including rainbow trout, Japanese flounder, carp, catfish, seabream and turbot (Hirono *et al.*, 2000; Laing *et al.*, 2000; Garcia-Castillo *et al.*, 2002, 2004; Zou *et al.*, 2002, 2003a, b; Park *et al.*, 2003; Saeij *et al.*, 2003; Ordas *et al.*, 2006). Furthermore, the recombinant protein has been expressed and functional studies suggest that teleost TNFα plays a key role in the induction of inflammatory responses in fish (Zou *et al.*, 2003), and may also be responsible for induction of ROI production in fish macrophages.

Another cytokine that plays an important role in the induction of macrophage antimicrobial responses in fish is IL-1 (reviewed by Bird *et al.*, 2002). Rainbow trout IL-1 has been cloned (Secombes *et al.*, 1999; Zou *et al.*, 1999a, b, 2000; Brubacher *et al.*, 2000), and subsequent studies have resulted in the cloning of the full-length gene in different fish species including sharks (Bird *et al.*, 2000; Fujiki *et al.*, 2000; Engelsma *et al.*, 2001, 2003; Pelegrin *et al.*, 2001; Scapigliati *et al.*, 2001). This cytokine has been shown to prime the respiratory burst response in mammalian macrophages and neutrophils (Ferrante, 1992; Sample and Czuprynski, 1994; Yagisawa *et al.*, 1995). Recently, recombinant trout IL-1 has been produced and functional studies performed (Hong *et al.*, 2001, 2003; Peddie *et al.*, 2001). The recombinant cytokine induced migration of head-kidney leukocytes (Peddie *et al.*, 2001) and was recently shown to increase protection of rainbow trout from infections with *Aeromonas salmonicida*, a finding that correlated with systemic IL-1β, COX-2, and lysozyme II gene expression (Hong *et al.*, 2003). The protective effects of IL-1 may also result from the ability to induce production of ROI by fish macrophages as seen in mammals.

The investigations of the mechanisms of activation of phagocytes have shifted from the use of relatively undefined substances such as MAF to the use of recombinant cytokines. One cytokine of particular interest with regards to the ROS is IFNγ (Nathan *et al.*, 1983). IFNg homologues have been recently identified in catfish, pufferfish and zebrafish but comprehensive functional analyses are yet to be performed (Zou *et al.*, 2004; Igawa *et al.*, 2006; Milev-Milovanovic *et al.*, 2006).

It has been shown that the respiratory burst response plays an important role in the destruction of several fish pathogens. *Renibacterium salmoninarum*, etiological agent of bacterial kidney disease, is susceptible to H_2O_2 killing by trout macrophages and addition of catalase to macrophage cultures inhibits killing of this fish pathogen *in vitro* (Hardie *et al.*, 1996). Graham and Secombes (1988) demonstrated that rainbow trout macrophages stimulated with crude MAF preparations could inhibit growth of the bacterium *Aeromonas salmonicida*. Subsequent work by this group, and others, demonstrated that killing of select pathogens correlated with the production of ROI, and the addition of scavengers of reactive oxygen, such as catalase, abolished the ability of macrophages to restrict the growth of pathogens (Hardie *et al.*, 1996; Campos-Perez *et al.*, 1997; Yin *et al.*, 1997).

Nitric Oxide Production (NO)

It was demonstrated in the early 1980s that nitrogen oxides were common by products of metabolism, and that the treatment of rats with bacterial endotoxin resulted in increased nitrate levels in body fluids (Green *et al.*, 1981a, b). Stuehr and Marletta (1985) demonstrated that endotoxin-stimulated murine macrophages produce both nitrate and nitrite, and subsequent studies showed that the production of these NO by-products by macrophages correlated with an increased cytotoxicity against tumors and pathogens (Hibbs, 1991; James, 1995).

Although constitutive nitric oxide synthase had been demonstrated from the central nervous system of fish (Brunung *et al.*, 1996; Holmquist and Ekstrom, 1997) only one report prior to 1995 had demonstrated that fish possess an inducible form of this enzyme. Schoor and Plum (1994) demonstrated inducible NO production, using enzyme histochemical techniques, from kidney homogenates obtained from channel catfish infected with the bacterium *Edwardsiella ictaluri*. Our laboratory subsequently demonstrated that NO production could be induced in a goldfish macrophage cell line stimulated with bacterial LPS (Wang *et al.*, 1995). Crude cytokine supernatants were also shown to contain a factor(s) that synergize with bacterial LPS to induce goldfish macrophages to produce NO (Neumann *et al.*, 1995), an effect since demonstrated in several fish species (Yin *et al.*, 1997; Mulero and Mesenguer, 1998; Tafalla and Novoa, 2000). Nucleotide sequences for goldfish, rainbow trout, carp, zebrafish, and Atlantic salmon inducible nitric oxide synthase (iNOS) have been identified (Laing *et al.* 1996, 1999; Campos-Perez *et al.* 2000; Laing Barroso *et al.* 2000; Saeij *et al.* 2000; Wang *et al.* 2001), and share approximately 60-70% homology with mammalian-derived macrophage iNOS. Rainbow trout head-kidney macrophages stimulated with 25-50 μg/ml LPS expressed maximal levels of iNOS between 2 and 6 h post-stimulation (Laing *et al.*, 1999). Furthermore, it was shown that the gills are an important site of iNOS expression in rainbow trout (Campos-Perez *et al.*, 2000). Following challenge with different fish pathogens iNOS message was rapidly upregulated in different tissues (Campos-Perez *et al.*, 2000; Acosta *et al.*, 2005; Tafalla *et al.*, 2005; Prabkaran *et al.*, 2006). For example, following challenge with *Renibacterium salmonarium*, iNOS message in the gills increased rapidly (i.e., between 3 and 6 h) and lasted for several days, while a delayed expression of iNOS was observed in the kidneys of challenged trout (i.e., after 24 h) that was rapidly downregulated (Campos-Perez *et al.*, 2000). Using a combination of biochemical, immunohistochemical, and immunoblotting analyses, iNOS immunoreactive cells in head-kidney tissues of rainbow trout were identified as heterophilic granulocytes, and iNOS positive macrophages and neutrophils were found in the liver (Barroso *et al.*, 2000) as well as retina of zebrafish (Shin *et al.*, 2000). In carp, induction of the iNOS gene was dependent on NF-kb and was observed following stimulation of carp phagocytes with LPS or *Trypanoplasma borreli*, which also correlated with the production of high levels of NO (Saeij *et al.*, 2000).

Unlike mammals where numerous studies have been conducted, relatively few studies have examined the NO-induced cytotoxic capability of fish macrophages *in vitro*. Yin *et al.* (1997) demonstrated that MAF activated catfish macrophages were bactericidal towards *Aeromonas hydrophila*, and that killing could be partially blocked using N^{G}MMLA, an inhibitor of NO production. Fish macrophages can be induced to produce NO in response to intracellular infection. For example, goldfish macrophages infected with *Leishmania major*, an obligate intracellular protozoan pathogen of mammalian macrophages, produce NO in the absence of any additional exogenous cytokine signals (Stafford *et al.*, 1999). Induction of this response appears to be mediated via the recognition of a foreign molecule by the macrophage, since phagocytosis of latex beads is insufficient for initiating NO production in goldfish macrophages (Neumann and Belosevic, 1996). This contrasts the scenario observed in mammalian macrophages, which require an accessory signal such as IFNγ for induction of the NO response (Green *et al.*, 1990). Similar effects were also reported for fish macrophages infected with the Gram-positive bacteria (Marsden *et al.*, 1994) as well as exposure to heat-killed

Trypanosoma danilewskyi and *Aeromonas salmonicida* (Stafford and Belosevic, 2003). Gilthead sea bream vaccinated against the pathogenic bacterium *Photobacterium damselae* have significantly higher levels of NO production than non-vaccinated individuals *in vivo* and *in vitro* (Acosta *et al.*, 2005). This heightened response correlated with greater protection from subsequent bacterial challenge. Using the sea bream/*P. damsela* model Acosta *et al.* (2005) reported that blocking the NO response using the iNOS inhibitor L-NAME significantly increased the susceptibility of fish to infection. Similarly, the bactericidal activity of catfish phagocytes against *Aeromonas hydrophila* was enhanced following vaccination and this bactericidal activity was partially blocked by the addition of N^G-MMLA another inhibitor of the NO pathway (Yin *et al.*, 1997). These authors also reported that the supernatants from immunized cell cultures exposed to the vaccinating strain induced greater NO response than supernatants collected from cells that were stimulated with a different bacterium (Yin *et al.*, 1997). Findings such as these demonstrate the importance of the NO response in the resistance of fish to certain pathogens.

Identification of the specific factors involved in the NO response has been facilitated by the availability of purified/recombinant immune mediators. In turbot, LPS in combination with a turbot IFN-αβ-like molecule induced the NO response in cells otherwise non-response to LPS alone. However, this stimulatory capacity was not present in all macrophage subpopulations. Human recombinant TNF-α when combined with LPS was able to induce a significant enhancement of the NO production of all macrophage subpopulations in turbot (Tafalla and Novoa, 2000). Cells treated with pentoxifylline an inhibitor of TNF-α (but not iNOS) were found to have significantly reduced NO production compared to controls (Saeij *et al.*, 2003). One of the important molecules capable of inducing NO response in fish macrophages is enzymatically cleaved transferrin (Stafford *et al.*, 2001; Stafford and Belosevic, 2003).

ROLE OF TRANSFERRIN IN INDUCTION OF MACROPHAGE ANTIMICROBIAL RESPONSES

Transferrin is a bi-lobed monomeric serum glycoprotein of approximately 70 to 80 kD and is responsible for the transport and delivery of iron to cells and is primarily produced in the liver (Ciechanover *et al.*, 1983; Dautry-Varsat *et al.*, 1983; Hopkins and Trowbridge 1983; Klausner *et al.*, 1983a, b; Ford, 2001). The N- and C- terminal lobes of transferrin have similar amino acid sequence, tertiary structure and are believed to have evolved as a result of gene duplication (Worwood, 1989; Baldwin, 1993). The two homologues globular lobes contain deep clefts capable of binding iron and are connected by a small peptide region (~15 amino acids) called the inter-domain bridge, which varies in length between different transferrin species (Anderson *et al.*, 1990; Retzer *et al.*, 1996). Transferrin is abundant in nature and has been identified in a wide range of organisms (i.e., insects, crustaceans, fish, and mammals) (Martin *et al.*, 1984; Bartfield and Law, 1990; Jamroz *et al.*, 1993; Yoshiga *et al.*, 1997). There is also extensive structural and sequence homology between transferrins from different species (Baldwin, 1993).

Binding of iron to transferrin creates a bacteriostatic environment by limiting the availability of iron to replicating pathogens. However, in addition to its primary described role as an iron-binding protein, transferrin appears to exhibit a variety of other biological functions. For example, transferrin induces neutrophilic end-stage maturation (Evans *et al.*, 1989), supports the growth and differentiation of the human promyelocytic cell line, HL-60 (Breitman *et al.*, 1980), and selectively stimulates cellular proliferation of prostatic carcinoma cells (Rossi and Zetter, 1992). Activation of casein kinase II, an enzyme involved in the regulation of cell growth, was shown to require the application of transferrin in combination with an insulin-like growth factor (Wang *et al.*, 1995a). Transferrin up-regulates chemokine synthesis by human proximal tubular epithelial cells (Tang *et al.*, 2002), and the addition of transferrin to rat cultured aortic smooth muscle cells induced a concentration- and dose-dependent increase in iNOS mRNA and nitrite accumulation. Elevated transferrin concentrations in cerebral spinal fluid after subarachnoid hemorrhage also increased iNOS mRNA expression by smooth muscle cells (Takenaka *et al.*, 2000). A recent study demonstrated that in addition to the binding of iron, transferrin functions as a protein-binding protein (Weinzmer *et al.*, 2001), and is one of the constituents secreted by platelets that can induce phagocytosis (Sakamoto *et al.*, 1997).

In chickens, ovatransferrin is a key inducer of cellular activation measured by its ability to induce the production of IL-6 and matrix metallopreoteinases as well as the induction of respiratory burst in macrophages (Xie *et al.*, 2001, 2002). Ibranim *et al.* (1998) have reported that ova transferrin can directly contribute to the killing of bacteria. This was demonstrated by the identification of a bactericidal domain in the amino-terminal half molecule (i.e., N-lobe, residues 1-332). The antibacterial properties of this domain were dependent on 3 intra-chain disulfide bonds and the protein sequence within the N-lobe demonstrated a marked sequence homology to insect defensins that contained 6 highly conserved cysteine residues. Therefore, ova transferrin is believed to be one of the key components found in inflammatory chicken serum that is capable of not only mediating immune cell functions but can also contribute to the direct killing of bacterial pathogens.

Teleost transferrin has also been described as an acute phase protein and increased levels of transferrin expression were observed following bacterial infections in rainbow trout (Bayne *et al.*, 2001). We have shown that transferrin can induce NO production by activated goldfish macrophages (Stafford *et al.*, 2001; Stafford and Belosevic, 2003). Transferrin must be cleaved in order for it to activate fish macrophages to produce NO. The native protein (~55-60 kD) undergoes proteolytic cleavage in goldfish leukocyte cultures stimulated with mitogens and/or mixed lymphocyte reactions (Stafford *et al.*, 2001). The resultant peptides of 33-37 kD synergized with LPS for induction of NO in goldfish macrophages. Products released from necrotic fish cells (i.e., macrophages and neutrophils) appear to play a major role in the cleavage of transferrin into its active NO-inducing form (Stafford and Belosevic, 2003). The connection between fish neutrophils, macrophages and transferrin provides an interesting model for understanding inflammation and regulation of the immune response of phagocytes in fish. During the initial phase of inflammation, vascular leakage of capillaries initiates swelling at the site of infection. In mammals, transferrin has been shown to leak into inflammatory sites during this early phase of inflammation (Bergman and Kolarz, 1976; Basran *et al.*, 1985; Colditz *et al.*, 1992; Raijmakers *et al.*, 1997).

In the goldfish, serum components have also been found to leak into the peritoneal cavity following induction of an inflammatory response (Chadzinska *et al.*, 2000). Neutrophils are one the first immune cells recruited to the site of inflammation, and their migration into the site may initiate the cleavage of transferrin via the production of neutrophil-derived proteases (i.e., elastase, gelatinase, matrix metalloproteases, etc.). We have recently shown that the intracellular contents of goldfish granulocytes were capable of cleaving transferrin (Stafford *et al.*, 2001). Monocytes that are subsequently recruited to the inflammatory site—and resident tissues macrophages—may then recognize the cleaved transferrin products as a signal for initiating the production of NO. This would be analogous to the recognition of endogenous proteins by TLRs in mammals (Beg, 2002). Although fish monocytes do not appear capable of producing NO (Neumann *et al.*, 2000a), transferrin may initiate the differentiation of these cells into the more mature macrophage phenotypes, as is the case in mammals (Bose and Farina, 1995). This may cause monocytes to become responsive to signals that initiate NO production. Differentiation of human monocytes into more mature phenotypes results in the acquired capacity to produce NO (Anderseen *et al.*, 1984). We have observed a similar effect in goldfish monocytes (Neumann *et al.*, 2000).

This novel finding implicates transferrin as primitive regulator of immune phagocyte function in lower vertebrates and possibly in higher vertebrates. Moreover, it is interesting to speculate that proteins homologous to transferrin may play an important role in the induction of the NO response in invertebrate immunocytes. It has been reported by several groups that invertebrate immunocytes possess the capacity to produce NO in response to immune challenge (Torreilles and Guerin, 1999). Since many invertebrates possess transferrin-like molecules (Baker and Lindley, 1992), it is conceivable that proteolytic cleavage of transferrin-like molecules may represent a primitive form of immunoregulation of innate immunity, and specifically, macrophage antimicrobial functions.

CONCLUSION

The recognition and elimination of invading pathogens is vital for host survival. Macrophages play a central role in host protection and cells functionally reminiscent of the vertebrate macrophage are present in virtually all metazoan organisms, attesting to the importance of these phagocytic cells in host defense in all multicellular organism. Macrophages contain a repertoire of non-self recognition receptors (i.e., PRRs) that recognize molecular patterns found on pathogens surfaces called PAMPs, and many of these innate immune receptors are highly conserved throughout evolution (i.e., Toll and TLRs). Recognition of PAMPs by PRRs leads to the rapid phagocytosis of the invading microbe followed by their eventual destruction using a variety of preformed enzymes or production of reactive intermediates (i.e., ROI and RNI) by inducible antimicrobial pathways.

Phagocytosis is the ancestral defense mechanism of all metazoan animals and is essential in preventing the dissemination of infectious agents. Many of the antimicrobial effector responses of vertebrate phagocytes are similar across diverse animal taxa. Inducible antimicrobial responses such as the respiratory burst pathway and production of NO have been demonstrated in fish phagocytes, and display biochemical and physiological similarities to homologous responses induced in mammalian phagocytes. Both respiratory burst activity and NO induction have been shown to be critical effector mechanisms in limiting the growth of fish pathogens and studies addressing the regulation of these responses in fish have provided some novel insights into how these mechanisms are regulated in vertebrates.

Acknowledgements

This work was supported by Natural Sciences and Engineering Council of Canada (NSERC) and Alberta Heritage Foundation for Medical Research (AHFMR) to M.B.

References

Acosta, F., F. Real, A.E. Ellis, C. Tabraue, D. Padilla and C.M. Ruiz de Galarreta. 2005. Influence of vaccination on the nitric oxide response of gilthead sea bream following infection with *Photobacterium damselae* subsp. *piscicida*. *Fish and Shellfish Immunology* 18: 31-38.

Acton, S., A. Rigotti, K.T. Landschulz, S. Xu, H.H. Hobbs and M. Krieger. 1996. Identification of scavenger receptor SR-BI as a high-density lipoprotein receptor. *Science* 271: 518-520.

Akira, S. 2003. Mammalian Toll-like receptors. *Current Opinion in Immunology* 15: 5-11.

Anderson, B.F., H.M. Baker, G.F. Noris, S.V. Rumball and E.N. Baker. 1990. Apolactoferrin structure demonstrates ligand-induced conformational change in transferrins. *Nature (London)* 344: 784-787.

Anderson, K.V., L. Bokla and C. Nusslein-Volhard. 1985. Establishment of dorsal-ventral polarity in the *Drosophila* embryo: The induction of polarity by the Toll gene product. *Cell* 42: 791-798.

Andreesen, R., J. Osterholz, H. Bodemann, K.J. Bross, U. Costabl and G.W. Lohr. 1984 Expression of transferrin receptors and intracellular ferritin during terminal differentiation of human monocytes. *Blut* 49: 195-202.

Aparicio, S., J. Chapman, E. Stupka, N. Putnam, J.M. Chia, P. Dehal, A. Christoffels, S. Rash, S. Hoon, A. Smit, M.D. Gelpke, J. Roach, T. Oh, I.Y. Ho, M. Wong, C. Detter, F. Verhoef, P. Prediki, A. Tay, S. Lucas, P. Richardson, S.F. Smith, M.S. Clark, Y.J. Edwards, N. Doggett, A. Zharkikh, S.V. Tavtigian, D. Pruss, M. Barnstead, C. Evans, H. Baden, J. Powell, G. Glusman, L. Rowen, L. Hood, Y.H. Tan, G. Elgar, T. Hawkins,

B. Venkatesh, D. Rokhsar and S. Brenner. 2002. Whole-genome shotgun assembly and analysis of the genome of Fugu fubripes. Science 297: 1301-1310.

Baker, E.N. and P.F. Lindley. 1992. New perspectives on the structure and function of transferrins. *Journal of Inorganic Biochemistry* 47: 147-160.

Baldwin, G.S. 1993. Comparison of transferrin sequences from different species. *Comparative Biochemistry and Physiology* B 106: 203-218.

Barreda, D., N.F. Neumann and M. Belosevic. 2000. Flow cytometric analysis of PKH26-labeled goldfish kidney derived macrophages. *Developmental and Comparative Immunology* 24: 308-317.

Barroso, J.B., A. Carreras, F.J. Esteban, M.A. Peinado, E. Martinez-Lara, R. Valderrama, A. Jimenez, J. Rodrigo and J.A. Lupianez. 2000. Molecular and kinetic characterization and cell type location of inducible nitric oxide synthase in fish. *American Journal of Physiology* 279: R650.

Bartfeld, N.S. and J.H. Law. 1990. Isolation and molecular cloning of transferrin from the tobacco hornworm, *Manduca sexta*. Sequence similarity to the vertebrate transferrins. *Journal of Biological Chemistry* 15: 21684-21691.

Basran, G.S., A.J. Byrne and J.G. Hardy. 1985. A noninvasive technique for monitoring lung vascular permeability in man. *Nuclear Medicine Communications* 6: 3-10.

Bayne, C.J., L. Gerwick, K. Fujiki, M. Nakao and T. Yano. 2001. Immune-relevant (including acute phase) genes identified in the livers of rainbow trout, *Oncorhynchus mykiss*, by means of suppression subtractive hybridization. *Developmental and Comparative Immunology* 25: 205-217.

Bazil, V., M. Baudys, I. Hilgert, I. Stefanova, M.G. Low, J. Zbrozek and V. Horejsi. 1989. Structural relationship between the soluble and membrane-bound forms of human monocyte surface glycoprotein CD14. *Molecular Immunology* 26: 657-662.

Beg, A.A. 2002. Endogenous ligands of Toll-like receptors: Implications for regulating inflammatory and immune responses. *Trends in Immunology* 23: 509-512.

Bergmann, H. and G. Kolarz. 1976. Pertechnetate uptake of joints in rheumatoid arthritis. *European Journal of Nuclear Medicine* 1: 205-210.

Bird, S., T. Wang, J. Zou, C. Cunningham and C.J. Secombes. 2000. The first cytokine sequence within cartilaginous fish: interleukin-1 in the small spotted catshark (*Scyliorhinus canicula*). *Journal of Immunology* 168: 3329-3340.

Bird, S., J. Zou, T. Wang, B. Munday, C. Cunningham and C.J. Secombes. 2002. Evolution of interleukin-1 beta. *Cytokine and Growth Factor Reviews* 13: 483-502.

Blackwell, J.M., T. Goswami, C.A. Evans, D. Sibthorpe, N. Papo, J.K. White, S. Searle, E.N. Miller, C.S. Peacock, H. Mohammed and M.A. Ibrahim. 2001. SLC11A1 (formerly NRAMP1) and disease resistance. *Cellular Microbiology* 3: 773-784.

Bosco, M.C., A. Rapisarda, S. Massazza, G. Melillo, H. Young and L. Varesio. 2000. The tryptophan catabolite picolinic acid selectively induces the chemokines macrophage inflammatory protein-1 alpha and -1 beta in macrophages. *Journal of Immunology* 164: 3283-3291.

Bose, M. and P. Farnia. 1995. Pro-inflammatory cytokines can significantly induce human mononuclear phagocytes to produce nitric oxide by a cell maturation-dependent process. *Immunology Letters* 48: 59-64.

Brattgjerd, S., O. Evensen and A. Lauve. 1994. Effect of injected yeast glucan on the activity of macrophages in Atlantic salmon, *Salmo salar* L., as evaluated by *in vitro* hydrogen peroxide production and phagocytic capacity. *Immunology* 83: 288-294.

Breitman, T.R., S.J. Collins and B.R. Keene. 1980. Replacement of serum by insulin and transferrin supports growth and differentiation of the human promyelocytic cell line, HL-60. *Experimental Cell Research* 126: 494-498.

Brittingham, A., C.J. Morrison, W.R. McMaster, B.S. McGwire, K.P. Chang and D.M. Mosser. 1995. Role of the *Leishmania* surface protease gp63 in complement fixation, cell adhesion, and resistance to complement-mediated lysis. *Journal of Immunology* 155: 3102-3111.

Brown, M.S. and J.L. Goldstein. 1983. Lipoprotein metabolism in the macrophage: Implications for cholesterol deposition in atherosclerosis. *Annual Review of Biochemistry* 52: 223-261.

Brubacher, J.L., C.J. Secombes, J. Zou and N.C. Bols. 2000. Constitutive and LPS-induced gene expression in a macrophage-like cell line from the rainbow trout (*Oncorhynchus mykiss*). *Developmental and Comparative Immunology* 24: 565-574.

Bruning, G., K. Hattwig and B. Mayer. 1996. Nitric oxide synthase in the peripheral nervous system of the goldfish, *Carassius auratus*. *Cell and Tissue Research* 284: 87-98.

Bullock, W.E. and S.D. Wright. 1987. Role of the adherence-promoting receptors, CR3, LFA-1, and p150,95, in binding of *Histoplasma capsulatum* by human macrophages. *Journal of Experimental Medicine* 165: 195-210.

Byrd, T.F. and M.A. Horwitz. 1989. Interferon gamma-activated human monocytes downregulate transferrin receptors and inhibit the intracellular multiplication of *Legionella pneumophila* by limiting the availability of iron. *Journal of Clinical Investigation* 83: 1457-1465.

Byrd, T.F. and M.A. Horwitz. 1991. Lactoferrin inhibits or promotes *Legionella pneumophila* intracellular multiplication in non-activated and interferon gamma-activated human monocytes depending upon its degree of iron saturation. Iron- lactoferrin and non-physiologic iron chelates reverse monocyte activation against *Legionella pneumophila*. *Journal of Clinical Investigation* 88: 1103-1112.

Byrd, T.F. and M.A. Horwitz. 2000. Aberrantly low transferrin receptor expression on human monocytes is associated with non-permissiveness for *Legionella pneumophila* growth. *Journal of Infectious Diseases* 181: 1394-1400.

Calduch-Giner, J.A., B.A. Sitja, P.P. Alvarez and S.J. Perez. 1997. Growth hormone as an *in vitro* phagocyte-activating factor in the gilthead sea bream (*Sparus aurata*). *Cell and Tissue Research* 287: 535-540.

Campos-Perez, J.J., A.E. Ellis and C.J. Secombes. 1997. Investigation of factors influencing the ability of *Renibacterium salmoninarum* to stimulate rainbow trout macrophage respiratory burst activity. *Fish and Shellfish Immunology* 7: 555-566.

Campos-Perez, J.J., M. Ward, R.S. Grabowski, A.E. Ellis and C.J. Secombes. 2000. The gills are an important site of iNOS expression in rainbow trout (*Oncorhynchus mykiss*) after challenge with Gram-negative pathogen *Renibacterium salmonarium*. *Immunology* 99: 153-161.

Carlin, J.M., E.C. Borden, P.M. Sondel and G.I. Byrne. 1989. Interferon-induced indoleamine 2,3-dioxygenase activity in human mononuclear phagocytes. *Journal of Leukocyte Biology* 45: 29-34.

Cassatella, M.A., R. Cappelli, B.V. Della, M. Grzeskowiak, S. Dusi and G. Berton. 1988. Interferon-gamma activates human neutrophil oxygen metabolism and exocytosis. *Immunology* 63: 499-506.

Chadzinska, M., A. Scislowska-Czarnecka and B. Plytycz. 2000. Inhibitory effects of morphine on some inflammation-related parameters in the goldfish *Carassius auratus* L. *Fish and Shellfish Immunology* 10: 531-542.

Chen, D.X. and A.J. Ainsworth. 1991. Assessment of metabolic activation of channel catfish peripheral blood neutrophils. *Developmental and Comparative Immunology* 15: 201-208.

Chen, H., G.C. Waldbieser, C.D. Rice, B. Elibol, W.R. Wolter and L.A. Hanson. 2002. Isolation and characterization of channel catfish natural resistance associated macrophage protein gene. *Developmental and Comparative Immunology* 26: 517-531.

Chung, S. and C.J. Secombes. 1987. Activation of rainbow trout macrophages. *Journal of Fish Biology* 31: 51-56.

Ciechanover, A., A.L. Schwartz, A. Duatry-Varsat and H.F. Lodish. 1983. Kinetics of internalization and recycling of transferrin and the transferrin receptor in a human hepatoma cell line. *Journal of Biological Chemistry* 258: 9681-9689.

Colditz, I.G., K.G. Altmann and D.L. Watson. 1992. Intradermal and percutaneous transudation of IgG1 and transferrin in sheep. *Immunology and Cell Biology* 70: 323-327.

Dalmo, R.A., J. Bogwald, K. Ingebrigtsen and R. Seljelid. 1996. The immunomodulatory effect of laminaran (beta(1,3)-D-glucan) on Atlantic salmon, *Salmo salar* L., anterior kidney leucocytes after intraperitoneal, peroral and peranal administration. *Journal of Fish Diseases* 19: 449-457.

Dautry-Varsat, A., A. Ciechanover and H.F. Lodish. 1983. pH and the recycling of transferrin during receptor-mediated endocytosis. *Proceedings of the National Academy of Sciences of the United States of America* 80: 2258-2262.

Diks, S.H., S.J. van Deventer and M.P. Peppelenbosch. 2001. Lipopolysaccharide recognition, internalisation, signalling and other cellular effects. *Journal of Endotoxin Research* 7: 335-348.

Dinauer, M.C. 1993. The respiratory burst oxidase and the molecular genetics of chronic granulomatous disease. *Critical Reviews in Clinical Laboratory Sciences* 30: 329-369.

Dorschner, M.O. and R.B. Philips. 1999. Comparative analysis of two NRAMP loci from rainbow trout. *DNA and Cell Biology* 18: 573-583.

Drapier, J.C., H. Hirling, J. Wietzerbin, P. Kaldy and L.C. Kuhn. 1993. Biosynthesis of nitric oxide activates iron regulatory factor in macrophages. *EMBO Journal* 12: 3643-3649.

Drickamer, K. and M.E. Taylor. 1993. Biology of animal lectins. *Annual Review of Cellular Biology* 9: 237-264.

Enane, N.A., K. Frenkel, J.M. O' Connor, K.S. Squibb and J.T. Zelikoff. 1993. Biological markers of macrophage activation: Applications for fish phagocytes. *Immunology* 80: 68-72.

Endemann, G., L.W. Stanton, K.S. Madden, C.M. Bryant, R.T. White and A.A. Protter. 1993. CD36 is a receptor for oxidized low-density lipoprotein. *Journal of Biological Chemistry* 268: 11811-11816.

Engelsma, M.Y., R.J.M. Stet, H. Schipper and B. Verburg-van Kemenade. 2001. Regulation of interleukin-1 beta RNA expression in the common carp. *Developmental and Comparative Immunology* 25: 195-203.

Engelsma, M.Y., R.J.M. Stet, J.P.J. Saeij and B.M.L. Verburg-van Kemenade. 2003. Differential expression and haplotypic variation of two interleukin-1 genes in the common carp (*Cyprinus carpio* L.). *Cytokine.* 15: 21-32.

Evans, W.H., S.M. Wilson, J.M. Bednarek, E.A. Peterson, R.D. Knight, M.G. Mage and L. McHugh. 1989. Evidence for a factor in normal human serum that induces human neutrophilic granulocyte end-stage maturation *in vitro. Leukemia Research* 13: 673-682.

Ezekowitz, R.A., K. Sastry, P. Bailly and A. Warner. 1990. Molecular characterization of the human macrophage mannose receptor: Demonstration of multiple carbohydrate recognition-like domains and phagocytosis of yeasts in Cos-1 cells. *Journal of Experimental Medicine* 172: 1785-1794.

Ferrante, A. 1992. Activation of neutrophils by interleukins-1 and -2 and tumor necrosis factors. *Immunology Series* 57: 417-436.

Forbes, J.R. and P. Gros. 2001. Divalent-metal transport by NRAMP proteins at the interface of host-pathogen interactions. *Trends in Microbiology* 9: 397-403.

Ford, M.J. 2001. Molecular evolution of transferrin: Evidence for positive selection in salmonids. *Molecular Biology and Evolution* 18: 639-647.

Francis, C.H. and A. Ellis. 1994. Production of a lymphokine (macrophage activating factor) by salmon (*Salmo salar*) leukocytes stimulated with outer membrane protein antigens of *Aeromonas salmonicida. Fish and Shellfish Immunology* 4: 489-497.

Fraser, I.P., H. Koziel and R.A. Ezekowitz. 1998. The serum mannose-binding protein and the macrophage mannose receptor are pattern recognition molecules that link innate and adaptive immunity. *Seminars in Immunology* 10: 363-372.

Fujiki, K., D. Shin, M. Nakao and T. Yano. 2000. Molecular cloning and expression analysis of carp (*Cyprinus carpio*) interleukin-1 beta, high affinity immunoglobulin E Fc receptor gamma subunit and serum amyloid A. *Fish and Shellfish Immunology* 10: 229-242.

Garcia-Castillo, J., P. Pelegrin, V. Mulero and J. Meseguer. 2002. Molecular cloning and expression analysis of tumor necrosis factor α from a marine fish reveal its constitutive expression and ubiquitous nature. *Immunogenetics* 54: 200-207.

Garcia-Castillo, J., E. Chaves-Pozo, P. Olivares, P. Pelegrin, J. Meseguer and V. Mulero. 2004. The tumor necrosis factor a of the bony fish seabream exhibits the *in vivo* pro-inflammatory and proliferative activities of its mammalian counterparts, yet it functions in a species-specific manner. *Cell and Molecular Life Science* 61: 1331-1340.

Govoni, G., S. Vidal, M. Cellier, P. Lepage, D. Malo and P. Gros. 1995. Genomic structure, promoter sequence, and induction of expression of the mouse Nramp1 gene in macrophages. *Genomics* 27: 9-19.

Graham, S. and C.J. Secombes. 1988. The production of a macrophage-activating factor from rainbow trout *Salmo gairdneri* leukocytes. *Immunology* 65: 293-297.

Graham, S. and C.J. Secombes. 1990a. Cellular requirements for lymphokine secretion by rainbow trout *Salmo gairdneri* leukocytes. *Developmental and Comparative Immunology* 14: 59-68.

Graham, S. and C.J. Secombes. 1990b. Do fish lymphocytes secrete interferon-gamma? *Journal of Fish Biology* 36: 563-573.

Green, L.C., D.L. Ruiz, D.A. Wagner, W. Rand, N. Istfan, V.R. Young and S.R. Tannenbaum. 1981a. Nitrate biosynthesis in man. *Proceedings of the National Academy of Sciences of the United States of America* 78: 7764-7768.

Green, L.C., S.R. Tannenbaum and P. Goldman. 1981b. Nitrate synthesis in the germ-free and conventional rat. *Science* 212: 56-58.

Green, S.J., R.M. Crawford, J.T. Hockmeyer, M.S. Meltzer and C.A. Nacy. 1990. *Leishmania major* amastigotes initiate the L-arginine-dependent killing mechanism in IFN-gamma-stimulated macrophages by induction of tumor necrosis factor-alpha. *Journal of Immunology* 145: 4290-4297.

Greenwalt, D.E., R.H. Lipsky, C.F. Ockenhouse, H. Ikeda, N.N. Tandon and G.A. Jamieson. 1992. Membrane glycoprotein CD36: A review of its roles in adherence, signal transduction, and transfusion medicine. *Blood* 80: 1105-1115.

Gruenheid, S., M. Cellier, S. Vidal and P. Gros. 1995. Identification and characterization of a second mouse Nramp gene. *Genomics* 25: 514-525.

Gruenheid, S., F. Canonne-Hergaux, S. Gauthier, D.J. Hackam, S. Grinstein and P. Gros. 1999. The iron transport protein NRAMP2 is an integral membrane glycoprotein that colocalizes with transferrin in recycling endosomes. *Journal of Experimental Medicine* 189: 831-841.

Gumulak-Smith, J., A. Teachman, A.H. Tu, J.W. Simecka, J.R. Lindsey and K. Dybvig. 2001. Variations in the surface proteins and restriction enzyme systems of *Mycoplasma pulmonis* in the respiratory tract of infected rats. *Molecular Microbiology* 40: 1037-1044.

Hackam, D.J., O.D. Rotstein, W. Zhang, S. Gruenheid, P. Gros and S. Grinstein. 1998. Host resistance to intracellular infection: Mutation of natural resistance-associated macrophage protein 1 (Nramp1) impairs phagosomal acidification. *Journal of Experimental Medicine* 188: 351-364.

Hamilton, T.A., P.W. Gray and D.O. Adams. 1984a. Expression of the transferrin receptor on murine peritoneal macrophages is modulated by *in vitro* treatment with interferon gamma. *Cellular Immunology* 89: 478-488.

Hamilton, T.A., J.E. Weiel and D.O. Adams. 1984b. Expression of the transferrin receptor in murine peritoneal macrophages is modulated in the different stages of activation. *Journal of Immunology* 132: 2285-2290.

Hardie, L.J., L.H. Chappell and C.J. Secombes. 1994. Human tumor necrosis factor alpha influences rainbow trout *Oncorhynchus mykiss* leukocyte responses. *Veterinary Immunology and Immunopathology* 40: 73-84.

Hardie, L.J., A.E. Ellis and C.J. Secombes. 1996. *In vitro* activation of rainbow trout macrophages stimulates inhibition of *Renibacterium salmoninarum* growth concomitant with augmented generation of respiratory burst products. *Diseases of Aquatic Organisms* 25: 175-183.

Hart, K. and M. Wilcox. 1993. A *Drosophila* gene encoding an epithelial membrane protein with homology to CD36/LIMP II. *Journal of Molecular Biology* 234: 249-253.

Haworth, R., N. Platt, S. Keshav, D. Hughes, E. Darley, H. Suzuki, Y. Kurihara, T. Kodama and S. Gordon. 1997. The macrophage scavenger receptor type A is expressed by activated macrophages and protects the host against lethal endotoxic shock. *Journal of Experimental Medicine* 186: 1431-1439.

Hayaishi, O., F. Hirata, T. Ohnishi, J.P. Henry, I. Rosenthal and A. Katoh. 1977. Indoleamine 2,3-dioxygenase: incorporation of 18O2—and 18O2 into the reaction products. *Journal of Biological Chemistry* 252: 3548-3550.

Hibbs, J.B., Jr. 1991. Synthesis of nitric oxide from L-arginine: A recently discovered pathway induced by cytokines with antitumour and antimicrobial activity. *Research Immunology* 142: 565-569.

Hirata, T., P.B. Bitterman, J.F. Mornex and R.G. Crystal. 1986. Expression of the transferrin receptor gene during the process of mononuclear phagocyte maturation. *Journal of Immunology* 4: 1339-1345.

Hirono, I., B. Nam, Y. Kurobe and T. Aoki. 2000. Molecular cloning, characterization and expression of tumor necrosis factor (TNF) cDNA and gene from Japanese flounder *Paralichthys olivaceus*. *Journal of Immunology* 165: 4423-4427.

Holmqvist, B. and P. Ekstrom. 1997. Subcellular localization of neuronal nitric oxide synthase in the brain of a teleost: An immunoelectron and confocal microscopical study. *Brain Research* 745: 67-82.

Hong, S., S. Peddie, J.J. Campos-Pérez, J. Zou and C.J. Secombes. 2003. The effect of intraperitoneally administered recombinant IL-1beta on immune parameters and resistance to *Aeromonas salmonicida* in the rainbow trout (*Oncorhynchus mykiss*). *Developmental and Comparative Immunology* 27: 801-812.

Hong, S., J. Zou, M. Crampe, S. Peddie, G. Scapigliati and C.J. Secombes. 2001. The production and bioactivity testing of rainbow trout (*Oncorhynchus mykiss*) recombinant IL-1beta. *Veterinary Immunology and Immunopathology* 81: 1-14.

Hopkins, C.R. and I.S. Trowbridge. 1983. Internalization and processing of transferrin and the transferrin receptor in human carcinoma A431 cells. *Journal of Cellular Biology* 97: 508-521.

Ibrahim, M.A., E. Iwamori, Y. Sugimoto and T. Aoki. 1998. Identification of a distinct antibacterial domain within the N-lobe of ovatransferrin. *Biochemical and Biophysical Acta* 1401: 289-303.

Igawa, D., M. Sakai and R. Savan. 2006. An unexpected discovery of two interferon gamma-like genes along with interleukin (IL)-22 and -26 from teleost: IL-22 and -26 genes have been described for the first time outside mammals. *Molecular Immunology* 43: 999-1009.

Itou, T., T. Iida and H. Kawatsu. 1996. Kinetics of oxygen metabolism during respiratory burst in Japanese eel neutrophils. *Developmental and Comparative Immunology* 20: 323-330.

Itou, T., T. Iida and H. Kawatsu. 1998. Evidence for the existence of cytochrome b558 in fish neutrophils by polyclonal anti-peptide antibody. *Developmental and Comparative Immunology* 22: 433-437.

Iyer, G.Y.N., M.F. Islam and J.H. Quastel. 1961. Biochemical aspects of phagocytosis. *Nature (London)* 192: 535-541.

Jacobs, A. 1977. Low molecular weight intracellular iron transport compounds. *Blood* 50: 433-439.

James, S.L. 1995. Role of nitric oxide in parasitic infections. *Microbiology Reviews* 59: 533-547.

Jamroz, R.C., J.R. Gasdaska, J.Y. Bradfield and J.H. Law. 1993. Transferrin in a cockroach: Molecular cloning, characterization, and suppression by juvenile hormone. *Proceedings of the National Academy of Sciences of the United States of America* 90: 1320-1324.

Jang, S.I., L.J. Hardie and C.J. Secombes. 1994. Effects of transforming growth factor-beta1 on rainbow trout *Oncorhynchus mykiss* macrophage respiratory burst activity. *Developmental and Comparative Immunology* 4: 315-323.

Jang, S.I., L.J. Hardie and C.J. Secombes. 1995a. Elevation of rainbow trout *Oncorhynchus mykiss* macrophage respiratory burst activity with macrophage-derived supernatants. *Journal of Leukocyte Biology* 57: 943-947.

Jang, S.I., V. Mulero, L.J. Hardie and C.J. Secombes. 1995b. Inhibition of rainbow trout phagocyte responsiveness to human tumor necrosis factor alpha (hTNF alpha) with monoclonal antibodies to the hTNF alpha 55 kDa receptor. *Fish and Shellfish Immunology* 5: 61-69.

Jørgensen, J.B. and B. Robertsen. 1995. Yeast beta-glucan stimulates respiratory burst activity of Atlantic salmon (*Salmo salar* L.) macrophages. *Developmental and Comparative Immunology* 19: 43-57.

Kent, S., E.D. Weinberg and P. Stuart-Macadam. 1994. The etiology of the anemia of chronic disease and infection. *Journal of Clinical Epidemiology* 47: 23-33.

Klausner, R.D., G. Ashwell, J.B. Harford and K.R. Bridges. 1983a. Binding of apotransferrin to K562 cells: Explanation of the transferrin cycle. *Proceedings of the National Academy of Sciences of the United States of America* 80: 2263-2266.

Klausner, R.D., J. van Renswoude, G. Ashwell, C. Kempf, A.N. Schechter, A. Dean and K.R. Bridges. 1983b. Receptor-mediated endocytosis of transferrin in K562 cells. *Journal of Biological Chemistry* 258: 4715-4724.

Kraal, G., L.J. van der Laan, O. Elomaa and K. Tryggvason. 2000. The macrophage receptor MARCO. *Microbes and Infection* 2: 313-316.

Krieger, M. and J. Herz. 1994. Structures and functions of multiligand lipoprotein receptors: Macrophage scavenger receptors and LDL receptor-related protein (LRP). *Annual Review of Biochemistry* 63: 601-637.

Laing, K.J., P.S. Grabowski, M. Belosevic and C.J. Secombes. 1996. A partial sequence for nitric oxide synthase from a goldfish (*Carassius auratus*) macrophage cell line. *Immunology and Cell Biology* 74: 374-379.

Laing, K.J., L.J. Hardie, W. Aartsen, P.S. Grabowski and C.J. Secombes. 1999. Expression of an inducible nitric oxide synthase gene in rainbow trout *Oncorhynchus mykiss*. *Developmental and Comparative Immunology* 23: 71-85.

Laing, K.J., T. Wang, J.J. Zou, J. Holland, I. Hirono, T. Aoki and C.J. Secombes. 2000. Cloning and expression analysis of *Oncorhynchus mykiss* tumor necrosis factor-alpha. *European Journal of Biochemistry* 268: 1315-1322.

Landmann, R., B. Muller and W. Zimmerli. 2000. CD14, new aspects of ligand and signal diversity. *Microbes and Infection* 2: 295-304.

Lee, G.R. 1983. The anemia of chronic disease. *Seminars in Hematology* 20: 61-80.

Lee, P.L., T. Gelbart, C. West, C. Halloran and E. Beutler. 1998. The human Nramp2 gene: Characterization of the gene structure, alternative splicing, promoter region and polymorphisms. *Blood Cells Molecules and Diseases* 24: 199-215.

Lemaitre, B., E. Nicolas, L. Michaut, J.M. Reichhart and J.A. Hoffmann. 1996. The dorsoventral regulatory gene cassette spatzle/Toll/cactus controls the potent antifungal response in *Drosophila* adults. *Cell* 86: 973-983.

Linehan, S.A., L. Martinez-Pomares and S. Gordon. 2000. Macrophage lectins in host defence. *Microbes and Infection* 2: 279-288.

MacKenzie, C.R., R. Langen, O. Takikawa and W. Daubener. 1999. Inhibition of indoleamine 2,3-dioxygenase in human macrophages inhibits interferon-gamma-induced bacteriostasis but does not abrogate toxoplasmastasis. *European Journal of Immunology* 29: 3254-3261.

MacKenzie, S., J.V. Planas and F.W. Goetz. 2003. LPS-stimulated expression of tumor necrosis factor-alpha mRNA in primary monocytes and *in vitro* differentiated macrophages. *Developmental and Comparative Immunology* 27: 393-400.

Malo, D., K. Vogan, S. Vidal, J. Hu, M. Cellier, E. Schurr, A. Fuks, N. Bumstead, K. Morgan and P. Gros. 1994. Haplotype mapping and sequence analysis of the mouse Nramp gene predict susceptibility to infection with intracellular parasites. *Genomics* 23: 51-61.

Marsden, M.J. and C.J. Secombes. 1997. The influence of vaccine preparations on the induction of antigen specific responsiveness in rainbow trout *Oncorhynchus mykiss*. *Fish and Shellfish Immunology* 7: 455-469.

Marsden, M.J., D. Cox and C.J. Secombes. 1994. Antigen-induced release of macrophage activating factor from rainbow trout *Oncorhynchus mykiss* leucocytes. *Veterinary Immunology and Immunopathology* 42: 199-208.

Martin, A.W., E. Huebers, H. Huebers, J. Webb and C.A. Finch. 1984. A mon-sited transferrin from a representative deuterostome: The ascidian *Pyura stolonifera* (subphylum Urochordata). *Blood* 64: 1047-1052.

Medzhitov, R. 2001. Toll-like receptors and innate immunity. *Nature Reviews Immunology* 1: 135-145.

Medzhitov, R. and C.A.J. Janeway. 1997a. Innate immunity: impact on the adaptive immune response. *Current Opinion in Immunology* 9: 4-9.

Medzhitov, R. and C.A.J. Janeway. 1997b. Innate immunity: The virtues of a nonclonal system of recognition. *Cell* 91: 295-298.

Medzhitov, R. and C.A.J. Janeway. 2000. Innate immune recognition: Mechanisms and pathways. *Immunological Reviews* 173: 89-97.

Melillo, G., G.W. Cox, D. Radzioch and L. Varesio. 1993. Picolinic acid, a catabolite of L-tryptophan, is a costimulus for the induction of reactive nitrogen intermediate production in murine macrophages. *Journal of Immunology* 150: 4031-4040.

Melillo, G., T. Musso, A. Sica, L.S. Taylor, G.W. Cox and L. Varesio. 1995. A hypoxia-responsive element mediates a novel pathway of activation of the inducible nitric oxide synthase promoter. *Journal of Experimental Medicine* 182: 1683-1693.

Milev-Milovanovic, I., S. Long, M. Wilson, E. Bengten, N.W. Miller and V.G. Chinchar. 2006. Identification and expression analysis of interferon gamma genes in channel catfish. *Immunogenetics* 58: 70-80

Mulero, V. and J. Meseguer. 1998. Functional characterization of a macrophage-activating factor produced by leucocytes of gilthead sea bream (*Sparus aurata* L.). *Fish and Shellfish Immunology* 8: 143-156.

Munoz, J., M.A. Esteban and J. Meseguer. 1999. *In vitro* culture requirements of sea bass (*Dicentrarchus labrax* L.) blood cells: Differential adhesion and phase contrast microscopic study. *Fish and Shellfish Immunology* 9: 417-428.

Munoz, P., J.A. Calduch-Giner, A. Sitja-Bobadilla, P. Alvarez-Pellitero and J. Perez-Sanchez. 1998. Modulation of the respiratory burst activity of Mediterranean sea bass (*Dicentrarchus labrax* L.) phagocytes by growth hormone and parasitic status. *Fish and Shellfish Immunology* 8: 25-36.

Nakagawara, A., N.M. DeSantis, N. Nogueira and C.F. Nathan. 1982. Lymphokines enhance the capacity of human monocytes to secret reactive oxygen intermediates. *Journal of Clinical Investigation* 70: 1042-1048.

Nathan, C.F., H.W. Murray, M.E. Wiebe and B.Y. Rubin. 1983. Identification of interferon-gamma as the lymphokine that activates human macrophage oxidative metabolism and antimicrobial activity. *Journal of Experimental Medicine* 158: 670-689.

Neumann, N.F. 1999. *Regulation of antimicrobial mechanisms of macrophages of the goldfish* (*Carassius auratus*). Ph.D. Thesis. University of Alberta, Edmonton, Alberta, Canada.

Neumann, N.F. and M. Belosevic. 1996. Deactivation of primed respiratory burst response of goldfish macrophages by leukocyte-derived macrophage activating factor(s). *Developmental and Comparative Immunology* 20: 427-439.

Neumann, N.F., D. Fagan and M. Belosevic. 1995. Macrophage activating factor(s) secreted by mitogen stimulated goldfish kidney leukocytes synergize with bacterial lipopolysaccharide to induce nitric oxide production in teleost macrophages. *Developmental and Comparative Immunology* 19: 473-482.

Neumann, N.F., D. Barreda and M. Belosevic. 1998. Production of a macrophage growth factor(s) by a goldfish macrophage cell line and macrophages derived from goldfish kidney leukocytes. *Developmental and Comparative Immunology* 4: 417-432.

Neumann, N.F., D. Barreda and M. Belosevic. 2000a. Generation and functional analysis of distinct macrophage sub-populations from goldfish (*Carassius auratus* L.) kidney leukocyte cultures. *Fish and Shellfish Immunology* 10: 1-20.

Neumann, N.F., J.L. Stafford and M. Belosevic. 2000b. Biochemical and functional characterization of macrophage stimulating factors secreted by mitogen-induced goldfish kidney leukocytes. *Fish and Shellfish Immunology* 10: 167-186.

Nickel, R., C. Ott, T. Dandekar and M. Leippe. 1999. Pore-forming peptides of *Entamoeba dispar*. Similarity and divergence to amoebapores in structure, expression and activity. *European Journal of Biochemistry* 265: 1002-1007.

Nikoskelainen, S., O. Kjellsen, E.M. Lilius and M.B. Schrøder. 2006. Respiratory burst activity of Atlantic cod (*Gadus morhua* L.) blood phagocytes differs markedly from that of rainbow trout. *Fish and Shellfish Immunology* 21: 199-208

Novoa, B., A. Figueras, I. Ashton and C.J. Secombes. 1996. *In vitro* studies on the regulation of rainbow trout (*Oncorhynchus mykiss*) macrophage respiratory burst activity. *Developmental and Comparative Immunology* 20: 207-216.

Nunez, M.T., V. Gaete, J.A. Watkins and J. Glass. 1990. Mobilization of iron from endocytic vesicles. The effects of acidification and reduction. *Journal of Biological Chemistry* 265: 6688-6692.

Okamura, Y., M. Watari, E.S. Jerud, D.W. Young, S.T. Ishizaka, J. Rose, J.C. Chow and I.J.F. Strauss. 2001. The extra domain A of fibronectin activates Toll-like receptor 4. *Journal of Biological Chemistry* 276: 10229-10233.

Ordas, M.C., M.M. Costa, F.J. Roca, G. Lopez-Castejon, V. Mulero, J. Meseguer, A. Figueras and B. Novoa. 2006. Turbot TNFα gene: Molecular characterization and biological activity of the recombinant protein. *Molecular Immunology* 44: 389-400.

Oshiumi, H., T. Tsujita, K. Shida, M. Matsumoto, K. Ikeo and T. Seya. 2003. Prediction of the prototype of the human Toll-like receptor gene family from the pufferfish, *Fugu rubripes*, genome. *Immunogenetics* 54: 791-800.

Overath, P., J. Haag, M.G. Mameza and A. Lischke. 1999. Freshwater fish trypanosomes: Definition of two types, host control by antibodies and lack of antigenic variation. *Parasitology* 119: 591-601.

Ozaki, Y., M.P. Edelstein and D.S. Duch. 1987. The actions of interferon and anti-inflammatory agents of induction of indoleamine 2,3-dioxygenase in human peripheral blood monocytes. *Biochemical and Biophysical Research Communications* 144: 1147-1153.

Park, C.I., T. Kurobe, I. Hirono and T. Aoki. 2003. Cloning and characterization of cDNAs for two distinct tumor necrosis factor receptor superfamily genes from Japanese flounder *Paralichthys olivaceus*. *Developmental and Comparative Immunology* 27: 365-375.

Payne, N.R. and M.A. Horwitz. 1987. Phagocytosis of *Legionella pneumophila* is mediated by human monocyte complement receptors. *Journal of Experimental Medicine* 166: 1377-1389.

Pearson, A.M. 1996. Scavenger receptors in innate immunity. *Current Opinion in Cell Biology* 8: 20-28.

Peddie, S., J. Zou, C. Cunningham and C.J. Secombes. 2001. Rainbow trout (*Oncorhynchus mykiss*) recombinant IL-1 beta and derived peptides induce migration of head-kidney leukocytes *in vitro*. *Fish and Shellfish Immunology* 11: 697-709.

Pelegrin, P., J. Garcia-Castillo, V. Mulero and J. Meseguer. 2001. Interleukin-1 isolated from a marine fish reveals up-regulated expression in macrophages following activation with lipopolysaccharide and lymphokines. *Cytokine* 16: 67-72.

Pfefferkorn, E.R. 1984. Interferon gamma blocks the growth of *Toxoplasma gondii* in human fibroblasts by inducing the host cells to degrade tryptophan. *Proceedings of the National Academy of Sciences of the United States of America* 81: 908-912.

Picard, V., G. Govoni, N. Jabado and P. Gros. 2000. Nramp 2 (DCT1/DMT1) expressed at the plasma membrane transports iron and other divalent cations into a calcein-accessible cytoplasmic pool. *Journal of Biological Chemistry* 275: 35738-35745.

Prabkaran, N., C. Binuramesh, D. Steinhagen and R.D. Michael. 2006. Immune response and disease resistance of *Oreochromis mossambicus* to *Aeromonas hydrophila* after exposure to hexavalent chromium. *Diseases of Aquatic Organisms* 68: 189-196.

Pugin, J., C.C. Schurer-Maly, D. Leturcq, A. Moriarty, R.J. Ulevitch and P.S. Tobias. 1993. Lipopolysaccharide activation of human endothelial and epithelial cells is mediated by lipopolysaccharide-binding protein and soluble CD14. *Proceedings of the National Academy of Sciences of the United States of America* 90: 2744-2748.

Qin, Q.W., M. Ototake, K. Noguchi, G. Soma, Y. Yokomizo and T. Nakanishi. 2001. Tumor necrosis factor alpha (TNFalpha)-like factor produced by macrophages in rainbow trout, *Oncorhynchus mykiss*. *Fish and Shellfish Immunology* 11: 245-256.

Raijmakers, P.G., A.B. Groeneveld, J.A. Rauwerda, G.J. Teule and C.E. Hack. 1997. Acute lung injury after aortic surgery: the relation between lung and leg microvascular permeability to 111indium-labelled transferrin and circulating mediators. *Thorax* 52: 866-871.

Reif, D.W. and R.D. Simmons. 1990. Nitric oxide mediates iron release from ferritin. *Archives of Biochemistry and Biophysics* 283: 537-541.

Retzer, M.D., A. Kabani, L.L. Button, R.H. Yu and A.B. Schryvers. 1996. Production and characterization of chimeric transferrins for the determination of the binding domains for bacterial transferrin receptors. *Journal of Biological Chemistry* 271: 1166-1173.

Rigotti, A., S.L. Acton and M. Krieger. 1995. The class B scavenger receptors SR-BI and CD36 are receptors for anionic phospholipids. *Journal of Biological Chemistry* 270: 16221-16224.

Robertsen, B. 1999. Modulation of the non-specific defence of fish by structurally conserved microbial polymers. *Fish and Shellfish Immunology* 9: 269-290.

Ross, G.D. 1989. Complement and complement receptors. *Current Opinion in Immunology* 2: 50-62.

Rossi, M.C. and B.R. Zetter. 1992. Selective stimulation of prostatic carcinoma cell proliferation by transferrin. *Proceedings of the National Academy of Sciences of the United States of America* 89: 6197-6201.

Saeij, J.P. and G.F. Wiegertjes. 1999. Identification and characterization of a fish natural resistance-associated macrophage protein (NRAMP) cDNA. *Immunogenetics* 50: 60-66.

Saeij, J.P., R.J. Stet, A. Groeneveld, L.B. Verburg-van Kemenade, W.B. van Muiswinkel and G.F. Wiegertjes. 2000. Molecular and functional chracterization of a fish inducible-type nitric oxide synthase. *Immunogenetics* 51: 339-346.

Saeij, J.P.J., R.J.M. Stet, B.J. de Vries, W.B. van Muiswinkel and G.F. Wiegertjes. 2003. Molecular and functional characterization of carp TNF: A link between TNF polymorphism and trypanotolerance? *Developmental and Comparative Immunology* 27: 29-41.

Sakamoto, H., N. Sakamoto, M. Oryu, T. Kobayashi, Y. Ogawa, M. Ueno and M. Shinnou. 1997. A novel function of transferrin as a constituent of macromolecular activators of phagocytosis from platelets and their precursors. *Biochemical and Biophysical Research Communications* 230: 270-274.

Sample, A.K. and C.J. Czuprynski. 1994. Bovine neutrophil chemiluminescence is preferentially stimulated by homologous IL-1, but inhibited by the human IL-1 receptor antagonist. *Veterinary Immunology and Immunopathology* 41: 165-172.

Scapigliati, G., F. Buonocore, S. Bird, J. Zou, P. Pelegrin, C. Falasca, D. Prugnoli and C.J. Secombes. 2001. Phylogeny of cytokines: molecular cloning and expression analysis of sea bass *Dicentrarchus labrax* interleukin-1 beta. *Fish and Shellfish Immunology* 11: 711-726.

Schlesinger, L.S. and M.A. Horwitz. 1991. Phagocytosis of *Mycobacterium leprae* by human monocyte-derived macrophages is mediated by complement receptors CR1 (CD35), CR3 (CD11b/CD18), and CR4 (CD11c/CD18) and IFN-gamma activation inhibits complement receptor function and phagocytosis of this bacterium. *Journal of Immunology* 147: 1983-1994.

Schoor, W.P. and J.A. Plum. 1994. Induction of nitric oxide synthase in channel catfish (*Ictalurus punctatus*) by *Edwardsiella ictaluri*. *Diseases of Aquatic Organisms* 19: 153-155.

Secombes, C.J., L.M. Laird and I.G. Priede. 1987. Lymphokine-release from rainbow trout leucocytes stimulated with concanavalin A. Effects upon macrophage spreading and adherence. *Developmental and Comparative Immunology* 11: 513-520.

Secombes, C.J., A.R. Cross, G.J. Sharp and R. Garcia. 1992. NADPH oxidase-like activity in rainbow trout *Oncorhynchus mykiss* (Walbaum) macrophages. *Developmental and Comparative Immunology* 16: 405-413.

Secombes, C.J., S. Bird, C. Cunningham and J. Zou. 1999. Interleukin-1 in fish. *Fish and Shellfish Immunology* 9: 335-343.

Segal, A.W. 1996. The NADPH oxidase and chronic granulomatous disease. *Molecular Medicine Today* 2: 129-135.

Shin, D.H., H.S. Lim, S.K. Cho, H.Y. Lee, H.W. Lee, K.H. Lee, Y.H. Chung, S.S. Cho, C. Ik Cha and D.H. Hwang. 2000. Immunocytochemical localization of neuronal and inducible nitric oxide synthase in the retina of zebrafish, *Brachydanio rerio*. *Neuroscience Letters* 292: 220-222.

Smiley, S.T., J.A. King and W.W. Hancock. 2001. Fibrinogen stimulates macrophage chemokine secretion through Toll-like receptor 4. *Journal of Immunology* 167: 2887-2894.

Solem, S.T., J.B. Jorgensen and B. Robertsen. 1995. Stimulation of respiratory burst and phagocytic activity in Atlantic salmon (*Salmo salar* L.) macrophages by lipopolysaccharide. *Fish and Shellfish Immunology* 5: 475-491.

Stafford, J.L. and M. Belosevic. 2003. Transferrin and the innate immune response of fish: Identification of a novel mechanism of macrophage activation. *Developmental and Comparative Immunology* 27: 539-554.

Stafford, J.L., N.F. Neumann and M. Belosevic. 1999. Inhibition of macrophage activity by mitogen-induced goldfish leukocyte deactivating factor. *Developmental and Comparative Immunology* 23: 585-596.

Stafford, J.L., N.F. Neumann and M. Belosevic. 2001. Products of proteolytic cleavage of transferrin induce nitric oxide response of goldfish macrophages. *Developmental and Comparative Immunology* 25: 101-115.

Stafford, J.L., P.E. McLauchlan, C.J. Secombes, A. Ellis and M. Belosevic. 2001. Generation of primary monocyte-like cultures from rainbow trout head kidney leukocytes. *Developmental and Comparative Immunology* 25: 447-459.

Stafford, J.L., K.K. Ellestad, K.E. Magor, M. Belosevic and B.G. Magor. 2003. A Toll-like receptor (TLR) gene that is upregulated in activated goldfish macrophages. *Developmental and Comparative Immunology* 27: 685-698.

Stuehr, D.J. and M.A. Marletta. 1985. Mammalian nitrate biosynthesis: Mouse macrophages produce nitrite and nitrate in response to *Escherichia coli* lipopolysaccharide. *Proceedings of the National Academy of Sciences of the United States of America* 82: 7738-7742.

Sun, H., B.N. Bristow, G. Qu and S.A. Wasserman. 2002. A heterotrimeric death domain complex in Toll signaling. *Proceedings of the National Academy of Sciences of the United States of America* 99: 12871-12876.

Tafalla, C. and B. Novoa. 2000. Requirements for nitric oxide production by turbot (*Scophthalmus maximus*) head kidney macrophages. *Developmental and Comparative Immunology* 24: 623-631.

Tafalla, C., B. Novoa and A. Figueras. 1999. Suppressive effect of turbot (*Scophthalmus maximus*) leukocyte-derived supernatants on macrophage and lymphocyte functions. *Fish and Shellfish Immunology* 9: 157-166.

Tafalla, C., J. Coll and C.J. Secombes. 2005. Expression of genes related to the early immune response in rainbow trout (*Oncorhynchus mykiss*) after viral haemorrhagic septicemia virus (VHSV) infection. *Developmental and Comparative Immunology* 29: 615-626.

Tahir, A. and C.J. Secombes. 1996. Modulation of dab (*Limanda limanda* L.) macrophage respiratory burst activity. *Fish and Shellfish Immunology* 6: 135-146.

Takenaka, K.V., N. Sakai, S. Murase, T. Kuroda, A. Okumura and M. Sawada. 2000. Elevated transferrin concentration in cerebral spinal fluid after subarachnoid hemorrhage. *Neurological Research* 22: 797-801.

Tang, S., J.C. Leung, A.W. Tsang, H.Y. Lan, T.M. Chan and K.N. Lai. 2002. Transferrin up-regulates chemokine synthesis by human proximal tubular epithelial cells: Implication on mechanism of tubuloglomerular communication in glomerulopathic proteinura. *Kidney International* 61: 1655-1665.

Taub, D.D., K. Conlon, A.R. Lloyd, J.J. Oppenheim and D.J. Kelvin. 1993. Preferential migration of activated CD4+ and CD8+ T cells in response to MIP-1 alpha and MIP-1 beta. *Science* 260: 355-358.

Taylor, M.J. and D. Hoole. 1995. The chemiluminescence of cyprinid leucocytes in response to zymosan and extracts of *Ligula intestinalis* (Cestoda). *Fish and Shellfish Immunology* 5: 191-198.

Taylor, M.W. and G.S. Feng. 1991. Relationship between interferon-gamma, indoleamine 2,3-dioxygenase, and tryptophan catabolism. *FASEB Journal* 5: 2516-2522.

Tobias, P.S., J. Mathison, D. Mintz, J.D. Lee, V. Kravchenko, K. Kato, J. Pugin and R.J. Ulevitch. 1992. Participation of lipopolysaccharide-binding protein in lipopolysaccharide-dependent macrophage activation. *American Journal of Respiratory Cell and Molecular Biology* 7: 239-245.

Torreilles, J. and M. Guerin. 1999. Production of peroxynitrite by zymosan stimulation of *Mytilus galloprovincialis* haemocytes *in vitro*. *Fish and Shellfish Immunology* 9: 509-518.

Ulevitch, R.J. and P.S. Tobias. 1995. Receptor-dependent mechanisms of cell stimulation by bacterial endotoxin. *Annual Review of Immunology* 13: 437-457.

Vaughan, R.B. 1965. The romantic rationalist: a study of Elie Metchnikoff. *Medical History* 9: 201-215.

Vega, M.A., B. Segui-Real, J.A. Garcia, C. Cales, F. Rodriguez, J. Vanderkerckhove and I.V. Sandoval. 1991. Cloning, sequencing, and expression of a cDNA encoding rat LIMP II, a novel 74-kDa lysosomal membrane protein related to the surface adhesion protein CD36. *Journal of Biological Chemistry* 266: 16818-16824.

Vidal, S.M., E. Pinner, P. Lepage, S. Gauthier and P. Gros. 1996. Natural resistance to intracellular infections: Nramp1 encodes a membrane phosphoglycoprotein absent in macrophages from susceptible (Nramp1 D169) mouse strains. *Journal of Immunology* 157: 3559-3568.

Wagner, D.A., V.R. Young and S.R. Tannenbaum. 1983. Mammalian nitrate biosynthesis: incorporation of 15NH3 into nitrate is enhanced by endotoxin treatment. *Proceedings of the National Academy of Sciences of the United States of America* 80: 4518-4521.

Wang, L.G., X.M. Liu, H. Wikiel and A. Bloch. 1995a. Activation of casein kinase II in ML-1 human myeloblastic leukemia cells requires IGF-1 and transferrin. *Journal of Leukocyte Biology* 57: 332-334.

Wang, R., N.F. Neumann, Q. Shen and M. Belosevic. 1995b. Establishment and characterization of macrophage cell line from the goldfish. *Fish and Shellfish Immunology* 5: 329-345.

Wang, T., M. Ward, P.S. Grabowski and C.J. Secombes. 2001. Molecular cloning, gene organization and expression of rainbow trout (*Oncorhynchus mykiss*) inducible nitric oxide synthase (iNOS) gene. *Biochemical Journal* 358: 747-755.

Wardrop, S.L. and D.R. Richardson. 2000. Interferon-gamma and lipopolysaccharide regulate the expression of Nramp2 and increase the uptake of iron from low relative molecular mass complexes by macrophages. *European Journal of Biochemistry* 267: 6586-6593.

Watanuki, H., K. Ota, A.C.M.A.R. Tassaka, T. Kato and M. Sakai. 2006. Immunostimulant effects of dietary *Spirulina platensis* on carp *Cyprinus carpio*. *Aquaculture* 258: 157-163.

Waterstrat, P.R., A.J. Ainsworth and G. Capley. 1991. *In vitro* responses of channel catfish, *Ictalurus punctatus*, neutrophils to *Edwardsiella ictaluri*. *Developmental and Comparative Immunology* 15: 53-63.

Weinberg, E.D. 1984. Iron withholding: A defense against infection and neoplasia. *Physiological Reviews* 64: 65-102.

Weinzimer, S.A., T.B. Gibson, P.F. Collett-Solberg, A. Khare, B. Liu and P. Cohen. 2001. Transferrin is an insulin-like growth factor-binding protein-3 binding protein. *Journal of Clinical Endocrinology and Metabolism* 86: 1806-1813.

Weis, W.I., M.E. Taylor and K. Drickamer. 1998. The C-type lectin superfamily in the immune system. *Immunological Reviews* 163: 19-34.

Weiss, G., B. Goossen, W. Doppler, D. Fuchs, K. Pantopoulos, G. Werner-Felmayer, H. Wachter and M.W. Hentze. 1993. Translational regulation via iron-responsive elements by the nitric oxide/NO-synthase pathway. *EMBO Journal* 12: 3651-3657.

Weiss, G., G. Werner-Felmayer, E.R. Werner, K. Grunewald, H. Wachter and M.W. Hentze. 1994. Iron regulates nitric oxide synthase activity by controlling nuclear transcription. *Journal of Experimental Medicine* 180: 969-976.

Weiss, G., S. Kastner, J. Brock, J. Thaler and K. Grunewald. 1997. Modulation of transferrin receptor expression by dexrazoxane (ICRF-187) via activation of iron regulatory protein. *Biochemical Pharmacology* 53: 1419-1424.

Weiss, G., C. Murr, H. Zoller, M. Haun, B. Widner, C. Ludescher and D. Fuchs. 1999. Modulation of neopterin formation and tryptophan degradation by Th1- and Th2-derived cytokines in human monocytic cells. *Clinical and Experimental Immunology* 116: 435-440.

Worwood, M. 1989. An overview of iron metabolism at a molecular level. *Journal of Internal Medicine* 226: 381-391.

Wright, S.D. 1995. CD14 and innate recognition of bacteria. *Journal of Immunology* 155: 6-8.

Wright, S.D., P.E. Rao, W.C. Van Voorhis, L.S. Craigmyle, K. Iida, M.A. Talle, E.F. Westberg, G. Goldstein and S.C. Silverstein. 1983. Identification of the C3bi receptor of human monocytes and macrophages by using monoclonal antibodies. *Proceedings of the National Academy of Sciences of the United States of America* 80: 5699-5703.

Xie, H., N.C. Rath, G.R. Huff, J.M. Balog and W.E. Huff. 2001. Inflammation-induced changes in serum modulate chicken macrophage function. *Veterinary Immunology and Immunopathology* 80: 225-235.

Xie, H., G.R. Huff, W.E. Huff, J.M. Balog and N.C. Rath. 2002. Effects of ovotransferrin on chicken macrophages and heterophil-granulocytes. *Developmental and Comparative Immunology* 26: 805-815.

Yagisawa, M., A. Yuo, S. Kitagawa, Y. Yazaki, A. Togawa and F. Takaku. 1995. Stimulation and priming of human neutrophils by IL-1 alpha and IL-1 beta: complete inhibition by IL-1 receptor antagonist and no interaction with other cytokines. *Experimental Hematology* 23: 603-608.

Yin, Z., T.J. Lam and Y.M. Sin. 1997. Cytokine-mediated antimicrobial immune response of catfish, *Clarias gariepinus*, as a defence against *Aeromonas hydrophila*. *Fish and Shellfish Immunology* 7: 93-104.

Yoshiga, T., V.P. Hernandez, A.M. Fallon and J.H. Law. 1997. Mosquito transferrin, an acute-phase protein that is upregulated upon infection. *Proceedings of the National Academy of Sciences of the United States of America* 94: 12337-12342.

Zhu, F.G. and D.S. Pisetsky. 2001. Role of the heat shock protein 90 in immune response stimulation by bacterial DNA and synthetic oligonucleotides. *Infection and Immunity* 69: 5546-5552.

Zou, J., C. Cunningham and C.J. Secombes. 1999a. The rainbow trout *Oncorhynchus mykiss* interleukin-1beta gene has a different organization to mammals and undergoes incomplete splicing. *European Journal of Biochemistry* 259: 901-908.

Zou, J., P.S. Grabowski, C. Cunningham and C.J. Secombes. 1999b. Molecular cloning of interleukin 1beta from rainbow trout *Oncorhynchus mykiss* reveals no evidence of an ice cut site. *Cytokine* 11: 552-560.

Zou, J., T. Wang, S. Hong, C. Cunningham and C.J. Secombes. 2000. Role of interleukin-1 beta in immune responses of rainbow trout. *Fish and Shellfish Immunology* 10: 289-299.

Zou, J., T. Wang, I. Hirono, T. Aoki, H. Inagawa, T. Honda, G.I. Soma, M. Ototake, T. Nakanishi, A.E. Ellis and C.J. Secombes. 2002. Differential expression of two tumor necrosis factor genes in rainbow trout, *Oncorhynchus mykiss*. *Developmental and Comparative Immunology* 26: 161-172.

Zou, J., C.J. Secombes, S. Long, N. Miller, L.W. Clem and V.G. Chinchar. 2003a. Molecular identification and expression analysis of tumor necrosis factor in channel catfish (*Ictalurus punctatus*). *Developmental and Comparative Immunology* 27: 845-858.

Zou, J., S. Peddie, G. Scapigliati, Y. Zhang, N.C. Bols, A.E. Ellis and C.J. Secombes. 2003b. Functional characterization of the recombinant tumor necrosis factors in rainbow trout, *Oncorhynchus mykiss*. *Developmental and Comparative Immunology* 27: 813-822.

Zou, J., Y. Yoshiura, J.M. Dijkstra, M. Sakai, M. Ototake and C.J. Secombes. 2004. Identification of an interferon gamma homologue in Fugu, *Takifugu rubripes*. *Fish and Shellfish Immunology* 17: 403-409.

CHAPTER

6

Immune Defence Mechanisms in the Sea Bass *Dicentrarchus labrax* L.

Francesco Buonocore and Giuseppe Scapigliati*

INTRODUCTION

SEA BASS AS A FISH MODEL

Immunological studies have been mostly performed in mammals and this could be taken as a basis for studies in other vertebrates. However, striking differences exist in manipulating homeothermic and ectothermic animals, for instance, mice and teleosts. Immunological studies in mice take advantage of characteristics such as: (i) better tolerance of external temperature ranges and a constant internal environment for studying substances administered to the animal, (ii) a minor sensitivity to stress and handling, (iii) presence of inbred strains to reduce individual variability in experiments, and (iv) the presence of markers for leucocyte populations,

Authors' address: Laboratory of Animal Biotechnology, Department of Environmental Sciences, University of Tuscia, Largo dell' Università, I-01100 Viterbo, Italy.

**Corresponding author:* E-mail: scapigg@unitus.it

subpopulations and molecules. These features are different when studying teleost fish, but in recent years, some difficulties have been diminished with the production of inbred strains of carp (Bandin *et al.*, 1997; Fischer *et al.*, 1998), trout (Ristow *et al.*, 1998), zebrafish (Link *et al.*, 2004) and medaka (Tsukamoto *et al.*, 2005) as also with the discovery of many genes homologous to known mammalian ones.

In addition, the zebrafish model is becoming a powerful system in studies of vertebrate immune development and disease, since it is possible to perform large-scale genetic screens on transparent, readily accessible embryos, in order to identify novel genes involved in pathogenesis, and to study the fate and effects of introduced substances (Trede *et al.*, 2004; van der Sar *et al.*, 2004; Langenau and Zon, 2005). However, with advances in current technology, a limitation of the zebrafish model is represented by the size of fish, since it is difficult to envisage *in vitro* and *in vivo* cell biology experiments with zebrafish leucocytes.

Despite these improvements, the availability of cellular markers for leucocytes in teleosts is, at present, scarce and heterogeneous. Most leucocyte markers are directed to B-cells, granulocytes and thrombocytes (for a review see Scapigliati *et al.*, 1999). On the other hand, although clear evidence of T-cell activities exist in teleosts, markers for T-cells are at present only available for two species: sea bass (Scapigliati *et al.*, 1995) and carp (Rombout *et al.*, 1998). The hypothesis that T-cell subpopulations exists in fish is supported by molecular and cell biology data in trout, since CD8 and CD4 co-receptors were cloned (Hansen and Strassburger, 2000; Suetake *et al.*, 2004) and biological activities were studied (Fischer *et al.*, 2006). The hypothesis is reinforced by data on the presence of the major histocompatibility complex (MHC) molecules class I and II on the surface of teleost cell populations (Rodrigues *et al.*, 1998; Stet *et al.*, 1998; Kruiswijk *et al.*, 2002).

The mAb DLT15 was the first marker for T-cells in fish, and gave the opportunity to describe for the first time the content of T-cells in lymphoid and non-lymphoid organs in a piscine model (Romano *et al.*, 1997).

Due to its importance in aquaculture, sea bass rearing is being continuously improved, mostly in southern European countries (Nehr *et al.*, 1996). A great deal of work has been performed on the prevention of the main pathologies that affect this species in aquaculture, which are vibriosis (Dec *et al.*, 1990), pasteurellosis (Bakopoulos *et al.*, 1997) and virosis (Sideris, 1997; Skliris and Richards, 1999). The current status of knowledge on sea bass immune system is henceforth reported.

INNATE IMMUNITY

Innate responses are rapidly mounted in response of invading organisms and include a complex network of molecules and cells that operate to kill and/or inactivate the putative pathogen. The leucocytes involved in innate responses are mononuclear phagocytes, polymorphonuclear leucocytes, and natural-killer cells. Molecules involved in innate responses are antibacterial peptides, lysozyme, transferrin, complement, reactive oxygen intermediates (ROI), cytokines, chemokines, and Toll-like receptors (TLR).

Cells

Widely employed methods to assess some cellular innate responses against invading organisms include measurement of ROI using nitroblue tetrazolium (NBT assay), and measurement of quantitative phagocytosis. These approaches were employed in a previous study, where the phagocytic activity of head-kidney adherent cells following stimulation by bacterial (*Aeromonas salmonicida*) and fungal (*Candida albicans*) pathogenic agents was studied using light microscopy and measuring ROI production (Bennani *et al.*, 1995). In this work, the ratio of macrophages to pathogenic agents and the amplitude of ROI response varied with the type of pathogenic agent, and opsonization by fish serum increased macrophage ROI response. Phagocytic responses of macrophages were further studied (Esteban and Meseguer, 1997) morphologically by analysing the influence of leucocyte source, bacterial species, presence or absence of a bacterial wall, bacterial status (live or dead), and bacterial opsonization. This work showed that peritoneal macrophages from sea bass showed a greater capacity to engulf bacteria than did those isolated from blood which, in turn, had greater engulfment properties than those isolated from head-kidney.

In another study, immune response of sea bass after intracoelomic immunization with the protozoan parasite *Sphaerospora dicentrarchi* was studied (Muñoz *et al.*, 2000). In this work, significant increases in serum lysozyme, NBT-producing cells, and antibody-secreting cells were detected.

The *in vitro* spontaneous cytotoxic reaction of head-kidney, blood or peritoneal exudate leucocytes against tumor target cells (NK activity) was studied by transmission and scanning electron microscopy, and effector cells exhibited ultrastructural features of either monocytic or lymphocytic lineages (Mulero *et al.*, 1994). The non-specific cell-mediated cytotoxicity process was further studied morphologically, indicating the fact that leucocytes were able to kill their targets by inducing necrosis and apoptosis, in a similar way to mammalian cytotoxic cells (Meseguer *et al.*, 1996). In another study, the spontaneous *in vitro* cytotoxic activity against tumor cell lines by unstimulated sea bass leucocytes was determined using the trypan blue exclusion test and lactate dehydrogenase release assay, and a high anti-tumor cell line activity of resident peritoneal leucocytes was found. A low activity was displayed by head-kidney and spleen cell populations, whereas blood leucocytes revealed no significant activity. Eosinophilic granule cells, isolated from a peritoneal wash, appeared to be responsible for most of the *in vitro* cytotoxic activity (Cammarata *et al.*, 2000). Some parasites can affect sea bass health in aquaculture, and although a specific humoral response against these organisms was reported recently (Muñoz *et al.*, 2000). These results show that immunization with the protozoan *Philasterides dicentrarchi* resulted mainly in the activation of the non-specific immune response measured by lysozyme activity, enhanced phagocytosis by macrophages and active ROI production.

Some diet-related changes in non-specific immune responses have been studied, and showed the modulation of responses to substances such as α-tocopherol and dietary oxidized fish (Obach *et al.*, 1993; Sitjà-Bobadilla and Pérez-Sánchez, 1999). In this last work, non-specific immune factors assayed included plasma lysozyme and complement activities, natural haemolysis against sheep red blood cells, and chemiluminescence response of head-kidney phagocytes as a measure of ROI activity.

Modulation of innate reactions to pathogens by immunostimulants and immunoadjuvants has been the subject of various studies. The effects of long-term oral administration of a combination of dietary glucans, alpha-tocopherol and ascorbic acid on the innate immune response has been reported (Bagni *et al.*, 2000). Alternative pathways of complement activation and lysozyme activity were both significantly enhanced in fish fed with glucans and elevated doses of vitamins. In this study it was also pointed out that in comparison to lysozyme activity, which showed marked individual variation, complement-mediated haemolytic activity was a more reliable indicator of immunocompetence.

A more recent study from the same group described several innate and acquired immune parameters in relation to short- and long-term feeding with yeast beta-glucan and alginic acid preparations (Bagni *et al.*, 2005). In this work, the serum complement, lysozyme, total proteins and heat shock protein (HSP) concentrations at various days after short-term and long-term feeding cycles were measured. Significant elevation of serum complement activity, of serum lysozyme, and gill and liver HSP concentration were observed at 30 days after the end of treatments. Over the long-term period, no significant differences were observed in innate and specific immune parameters.

Taken together, the results of this elegant work suggested the potential of alginic acid and beta-glucans to activate some innate immune responses in sea bass, and particularly under conditions of immunodepression related to environmental stress.

Yeast beta-glucans were also employed to stimulate *in vitro* leucocytes isolated from different tissues (Vazzana *et al.*, 2003), and in this study it was solely observed an enhanced chemiluminescence activity after stimulation.

Key molecules of innate defence reactions against bacteria are TLRs, and recently some of the TLR mammalian counterparts have been cloned in zebrafish (Phelan *et al.*, 2005), trout (Rodriguez *et al.*, 2005) and sea bream (Genebank accession no AY751797), but no sequences are at present available for sea bass.

Molecules

In sea bass, the knowledge on cells and molecules of innate responses is not extensive. Antibacterial peptides are produced by virtually any animal species (for a review see Reddy *et al.*, 2004), but despite their importance, no information is available on sea bass. The only study regarding an antibacterial peptide of the hepcidin family produced in response to Gram-negative bacteria was performed in an evolutionary relative of sea bass (Shike *et al.*, 2002).

The presence in teleosts of typical inflammatory activities involving neutrophils and macrophages, as well as the presence of pro-inflammatory cytokines, has been established. Efforts in finding fish homologues of mammalian cytokine genes led to the cloning of important molecules such as IL-1β, TNFα, IL-8 and IL-18. The common employed strategy was to use degenerate primers designed to amplify evolutionarily-conserved regions in the target molecule and preparing cDNA from cells likely able to express it at high levels. This approach has been successful for the cloning of IL-1β from rainbow trout (*Oncorhynchus mykiss*) (Zou *et al.*, 1999a), the first non-mammalian sequence obtained, and subsequently from sea bass (*Dicentrarchus labrax*) (Scapigliati *et al.*, 2001). These cloned genes for pro-inflammatory cytokines were studied to assess the expression patterns induced by LPS in various organs and tissues and to clarify their gene structure, but the knowledge on the biological activity of recombinant mature peptides is available only for few molecules.

In this respect, sea bass is the fish species in which much work on IL-1β bioactivity has been done (Scapigliati *et al.*, 2006). Fish IL-1β lacks the sequence coding for ICE cleavage site (Zou *et al.*, 1999a), as it happens in other non-mammalian vertebrates. However, by alignment with known sequences the putative initiation site of the mature peptide can be predicted. In sea bass, a predicted putative mature peptide starting at Ala^{86} was designed and the availability of a recombinant IL-1β (rIL-1β) molecule has allowed the characterization of *in vitro* and *in vivo* IL-1β biological activities in this teleost species (Buonocore *et al.*, 2003, 2005). For many years since its discovery, the main way to test the presence of IL-1-containing preparations was the lymphocyte-activating factor (LAF) assay which employed murine thymocytes from young animals (Krakauer and Mizel, 1982). The LAF-assay has been employed to test the effects of rIL-1β on sea bass thymocytes, and the results obtained are shown in recent studies (Buonocore *et al.*, 2005; Scapigliati *et al.*, 2006). The data show a dose-dependent increase of proliferation induced by rising concentrations of the cytokine. The highest quantity of rIL-1β employed (50 ng/ml) induced some downregulation of proliferation statistically not different by values obtained using Con-A only and IL-1β at 0.5 ng/ml, probably due to toxicity for the cells. No synergistic effect of IL-1 and Con-A was observed. Using thymocytes from non-inbred fish, a mean of one fish out of three responded to rIL-1β (Scapigliati *et al.*, 2006).

The bioactivity of sea bass rIL-1β was also tested for its effects on cell function using the D10.G4.1 cells, an IL-1-dependent murine Th2 cell line, and a clear induction of proliferation was evident with a dose-dependent increase at different IL-1β doses. Very interestingly, whilst proliferation was still increasing with the highest dose of sea bass rIL-1β tested (30 ng/ml) relative to the proliferation induced with human rIL-1β (14 pg/ml), it was clearly noted that a far greater concentration of sea bass rIL-1β was needed. This discrepancy has been explained when the 3D-structure of fish IL-1R become available (Scapigliati *et al.*, 2004).

The phagocytosis of foreign substances is affected by IL-1β (Crampe *et al.*, 1997; Hong *et al.*, 2001), and this assay has been adapted in sea bass to test rIL-1β (Buonocore *et al.*, 2005). The results from these experiments clearly showed that added rIL-1β increased both the percent phagocytosis and the phagocytic index. Some experiments already performed in trout were adapted to sea bass (Buonocore *et al.*, 2005), thus it was shown that rIL-1β increased COX-2 transcription in a dose-dependent manner (50 ng/ml of rIL-1β produced an effect on the expression that was quite similar to that seen with 5 μg/ml of LPS), and that rIL-1β increased gene expression through the autocrine induction of IL-1, as observed in mammals (Dinarello *et al.*, 1987; Warner *et al.*, 1987). Cytokines can have an immunoadjuvant effect when administered together with vaccine formulations (Tagliabue and Boraschi, 1993), and, taking advantage of preliminary observations performed in carp (Yin and Kwang, 2000), sea bass rIL-1β was tested as an immuno-adjuvant in vaccination experiments. The vaccine selected was a commercial and efficient bacterin preparation against the sea bass pathogen *Vibrio anguillarum*, and was administered intraperitoneally. The results obtained with these experiments indicated that the recombinant molecule, when present together with the antigenic preparation, enhanced the serum antibody response, as seen by the increased production of specific anti-*Vibrio* immunoglobulins (Buonocore *et al.*, 2004).

Cytokines are known to exert their effects through specific receptors on the cell surface of target cells, and binding of cytokines to these receptor induce physiological activities mediated by intracellular cascades of second messengers. In teleost fish, virtually nothing is known on the effects induced by cytokines on second messengers, but considering that specific IL-1R have been identified in fish (Scapigliati *et al.*, 2004), it is likely to speculate that they may transduce intracellular signaling.

On this basis, 100 ng/ml of rIL-1β were added to head-kidney leucocytes of sea bass previously loaded with the Ca^{++}-testing molecule FURA-2, and a clear and reproducible increase in Ca^{++} concentration inside cells was observed (Scapigliati *et al.*, 2006). This observation is interesting and might reflect acquired differences of IL-1-driven, signaling during evolution, since in mammals IL-1b did not directly affects Ca^{++} metabolism in leucocytes (Georgilis *et al.*, 1987; Rosoff *et al.*, 1988).

Another molecule related to the immune system is COX-2 that is involved in fundamental processes of vertebrate innate immunity and inflammatory processes (Yu *et al.*, 1998; Mitchell *et al.*, 2002; Steer and Corbett, 2003). Macrophages are the cell population mainly involved in the production of COX-2 upon activation (Patel *et al.*, 1999; Joo *et al.*, 2004), and they have been shown to perform this activity also in teleost fishes (Brubacher *et al.*, 2000). Indeed, macrophages play a major role in the first-line innate defence against pathogens, and the intimate involvement of COX-2 in these activities has previously been shown (Zou *et al.*, 1999b; von Aulock *et al.*, 2003). Sea bass COX-2 full-length cDNA (2350 bp) was predicted to contain 596 amino acids that include a classic 19-aa signal peptide. This cDNA is much shorter than its mammalian counterparts (usually 4.0-4.5 kb), but its size is in accordance with that of the other known fish COX-2 sequences. There is high conservation of amino acids essential for its biochemical function. These include 12 conserved cysteines and fundamental residues for catalysis (Arg-106, Tyr-340 and Tyr-369), aspirin acetylation (Ser-514) and haem coordination (His-193 and His-372).

The COX-2 molecule was not constitutively expressed in sea bass head-kidney and was induced by LPS, a virulence factor of many pathogens, which may have a Gram negative like action, by a pro-inflammatory molecule like sea bass IL-1β (Buonocore *et al.*, 2004) and by i.p. injection with a common pathogen, *Vibrio anguillarum* (Buonocore *et al.*, 2005). These results confirm what has been found with other mammalian and bird COX-2 genes (Xie *et al.*, 1991; Janicke *et al.*, 2003) and suggest that COX-2 is involved in inflammatory processes in sea bass.

ACQUIRED IMMUNITY

Immunoglobulins and B-cells

Antibody responses are mounted within bony fish after immunization with particular antigens (Kaattari and Piganelli, 1996; Manning and Nakanishi, 1996; Joosten, 1997a; Palm *et al.*, 1998; Meloni and Scapigliati, 2000) and in response to vaccination (Marsden *et al.*, 1996; Boesen *et al.*, 1997; Marsden and Secombes, 1997; Palm *et al.*, 1998; Lange *et al.*, 2001). In higher vertebrates, there are several Ig classes, determined by the type of heavy chain they possess, that have different structure and function. It is evident from molecular and biochemical studies that teleost species looked at to date produce predominantly an antibody with one type of heavy chain (μ), called IgM (Warr, 1995). Indeed, IgM is the predominant

B-cell antigen receptor conserved among vertebrates (Marchalonis *et al.*, 1992; Scapigliati *et al.*, 1997), and is presently considered to be the only class of immunoglobulin universally found in all jawed vertebrates (Bengten *et al.*, 2000). Recently, other classes than IgM have been found in fish, namely IgD (Wilson *et al.*, 1997; Stenvik and Jörgensen, 2000), IgT (Hansen *et al.*, 2005), and IgZ (Danilova *et al.*, 2005). Fish IgM is composed of heavy and light polypeptide m chains, each with an antigen binding (variable) region and a more constant class-specific region. These chains combine in equimolar amounts to give a complex polymeric molecule, usually a tetramer in teleosts, although it can also be monomeric (Warr, 1995). Both membrane-bound and secreted forms of immunoglobulin exist in fish (Clem *et al.*, 1977). The membrane receptor form of Ig differs from that of the secreted form, and it is able to activate intracellular second messenger pathways in response to immunoglobulin cross-linking (van Ginkel *et al.*, 1994), and in response to antigen binding (MacDougal *et al.*, 1999). The conspicuous switch from IgM to a low molecular weight immunoglobulin (IgG in mammals) that occurs in the course of an immune response in higher vertebrates is not, however, manifested in fish. Fish antibodies are of lower affinity and diversity than those of mammals, with only a low level of heterogeneity in the antibodies produced in response to a single, defined hapten (Du Pasquier, 1982).

All jawed vertebrates (Gnathostomata) have B cells. With the exception of some B cells in cartilaginous fish that express germ-line joined Ig genes, all B cells, irrespective of the organization of their Ig genes rearrange the Ig-gene segments somatically (Du Pasquier, 1993). B lymphocytes represent about 15% of the human circulating lymphoid pool and are the only cells which produce antibodies. They are classically defined by the presence of endogenously produced Ig and by the presence of surface-associated Ig inserted into the plasma membrane which functions as an antigen receptor (BcR). Interaction of antigens with these membrane antibody molecules initiates the sequence of B-cell activation, which leads to the development of effector cells, or plasma cells, that actively secrete antibody molecules. Upon direct or cell-mediated contact with an antigen, B-cells change their biological status from a 'cognitive phase' to an 'effector phase', and the binding of antigen to membrane Ig on B cells is the initiating event in B lymphocyte activation and, therefore, in humoral immunity.

In sea bass several studies reported on the antibody response against various antigens, and the knowledge on antigen-induced humoral immunity is increasing. Here, we will summarize the main obtained results.

Total Serum Immunoglobulins

A previous study was addressed to evaluate the content of serum total antibody concentration with respect to age, gender, and water oxygenation levels (Scapigliati *et al.*, 1999). A total of 586 sera were tested by ELISA assays and the results have shown that the total Ig content increased in sea bass with age/size. These data are in agreement with observations of other workers, which reported an increase of serum Ig in relation to size in catfish (Klesius, 1990), trout (Sànchez *et al.*, 1993), and turbot (Estèvez *et al.*, 1995). In these previous works, however, the size of the fish was the main consideration, and no information on age of the animals was given. It is important to note that, in our knowledge, 10-year-old sea bass were never sampled previously. More recently, some authors reported on the serum antibody response of red drum against bacteria in relation to the age of fish (Evans *et al.*, 1997). Although not necessarily reflective of total serum Ig concentration, they showed that the percentage of red drum within a natural population, which exhibited serum antibody responses against indigenous bacteria, increased with the age of the animals. It is matter of speculation whether the rise of Ig with age depends on the non-specific accumulation of the molecule, or from an increase of Ig specific for diverse antigens encountered during the life of the animal.

By analysing sera for total Ig content with respect to the period of sampling, the results showed that winter was the season in which a higher serum Ig concentration was present. In teleost fish, some size-related and season-related variations of physiological activities have been described (Zapata *et al.*, 1992). For instance, seasonal changes in the humoral immune response and the lymphoid tissues were observed in *Sebastiscus marmoratus* (Nakanishi, 1986), a seasonal trend in serum lysozyme activity and serum total protein content has been reported for the dab (Hutchinson and Manning, 1996) and, more recently, some authors reported on the serum antibody response of red drum against bacteria in relation to the age of fish (Evans *et al.*, 1997). It should be noted that in humans seasonal variation in total IgG and IgM were reported in children exposed to air-borne lead particles (Wagnerova *et al.*, 1986), and serum IgE levels were found to be related to the menstrual cycle (Vellutini *et al.*,

1997). Other parameters investigated were the immune response to *Chlamidia pneumoniae* in relation to gender, (von Hertzen *et al.*, 1998), and serum antibody levels for commensal oral bacteria in relation to age (Percival *et al.*, 1996). In dogs, the influence of age on serum total IgE content was investigated (Racine *et al.*, 1999) and it was shown to increase with age.

For a better understanding of the data obtained in sea bass, it should be observed that the presence of Ig or Ig-like molecules have been demonstrated with various methods in the eggs and/or newborn fry of several species, and the uptake of Ig-like in the eggs studied in detail in sea bream and sea bass (Picchietti *et al.*, 2002, 2004). Winter is the spawning period for sea bass and it is, therefore, conceivable that the higher Ig content detected in winter in blood could be a consequence of storing some Ig in developing eggs during oogenesis. Indeed, in sea bass, a concentration of 3.85 ± 0.9 μg/ml of purified Ig (corresponding to 6 μg/g egg weight) was shown, and immunocytochemical analysis of paraffin sections from 5-day post-hatched embryos revealed an accumulation of material immunoreactive with the anti-Ig mAb DLIg3 in the yolk sac (Scapigliati *et al.*, 1999).

The effect of water oxygenation levels was also investigated, and fish Ig levels were significantly higher for fish maintained in hyperoxygenated water, than for those reared in running non-hyperoxygenated seawater. To better understand these data, it is important to note that a slight difference in organ distribution of B-cells and T-cells in lymphoid organs of sea bass reared at different water oxygenation conditions has been observed (Abelli *et al.*, 1998). Further experiments are required to explain these data and relate observed differences to other physiological parameters, perhaps the increase of some metabolic activities induced by water hyperoxygenation. For instance, it has been described that a low level of oxygen in water can induce a rise in the concentration of cortisol and plasma catecholamines, which, in turn, can induce immunodepression (Sorensen and Weber, 1995). In stress-induced immunosuppression the action of cortisol appeared to operate via specific receptors on leucocytes (Schreck, 1996). Indeed, the immuno-neuro-endocrine system may be important among the variety of mechanisms involved in the response of fish to stress factors such as temperature, salinity, season and crowding (Schreck, 1996; Bly *et al.*, 1997; Pickering, 1999).

Taken together, the results obtained by analysing total Ig content showed that in sea bass, like in mammals, there is a modulation of Ig concentration not directly related to an evident immune function. This suggests a conservation of this feature in vertebrates and stimulates further research for its understanding.

Antigen-specific Immunoglobulins

Teleost fish display a primary and a secondary humoral response upon antigen administration, although a difference relative to mammals is that a shift in Ig class is absent (Van Muiswinkel, 1995). Sea bass is a teleost species susceptible to many pathogens, the most studied pathology being vibriosis and a septicaemia caused by *Vibrio anguillarum* serotypes 01 and 02 and *Photobacterium damselae* subsp. *piscicida*. With regard to *V. anguillarum*, during experimental or 'in field' vaccination trials, it is usual to employ a bacterin suspension (killed bacteria) administered intraperitoneally, orally or by immersion. In an early work (Dec *et al.*, 1990), a comparison of oral vaccination of sea bass with respect to intraperitoneal administration showed the inefficiency of oral treatment, as measured by both antigen-specific Ig serum titers and bacteriostatic activities. However, passive immunization using serum from orally vaccinated fish (2 months after vaccination) conferred some protection against a challenge carried out by inoculation of virulent *Vibrio*. Subsequently, the relationships between the levels of total protein, Ig and antibody activity in serum of sea bass broodstock, following one or two intraperitoneal injections of heat-killed *V. anguillarum*, were investigated (Coeurdacier *et al.*, 1997). Results from this work showed that *Vibrio* injection did not modify serum protein levels, and that Ig production was significantly higher in immunized animals, whereas no significant difference was found between males and females in antigen-specific antibody levels.

With regard to the Gram-negative *P. damselae*, previously classified as *Pasteurella piscicida*, many studies have addressed the effects of this organism on the immune system in relation to the preparation of effective vaccines. In previous work (Bakopoulos *et al.*, 1997), fish were injected with live as well as heat-killed bacteria, and serum antibody titers determined by western blot analysis. Great variation among the sera was evident with reference to the recognition of antigens in the high molecular weight group, whereas lipopolysaccharide and/or lipoprotein situated in the low molecular weight group products appeared to be the most immunogenic material in the bacterial cell. Western blot analysis was also employed to assess the presence of *Pasteurella* antigens in organs of sea bass (Pretti *et al.*, 1999), and showed variability in the molecular weight of antigens recognized by immune sera. This reported variability was further studied (Mazzolini *et al.*, 1998) using different antigen preparations administered intraperitoneally. Interestingly, bacterial extracellular products carried out most of the toxic activity.

As already reported, the main sea bass infective pathologies are vibriosis and pasteurellosis, caused by gram-negative bacteria. In particular, the pasteurellosis, present in the Mediterranean Sea since about 1990 (Barriero *et al.*, 1991) and caused by *Photobacterium damselae* sub. *piscicida* (*Phda*), is particularly virulent and population of resistant sea bass have been not yet naturally selected. Some studies showed that *Phda* is an intracellular bacteria, colonizes mucosal epithelium of gills and intestine (Magarinos *et al.*, 1996; Lòpez-Dòriga *et al.*, 2000), expresses some polysaccharides (Bonet *et al.*, 1994) and activates innate defenses (Santarem and Figueras, 1995). *Phda* expresses its pathogenic effect after contact with host mucosal epithelia (epidermis, gills, intestine). As result, a vaccination that stimulates the mucosal immunity system (gills and/or intestine), may represent the best way to stimulate the immune system and to give protection towards further exposure to pathogen.

The production of antigenic formulations to be used as a potential vaccine anti *Phda* was tested in a Japanese yellowfish (Kawakami *et al.*, 1997), and in the sea bass (Mazzolini *et al.*, 1998; Dos Santos *et al.*, 2001). In another work, intraperitoneal inoculation of *Phda* did induced production of specific antibodies in serum, even if with low titer (Bakopoulos *et al.*, 1997), and immune positivity to the bacterium in liver (Pretti *et al.*, 1999). Despite these efforts, the way to vaccinate sea bass by immersion against pasteurellosis is still under improvement.

Vaccination of sea bass is usually performed by immersion in juveniles of 2-4 gr using killed bacteria because it was demonstrated that at this size the immune system is complete in its humoral and cellular component (Abelli *et al.*, 1996; Picchietti *et al.*, 1997; Dos Santos *et al.*, 2001) and it can react to the vaccination without side effects like tolerance (Petrie-Hanson and Ainsworth, 1997). Nowadays, the most common methods of vaccination by immersion of teleosts consist in maintaining a high concentration of juvenile fish immersed in a solution with water and vaccine for a few minutes. An efficient stimulation of the mucosal immune system through surfaces in direct contact with the antigen, as gills

(Davidson *et al.*, 1997, Moore *et al.*, 1998), intestine and epidermis (Jenkins *et al.*, 1994; Joosten, 1995; Lumsden *et al.*, 1995; Robohm and Koch, 1995) seems to play a major role to confer a significant protection.

Some molecules have been tested as putative adjuvants for their ability to increase the antigenic capability in mucosal environments, especially in mammals. In fish, during a mucosal vaccination of catfish it was used a cholera toxin as adjuvant, with encouraging results about protection against live pathogen challenge (Hebert *et al.*, 2000). Recently, the *Escherichia coli* heat-sensible enterotoxin (LTK63) was demonstrated in humans to be very efficient in increasing antibody response of antigen supplied through mucosal vaccination (Lycke and Holmgren, 1986; Czerkinsky *et al.*, 1989; Vervelde *et al.*, 1998). The LTK63 was produced by a natural mutant strain that has lost its ability for ADP-rybosylation and thereby its toxicity (Neidleman, *et al.*, 2000). This mutant toxin was produced as recombinant peptide, tested for its potential immunoadjuvant capability in sea bass juveniles, and the results presented in this work.

This experimental work has been organized by using young fish (ca. 5 g) deriving from a genetically wild broodstock and maintained in a fish farm in a controlled environment at 16-18°C. Fish were maintained in 1000-liter tanks with running filtered seawater and were treated by immersing batches of 25 fishes for 2 minutes in 30 liters of an experimental vaccine anti-Phda (AVL/Schering Plough) used following the suggested dilution (1:10 in sea water). The animals were treated with and without LTK63 adjuvant (500 ng/ml), whereas controls received a mock stimulation, then they were transferred again in respective tanks and maintained in aquacultured normal condition. After 21 days the same treatment was repeated with the same vaccine dose, but without adjuvant. The fish were sampled at day 0, day 21, day 40 and day 70 after initial treatment, and blood was taken out by cutting tail after lethal anaesthesia with 1 g/l of MS-622. As the fish were very small, it was necessary to make pools of blood from 4-5 animals. The sera were than frozen at –20°C.

The presence of anti-Phda specific antibody was tested in pools of sera by ELISA using a polyclonal anti-sea bass IgM as previously described (Scapigliati *et al.*, 1999), and the presence of anti-Phda specific antibody produced *in vitro* by head kidney cells of individual fish was tested, as previously described (Meloni and Scapigliati, 2000).

At 21 days, the amount of specific antibody was barely detectable, with no significant differences between experimental fish groups

(Fig. 6.1a), at 40 days was present an evident antibody titer against Phda in vaccinated animals, and this titer was raised in the experimental group treated with the LTK63 immunoadjuvant (Fig. 6.1b). At 70 days, the specific antibody titer against Phda was even higher in vaccinated animals, but LTK63 did not showed a further increase (Fig. 6.1c). These results clearly showed that the vaccination without boosting does not induce a relevant antibody titer, whereas a vaccine boosting after 21 days induces a significant antibody titer anti-Phda. Interestingly, the immunoadjuvant LTK63 supplied with the vaccine anti-Phda increased the amount of specific immunoglobulins measured at day 40, but did not had an effect day 70. At this time, the specific anti-Phda titer was still high.

The presence of anti-Phda specific antibody was also tested in same fish through an *in vitro* production of antibody by head-kidney cells and gills cells, and the results are shown in Fig. 6.2. In head-kidney (Fig. 6.2a) the presence of specific antibody was evident in all samplings, and increased steadily from day 21 till day 70. The fish group treated with LTK63 did not showed differences in this assay. In cells from gills an intense production of Ig was detected (Fig. 6.2b) at all sampling dates, and also in this organ no additive effects induced by LTK63 were observed.

To study humoral reactions of the sea bass, some mAbs have been prepared against Ig and Ig-bearing cells (Scapigliati *et al.*, 1999). All these mAbs were prepared using Ig as the immunogen purified with various biochemical methods (Estévez *et al.*, 1994; Palenzuela *et al.*, 1996). Most of these mAbs recognized the Ig heavy chain, whereas a few were obtained against the light chain (Romestand *et al.*, 1995; Scapigliati *et al.*, 1996; Dos Santos *et al.*, 1997). In particular, in the first report (Romestand *et al.*, 1995), the obtained mAb were employed to set up an ELISA assay to detect total and antigen-specific serum Ig, whereas the organ distribution of Ig-bearing cells was not studied. Shortly after, a mAb raised against the light chain of the Ig molecule was selected to be effective in recognizing the immunization antigen both in denatured and native form (Scapigliati *et al.*, 1996). In this work, sea bass Ig were single-step purified from whole serum by affinity chromatography on protein A-sepharose and used as immunogen in mice. Among the positive hybridomas obtained, some clones were selected according to their ability to recognize either the Ig light chain (DLIg3) or the heavy chain (DLIg13 and DLIg14). DLIg3 stained in IIF and flow cytometric analysis 21% of PBL, 3% of thymocytes, 30% of splenocytes, 33% of head-kidney leucocytes, and 2% of

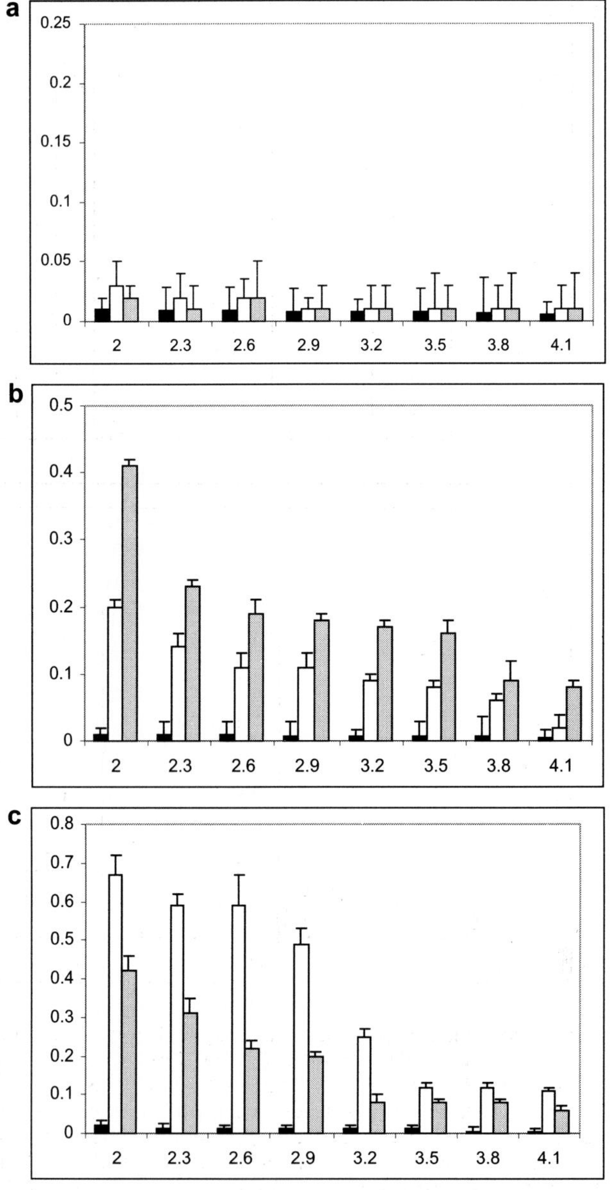

Fig. 6.1 ELISA assay. Figure 6.1 shows the results of indirect ELISA assay on pooled sera of young sea bass vaccinated against *Photobacterium damselae* bacterin and tested at day 21 (a), 49 (b), and 70 (c) after vaccination. Control fish are shown by black bars, vaccinated fish by empty bars, and LTK63-treated fish by grey bars. From each fish group not less than 20 pooled sera were analysed in duplicates, displayed values are the mean ± standard deviation of absorbance measured at 492 nm (in the Y-axis), at various serum dilutions (represented in the X-axis by the log of the reciprocal serum dilution).

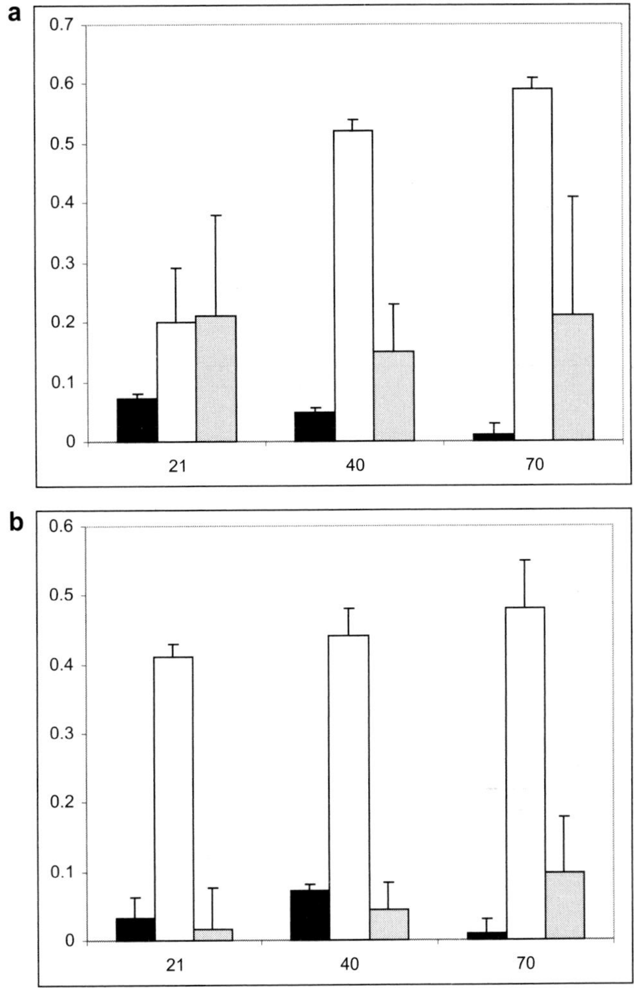

Fig. 6.2 *In vitro* production of anti-Phda immunoglobulins. Cells obtained from head kidney (a) and gills (b) at indicated days (in the X-axis) were incubated with the immunisation antigen (*Photobacterium damselae* bacterin) adsorbed onto plastic for 48 hours at 18°C. Cells were then removed and specific antibody detected by ELISA. Control fish are shown by black bars, vaccinated fish by empty bars, and LTK63-treated fish by grey bars. From each fish group, 5 individuals were analysed in duplicates, displayed values are the mean ± standard deviation of absorbance measured at 492 nm (in the Y-axis).

gut-associated lymphoid tissue (GALT) (Romano *et al.*, 1997). DLIg13 and DLIg14 were unable to stain living cells by IIF and FACS, but recognized fixed cells following avidin-biotin complex (ABC)-immunoperoxidase staining of sections from spleen, head-kidney and midgut. The mAb DLIg3 (IgG class) was the most interesting mAb obtained, since it worked in all systems used, and was used to set up a sensitive ELISA assay (detection limit 1.2 ng/ml) to detect and quantify purified and serum Ig. In later work (Dos Santos *et al.*, 1997), three anti-Ig mAbs were selected (WDI 1-3) with criteria based on ELISA, Western blot, IIF and flow cytometric analysis. All mAb were found to belong to the IgG class, and were effective in detecting antigen-specific antibody titers in ELISA. Under reducing conditions WDI 1 recognizes the heavy chain and both WDI 2 (slightly) and WDI 3 (strongly) recognize the light chain. The average percentages of surface Ig-positive cells in PBL, head-kidney, spleen, thymus and gut for the different mAbs were similar to that previously reported, so confirming the estimation of B-cells in sea bass organs. mAbs DLIg3 and WDI 1-3 were also employed in immunogold labelling, and showed specificity for subpopulations of lymphoid cells (B-cells, plasma cells and macrophages) in both PBL and lymphoid tissues (Dos Santos *et al.*, 1997; Romano *et al.*, 1997).

T-cells

The existence of T-cell populations has been known within bony fish since the 1970s (Stolen and Makela, 1975), as demonstrated "*in vitro*" by the proliferation induced with mitogens (Etlinger *et al.*, 1976; Sizemore *et al.*, 1984) and non-self antigens (Marsden *et al.*, 1996), by the responses in the mixed-leukocyte reaction (Miller *et al.*, 1985), and by the function as helper cells in antibody production against thymus-dependent antigens (Miller *et al.*, 1987). The importance of antigen processing and presentation in the generation of secondary *in vitro* immune responses in the channel catfish to both simple and complex T-dependent antigens were shown by Vallejo *et al.* (1991). In this work, employing functionally active long-term monocyte lines as antigen-presenting cells 'putative restriction' of immune responses by PBL as responders was revealed, and these results provided evidence that alloantigens, presumably MHC or MHC-like molecules, could drive antigen presentation and restriction of teleost immune responses similar to the situation in mammals. Another T-cell activity shown to take place in teleosts is the graft-versus-host reaction (GVHR). A model system of clonal triploid ginbuna and tetraploid ginbuna-goldfish hybrids was employed to demonstrate GVHR in carp (Nakanishi and Ototake, 1999). In this work, sensitized triploid cells were injected into tetraploid recipients and a typical GVHR was induced, leading to the death of recipients within one month, thereby suggesting the presence of allo-reactive Tc in teleosts. An evidence of an *in vivo* T-cell activity in bony fish was provided by Abelli *et al.* (1999).

The sea bass is, at present, the only marine species for which a specific anti putative T-cell marker is available. The mAb DLT15, specific for thymocytes and peripheral T-cells, was obtained by immunizing mice with paraformaldehyde-fixed thymocytes from juvenile sea bass (Scapigliati *et al.*, 1995). This antibody (IgG_3 subclass) is able to recognize its antigen(s) both in living cells and in tissue sections, and its use in IIF and cytofluorimetric analysis of leucocytes enriched over Percoll permitted the first evaluation of a T-cell population in sea bass, consisting of 3% of PBL, 9% of splenocytes, 4% of head-kidney cells, 75% of thymocytes, 51% of GALT, and 60% of gill-associated lymphoid tissue (Romano *et al.*, 1997). In view of the desire for oral delivery of antigens, gut-associated lymphoid tissue has been the subject of particular research, since it revealed a striking abundance of T-cells (Abelli *et al.*, 1997), and a remarkable precocity of their appearance during development (Picchietti *et al.*, 1997). DLT15 was used in immunocytochemistry to show for the first time in a piscine system T-cell activity *in vivo,* where muscle transplants have been grafted onto allogenic recipient fish (Abelli *et al.*, 1999). The immunocytochemical analysis with DLT15 of rejected sea bass muscle allografts showed that many cells infiltrating the tissue were DLT15-positive. Another important use of DLT15 was to purify immunoreactive cells from sea bass organs, mainly from blood and gut-associated lymphoid tissue (Scapigliati *et al.*, 2000). The purification was performed using labelled immunobeads with leucocyte fractions enriched by Percoll density gradient centrifugation, and the purity of DLT15-purified cells was 90% for gut-associated lymphoid tissue, and 80% for blood leucocytes.

Some information is available on '*in vitro* cell biology responses of sea bass T-cells. Previous studies reported that leucocytes from head-kidney proliferated poorly in response to mitogens, which in mammals, are specific for T-cells, such as concanavalin-A or phytohemagglutinin (Volpatti *et al.*, 1996; Galeotti *et al.*, 1999).

Recently, some studies investigated in sea bass some '*in vitro*' responses of T-cells, such as thymocyte proliferation, and allorecognition. The presence of IL-1-containing preparations can be tested in mammals employing the lymphocyte-activation assay (LAF) with murine thymocytes from young animals (Krakauer *et al.*, 1982; Oppenheim *et al.*,

1982). The LAF-assay has been adapted to test the effects of rIL-1β on sea bass thymocytes, and the obtained results were reported in recent works (Buonocore *et al.*, 2005; Scapigliati *et al.*, 2006). The data showed a dose-dependent increase of proliferation induced by rising concentrations of the cytokine from 0.2 to 50 ng/ml. At difference with mammals, no synergistic effect between IL-1 and lectins in inducing thymocyte proliferation was observed.

Another recent investigation performed on sea bass reported the first direct quantitative determination of an '*in vitro*' T-cell activity from primary cultures of leucocytes in a teleost species (Meloni *et al.*, 2006). In this study, a number of cellular activities of sea bass PBL against allogeneic PBL inactivated by irradiation were studied. The number of T-cells and B-cells were evaluated after two weeks of incubation by using specific mAbs in immunofluorescence and flow cytometry. The results showed an increase of T cells in a one-way mixed leucocyte reaction (MLR), whereas the percentage of B cells remained similar to that in control PBL. The increase of T-cells in MLR cultures was also confirmed using RT-PCR through the analysis of the T-cell receptor β (TcR) mRNA expression. As previously reported in mammals (Kronke *et al.*, 1984), the addition of cyclosporin-A to the MLR caused a significant decrease in T-cell proliferation, thus suggesting a close similarity between fish and mammalian T-cell behaviour. Leucocytes from MLR cultures displayed an enhanced cytotoxic activity against xenogeneic target cells with respect to control PBL, raising the possibility of the presence of cytotoxic-like T lymphocytes. Additionally, this work also measured the cellular activation of PBL induced by allorecognition during a MLR. This latter was done by measuring with FURA-2 AM the mobilization of intracellular Ca^{++}, induced by adding anti- lymphocyte mAbs (DLT15 and DLIg3), that resulted affected in MLR with respect to controls, thus suggesting the presence of activated lymphocytes.

Immunoregulatory Molecules

T-cell Receptor

At contrast with B lymphocytes, T lymphocytes have antigen receptors which are not secreted and are membrane molecules distinct but structurally related to the antibody molecule. In mammals, the T lymphocyte family is subdivided into functionally distinct cell populations with different cell-associated macromolecules, the best defined of which are helper T-cells (Th) and cytolytic (or cytotoxic) (Tc) T-cells. Th and Tc recognize only peptide antigens that are associated with MHC proteins expressed on the surface of accessory cells. As a results T-cells recognize and respond to cell surface-associated but not soluble protein antigens, in contrast to B-cells that can respond to proteins, nucleic acids, polysaccharides, lipids, and small chemicals (Babbitt *et al.*, 1985). The development of technologies for propagating monoclonal T-cell populations *in vitro*, including hybridomas and antigen-specific T-cell clones (Kappler *et al.*, 1981) was a milestone for the study of TcR. All the cells in a clonal T-cell population that are derived from a single cell are genetically identical, and therefore express identical TcR different from the receptors produced by all other clones. Therefore, TcR produced by any clone express unique determinants, or idiotypes, in their antigen-binding regions that are not present on the antigen receptors of any other clone. This feature has been the basis for a successful attempt in studying T-cell acitivity without the use of specific markers, namely the molecular study of antigen-induced V-D-J rearrangements of the TcRβ chain with the immunoscope methodology (Boudinot *et al.*, 2001, 2002). In these works, the DNA sequences originating by V-D-J rearrangements have been amplified by PCR using specific primers, and the pools of amplified sequences separated by chromatography. In immunized fish, a rapid induction of rearrangements was detected as a rise in the number of detected chromatographic peaks.

The cloning of TcR genes revealed that they, like Ig genes, utilize somatic rearrangement as a mechanism of generating diversity with variable (V), constant (C), and joining (J) segments remarkably similar in size and organization to those found in Ig genes. The antigen-MHC receptor on the majority of T-cells, including MHC-restricted helper T-cells and cytotoxic T lymphocytes, is a heterodimer consisting of two polypeptide chains, designated α and β, covalently linked by a disulfide bond. In mammals, the α chain is a 40-50 kDa acidic glycoprotein, and the β chain is a 40-45 kDa uncharged or basic glycoprotein. The αβ heterodimer recognizes complexes of processed peptides, generated from foreign protein antigens, bound to self MHC molecules (Abbas *et al.*, 1991). This heterodimer provides T-cells with the ability to recognize antigen-MHC complexes, but both the cell-surface expression of TcR molecules and their function in activating T-cells are dependent on a group of associated proteins that form the CD3 complex (Van Wauwe *et al.*, 1984; Tsoukas *et al.*, 1985). The CD3 complex in mammals consists of at least five distinct integral membrane proteins of 16-28 kDa non-covalently associated one another and with the TcRαβ heterodimer. The γδ TcR is another type of TcR present on the surface of a subset of αβ-negative peripheral T-cells and immature thymocytes (Heilig and Tonegawa, 1986; Koning *et al.*, 1988). In addition to mammals, the αβ- and γδ- TcR have been cloned in other vertebrate species, as seen in birds (Sowder *et al.*, 1988), and in amphibians (Chretien *et al.*, 1997; Haire *et al.*, 2002). As above reported, T-cells of vertebrates must have the TcR complex on their cell surface and, consequently, the sea bass leucocytes immunopurified with mAb DLT15 should be enriched in cells expressing mRNAs coding for αβ and/or γδ chains of TcR. This approach was addressed to identify and clone the sea bass TcR by homology cloning using degenerate oligonucleotide primers derived from the peptide sequences MYWY and VYFCA of the rainbow trout TcRβ (Partula *et al.*, 1995), thus allowing the molecular cloning of the sea bass TcRβ-chain V_β region (Scapigliati *et al.* 2000).

CD8 Co-receptor

In mammals, the recognition of peptides presented by MHC molecules on antigen presenting cells to cytotoxic and helper T-cells is mediated by a number of receptor-ligand or receptor-counter receptor interactions. This process involves the T-cell coreceptors CD8 and CD4 which bind to MHC class I and II molecules, respectively. Cytotoxic reactions and cellular equivalents of $CD8^+$ cytotoxic cells have been reported recently in fish (Fisher *et al.*, 2003, 2006; Secombes *et al.*, 2005) and the first fish CD8α gene was cloned in rainbow trout a few years ago (Hansen and Strassburger, 2000).

Sea bass CD8α has been recently cloned (Buonocore *et al.*, 2006) and its size was in accordance with other fish and mammalian counterparts. The overall structure seems well conserved during evolution, although some differences are evident. Most cysteine residues are present in all CD8α sequences, but an extra cysteine in the IgSf V domain was found only in mammals and the second cysteine of the CXCP motif in the cytoplasmic tail, responsible in mammals for $p56^{lck}$ binding, was lacking in all the fish sequences. This latter finding is quite relevant, as the association of CD8α with $p56^{lck}$ leads to the phosphorylation of the T-cell receptor by tyrosine kinases as shown in higher vertebrates. The major source of CD8α expression in sea bass was the thymus, in agreement with the finding that this organ is a main site for T cell lymphopoiesis in all vertebrate species (Hansen and Zapata, 1998). The availability of known 3D structures for human and mouse CD8α Ig-like domains allowed the creation of a sea bass protein model by comparative modelling. The analyzed N-terminal CD8α region contains two cysteine residues that are fundamental for the folding of the IgSf V domain in mammals and the distance of these residues in the sea bass model is sufficient to allow disulfide bond formation. The comparison of secondary structures suggests that all b-strands in this region are well conserved among species, although some little differences can be appreciated.

The MHC class I and II molecules have not been yet characterized in sea bass although a short MHC II class II sequence is available (Venkatesh *et al.*, 1999).

ONTOGENY OF THE IMMUNE SYSTEM

In contrast to higher vertebrates, most fish species are free-living organisms already at the embryonic stage of life. Living in an aquatic environment, they must have defense mechanisms to protect themselves against a variety of microorganisms. Consequently, during a rather long period, the fish are dependent on their innate immune system, and it is expected that this develops at a very early embryonic age. A very good review on the ontogenesis of fish leucocytes has been recently published (Rombout *et al.*, 2005), In this work emerged that best characterized fish species are two cyprinids (zebrafish, carp) and sea bass, and that timing of leucocyte appearance may vary considerably among species. This review clearly shows that, dependent on the species, young fish use innate mechanisms during the first weeks/months of their development, and this may find application for the defense of farmed fish against pathogens at early age. It is evident that T-cells develop much earlier than B-cells in all species investigated. Extra-thymic origin of T cells mostly of the gd phenotype was reported in mammals, and the data obtained on zebrafish, carp, and sea bass reported in the review, indicate that a similar process takes place in bony fish. This is an important finding, since thymus and gut-associated lymphoid tissue (GALT) are evolutionary related and GALT possibly precedes thymus during evolution.

Acknowledgements

Experimental original work presented was supported by the Italian Ministero delle Politiche Agricole, 6° Piano Triennale, progetto di Ricerca

6C59. The EU Integrated Project IMAQUANIM CT-2005-007103 supported some of cited work. Authors are also indebted to Dr G. Del Giudice and Dr P. Ruggiero (Chiron Vaccines, Siena, Italy) for the supply of the LTK63 molecule, and to Dr C. Magugliani (Nuova Azzurro, Civitavecchia, Italy) for the work in fish farm.

References

Abbass, I.M. and A.A. Razak. 1991. Cadmium, selenium, and tellurium chelators in *Aspergillus terreus*. *Biological Trace Element Research* 28: 173-179.

Abelli, L., M.R. Baldassini, R. Meschini and L. Mastrolia. 1998. Apoptosis of thymocytes in developing sea bass *Dicentrarchus labrax* (L.). *Fish and Shellfish Immunology* 8: 13-24.

Abelli, L., M.R. Baldassini, L. Mastrolia and G. Scapigliati. 1999. Immunodetection of lymphocyte subpopulations involved in allograft rejection in a teleost, *Dicentrarchus labrax* (L.). *Cell Immunology* 1: 152-160.

Abelli, L., S. Picchietti, N. Romano, L. Mastrolia and G. Scapigliati. 1996. Immunocytochemical detection of a thymocyte antigenic determinant in developing lymphoid organs of sea bass *Dicentrarchus labrax* (L.). *Fish and Shellfish Immunology* 6: 493-505.

Babbitt, B.P., P.M. Allen, G. Matsueda, E. Haber and E.R. Unanue. 1985. Binding of immunogenic peptidesd to Ia histocompatibility molecules. *Nature (London)* 317: 359-361.

Bagni, M., L. Archetti, M. Amadori and G. Marino. 2000. Effect of long-term oral administration of an immunostimulant diet on innate immunity in sea bass *Dicentrarchus labrax*. *Journal of Veterinary Medicine* B 47: 745-751.

Bagni, M., N. Romano, M.G. Finoia, L. Abelli, G. Scapigliati, P.G. Tiscar, M. Sarti and G. Marino. 2005. Short- and long-term effects of a dietary yeast beta-glucan (Macrogard) and alginic acid (Ergosan) preparation on immune response in sea bass *Dicentrarchus labrax*. *Fish and Shellfish Immunology* 18: 311-325.

Bakopoulos, V., D. Volpatti, A. Adams, M. Galeotti and R. Richards. 1997. Qualitative differences in the immune response of rabbit, mouse and sea bass, *Dicentrarchus labrax* L. to *Photobacterium damsela* subsp. *piscicida*, the causative agent of fish Pasteurellosis. *Fish and Shellfish Immunology* 7: 161-174.

Bandin, I., C.P. Dopazo, W.B. Vanmuiswinkel and G.F. Wiegertjes. 1997. Quantitation of antibody secreting cells in high and low antibody responder inbred carp *Cyprinus carpio* L. strains. *Fish and Shellfish Immunology* 7: 487-501.

Barriero, S., J.F. Casal, A. Figueras, B. Margariños and J.L. Barja. 1991. Pasteurellosis in cultured gilthead sea bream, *Sparus auratus*: first report in Spain. *Aquaculture* 99: 1-15.

Bengten, E., M. Wilson, N. Miller, L.W. Clem, L. Pilstrom and G.W. Warr. 2000. Immunoglobulin isotypes: Structure, function, and genetics. *Current Topics in Microbiology and Immunology* 248: 189-219.

Bennani, N., A. Schmid-Alliana and M. Lafaurie. 1995. Evaluation of phagocytic activity in a teleost fish, *Dicentrarchus labrax*. *Fish and Shellfish Immunology* 5: 237-246.

Bly, J.E., S.M.A. Quiniou and L.W. Clem. 1997. Environmental effects on fish immune mechanisms. In: *Fish Vaccinology*, R. Gudding, A. Lillehaug, P.J. Midtlyng and F. Brown (eds.). Development Biology Standard, S. Karger, Basel, 90: 33-43.

Boesen, H.T., K. Pedersen, C. Koch and J.L. Larsen. 1997. Immune response of rainbow trout *Oncorhynchus mykiss* to antigenic preparations from *Vibrio anguillarum* serogroup O1. *Fish and Shellfish Immunology* 7: 543-553.

Bonet, R., B. Magarinos, J.L. Romalde, M.D. Simonpujol, A.E. Touranzo and F. Congregado. 1994. Capsular polysaccharide expressed by *Pasteurella piscicida* grown in vitro. *FEMS Microbiology Letters* 124: 285-289.

Boudinot, P., S. Boubekeur and A. Benmansour. 2001. Rhabdovirus infection induces public and private T cell responses in teleost fish. *Journal of Immunology* 167: 6202-6209.

Boudinot, P., S. Boubekeur and A. Benmansour. 2002. Primary structure and complementarity-determining region (CDR) 3 spectratyping of rainbow trout TCRbeta transcripts identify ten Vbeta families with Vbeta6 displaying unusual CDR2 and differently spliced forms. *Journal of Immunology* 169: 6244-6252.

Brubacher, J.L., C.J. Secombes, J. Zou and N.C. Bols. 2000. Constitutive and LPS-induced gene expression in a macrophage-like cell line from the rainbow trout *Oncorhynchus mykiss*. *Developmental and Comparative Immunology* 24: 565-574.

Buònocore, F., M. Forlenza, C.J. Secombes, J. Zou, M. Mazzini and G. Scapigliati. 2003a. Expression in *E. coli* and purification of sea bass *Dicentrarchus labrax* interleukin-1, a possible immuno-adjuvant in aquaculture. *Marine Biotechnology* 5: 214-221.

Buonocore, F., D. Prugnoli, C. Falasca, C.J. Secombes and G. Scapigliati. 2003b. Peculiar gene organization and incomplete splicing of sea bass *Dicentrarchus labrax* L. interleukin-1β. *Cytokine* 21: 257-264.

Buonocore, F., M. Mazzini, M. Forlenza, E. Randelli, C.J. Secombes, J. Zou and G. Scapigliati. 2004. Expression in *Escherichia coli* and purification of sea bass *Dicentrarchus labrax* interleukin 1β, a possible immuno-adjuvant in aquaculture. *Marine Biotechnology* 6: 53-59.

Buonocore, F., E. Randelli, D. Casani, M. Mazzini, I. Cappuccio, C.J. Secombes and G. Scapigliati. 2005. cDNA cloning and expression analysis of a cyclooxygenase-2 from sea bass *Dicentrarchus labrax* L. after vaccination. *Aquaculture* 245: 301-310.

Buonocore, F., E. Randelli, S. Bird, C.J. Secombes, S. Costantini, A. Facchiano, M. Mazzini and G. Scapigliati. 2006. The CD8α from sea bass *Dicentrarchus labrax* L.: Cloning, expression and 3D modelling. *Fish and Shellfish Immunology* 20: 637-646.

Cammarata, M., M. Vazzana, M. Cervello, V. Arizza and N. Parrinello. 2000. Spontaneous cytotoxic activity of eosinophilic granule cells separated from the normal peritoneal cavity of *Dicentrarchus labrax*. *Fish and Shellfish Immunology* 10: 143-154.

Chretien, I., A. Marcuz, J. Fellah, J. Charlemagne and L. Du Pasquier. 1997. The T-cell receptor beta genes of *Xenopus*. *European Journal of Immunology* 27: 763-771.

Clem, L.W., W.E. McLean, V.T. Shankey and M.A. Cuchens. 1977. Phylogeny of lymphocyte heterogeneity. I. Membrane immunoglobulins of teleost lymphocytes. *Developmental and Comparative Immunology* 1: 105-118.

Coeurdacier, J.L., J.F. Pepin, C. Fauvel, P. Legall, A.F. Bourmaud and B. Romestand. 1997. Alterations in total protein, IgM and specific antibody activity of male and female sea bass *Dicentrarchus labrax* L. sera following injection with killed *Vibrio anguillarum*. *Fish and Shellfish Immunology* 7: 151-160.

Crampe, M., S.R. Farley, A. Langston and A.L. Pulsford. 1997. Measurement of phagocytosis: A comparative evaluation of microscopic and microplate methods. In: *Methodology in Fish Diseases Research*. Fisheries Research Services, Aberdeen, pp. 81-89.

Czerkinsky, C., M.W. Russell, N. Lycke, M. Lindblad and J. Holmgren. 1989. Oral administration of a streptococcal antigen coupled to cholera toxin B subunit evokes strong antibody responses in salivary glands and extramucosal. *Infectious Immunology* 57: 1072-1077.

Danilova, N., J. Bussmann, K. Jekosch and L.A. Szteiner. 2005. The immunoglobulin heavy-chain locus in zebrafish: Identification and expression of a previously unknown isotype, immunoglobulin. *Immunoglobulin Z. Nature Immunology* 6: 295-302.

Davidson, G.A., S.H. Lin, C.J. Secombes and A.E. Ellis. 1997. Detection of specific and 'constitutive' antibody secreting cells in the gills, head kidney and peripheral blood leucocytes of dab *Limanda limanda*. *Veterinary Immunology and Immunopathology* 58: 363-374.

Dec, C., P. Angelidis and F. Baudin-Laurencin. 1990. Effects of oral vaccination against vibriosis in turbot, *Scophthalmus maximus* L., and sea bass *Dicentrarchus labrax* L. *Journal of Fish Diseases* 13: 369-376.

Dinarello, C.A., T. Ikejima, S.J. Warner, S.F. Orencole, G. Lonnemann, J.G. Cannon and P. Libby. 1987. Interleukin 1 induces interleukin 1. I. Induction of circulating interleukin 1 in rabbits *in vivo* and in human mononuclear cells *in vitro*. *Journal of Immunology* 139: 1902-1910.

Dos Santos, N.M., N. Taverne, A.J. Taverne-Thiele, M. Desousa and J.H.W.M. Rombout. 1997. Characterisation of monoclonal antibodies specific for sea bass (*Dicentrarchus labrax* L.) IgM indicates the existence of B cell subpopulations. *Fish and Shellfish Immunology* 7: 175-191.

Dos Santos, N.M., T. Hermesen, J.H. Rombout, L. Pilstrom and R.J. Stet. 2001. Ig light chain variability in DNP(494)-KLH immunised sea bass (*Dicentrarchus labrax* L.): Evidence for intra-molecular induced suppression. *Developmental and Comparative Immunology* 25: 387-401.

Du Pasquier, L. 1982. Antidody diversity in lower vertebrates—Why is it so restricted? *Nature (London)* 296: 311-313.

Du Pasquier, L. 1993. Phylogeny of B-cell development. *Current Opinion in Immunology* 5: 185-193.

Esteban, M.A. and J. Meseguer. 1997. Factors influencing phagocytic response of macrophages from the sea bass *Dicentrarchus labrax* L.: An ultrastructural and quantitative study. *Anatomical Record* 248: 533-541.

Estévez, J., J. Leiro, M.T. Santamarina and F.M. Ubeira. 1995. A sandwich immunoassay to quantify low levels of turbot *Scophthalmus maximus* immunoglobulins. *Veterinary Immunology and Immunopathology* 45: 165-174.

Estévez, J., J. Leiro, M.T. Santamarina, J. Domínguez and F.M. Ubeira. 1994. Monoclonal antibodies to turbot *Scophthalmus maximus* immunoglobulins: Characterization and applicability in immunoassays. *Veterinary Immunology and Immunopathology* 41: 353-366.

Etlinger, H.M., H.O. Hodgins and J.M. Chiller. 1976. Evolution of the lymphoid system. Evidence for lymphocyte heterogeneity in rainbow trout revealed by the organ distribution of mitogenic responses. *Journal of Immunology* 116: 1547-1553.

Evans, M.R., S.J. Larsen, G.H.M. Riekerk and K.G. Burnett. 1997. Patterns of immune response to environmental bacteria in natural populations of the red drum, *Sciaenops ocellatus* (Linnaeus). *Journal of Experimental Marine Biology and Ecology* 208: 87-105.

Fischer, U., M. Ototake and T. Nakanishi. 1998. In vitro cell-mediated cytotoxicity against allogeneic erythrocytes in ginbuna crucian carp and goldfish using a non-radioactive assay. *Developmental and Comparative Immunology* 22: 195-206.

Fischer, U., K. Utke, M. Ototake, J.M. Dijkstra and B. Kollner. 2003. Adaptive cell-mediated cytotoxicity against allogeneic targets by CD8-positive lymphocytes of rainbow trout *Oncorhynchus mykiss*. *Developmental and Comparative Immunology* 27: 323-337.

Fisher, U., K. Utke, T. Somamoto, B. Kollner, M. Ototake and T. Nakanishi. 2006. Cytotoxic activities of fish leucocytes. *Fish and Shellfish Immunology* 20: 209-226.

Galeotti, M., D. Volpatti and M. Rusvai. 1999. Mitogen induced *in vitro* stimulation of lymphoid cells from organs of sea bass *Dicentrarchus labrax* L. *Fish and Shellfish Immunology* 9: 227-232.

Georgilis, K., C. Schaefer, C.A. Dinarello and M.S. Klempner. 1987. Human recombinant interleukin 1 beta has no effect on intracellular calcium or on functional responses of human neutrophils. *Journal of Immunology* 138: 3403-3407.

Haire, R.N., M.K. Kitzan Haindfield, J.B. Turpen and G.W. Litman. 2002. Structure and diversity of T-lymphocyte antigen receptors alpha and gamma in *Xenopus*. *Immunogenetics* 54: 431-438.

Hansen, J.D. and A.G. Zapata. 1998. Lymphocyte development in fish and amphibians. *Immunological Reviews* 166: 199-220.

Hansen, J.D. and P. Strassburger. 2000. Description of an ectothermic TCR coreceptor, CD8α, in rainbow trout. *Journal of Immunology* 164: 3132-3139.

Hansen, J.D., E.D. Landis and R.B. Phillips. 2005. Discovery of a unique Ig heavy-chain isotype (IgT) in rainbow trout: Implications for a distinctive B cell developmental pathway in teleost fish. *Proceedings of the National Academy of Sciences of the United States of America* 102: 6919-6924.

Hebert, P., J. Ainsworth and B. Boyd. 2000. Cholera toxin has adjuvant properties in channel catfish when injected intraperitoneally. *Fish and Shellfish Immunology* 10: 469-474.

Heilig, J.S. and S. Tonegawa. 1986. Diversity of murine gamma genes and expression in fetal and adult T lymphocytes. *Nature (London)* 322: 836-840.

Hong, S., J. Zou, M. Crampe, S. Peddie, G. Scapigliati, N. Bols, C. Cunningham and C.J. Secombes. 2001. The production and bioactivity of rainbow trout *Oncorhynchus mykiss* recombinant IL-1β. *Veterinary Immunology and Immunopathology* 81: 1-14.

Hutchinson, T.H. and M.J. Manning. 1996. Seasonal trends in serum lysozyme activity and total protein concentration in dab *Limanda limanda* L. sampled from Lyme Bay, UK. *Fish and Shellfish Immunology* 6: 473-482.

Janicke, H., P.M. Taylor and C.E. Bryant. 2003. Lipopolysaccharide and interferon gamma activate nuclear factor kappa B and induce cyclo-oxygenase-2 in equine vascular smooth muscle cells. *Research in Veterinary Science* 75: 133-140.

Jenkins, P.G., A.B. Wrathmell, J.E. Harris and A.L. Pulsford. 1994. Systemic and mucosal immune responses to enterically delivered antigen in *Oreochromis mossambicus*. *Fish and Shellfish Immunology* 4: 255-271.

Joo, M., G.Y. Park, J.G. Wright, T.S. Blackwell, M.L. Atchison and J.W. Christman. 2004. Transcriptional regulation of the cyclooxygenase-2 gene in macrophages by PU.1. *Journal of Biological Chemistry* 279: 6658-6665.

Joosten, E.P.H.M. 1997. *Immunological aspects of oral vaccination in fish*. Ph.D. Thesis. Wageningen Agricultural University, Wageningen, The Netherlands.

Joosten, E.P.H.M., M. Aviles-Trigueros, P. Sorgeloos and J.H.W.M. Rombout. 1995. Oral vaccination of juvenile carp *Cyprinus carpio* and gilthead seabrea *Sparus aurata* with bioencapsulated *Vibrio anguillarum* bacterin. *Fish and Shellfish Immunology* 5: 289-299.

Kaattari, S.L. and J.D. Piganelli. 1996. The specific immune system: Humoral defence. In: *The Fish Immune System: Organism, Pathogen and Environment,* G.K. Iwama and T. Nakanishi (eds.). Academic Press, San Diego, pp. 207-254.

Kappler, J.W., B. Skidmore, J. White and P. Marrack. 1981. Antigen-inducible, H-2 restricted, interleukin-2 producing T-cell hybridomas: lack of independent antigen and H-2 recognition. *Journal of Experimental Medicine* 153: 1198-1214.

Kawakami, Y., H.N. Shinohara, Y. Fukuda, H. Yamashita, H. Kihara and M. Sakai. 1997. The efficacy of lipopolysaccharide mixed chloroform-killed cell (LPS-CKC) bacterin of *Pasteurella piscicida* on Yellowtail, *Seriola quinqueradiata*. *Aquaculture* 154: 95-105.

Klesius, P.H. 1990. Effect of size and temperature on the quantity of immunoglobulin in channel catfish *Ictalurus punctatus*. *Veterinary Immunology and Immunopathology* 24: 187-196.

Koning, F., A.M. Kruisbeek, W.L. Maloy, S. Marusic-Galesic, D.M. Pardoll, E.M. Shevach, G. Stingl, R. Valas, W.M. Yokoyama and J.E. Coligan. 1988. T cell receptor gamma/delta chain diversity. *Journal of Experimental Medicine* 167: 676-681.

Krakauer, T., D. Mizel and J.J. Oppenheim. 1982. Independent and synergistic thymocyte proliferative activities of PMA and IL1. *Journal of Immunology* 129: 939-941.

Kronke, M., W.J. Leonard, J.M. Depper, S.K. Arya, F. Wong-Staal, R.C. Gallo, T.A. Waldmann and W.C. Greene. 1984. Cyclosporin A inhibits T-cell growth factor gene expression at the level of mRNA transcription. *Proceedings of the National Academy of Sciences of the United States of America* 81: 5214-5218.

Kruiswijk, C.P., T.T. Hermsen, A.H. Westphal, H.F. Savelkoul and R.J. Stet. 2002. A novel functional class I lineage in zebrafish *Danio rerio*, carp *Cyprinus carpio*, and large barbus *Barbus intermedius* showing an unusual conservation of the peptide binding domains. *Journal of Immunology* 169: 1936-1947.

Lange, S., B.K. Gudmundsdottir and B. Magnadottir. 2001. Humoral immune parameters of cultured Atlantic halibut *Hippoglossus hippoglossus* L. *Fish and Shellfish Immunology* 11: 523-535.

Langenau, D.M. and L.I. Zon. 2005. The zebrafish: A new model of T-cell and thymic development. *National Review of Immunology* 5: 307-317.

Link, B.A., M.P. Gray, R.S. Smith and S.W. John. 2004. Intraocular pressure in zebrafish: Comparison of inbred strains and identification of a reduced melanin mutant with raised IOP. *Investigations in Ophthalmology and Visual Science* 45: 4415-4422.

Lòpez-Dòriga, M.V., A.C. Barnes, N.M.S. Dos Santos and A.E. Ellis. 2000. Invasion of fish epithelial cells by *Photobacterium damselae* ssp. *piscicida*: Evidence for receptor specificity, and effect of capsule and serum. *Microbiology* 146: 21-30.

Lumsden, J.S., V.E. Ostland, D.D. MacPhee and H.W. Ferguson. 1995. Production of gill-associated and serum antibody by rainbow trout *Oncorhynchus mykiss* following immersion immunization with acetone-killed *Flavobacterium branchiophilum* and the relationship to protection from experimental challenge. *Fish and Shellfish Immunology* 5: 151-165.

Lycke, N. and J. Holmgren. 1986. Strong adjuvant properties of cholera toxin on gut mucosal immune responses to orally presented antigens. *Immunology* 59: 301-308.

MacDougal, K.C., P.A. Mericko and K.G. Burnett. 1999. Antigen receptor-mediated activation of extracellular related kinase (ERK) in B lymphocytes of teleost fishes. *Developmental and Comparative Immunology* 23: 221-230.

Magarinos, B., J.L. Romalde, M. Noya, J.L. Barja and A.E. Toranzo. 1996. Adherence and invasive capacities of the fish pathogen *Pasteurella piscicida*. *FEMS Microbiology Letters* 138: 29-34.

Manning, M. and T. Nakanishi. 1996. The specific immune system: Cellular defenses. In: *The Fish Immune System: Organism, Pathogen and Environment*, G.K. Iwama and T. Nakanishi (eds.). Academic Press Inc, San Diego, pp. 159-205.

Marchalonis, J.J., S.F. Schluter, H.Y. Yang, V.S. Hohman, K. McGee and L. Yeaton. 1992. Antigenic cross-reactions among immunoglobulin of diverse vertebrates (elasmobranchs to man) detected using xenoantisera. *Comparative Biochemistry and Physiology* A101: 675-687.

Marsden, M. and C.J. Secombes. 1997. The influence of vaccine preparations on the induction of antigen specific responsiveness in rainbow trout, *Oncorhynchus mykiss*. *Fish and Shellfish Immunology* 7: 455-469.

Marsden, M.J., L.M. Vaughan, T.J. Foster and C.J. Secombes. 1996. A live (Delta aroA) *Aeromonas salmonicida* vaccine for furunculosis responses in rainbow trout *Oncorhynchus mykiss*. *Infection and Immunology* 64: 3863-3869.

Mazzolini, E., A. Fabris, G. Ceschia, D. Vismara, A. Magni, A. Amadei, A. Passera, L. Danielis and G. Giorgetti. 1998. Pathogenic variability of *Pasteurella piscicida* during *in vitro* cultivation as a preliminary study for vaccine production. *Journal of Applied Ichthyology—Zeitschrift für Angewandte Ichthyologie* 14: 265-268.

Meloni, S. and G. Scapigliati. 2000. Evaluation of immunoglobulins produced in vitro by head kidney leucocytes of sea bass *Dicentrarchus labrax* by immunoenzymatic assay. *Fish and Shellfish Immunology* 10: 95-99.

Meloni, S., G. Zarletti, S. Benedetti, E. Randelli, F. Buonocore and G. Scapigliati. 2006. Cellular activities during a mixed leucocyte reaction in the teleost sea bass *Dicentrarchus labrax*. *Fish and Shellfish Immunology* 20: 739-749.

Meseguer, J., M.A. Esteban and V. Mulero. 1996. Nonspecific cell-mediated cytotoxicity in the seawater teleosts *Sparus aurata* and *Dicentrarchus labrax*: Ultrastructural study of target cell death mechanisms. *Anatomical Record* 244: 499-505.

Miller, N.W., R.C. Sizemore and L.W. Clem. 1985. Phylogeny of lymphocyte heterogeneity: The cellular requirements for in vitro antibody responses of channel catfish leukocytes. *Journal of Immunology* 134: 2884-2888.

Miller, N.W., J.E. Van Ginkel, F. Ellsaesser and L.W. Clem. 1987. Phylogeny of lymphocyte heterogeneity: Identification and separation of functionally distinct subpopulations of channel catfish lymphocytes with monoclonal antibodies. *Developmental and Comparative Immunology* 11: 739-748.

Mitchell, R.A., H. Liao, J. Chesney, G. Fingerle-Rowson, J. Baugh, J. David and R. Bucala. 2002. Macrophage migration inhibitory factor (MIF) sustains macrophage pro-inflammatory function by inhibiting p53: Regulatory role in the innate immune response. *Proceedings of the National Academy of Sciences of the United States of America* 99: 345-350.

Moore, J.D., M. Ototake and T. Nakanishi. 1998. Particulate antigen uptake during immersion immunisation of fish: The effectiveness of prolonged exposure and the roles of skin and gill. *Fish and Shellfish Immunology* 8: 393-407.

Mulero, V., M.A. Esteban, J. Muñoz and J. Meseguer. 1994. Non-specific cytotoxic response against tumor target cells mediated by leucocytes from seawater teleosts, *Sparus aurata* and *Dicentrarchus labrax:* An ultrastructural study. *Archives of Histology and Cytology* 57: 351-358.

Muñoz, P., A. Sitja-Bobadilla and P. Alvarez-Pellitero. 2000. Cellular and humoral immune response of European sea bass *Dicentrarchus labrax* L. (Teleostei: Serranidae) immunized with *Sphaerospora dicentrarchi* Myxosporea: Bivalvulida. *Parasitology* 120: 465-477.

Nakanishi, T. 1986. Seasonal changes in the humoral immune response and the lymphoid tissues of the marine teleost, *Sebastiscus marmoratus*. *Veterinary Immunology and Immunopathology* 12: 213-221.

Nakanishi, T. and M. Ototake. 1999. The graft-versus-host reaction (GVHR) in the ginbuna crucian carp, *Carassius auratus langsdorfii*. *Developmental and Comparative Immunology* 23: 15-26.

Nehr, O., I.P. Blancheton and E. Alliot. 1996. Development of an intensive culture system for sea bass *Dicentrarchus labrax* larvae in sea enclosures. *Aquaculture* 142: 43-58.

Neidleman, J.A., M. Vajdy, M. Ugozzoli, G. Ott and D. O'Hagan. 2000. Genetically detoxired mutants of heat-labile enterotoxin from *Escherichia coli* are effective adjuvants for induction of cytotoxic T-cell responses against HIV-1 gag-p55. *Immunology* 101: 154-160.

Obach, A., C. Quentel and F.B. Laurencin. 1993. Effects of alpha-tocopherol and dietary oxidized fish oil on the immune response of sea bass *Dicentrarchus labrax*. *Diseases of Aquatic Organisms* 15: 175-185.

Oppenheim, J.J., B.M. Stadler, R.P. Siraganian, M. Mage and B. Mathieson. 1982. Lymphokines: Their role in lymphocyte responses. Properties of interleukin 1. *Federal Proceedings* 41: 257-262.

Palenzuela, O., A. Sitjà-Bobadilla and P. Alvarez-Pellitero. 1996. Isolation and partial characterization of serum immunoglobulins from sea bass *Dicentrarchus labrax* L. and gilthead sea bream *Sparus aurata* L. *Fish and Shellfish Immunology* 6: 81-94.

Palm, R.C., M.L. Landolt and R.A. Busch. 1998. Route of vaccine administration: Effects on the specific humoral response in rainbow trout *Oncorhynchus mykiss*. *Diseases of Aquatic Organisms* 33: 157-166.

Partula, S., A. De Guerra, J.S. Fellah and J. Charlemagne. 1995. Structure and diversity of the T cell antigen receptor-chain in a teleost fish. *Journal of Immunology* 155: 699-706.

Patel, R., M.G. Attur, M. Dave, S.B. Abramson and A.R. Amin. 1999. Regulation of cytosolic COX-2 and prostaglandin E2 production by nitric oxide in activated murine macrophages. *Journal of Immunology* 162: 4191-4197.

Percival, R.S., P.D. Marsh and S.J. Challacombe. 1996. Serum antibodies to commensal oral and gut bacteria vary with age. *FEMS Immunology and Medical Microbiology* 15: 35-42.

Petrie-Hanson, L. and A.J. Ainsworth. 1997. Humoral immune responses of channel catfish fry and secreting cells in the gills, head-kidney and peripheral blood leucocytes of dab *Limanda limanda*. *Veterinary Immunology and Immunopathology* 58: 363-374.

Phelan, P. E., M.T. Mellon and C.H. Kim. 2005. Functional characterization of full-length TLR3, IRAK-4, and TRAF6 in zebrafish (*Danio rerio*). *Molecular Immunology* 42: 1057-1071.

Picchietti, S., F.R. Terribili, L. Mastrolia, G. Scapigliati and L. Abelli. 1997. Expression of lymphocyte antigenic determinants in developing GALT of the sea bass *Dicentrarchus labrax* L. *Anatomy and Embryology* 196: 457-463.

Picchietti, S., G. Scapigliati, M. Fanelli, F. Barbato, S. Canese, L. Mastrolia, M. Mazzini and L. Abelli. 2002. Sex-related variation in serum immunoglobulins during reproduction in gilthead seabream and evidence for a transfer from the female to the eggs. *Journal of Fish Biology* 59: 1503-1511.

Picchietti, S., A.R. Taddei, G. Scapigliati, F. Buonocore, A.M. Fausto, N. Romano, M. Mazzini, L. Mastrolia and L. Abelli. 2004. Immunoglobulin protein and gene transcripts in ovarian follicles throughout oogenesis in the teleost *Dicentrarchus labrax*. *Cell and Tissue Research* 315: 259-270.

Pickering, A.D. 1999. Stress responses of farmed fish. In: *Biology of Farmed Fish*, K.D. Black and A.D. Pickering (eds.). Sheffield Academic Press, Sheffield, UK, pp. 222-255.

Pretti, C., M.T.A. Milone and A.M. Cognetti Varriale. 1999. Fish pasteurellosis: Sensitivity of Western blotting analysis on the internal organs of experimentally infected sea bass *Dicentrarchus labrax*. *Bulletin of the European Association of Fish Pathologists* 19: 120-122.

Racine, B.P., E. Marti, A. Busato, R. Weilenmann, S. Lazary and M.E. Griot-Wenk. 1999. Influence of sex and age on serum total immunoglobulin E concentration in Beagles. *American Journal of Veterinary Research* 60: 93-97.

Reddy, K.V., R.D. Yedery and C. Aranha. 2004. Antimicrobial peptides: premises and promises. *International Journal of Antimicrobial Agents* 24: 536-547.

Ristow, S.S., L.D. Grabowski, C. Ostberg, B. Robison and G.H. Thorgaard. 1998. Development of long-term cell lines from homozygous clones of rainbow trout. *Journal of Aquatic Animal Health* 10: 75-82.

Robohm, R.A. and R.A. Koch. 1995. Evidence for oral ingestion as the principal route of antigen entry in bath-immunized fish. *Fish and Shellfish Immunology* 5: 137-150.

Rodrigues, P.N., T.T. Hermsen, A. van Maanen, A.J. Taverne-Thiele, J.H. Rombout, B. Dixon and R.J. Stet. 1998. Expression of MhcCyca class I and class II molecules in the early life history of the common carp *Cyprinus carpio* L. *Developmental and Comparative Immunology* 22: 493-506.

Rodriguez, M.F., G.D. Wiens, M.K. Purcell and Y. Palti. 2005. Characterization of Toll-like receptor 3 gene in rainbow trout *Oncorhynchus mykiss. Immunogenetics* 57: 510-519.

Romano, N., L. Abelli, L. Mastrolia and G. Scapigliati. 1997. Immunocytochemical detection and cytomorphology of lymphocyte subpopulations in a teleost fish *Dicentrarchus labrax* L. *Cell and Tissue Research* 289: 163-171.

Rombout, J.H., P.H. Joosten, M.Y. Engelsma, A.P. Vos, N. Tarvene and J.J. Taverne-Thiele. 1998. Indications for a distinct putative T cell population in mucosal tissue of carp (*Cyprinus carpio* L.). *Developmental and Comparative Immunology* 22: 63-77.

Rombout, J.H., H.B. Huttenhuis, S. Picchietti and G. Scapigliati. 2005. Phylogeny and ontogeny of fish leucocytes. *Fish and Shellfish Immunology* 19: 441-955.

Romestand, B., G. Breuil, C.A.F. Bourmaud, J.L. Coeurdacier and G. Bouix. 1995. Development and characterisation of monoclonal antibodies against seabass immunoglobulins *Dicentrarchus labrax* L. *Fish and Shellfish Immunology* 5: 347-357.

Rosoff, P.M., N. Savage and C.A. Dinarello. 1988. Interleukin-1 stimulates diacylglycerol production in T lymphocytes by a novel mechanism. *Cell* 54: 73-81.

Sánchez, C.P., A. Lopez-Fierro Zapata and J. Dominguez. 1993. Characterisation of monoclonal antibodies against heavy and light chains of trout immunoglobulins. *Fish and Shellfish Immunology* 3: 237-251.

Santarem, M.M. and A. Figueras. 1995. Leucocyte numbers and phagocytic activity in turbot *Scophthalmus maximus* following immunization with *Vibrio damsela* and *Pasteurella piscicida* O-antigen bacterins. *Diseases of Aquatic Organisms* 23: 213-220.

Scapigliati, G., N. Romano and L. Abelli. 1999. Monoclonal antibodies in teleost fish immunology: Identification, ontogeny and activity of T- and B-lymphocytes. *Aquaculture* 172: 3-28.

Scapigliati, G., F. Buonocore and M. Mazzini. 2006. Biological activity of cytokines: an evolutionary perspective. *Current Pharmaceutical Design* 12: 3071-3081.

Scapigliati, G., M. Mazzini, L. Mastrolia, N. Romano and L. Abelli. 1995. Production and characterisation of a monoclonal antibody against the thymocytes of the sea bass *Dicentrarchus labrax* L. (Teleostea, Percicthydae). *Fish and Shellfish Immunology* 5: 393-405.

Scapigliati, G., N. Romano, S. Picchietti, M. Mazzini, L. Mastrolia, D. Scalia and L. Abelli. 1996. Monoclonal antibodies against sea bass *Dicentrarchus labrax* L. immunoglobulins: Immunolocalization of immunoglobulin-bearing cells and applicability in immunoassays. *Fish and Shellfish Immunology* 6: 383-401.

Scapigliati, G., F. Chausson, E.L. Cooper, D. Scalia and M. Mazzini. 1997. Qualitative and quantitative analysis of serum immunoglobulins of four Antarctic fish species. *Polar Biology* 18: 209-213.

Scapigliati, G., N. Romano, L. Abelli, S. Meloni, A.G. Ficca, F. Buonocore, S. Bird and C.J. Secombes. 2000. Immunopurification of T-cells from sea bass *Dicentrarchus labrax* L. *Fish and Shellfish Immunology* 10: 329-341.

Scapigliati, G., F. Buonocore, S. Bird, J. Zou, P. Pelegrin, C. Falasca, D. Prugnoli and C.J. Secombes. 2001. Phylogeny of cytokines: Molecular cloning and expression analysis of sea bass *Dicentrarchus labrax* interleukin-1 beta. *Fish and Shellfish Immunology* 11: 711-726.

Scapigliati, G., S. Costantini, G. Colonna, A. Facchiano, F. Buonocore, P. Bossù, C. Cunningham, J.W. Holland and C.J. Secombes. 2004. Modelling of fish interleukin-1 and its receptor. *Developmental and Comparative Immunology* 28: 429-441.

Schreck, C.B. 1996. Immunomodulation: endogenous factors. In: *The Fish Immune System: Organisms, Pathogen and Environment,* G.K. Iwama and T. Nakanishi (eds.). Academic Press, San Diego, pp. 311-337.

Secombes, C.J., S. Bird and J. Zou. 2005. Adaptive immunity in teleosts: cellular immunity. *Developmental Biology* (Basel) 121: 25-32.

Shike, H., X. Lauth, M.E. Westerman, V.E. Ostland, J.M. Carlberg, J.C. Van Olst, C. Shimizu, P. Bulet and J.C. Burns. 2002. Bass hepcidin is a novel antimicrobial peptide induced by bacterial challenge. *European Journal of Biochemistry* 269: 2232-2237.

Sideris, D.C. 1997. Cloning, expression and purification of the coat protein of encephalitis virus (DIEV) infecting *Dicentrarchus labrax. Biochemistry, Molecular Biology International* 42: 409-417.

Sitjà-Bobadilla, A. and J. Pérez-Sánchez. 1999. Diet-related changes in non-specific immune response of European sea bass *Dicentrarchus labrax* L. *Fish and Shellfish Immunology* 9: 637-640.

Sizemore, R.G., N.W. Miller, M.A. Cuchens, C.J. Lobb and L.W. Clem. 1984. Phylogeny of lymphocyte heterogeneity: The cellular requirements for *in vitro* mitogenic responses of channel catfish leukocytes. *Journal of Immunology* 133: 2920-2924.

Skliris, G.P. and R.H. Richards. 1999. Induction of nodavirus disease in sea bass, *Dicentrarchus labrax*, using different infection models. *Virus Research* 63: 85-93.

Sorensen, B. and R.E. Weber. 1995. Effects of oxygenation and the stress hormones adrenaline and cortisol on the viscosity of blood from the trout *Oncorhynchus mykiss. Journal of Experimental Biology* 198: 953-959.

Sowder, J.T., C.L. Chen, L.L. Ager, M.M. Chan and M.D. Cooper. 1988. A large subpopulation of avian T cells express a homologue of the mammalian T gamma/delta receptor. *Journal of Experimental Medicine* 167: 315-322.

Steer, S.A. and J.A. Corbett. 2003. The role and regulation of COX-2 during viral infection. *Viral Immunology* 16: 447-460.

Stenvik, J. and T.O. Jörgensen. 2000. Immunoglobulin D (IgD) of Atlantic cod has a unique structure. *Immunogenetics* 51: 452-461.

Stet, R.J.M., C.P. Kruiswijk, J.P.J. Saeji and G.F. Wiegertjes. 1998. Major histocompatibility genes in ciprinid fishes: Theory and practice. *Immunological Reviews* 166: 301-316.

Stolen, S.S and O. Makela. 1975. Carrier preimmunisation in the anti-hapten response of a marine fish. *Nature (London)* 254: 718-719.

Suetake, H., K. Arakia and Y. Suzuki. 2004. Cloning, expression, and characterization of fugu CD4, the first ectothermic animal CD4. *Immunogenetics* 56: 368-374.

Tagliabue, A. and D. Boraschi. 1993. Cytokines as vaccine adjuvants: Interleukin 1 and its synthetic peptide. *Vaccine* 11: 594-595.

Trede, N.S., D.M. Langenau, D. Traver, A.T. Look and L.I. Zon. 2004. The use of zebrafish to understand immunity. *Immunity* 20: 367-379.

Tsoukas, C.D., B. Landgraf, J. Bentin, M. Valentine, M. Lotz, J.H. Vaughan and D.A Carson. 1985. Activation of resting T lymphocytes by anti-CD3 (T3) antibodies in the absence of monocytes. *Journal of Immunology* 135: 1719-1723.

Tsukamoto, K., S. Hayashi, M.Y. Matsuo, M.I. Nonaka, M. Kondo, A. Shima, S. Asakawa, N. Shimizu and M. Nonaka. 2005. Unprecedented intraspecific diversity of the MHC class I region of a teleost medaka, *Oryzias latipes*. *Immunogenetics* 57: 420-431.

Vale, A., A. Afonso and M.T. Silva. 2002. The professional phagocytes of sea bass *Dicentrarchus labrax* L.: Cytochemical characterisation of neutrophils and macrophages in the normal and inflamed peritoneal cavity. *Fish and Shellfish Immunology* 13: 183-198.

Vale, A., F. Marques and M.T. Silva. 2003. Apoptosis of sea bass *Dicentrarchus labrax* L. neutrophils and macrophages induced by experimental infection with *Photobacterium damselae* subsp. *piscicida*. *Fish and Shellfish Immunology* 15: 129-144.

Vallejo, A.N., N.W. Miller and L.W. Clem. 1991. Phylogeny of immune recognition: Role of alloantigens in antigen presentation in channel catfish immune responses. *Immunology* 74: 165-168.

Van der Sar, A.M., B.J. Appelmelk, C.M. Vanderbrouke-Grauls and W. Bitter. 2004. A star with stripes: Zebrafish as an infection model. *Trends in Microbiology* 12: 451-457.

Van Ginkel, F.W., N.W. Miller, M.A. Cuchens and L.W. Clem. 1994. Activation of channel catfish B cells by membrane immunoglobulin cross-linking. *Developmental and Comparative Immunology* 18: 97-107.

Van Muiswinkel, W.B. 1995. The Piscine immune system: innate and acquired immunity. In: *Fish Diseases and Disorders*, P. T. K. Woo (ed.). Department of Zoology, University of Guelph, Canada, Vol. 1: Protozoan and Metazoan Infections: pp. 729-749.

Van Wauwe, J.P., J. Gossens and G. Van Nyen. 1984. Inhibition of lymphocyte proliferation by monoclonal antibody directed against the T3 antigen on human T cells. *Cell Immunology* 86: 525-534.

Vazzana, M., D. Parrinello and M. Cammarata. 2003. Chemiluminescence response of beta-glucan stimulated leukocytes isolated from different tissues and peritoneal cavity of *Dicentrarchus labrax*. *Fish and Shellfish Immunology* 14: 423-434.

Vellutini, M., G. Viegi, D. Parrini, M. Pedreschi, S. Baldacci, P. Modena, P. Biavati, M. Simoni, L. Carrozzi and C. Giuntini. 1997. Serum immunoglobulins E are related to menstrual cycle. *European Journal of Epidemiology* 13: 931-935.

Venkatesh, B., Y. Ning and S. Brenner. 1999. Late changes in spliceosomal introns define clades in vertebrate evolution. *Proceedings of the National Academy of Sciences of the United States of America* 96: 10267-10271.

Vervelde, L., E.M. Janse, A.N. Vermeulen and S.H. Jeurissen. 1998. Induction of a local and systemic immune response using cholera toxin as vehicle to deliver antigen in the lamina propria of the chicken intestine. *Veterinary Immunology and Immunopathology* 16: 261-272.

Volpatti, D., M. Rusvai, L. D'Angelo, M. Manetti and M. Galeotti. 1996. Evaluation of sea bass *Dicentrarchus labrax* lymphocytes proliferation after in vitro stimulation with mitogens. *Bollettino della Societa Italiana di Patologia Ittica* 8: 22-26.

Von Aulock, S., C. Hermann and T. Hartung. 2003. Determination of the eicosanoid response to inflammatory stimuli in whole blood and its pharmacological modulation ex vivo. *Journal of Immunological Methods* 277: 53-63.

Von Hertzen, L., H.M. Surcel, J. Kaprio, M. Koskenvuo, A. Bloigu, M. Leinonen and P. Saikku. 1998. Immune responses *to Chlamydia pneumoniae* in twins in relation to gender and smoking. *Journal of Medical Microbiology* 47: 441-446.

Wagnerova, M., V. Wagner, Z. Madlo, V. Zavazal, D. Wokounova, J. Kriz and O. Mohyla. 1986. Seasonal variations in the level of immunoglobulins and serum proteins of children differing by exposure to air-borne lead. *Journal of Hygiene, Epidemiology, Microbiology and Immunology* 30: 127-138.

Warner, S.J.C., K.R. Auger and P. Libby. 1987. Interleukin 1 induces interleukin 1. II. Recombinant human interleukin 1 induces interleukin 1 production by adult human vascular endothelial cells. *Journal of Immunology* 139: 1911-1917.

Warr, G. 1995. The immunoglobulin genes of fish. *Developmental and Comparative Immunology* 19: 1-12.

Wilson, M., E. Bengten, N.W. Miller, L.W. Clem, L. Du Pasquier and G.W. Warr. 1997. A novel chimeric Ig heavy chain from a teleost fish shares similarities to IgD. *Proceedings of the National Academy of Sciences of the United States of America* 94: 4593-4597.

Xie, W., J.G. Chipman, D.L. Robertson, R.L. Erikson and D.L. Simmons. 1991. Expression of a mitogen-responsive gene encoding prostaglandin synthase is regulated by mRNA splicing. *Proceedings of the National Academy of Sciences of the United States of America* 88: 2692-2696.

Yin, Z. and J. Kwang. 2000. Carp interleukin-1β in the role of an immuno-adjuvant. *Fish and Shellfish Immunology* 10: 375-378.

Yu, C.L., M.H. Huang, Y.Y. Kung, C.Y. Tsai, S.T. Tsai, D.F. Huang, K.H. Sun, S.H. Han and H.S. Yu. 1998. Interleukin-13 increases prostaglandin E2 (PGE2) production by normal human polymorphonuclear neutrophils by enhancing cyclooxygenase 2 (COX-2) gene expression. *Inflammation Research* 47: 167-173.

Zapata, A.G., A. Varas and M. Torroba. 1992. Seasonal variations in the immune system of lower vertebrates. *Immunology Today* 13: 142-147.

Zou, J., P.S. Grabowski, C. Cunningham and C.J. Secombes. 1999a. Molecular cloning of interleukin 1β from rainbow trout *Oncorhynchus mykiss* reveals no evidence of an ICE cut site. *Cytokine* 11: 552-560.

Zou, J., N.F. Neumann, J.W. Holland, M. Belosevic, C. Cunningham, C.J. Secombes and A.F. Rowley. 1999b. Fish macrophages express a cyclo-oxygenase-2 homologue after activation. *Biochemical Journal* 340: 153-159.

CHAPTER

7

Immunoglobulin Genes of Teleosts: Discovery of New Immunoglobulin Class

Ram Savan* and Masahiro Sakai§

INTRODUCTION

Immunoglobulins form the humoral component of the adaptive immune system. An immunoglobulin molecule is made up of a heterodimer of two immunoglobulin heavy (IGH) and two light chains (IGL). The heavy and light chains are composed of variable and constant domains. In mammals, immunoglobulin heavy chain genes are classified mainly into five isotypes: IgM (μ), IgD (δ), IgG (γ), IgA (α) and IgE (ε). However, not all mammalian Ig homologs have been found in other vertebrates. Among the Ig isotypes, definitive orthologs of mammalian IgM and IgD have been reported in most of the non-mammalian vertebrates (Dahan *et al.*, 1983; Kokubu *et al.*, 1988; Schwager *et al.*, 1988a, b; Wilson *et al.*, 1997;

Authors' addresses: Faculty of Agriculture, University of Miyazaki, Gakuen kibanadai nishi 1-1, Miyazaki, 889-2192, Japan.
**Present address*: National Cancer Institute, Frederick, MD 21701, USA.
Corresponding authors: E-mail: *savanr@mail.nih.gov; §m.sakai@cc.miyazaki-u.ac.jp

Ota *et al.*, 2003; Ohta and Flajnik, 2006; Zhao *et al.*, 2006). However, few of the Ig heavy chain isotypes discovered in non-mammalian vertebrates have been assigned 'non-conventional' nomenclature since they were not readily classifiable to mammalian Igs based on the structure and genomic organization. IgX (ξ) (similar to IgA) and IgY (ψ) (ortholog of IgG) have been reported in birds, reptiles and amphibians. Recently, IgF (similar to IgY) has been discovered from frog (*Xenopus laevis*) (Zhao *et al.*, 2006).

The chromosomal organization of immunoglobulin loci has been thoroughly described in mammals, wherein the variable heavy chain (V_H) segments are followed by the diversity (D_H), the joining (J_H) and constant (C_H) domain exons. This arrangement is called the classical 'translocon' type. In mammals, the μ constant domain exons are located closest to the J_H segments and followed in order by δ, γ, ε, and α constant domains. In amphibians, the known constant Ig domains are IGM, IGD, IGX, IGY, and IGF and these are present in the same order (Ohta and Flajnik, 2006; Zhao *et al.*, 2006). Although, the overall translocon type arrangement of the IGH loci is found conserved in birds and amphibians, some species-specific variations in the order and orientation of the genes have been observed. Variations in the transcriptional orientation of constant genes have been reported in duck IGH locus; the α (in inverted transcriptional orientation) is located between μ and γ regions (Zhao *et al.*, 2000).

Among fish, the cartilaginous and bony fishes have distinct genomic organization of IGH loci. Cartilaginous fish have a distinct 'multi-cluster' type of organization of IGH loci; each cluster is a unit of V_H, D_H, J_H and C_H segments, wherein the gene rearrangements are restricted to individual clusters (Du Pasquier and Flajnik, 1999). Among teleosts, catfish (*Ictalurus punctatus*) has been one of the first and widely studied IGH loci (Wilson *et al.*, 1990, 1997; Ghaffari and Lobb, 1991, 1992, 1999; Warr *et al.*, 1991, 1992; Ventura-Holman *et al.*, 1996; Bengten *et al.*, 2002, 2006; Ventura-Holman and Lobb, 2002). Studies in Atlantic salmon, *Salmo salar* (Hordvik *et al.*, 1999), Atlantic cod, *Gadus morhua* (Stenvik and Jorgensen, 2000), Atlantic halibut, *Hippoglossus hippoglossus* (Hordvik, 2002), Japanese flounder, *Paralichthys olivaceus* (Srisapoome *et al.*, 2004) and fugu, *Takifugu rubripes* (Saha *et al.*, 2004) have significantly contributed to the understanding of IGH loci in teleosts. The genome projects of fugu (Aparicio *et al.*, 2002) and zebrafish (*Danio rerio*; http://www.sanger.ac.uk/Projects/D_rerio/) have significantly aided in the characterization of the teleost IGH loci (Sakai and Savan, 2004; Danilova *et al.*, 2005; Savan *et al.*, 2005b). These and other studies [(rainbow trout, *Oncorhynchus mykiss* (Hansen *et al.*, 2005)] give an emerging consensus that the IGH loci in bony fishes shares a slightly different genomic organization compared to other vertebrates. In this chapter, we will discuss the IGH loci comparing similarities and differences across teleosts.

IMMUNOGLOBULIN HEAVY CHAIN LOCI IN TELEOSTS

The Heavy Chain Variable (VH) Region

The VH, DH, and JH segments in the variable region provide structural diversity necessary for the recognition of antigens. These gene segments undergo processes such as somatic recombination, junctional diversity and somatic mutation. A 110 residue VH region is composed of four framework (FR) regions and three complementarity determining regions (CDR).

The VH genes are classified into families based on the nucleotide identities and genes with 80% nucleotide identity belong to the same family (Bordeur and Riblet, 1984). Furthermore, Ota and Nei (1994) classified VH genes across vertebrates into five major groups A to E. Among these groups, bony fish VH genes were represented in groups C and D.

In teleosts, extensive VH gene repertoire analysis has been conducted through cDNA cloning. Thirteen VH cDNA clones from Atlantic cod were identified which were classified into three VH families (Bengten *et al.*, 1994). The studies on expressed VH genes from Arctic char (*Salvelinus alpinus*) and trout were classified into eight and eleven families, respectively (Roman *et al.*, 1996; Andersson and Matsunaga, 1998). In zebrafish, 74 cDNA clones identified as VH genes were classified into four families (Danilova *et al.*, 2000). To investigate the diversity of immunoglobulin heavy chain variable domain in cold-adapted teleosts emerald rock cod *Trematomus bernacchii*, 45 cDNA clones from spleen library were analyzed (Coscia and Oreste, 2003). These VH sequences were divided into two gene families, which clustered into groups D and C. In catfish, the VH genes were classified into 13 families (Ghaffari and Lobb, 1991; Warr *et al.*, 1991; Ventura-Holman *et al.*, 1996; Yang *et al.*, 2003).

Genomic Structure of VH

Similar to VH genes in mammals, the genomic structure of the fish variable gene segment is composed of a short 5′UTR (un-translated region), a leader peptide (L) interrupted by a short intron (L1-intron-L2), a variable segment ending with a heptamer-spacer-nonamer recombination signal sequence (RSS) at the 3′UTR. The variable domains invariantly harbor two cysteines (IMGT numbering Cys^{23} and Cys^{104}) important for intra-domain-disulphide bridge, Typ^{41} residue in FR2 region and the *Tyr-Tyr-Cys* (YYC) residues in the FR3 region.

The VH segments from zebrafish IGH locus were identified from a genome contig (ZV5 No. zK148A13.00312) of 11.04 Mb in length (Sakai and Savan, 2004; Danilova *et al.*, 2005). These VH segments are present upstream of novel ζ domains and in the same transcriptional orientation (Fig. 7.1). A total of 49 VH segments were found in a 94.2 kb (94,232 bp) segment of this contig (Fig. 7.1; Table 7.1). The region contained 49 VH segments divided into 12 families including 10 pseudo-genes (Table 7.1). Among the 49 VH segments identified, 39 segments fulfilled the requirement to be designated as functional and 10 were classified as pseudogenes (Brodeur *et al.*, 1988; Kabat *et al.*, 1991) (Table 7.1). Out of the 10 pseudogenes, VH12, VH37, and VH16 had truncation's and others encountered premature-termination in the VH region. A highly conserved octamer motifs (ATGCAAT) was present in 34 VH genes and this motif was present upstream of the leader region of VH segment.

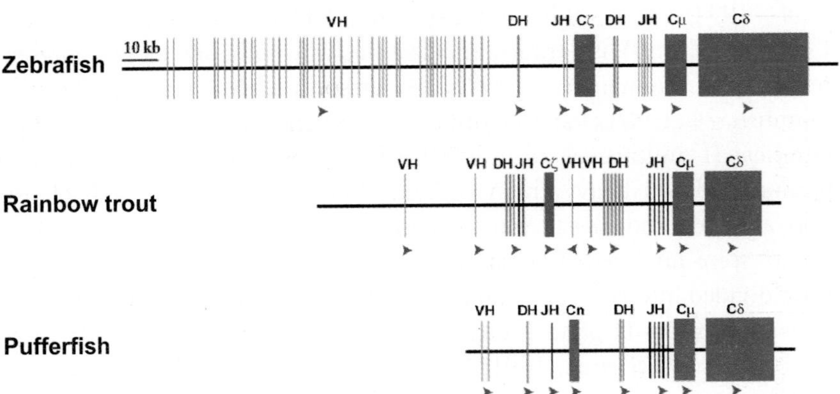

Fig. 7.1 IGH loci in teleosts. A genomic map of IGH locus showing the μ, δ and novel Ig constant domains from pufferfish, zebrafish and rainbow trout. The arrowheads represent the transcriptional orientation of the gene segments.

Table 7.1 Summary of VH segments in zebrafish.

Name	*Octamer*	*bp*	*TATA*	*bp*	ATG	*gt/ag*	*Heptermer*	*bp*	*Nonamer*	*hits to VH of Dr*	*Defects in the psuedogene*	
VH1	ATACAAAT	23	CCTATACATG	18	+	+	CACAGCG	23	ACAAATACT	CA473026	–	F
VH2	ATGCAAAT	–	–	–	+	+	CACCGTG	23	ACAAAAACT	–	–	ORF
VH3	ATGCAAAT	24	AGTATTTAAG	43	+	+	CACAGTG	23	ACAAAAACC	–	–	ORF
VH4	ATGTAAAT	9	CATATTTATA	62	+	+	CACAATG	23	GCAAAAACA	–	–	ORF
VH5	ATGCAAT	24	AGTATTTAAG	43	+	+	CACAGTG	23	ACAAAAACC	–	deletion VH(56)	P
VH6	ATGCAAAT	24	GTTATAACGG	33	+	+	CACAGTG	23	GCAAAAACA	AW184426	–	F
VH7	ATGTAAAT	–	–	–	+	+	CACTGTG	23	ACAAAAACC	–	–	ORF
VH8	ATGCAAAT	40	GATAGAAAGA	31	+	+	CACAGAG	23	TCAAAAACA	–	–	ORF
VH9	ATGTAAAT	–	–	–	+	+	CACAATG	24	ACAAAAACA	–	–	ORF
VH10	ATGCAAAT	24	GTTATAATGG	33	+	+	CACAGTG	23	TCAAAAACC	CK239518	–	F
VH11	ATGCAAAT	–	–	–	+	+	CACAGTG	23	ACAAAAACC	CK240200	–	F
VH12	–	–	–	–	–	–	CACAGTC	21	TCAAATACT	–	truncated	P
VH13	–	–	–	–	–	–	CACAATT	23	ACAAAAACA	–	–	ORF
VH14	ATGCAAAT	24	GTTATAACGG	33	+	+	CACAGTG	23	TCAAAAACC	CK239518	–	F
VH15	ATGCAAAT	–	–	–	+	+	CACAGTG	23	ACAAAAACC	CK240200	–	F
VH16	–	–	–	–	+	–	TGCAGTG	23	ACAAACACA	–	truncated(48–102)	P
VH17	ATGCAAAT	24	AGTATTTAAG	42	+	+	CACAGTG	23	ACAATAACC	CK240200	–	F
VH18	–	–	–	–	–	–	CACAGTG	23	ACAAAAAAA	–	–	ORF
VH19	ATGCAAAT	14	TTTATTAAGT	43	+	+	CACAGTG	23	TCAAAAACT	CK239518	–	F
VH20	ATGCAAAT	15	TGTACAAAAG	55	+	+	CACAGTG	22	ACAAAAACC	–	–	ORF
VH21	ATGCAAAT	15	TTATAAGTTGT	62	+	+	CACTGTG	23	ACAAAAACC	–	–	ORF
VH22	ATGTAAAT	–	–	–	+	+	CACAGTG	23	ACAATAATG	–	–	ORF
VH23	ATGCAAAT	66	ATGATAAAGG	51	+	+	CACAGCA	23	ACAATAAAT	–	–	ORF
VH24	ATGCAAAT	14	CATTTAAGCA	65	+	+	CACAGCA	23	ACAAAAACC	–	–	ORF

(*Table 7.1 contd.*)

(Table 7.1 contd.)

VH25	ATGCAAAC	15	CCTATATAAA	39	+	+	CACTGTG	23	AGGTAAACC	–	–	ORF
VH26	ATGCAAAT	10	CTTAAAACCT	49	+	+	CACAGTG	24	ACAAAAAAC	–	–	ORF
VH27	ATGCAAAT	16	CATATTTAAA	36	+	+	CACTGTG	23	TCAAAAACT	–	–	F
VH28	ATGCAAAT	17	TTTATAAGCC	61	+	+	–	–	–	–	Ter in VH region from 80	P
VH29	–	–	–	–	–	–	–	–	TCAAAAACT	–	fragment of VH only	P
VH30	ATGCAAAC	–	–	–	+	+	CACAGTG	23	ACAAGAACT	–	–	ORF
VH31	ATGCAAAT	–	–	–	+	+	CACAGTT	22	ACAAAAATA	AF273897	–	F
VH32	ATGCAAAC	85	CATATTCTTT	10	+	+	CACTGTG	23	ACAAGAACT	AF273879	–	F
VH33	ATGCAAAT	50	GTTATAAACT	63	+	+	CACAGTT	22	ACAAAAACA	AF273880	–	F
VH34	ATGCAAAC	38	CATATTTAAA	48	+	+	CACAATA	22	ACATAAACC	–	–	ORF
VH35	ATGCAAAC	13	CCTTTAAAAG	96	+	+	CACAGTG	22	ACAAAATCT	AF273883	–	F
VH36	–	–	–	–	+	+	CACAGAT	23	ACAAGAACT	–	Ter in VH region from 94	P
VH37	TTGCAAAAT	60	CTTATTTAAC	141	+	+	CACTGTG	23	ACAAAAACC	–	truncated from 100	P
VH38	ATGCAAAT	31	ATGATAAAGG	38	+	+	CACAGCA	23	ACAAAAACA	–	–	ORF
VH39	–	–	–	–	+	+	CACAGAT	23	ACAAGAACT	–	–	ORF
VH40	ATGCAAAT	10	CTTAAAAACT	49	+	+	CACAGTG	22	GCACAAAAT	–	Ter in VH region from 55	P
VH41	ATGCAAAT	16	CATATTTAAA	39	+	+	CACTGTG	23	TCAAAAACA	–	–	ORF
VH42	ATGCAAAC	17	TTTATAAGCC	69	+	+	TGCAGTG	23	ACATAAATG	–	Ter in VH region from 3	P
VH43	ATGCAAAT	17	TGTGTAAATA	55	+	+	CACAGTG	22	ACATAAACC	–	–	ORF
VH44	ATGCAAAT	–	–	–	+	+	CACAGTG	23	ACAAAAACA	–	–	ORF
VH45	ATGCAAAT	13	TGTATATAAC	72	+	+	CACAGCG	23	ACAAGAACT	–	–	ORF
VH46	–	–	–	–	+	+	GACACTC	24	TCAAAAACC	–	–	ORF
VH47	ATGCAAAT	41	TGTCTATATG	36	+	+	CACAGTA	22	ACAAAAACA	CD758662	–	F
VH48	ATGCAAAT	–	–	–	+	–	CACAATG	23	GCACCTTAA	–	Ter in VH region from 94	P
VH49	ATTCAAAT	18	TCTATTCTCT	82	+	+	CACAGTC	22	TCAAAAACA	–	–	ORF

P: Psuedogene; ORF: Open-reading frame; F: Functional gene

Among the rest, nine and eight genes had loosely defined or without octamer motifs, respectively. TATA box were also identified in 31 VH genes, with some sequences harboring two such motifs. The recombination signal sequences (RSS; heptamer) essential for the V-DJ rearrangement are present at the 3′UTR of the VH. Among the 49 VH sequences, twenty of them harbored heptamers (CACAGTG) that were highly conserved; three RRS sequences did not show conservation of the first three nucleotides (CAC) and two heptamers did not harbor the RSS motifs. Nonamer motifs, however less conserved, were present in 48 VH sequences. The heptamer and nonamer sequences were separated by 23±1 nucleotides. The identified VH gene segments could be classified into 12 families (Table 7.2). However, studies by Steiners group (Danilova *et al.*, 2005) reported 39 functional genes and 8 pseudogenes from their zebrafish IGH locus analysis. Recently, Bengten and co-workers (Bengten *et al.*, 2006) reported 55 genes on the catfish IGH locus, among which, 27 genes were classified as functional. The 28 pseudogenes either had inframe-stop codons, frame shifts or 5′ or 3′ gene fragments. Compared to zebrafish, catfish has a low ratio of functional genes. Peixoto and Brenner (2000) characterized 50 kb of fugu VH locus and two VH families were identified from eight full-length VH genes.

Table 7.2 Classification of zebrafish VH segments into families.

Families of IGHV segments	*VH segments*
I	1-1[a], 1-2, 1-3; 1-5, 1-11, 1-15, 1-17
II	2-4, 2-6, 2-16, 2-18
III	3-7, 3-21
IV	4-8, 4-10, 4-14, 4-19
V	5-9, 5-13
VI	6-20, 6-22, 6-23, 6-24, 6-38
VII	7-25, 7-39, 7-27, 7-41
VIII	8-30, 8-32, 8-45
IX	9-31, 9-33, 9-49, 9-46
X	10-37, 10-44
XI	11-26, 11-47, 11-43, 11-28, 11-42, 11-43, 11-35
XII	12-36, 12-48

[a]The first number corresponds to the family and the second to the VH segment listed in Table 7.1

Diversity Heavy (DH) Region

The structure of the DH region in fish is similar to its mammalian counterparts. The DH sequences are flanked with heptamer and nonamer motifs separated by conventional 12±1 bp spacers.

In zebrafish, nine DH segments have been identified by the analysis of IgZ and IgM transcripts. Among those identified DH segments, four are placed upstream of ζ region and five are located upstream of μ region (Fig. 7.1). The DH segments are composed of 10 to 42 bp coding nucleotides and 12 bp recombination signal sequences (RSS) elements with conserved heptamers and nonamers. In fugu, we have identified five DH-gene segments in fugu IGH locus, these segments are placed upstream of the novel Ig (one DH-gene segment) and μ (four DH-gene segments) domains. The DH1 is present 1.2 kb upstream of the new IgH CH1 domain and is a 15-nt-long segment open in all the three reading frames. DHμ1 to DHμ4 segments span a 1.5 kb region present 3.7 kb upstream of μ exon. In rainbow trout, Hansen and co-workers (2005) report at least three and six DH-gene segments upstream of the τ and μ regions, respectively. In catfish, five DH-gene segments have been identified upstream of JH cluster and the sixth DH-gene segment is located between VH11 and VH1 gene segments. Furthermore, two additional D-like segments have also been reported to be present in the VH locus (Bengten *et al.*, 2006).

Joining Heavy (JH) Region

The JH gene segments have a characteristic WGXG motif. The JH segments have an upstream RSS, which includes a T-rich nonamer, a 22 to 23 spacer, and a heptamer.

In zebrafish, a total of 7 JH segments have been identified. Two JH gene segments are placed 5′ of ζ1 exon and the remaining five are present upstream of μ1 (Fig. 7.1). These segments are composed of 15 to 17 amino acids. Similarly, in fugu, five and one JH gene segments are found upstream of the new Ig C1 and μ1 exons, respectively (Saha *et al.*, 2004; Savan *et al.*, 2005b). In rainbow trout, two and five JH gene segments were found upstream of τ1 and μ1 exons, respectively (Hansen *et al.*, 2005). Hayman and Lobb (2000) reported a total of 9 JH (JH1 to JH9) gene segments upstream of μ1 exon, tightly clustered within a region spanning about 2.2 kb in catfish. Recently, Bengten and co-workers (2006) have shown an additional three JH-gene segments, thereby increasing the total JH segments to 13 in catfish. Among these, two JH segments (JH13 and JH14) are present in the JH cluster upstream of μ. One JH-gene segment (JH12) is present along with DH-gene cluster in catfish. Atlantic salmon possesses five JH segments located 0.5 to 1.6 kb upstream of μ1 exon (Hordvik *et al.*, 1997).

Constant Regions on the Teleosts IgH Loci

Structure of IgZ/IgT/IgH Region

Recently, novel Ig heavy chain region on the teleost IGH loci were discovered (Sakai and Savan, 2004; Danilova *et al.*, 2005; Hansen *et al.*, 2005; Savan *et al.*, 2005b) (Fig. 7.1; Table 7.3). As this Ig class did not share similarities to any known Ig isotype and was identified during the same time, it was independently named as IgZ (zebrafish), IgT (rainbow trout) and IgH (fugu). However, some typical similarities exist among these Ig isotypes: (1) the genomic position (relative to IgM) of these constant domains are similar in the IGH loci of these fish; (2) the novel Ig isotype is sandwiched between two D and J clusters upstream of μ region; and (3) the zebrafish and trout Ig isotype is composed of four constant domain encoding exons and two membrane coding exons (TM1

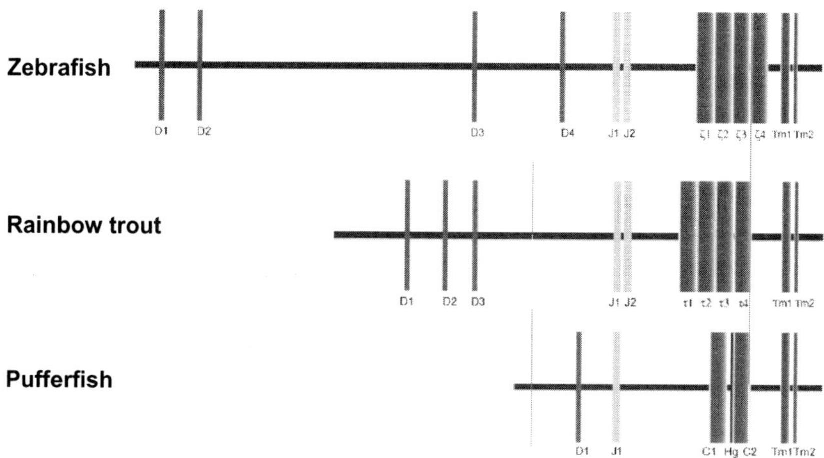

Fig. 7.2 The genomic organization of novel Igs. The fugu IgH comprises two exons encoding for the new constant and membrane coding exons (Tm1 & Tm2). Zebrafish and rainbow trout genes are composed of four constant domains and two membrane-coding exons.

Table 7.3 List of teleost species where Ig heavy chain genes have been cloned.

Species	*IgM*	*References*	*IgD*	*References*	*Novel IgH*	*References*
Zebrafish	+[1]/+[2]		+/+	Zimmerman and Steiner 2006[3]	+/+	Danilova *et al.* (2005)
Catfish	+/+	Wilson *et al.* (1990)	+/+	Wilson *et al.* (1997); Bengten *et al.* (2002)	–/–	
Common carp	+/+	Nakao *et al.* (1998)	+/–	CA966291[4] (unpublished)	+/+	Savan *et al.* (2005a)
Rainbow trout	+/+	Hansen *et al.* (2005)	+/+	Hansen *et al.* (2005)	+/+	Hansen *et al.* (2005)
Fugu	+/+	Saha *et al.* (2005)	+/+	Saha *et al.* (2004)	–/–	Savan *et al.* (2005b)
Japanese flounder	+/+	Aoki *et al.* (2000); Srisapoome *et al.* (2004)	+/+	Aoki *et al.* (2000); Hirono *et al.* (2003); Srisapoome *et al.* (2004)	–/–	–
Atlantic salmon	+/+	Hordvik *et al.* (1992)	+/+	Hordvik *et al.* (1999); Hordvik *et al.* (2002)	–/–	–
Atlantic halibut	+/–	AAF69488[5] (unpublished)	+/+	Hordvik (2002)	–/–	–
Atlantic cod	+/+	Bengten *et al.* (1991)	+/+	Stenvik and Jorgensen (2000)	–/–	–
Tilapia	+/–	AY522596 (unpublished)	–/–	–	–/–	–
Yellow tail	+/–	Savan, R. & Sakai, M., pers. comm.	–/–	–		–
Grouper	+/–	Cheng *et al.* (2006)	–/–	–	–/–	–
Brown trout	+/–	Hordvik *et al.* (2002)	+/–	Hordvik *et al.* (2002)	–/–	–
Wolf fish	+/–	Espelid *et al.* (2001)	–/–	–	–/–	–
Blackfin icefish	+/–	Ota *et al.* (2003)	–/–	–	–/–	–
Chinese perch	+/–	AAQ14845[5] (unpublished)	–/–	–	–/–	–
Haddock	+/–	CAH04752[5] (unpublished)	–/–	–	–/–	–

[1]Genomic information published;
[2]cDNA information available or published;
[3]A.M. Zimmerman and L.A. Steiner. The zebrafish IgD gene: exon usage, chimerism and quantitative expression profiles during development. Abstract in 10th ISDCI conference, July 1–6th 2006 Charleston, SC, USA;
[4]Expressed sequence tag of IgD (incomplete sequence);
[5]Accession number of a single representative sequence has been shown.

and TM2) (Fig. 7.2). Surprisingly, the fugu IgH is a two constant domain isotype (Savan *et al.*, 2005b). Furthermore, a hinge-like region between CH1 and CH2 was identified first from the expressed gene and later confirmed on the genomic sequence. This hinge segment is present within CH2 exon. A preliminary investigation by computational modeling predicts the presence of a hinge (Savan *et al.*, 2005b). This hinge region is composed of 21 amino acid residues with repeats of Val-Lys-Pro-Thr (VKPT). A similar pattern of fused hinge region has been seen only in mammalian IgA. Recently, in frog, a novel two-domain Ig isotype called IgF was identified (Zhao *et al.*, 2006). This isotype has a hinge region between CH1 and CH2, which is encoded by a separate exon. Although there is low similarity among hinge sequences across species, hinge might have evolved early in vertebrates (Zhao *et al.*, 2006).

The exons encoding the novel heavy chain constant domains harbored typical immunoglobulin domain motifs (**[FY]-x-C-x[V]-x-H**). The first domain (CH1) harbored three cysteine residues. The first Cys residue is important for disulphide linkage with V_HL chain and the other two Cys are involved in intra-domain disulphide linkages to form the core Ig-domain loop. The remaining constant domains harbor two conserved Cys residues, each required for intra-domain disulphide linkages to form Ig-domain loops. All the constant domains had splice sites (GT/AG) conserved at the intron/exon junctions.

In phylogenetic studies, the novel Ig isotypes formed a distinct cluster apart from the known Igs (Danilova *et al.*, 2005; Savan *et al.*, 2005b). Zebrafish, carp and trout ζ/τ isotypes clearly belong to a single Ig class and might be found in other teleosts. The structural differences of the novel fugu Ig isotype compared to ζ/τ might have functional consequences that remain to be clarified. The carp μ-ζ chimera is an interesting Ig that needs to be cloned in other fish species.

Structural Variations and Expression of the Novel Ig Isotype

The functional data on novel isotypes is limited as they are just beginning to be identified in other teleosts (Fig. 7.3). Here we will discuss the investigations conducted in zebrafish, rainbow trout, fugu and common carp (*Cyprinus carpio* L.). The IgT/Z isotype is expressed as secretory and membrane forms in zebrafish (Danilova *et al.*, 2005), rainbow trout (Hansen *et al.*, 2005) and common carp (Savan, R., Sogabe, K.,

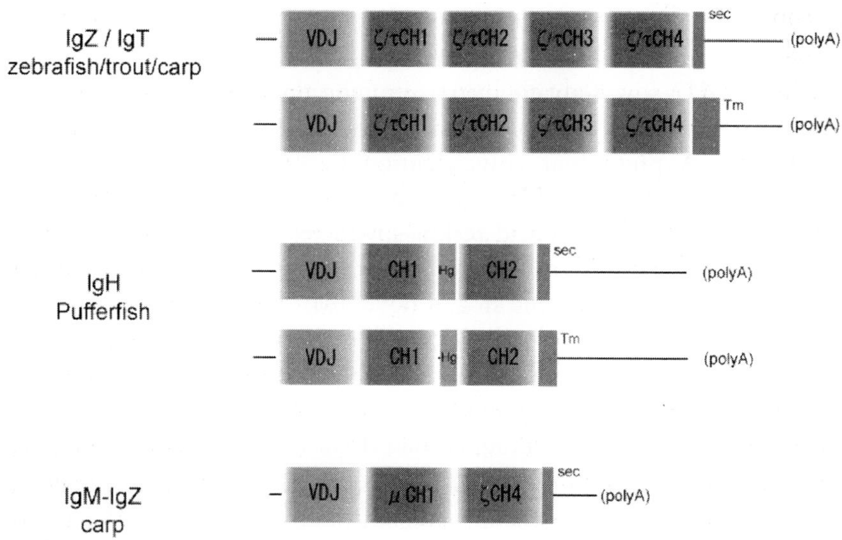

Fig. 7.3 Schematic representation of the transcripts of novel IgH chain isotypes cloned in teleosts.

Kemenade, L. and Sakai, M., pers. comm.). Furthermore, an unusual IgM-IgZ chimeric isotype, expressed in membrane and secretary forms, has been cloned from common carp (Savan *et al.*, 2005a). This chimera is composed of VH, μ1 (CH1) and ζ4 (CH2) Ig domains. This new IgM-IgZ isotype in fish has a novel structure, wherein it harbors only two constant domains, with μ1 as the first domain and ζ4 as the second domain. It would be interesting to find out the type of a recombination event occurring for the formation of a μ-ζ chimera. At present, we think in carp there might be two IGH loci in tandem (as seen in catfish; Bengten *et al.*, 2002) or a rearranged IGH locus on a separate chromosome. Furthermore, the prevalence of this isotype in other teleosts and its function needs to be addressed.

In fugu, when a tissue wide gene expression was conducted by RT-PCR, spleen, anterior kidney (mammalian equivalent of bone marrow), liver, intestine and gill tissues were positive for the s- and m-forms of novel Ig. By *in situ* hybridization, lymphocytes strongly expressing novel Ig were detected. Further confirmation of strong expression of the novel Ig in gill epithelial cells, and goblet cells in the intestinal epithelium was observed. The production of fugu Ig on mucosal surfaces predicts a role in mucosal immunity. In zebrafish, this gene was expressed in thymus, pronephros and mesonephros. The zebrafish, IgZ was largely restricted to the primary lymphoid tissues.

In an ontogeny study conducted in fugu, the novel membrane form was first detected at 4 days post fertilization (dpf), while the secretory form was expressed 1 dph (days post-hatching), i.e., 8 dpf. In the same study, IgM was first expressed at 4 dpf. In case of trout, IgZ and IgM were expressed during the same period. Contrastingly, in zebrafish, IgZ was expressed prior to IgM. Although further studies are needed in other teleosts to study the ontogeny of Ig isotypes, the above results indicate that the novel Ig is expressed during the same time or earlier than IgM.

The structure of IGH locus in teleosts is similar to the tandemly arranged TCRa/TCRd loci (Danilova *et al.*, 2005). As the TCRa/TCRd loci rearranges to produce αβ and γδ T-cells, the arrangement of the IGH loci in fish ensures that IgM and IgZ are mutually exclusive. Furthermore, Danilova *et al.* (2005) suggest that the time and tissue specificity of IgM and IgZ expression might indicate that these isotypes are produced in separate B-cell lineages.

Structure of IgM Region

IgM is found in most of the vertebrates and is presumed to be evolutionarily ancient. In vertebrates, this isotype exists as secretory (sIgM) and membrane (mIgM) bound forms. IgM is expressed on B-cells (membrane-bound form) and this is the first isotype expressed during B-cell development. The secretory form (produced by plasma cells upon B-cells differentiation) of IgM is known to exist in multimeric forms. Predominant among those are monomers (Elasmobranchs), tetramers (in teleosts) and pentameric (vertebrates) forms. With an exception to human IgM hexamer, the multimeric form of IgM in mammals is formed in association with J chain.

IgM is the most widely cloned and characterized immunoglobulin in fish (Wilson *et al.*, 1990; Bengten *et al.*, 1991; Lundqvist *et al.*, 1998; Espelid *et al.*, 2001; Saha *et al.*, 2004; Srisapoome *et al.*, 2004; Cheng *et al.*, 2006) (Table 7.3). The μ region is composed of four constant-coding exons (μCH1 to μCH4) and two membrane-coding exons (μTM1 and μTM2). This genomic structure of μ gene is common in all vertebrates. The membrane (μm) and secretory (μs) forms of IgM transcript are generated from the same genomic region by alternative RNA processing. However, the processing of the μm in teleosts is distinct from other vertebrates. Most of the vertebrates have a cryptic donor splice site at the μ4 exon, which splices directly to TM1 exon. However, in teleosts, the cryptic site in the μ4 is absent and the splicing takes place between μ3 exon and TM1. However, this evolutionary adaptation, deletion of CH4 domain in μm has been shown to have no effect on its function.

IgM is the major immunoglobulin isoform expressed in teleosts. IgM positive cells have been found in spleen, head kidney, and kidney confirming that these tissues are major sites of antibody production in fish (Saha *et al.*, 2005). The pronephros (head-kidney) is known to be the primary organ for B and plasma cells (Rijkers *et al.*, 1980; Razquin *et al.*, 1990; Zapata and Amemiya, 2000). However, B and plasma cells have also been found in thymus, which is a T-cell maturation site (Schroder *et al.*, 1998; Grontvedt and Espelid, 2003). The expression of IgM positive cells is also seen in mucosal organs such as the skin, gills, and intestine. Danilova and Steiner (2002) reported IgM^+ B cells from pancreas.

Delta (δ) Region

In the IGH loci of vertebrates, IgD is located downstream of IgM. In mammals, IgD exists as secreted and membrane-bound forms. In B-cells, IgM and IgD are co-expressed and their expression is regulated through RNA processing. IgD is made up of two and three constant domains in human and mouse, respectively. The number of hinge encoding exons varies between human (one) and mouse (two). Among the lower vertebrates, IgD was first discovered in catfish (Wilson *et al.*, 1997). The fish IgD has the first constant domain of μ1 followed by seven constant domains encoded by δ gene. This chimeric IgD molecule is formed by splicing of m1 into the δ1 exon. Reports of cloning of IgD from birds, cartilaginous fish (IgW is homologous to IgD), lungfish and recently from frog suggests that this isotype along with IgM are primordial (Ohta and Flajnik, 2006). However, the function of this isotype is still not clear.

After the first report of IgD from catfish (Wilson *et al.*, 1997), this isotype has been reported in other teleosts like Atlantic cod (Stenvik and Jorgensen, 2000), Atlantic salmon, Atlantic halibut (Hordvik, 2002), Japanese flounder (Hirono *et al.*, 2003) fugu (Saha *et al.*, 2004) and zebrafish (Sakai and Savan, 2004; Danilova *et al.*, 2005) (Table 7.3). Compared to mammalian IgD, teleostean orthologue has a unique structural organization. In a majority of teleosts, IgD is composed of seven constant region-encoding exons (δCH1 to δCH7) and two exons coding for the membrane region (δTM1 and δTM). However, species-specific exceptions to this basic structure can be seen in teleosts (Table 7.4). Tandem duplications of δ1-δ2 have been seen in Atlantic cod and similar tandem duplications of δ2-δ3-δ4 has been reported from Atlantic halibut, Atlantic salmon and recently also from catfish. In fugu and spotted green pufferfish, an unusual duplication of six δ domains can be seen (Saha *et al.*, 2004; Savan, R. and Sakai, M., pers. comm.). Contrastingly, δ domains are not duplicated in Japanese flounder (Srisapoome *et al.*, 2004). In zebrafish, δ region consists of 16 δ domains and two membrane-coding exons. Here, δ2-δ3-δ4 domains are repeated four times in tandem and the first repeat has a stop codon within the δ2 domain. The expression of this gene is not yet confirmed in zebrafish.

Table 7.4 Genomic organization of the δ domains in teleosts.

Species	*Genomic structure*	*References*
Pufferfish	$(\delta1\text{-}\delta2\text{-}\delta2\text{-}\delta3\text{-}\delta4\text{-}\delta5\text{-}\delta6)_2\text{-}\delta7$-TM1-TM2	Saha *et al.* (2004)
Catfish	$\delta1\text{-}\delta2\text{-}(\delta2\text{-}\delta3\text{-}\delta4)_3\text{-}\delta5\text{-}\delta6\text{-}\delta7$-TM1-TM2	Bengten *et al.* (2006)
Atlantic cod	δ1-δ2-δγ-δ1-δ2-Ψδψ-δ7-TM1-TM2-//-Ψδ7	Stenvik and Jorgensen (2000)
Japanese flounder	δ1-δ2-δ2-δ3-δ4-δ5-δ6-δ7-TM	Srisapoome *et al.* (2004)
Atlantic salmon	$\delta1\text{-}\delta2\text{-}(\delta2\text{-}\delta3\text{-}\delta4)_2\text{-}\delta5\text{-}\delta6\text{-}\delta7$-TM1-TM2	Hordvik *et al.* (1999)
Zebrafish	$\delta1\text{-}\delta2\text{-}(\delta2\text{-}\delta3\text{-}\delta4)_4\text{-}\delta5\text{-}\delta6\text{-}\delta7$-TM1-TM2	Sakai and Savan (2004) Danilova *et al.* (2005)

In fish, however, the secretory type of IgD has so far been reported only in catfish (Bengten *et al.*, 2002). In fugu, IgD gene is expressed in the spleen and head kidney. The expression pattern is similar to IgM. The functional role of IgD remains to be examined in mice and humans. The structural variations of the IgD in vertebrates indicate that this isotype may differ in their biological properties (Zhao *et al.*, 2002).

CONCLUSION

Although, the basic structure of the IGH loci is conserved, species-specific variations exist across teleosts. In general, teleosts have three major Ig heavy chain isotypes. Analysis of IGH genomic region suggests no additional discovery of IgH genes will be forthcoming, at least not from downstream region of δ gene (Bengten *et al.*, 2006). The novel Ig isotype, along with IgM and IgD, will prove to be good tools to study the immune responses and biology of B-cells in fish.

References

Andersson, E. and T. Matsunaga. 1998. Evolutionary stability of the immunoglobulin heavy chain variable region gene families in teleost. *Immunogenetics* 47: 272-277.

Aparicio, S., J. Chapman, E. Stupka, N. Putnam, J.M. Chia, P. Dehal, A. Christoffels, S. Rash, S. Hoon, A. Smit, M.D. Gelpke, J. Roach, T. Oh, I.Y. Ho, M. Wong, C. Detter, F. Verhoef, P. Predki, A. Tay, S. Lucas, P. Richardson, S.F. Smith, M.S. Clark, Y.J. Edwards, N. Doggett, A. Zharkikh, S.V. Tavtigian, D. Pruss, M. Barnstead, C. Evans, H. Baden, J. Powell, G. Glusman, L. Rowen, L. Hood, Y. H. Tan, G. Elgar, T. Hawkins, B. Venkatesh, D. Rokhsar and S. Brenner. 2002. Whole-genome shotgun assembly and analysis of the genome of *Fugu rubripes*. *Science* 297: 1301-1310.

Bengten, E., T. Leanderson and L. Pilstrom. 1991. Immunoglobulin heavy chain cDNA from the teleost Atlantic cod (*Gadus morhua* L.): Nucleotide sequences of secretory and membrane form show an unusual splicing pattern. *European Journal of Immunology* 21: 3027-3033.

Bengten, E., S. Stromberg and L. Pilstrom. 1994. Immunoglobulin VH regions in Atlantic cod (*Gadus morhua* L.): Their diversity and relationship to VH families from other species. *Developmental and Comparative Immunology* 18: 109-122.

Bengten, E., S.M. Quiniou, T.B. Stuge, T. Katagiri, N.W. Miller, L.W. Clem, G.W. Warr and M. Wilson. 2002. The IgH locus of the channel catfish, *Ictalurus punctatus*, contains multiple constant region gene sequences: Different genes encode heavy chains of membrane and secreted IgD. *Journal of Immunology* 169: 2488-2497.

Bengten, E., S. Quiniou, J. Hikima, G. Waldbieser, G.W. Warr, N.W. Miller and M. Wilson. 2006. Structure of the catfish IGH locus: Analysis of the region including the single functional IGHM gene. *Immunogenetics* 58: 831-844.

Brodeur, P.H., G.E. Osman, J.J. Mackle and T.M. Lalor. 1988. The organization of the mouse Igh-V locus. Dispersion, interspersion, and the evolution of VH gene family clusters. *Journal of Experimental Medicine* 168: 2261-2278.

Cheng, C.A., J.A. John, M.S. Wu, C.Y. Lee, C.H. Lin, C.H. Lin and C.Y. Chang. 2006. Characterization of serum immunoglobulin M of grouper and cDNA cloning of its heavy chain. *Veterinary Immunology and Immunopathology* 109: 255-265.

Coscia, M.R. and U. Oreste. 2003. Limited diversity of the immunoglobulin heavy chain variable domain of the emerald rockcod *Trematomus bernacchii*. *Fish and Shellfish Immunology* 14: 71-92.

Dahan, A., C.A. Reynaud and J.C. Weill. 1983. Nucleotide sequence of the constant region of a chicken mu heavy chain immunoglobulin mRNA. *Nucleic Acids Research* 11: 5381-5389.

Danilova, N., V.S. Hohman, E.H. Kim and L.A. Steiner. 2000. Immunoglobulin variable-region diversity in the zebrafish. *Immunogenetics* 52: 81-91.

Danilova, N., J. Bussmann, K. Jekosch and L.A. Steiner. 2005. The immunoglobulin heavy-chain locus in zebrafish: Identification and expression of a previously unknown isotype, immunoglobulin. *Nature Immunology* 6: 295-302.

Danilova, N. and L.A. Steiner. 2002. B cells develop in the zebrafish pancreas. *Proceedings of the National Academy of Sciences of the United States of America* 99: 13711-13716.

Du Pasquier, L. and M.F. Flajnik. 1999. *Origin and Evolution of the Vertebrate Immune System.* Lipponcott-Raven, Philadelphia.

Espelid, S., M. Halse, S.T. Solem and T.O. Jorgensen. 2001. Immunoglobulin genes and antibody responses in the spotted wolffish (*Anarhichas minor* Olafsen). *Fish and Shellfish Immunology* 11: 399-413.

Ghaffari, S.H. and C.J. Lobb. 1991. Heavy chain variable region gene families evolved early in phylogeny. Ig complexity in fish. *Journal of Immunology* 146: 1037-1046.

Ghaffari, S.H. and C.J. Lobb. 1992. Organization of immunoglobulin heavy chain constant and joining region genes in the channel catfish. *Molecular Immunology* 29: 151-159.

Ghaffari, S.H. and C.J. Lobb. 1999. Structure and genomic organization of a second cluster of immunoglobulin heavy chain gene segments in the channel catfish. *Journal of Immunology* 162: 1519-1529.

Grontvedt, R.N. and S. Espelid. 2003. Immunoglobulin producing cells in the spotted wolffish (*Anarhichas minor* Olafsen): Localization in adults and during juvenile development. *Developmental and Comparative Immunology* 27: 569-578.

Hansen, J.D., E.D. Landis and R.B. Phillips. 2005. Discovery of a unique Ig heavy-chain isotype (IgT) in rainbow trout: Implications for a distinctive B cell developmental pathway in teleost fish. *Proceedings of the National Academy of Sciences of the United States of America* 102: 6919-6924.

Hayman, J.R. and C.J. Lobb. 2000. Heavy chain diversity region segments of the channel catfish: Structure, organization, expression and phylogenetic implications. *Journal of Immunology* 164: 1916-1924.

Hirono, I., B.H. Nam, J. Enomoto, K. Uchino and T. Aoki. 2003. Cloning and characterisation of a cDNA encoding Japanese flounder *Paralichthys olivaceus* IgD. *Fish and Shellfish Immunology* 15: 63-70.

Hordvik, I. 2002. Identification of a novel immunoglobulin delta transcript and comparative analysis of the genes encoding IgD in Atlantic salmon and Atlantic halibut. *Molecular Immunology* 39: 85-91.

Hordvik, I., C. De Vries Lindstrom, A.M. Voie, A. Lilybert, J. Jacob and C. Endresen. 1997. Structure and organization of the immunoglobulin M heavy chain genes in Atlantic salmon, *Salmo salar*. *Molecular Immunology* 34: 631-639.

Hordvik, I., J. Thevarajan, I. Samdal, N. Bastani and B. Krossoy. 1999. Molecular cloning and phylogenetic analysis of the Atlantic salmon immunoglobulin D gene. *Scandinavian Journal of Immunology* 50: 202-210.

Kabat, E.A., T.T. Wu, H.M. Perry, K.S. Gottesman and C. Foeller. 1991. Sequences of Proteins of Immunological Interest. National Institutes of Health, Bethesda, MD.

Kokubu, F., K. Hinds, R. Litman, M.J. Shamblott and G.W. Litman. 1988. Complete structure and organization of immunoglobulin heavy chain constant region genes in a phylogenetically primitive vertebrate. *EMBO Journal* 7: 1979-1988.

Lundqvist, M.L., S. Stromberg and L. Pilstrom. 1998. Ig heavy chain of the sturgeon *Acipenser baeri*: cDNA sequence and diversity. *Immunogenetics* 48: 372-382.

Ohta, Y. and M. Flajnik. 2006. IgD, like IgM, is a primordial immunoglobulin class perpetuated in most jawed vertebrates. *Proceedings of the National Academy of Sciences of the United States of America* 103: 10723-10728.

Ota, T., J.P. Rast, G.W. Litman and C.T. Amemiya. 2003. Lineage-restricted retention of a primitive immunoglobulin heavy chain isotype within the Dipnoi reveals an evolutionary paradox. *Proceedings of the National Academy of Sciences of the United States of America* 100: 2501-2506.

Peixoto, B.R. and S. Brenner. 2000. Characterization of approximately 50 kb of the immunoglobulin VH locus of the Japanese pufferfish, *Fugu rubripes*. *Immunogenetics* 51: 443-451.

Razquin, B.E., A. Castullo, P. Lopezfierro, F. Alvarez, A. Zapata and A.J. Villena. 1990. Ontogeny of IGM-producing cells in the lymphoid organs of rainbow trout, *Salmo gairdneri* Richardson—An immunohistochemical and enzyme-istochemical study. *Journal of Fish Biology* 36: 159-173.

Rijkers, G.T., E.M. Frederix-Wolters and W.B. van Muiswinkel. 1980. The immune system of cyprinid fish. Kinetics and temperature dependence of antibody-producing cells in carp (*Cyprinus carpio*). *Immunology* 41: 91-97.

Roman, T., E. Andersson, E. Bengten, J. Hansen, S. Kaattari, L. Pilstrom, J. Charlemagne and T. Matsunaga. 1996. Unified nomenclature of Ig VH genes in rainbow trout (*Oncorhynchus mykiss*): Definition of eleven VH families. *Immunogenetics* 43: 325-326.

Saha, N.R., H. Suetake, K. Kikuchi and Y. Suzuki. 2004. Fugu immunoglobulin D: a highly unusual gene with unprecedented duplications in its constant region. *Immunogenetics* 56: 438-447.

Saha, N.R., H. Suetake and Y. Suzuki. 2005. Analysis and characterization of the expression of the secretory and membrane forms of IgM heavy chains in the pufferfish, *Takifugu rubripes*. *Molecular Immunology* 42: 113-124.

Sakai, M. and R. Savan. 2004. *Characterization of zebrafish immunoglobulin heavy chain (IGH) locus*. Proceedings of JSPS-NRCT International Symposium, Kasetsart University, Thailand.

Savan, R., A. Aman, M. Nakao, H. Watanuki and M. Sakai. 2005a. Discovery of a novel immunoglobulin heavy chain gene chimera from common carp (*Cyprinus carpio* L.) *Immunogenetics* 57: 458-463.

Savan, R., A. Aman, K. Sato, R. Yamaguchi and M. Sakai. 2005b. Discovery of a new class of immunoglobulin heavy chain from fugu. *European Journal of Immunology* 35: 3320-3331.

Schroder, M.B., A.J. Villena and T.O. Jørgensen. 1998. Ontogeny of lymphoid organs and immunoglobulin producing cells in Atlantic cod (*Gadus morhua* L.). *Developmental and Comparative Immunology* 22: 507-517.

Schwager, J., D. Grossberger and L. Du Pasquier. 1988a. Organization and rearrangement of immunoglobulin M genes in the amphibian *Xenopus*. *EMBO Journal* 7: 2409-2415.

Schwager, J., C.A. Mikoryak and L.A. Steiner. 1988b. Amino acid sequence of heavy chain from *Xenopus laevis* IgM deduced from cDNA sequence: Implications for evolution of immunoglobulin domains. *Proceedings of the National Academy of Sciences of the United States of America* 85: 2245-2249.

Srisapoome, P., T. Ohira, I. Hirono and T. Aoki. 2004. Genes of the constant regions of functional immunoglobulin heavy chain of Japanese flounder, *Paralichthys olivaceus*. *Immunogenetics* 56: 292-300.

Stenvik, J. and T.O. Jorgensen. 2000. Immunoglobulin D (IgD) of Atlantic cod has a unique structure. *Immunogenetics* 51: 452-461.

Ventura-Holman, T., S.H. Ghafari and C.J. Lobb. 1996. Characterization of a seventh family of immunoglobulin heavy chain VH gene segments in the channel catfish, *Ictalurus punctatus*. *European Journal of Immunogenetics* 23: 7-14.

Ventura-Holman, T. and C.J. Lobb. 2002. Structural organization of the immunoglobulin heavy chain locus in the channel catfish: The IgH locus represents a composite of two gene clusters. *Molecular Immunology* 38: 557-564.

Warr, G.W., D.L. Middleton, N.W. Miller, L.W. Clem and M.R. Wilson. 1991. An additional family of VH sequences in the channel catfish. *European Journal of Immunogenetics* 18: 393-397.

Warr, G.W., N.W. Miller, L.W. Clem and M.R. Wilson. 1992. Alternate splicing pathways of the immunoglobulin heavy chain transcript of a teleost fish, *Ictalurus punctatus*. *Immunogenetics* 35: 253-256.

Wilson, M.R., A. Marcuz, F. van Ginkel, N.W. Miller, L.W. Clem, D. Middleton and G.W. Warr. 1990. The immunoglobulin M heavy chain constant region gene of the channel catfish, *Ictalurus punctatus*: An unusual mRNA splice pattern produces the membrane form of the molecule. *Nucleic Acids Research* 18: 5227-5233.

Wilson, M.R., E. Bengten, N.W. Miller, L.W. Clem, L. Du Pasquier and G.W. Warr. 1997. A novel chimeric Ig heavy chain from a teleost fish shares similarities to IgD. *Proceedings of the National Academy of Sciences of the United States of America* 94: 4593-4597.

Yang, F., T. Ventura-Holman, G.C. Waldbieser and C.J. Lobb. 2003. Structure, genomic organization, and phylogenetic implications of six new VH families in the channel catfish. *Molecular Immunology* 40: 247-260.

Zapata, A. and C.T. Amemiya. 2000. Phylogeny of lower vertebrates and their immunological structures. *Current Topics in Microbiology and Immunology* 248: 67-107.

Zhao, Y., I. Kacskovics, Q. Pan, D.A. Liberles, J. Geli, S.K. Davis, H. Rabbani and L. Hammarstrom. 2002. Artiodactyl IgD: The missing link. *Journal of Immunology* 169: 4408-4416.

Zhao, Y., Q. Pan-Hammarstrom, S. Yu, N. Wertz, X. Zhang, N. Li, J.E. Butler and L. Hammarstrom. 2006. Identification of IgF, a hinge-region-containing Ig class, and IgD in *Xenopus tropicalis*. *Proceedings of the National Academy of Sciences of the United States of America* 103: 12087-12092.

CHAPTER

8

Antimicrobial Peptides of the Innate Immune System

Valerie J. Smith[1] and Jorge M.O. Fernandes[2]

INTRODUCTION

The *milieu* inhabited by fish, whether pelagic or demersal, marine or freshwater, teems with a huge diversity of microorganisms. For them, the fish body is a stable, nutrient-rich habitat, ideal for exploitation. Some, including commensals and opportunists, lodge on the skin surface or colonize the gut, urogenital tract or gills. Others, more usually pathogens, invade the blood, organs and internal tissues, their entry facilitated by the lack of keratin in the skin, which, if present, would confer a degree of imperviousness to abrasion and proteolytic attack. For fish, therefore, wounds and injuries may not represent the only portal for microbes to gain

Authors' addresses: [1]Comparative Immunology Group, Gatty Marine Laboratory, School of Biology, University of St Andrews, St Andrews, Fife, KY16 8LB, Scotland, UK.
E-mail: vjs1@st-andrews.ac.uk
[2]Marine Molecular Biology and Genomics Research Group, Faculty of Biosciences and Aquaculture, Bodø University College, N-8049 Bodø, Norway.
E-mail: Jorge.Fernandes@hibo.no
Corresponding authors: vjs1@st-andrews.ac.uk *(VJS)* and jorge.fernandes@hibo.no *(JMOF)*.

access to the rich pickings of the fish body. Copious mucus secretion on mucosal surfaces helps to some extent in the mechanical removal of microorganisms, but it is costly to produce and may itself promote the growth of bacteria as a nutrient. Wherever microbes colonize, they pose a threat to the well-being of the fish host, draining it of its resources, causing disease or imposing a metabolic cost in terms of immunological reaction. Control of microbial growth—not only on exposed surfaces but also within the tissues—is thus crucial not only to survival but also to the fitness, growth, and reproductive success of fish.

Clearly, both the internal body fluids and the mucosal secretions need to contain microbicidal agents that can act as general disinfectants. Such compounds are well recognized as key components of the systemic and mucosal defence systems of other animals and are especially important to animals that lack specific 'adaptive type' immunity or those that, like fish, possess one that may be age or temperature dependent, slow to develop and short-lived.

Earlier, reviews of the microbicidal agents produced by fish have been given by Yano (1996) and Ellis (1999, 2001), although these researches have focused on factors such as lytic enzymes (e.g., lysozyme), agglutinating factors, oxyradicals and protease inhibitors. The present review concentrates on another group of agents, the antimicrobial peptides (AMPs). These were discovered some 25 years ago by Hans Boman's group (Hultmark *et al.*, 1982), Robert Lehrer *et al.* (Selsted *et al.*, 1985) and Michael Zasloff (1987) in insects, mammals and amphibians, respectively. Now, over 700 AMPs have been found from acoelomates, amphibians, mammals and higher plants as well as protostome and deuterostome invertebrates, but reports for teleosts did not appear until about 20 years ago (Lazarovici *et al.*, 1986; Thompson *et al.*, 1986) and it took yet another decade for research to really gain momentum. In recent years, the number of AMPs found in fish has grown considerably (Fig. 8.1).

In this chapter, we will describe the diversity of AMPs and other low molecular weight microbicidal proteins presently known for fish, and consider their structure, range of activity, sites of expression and modes of action. We will also discuss the extent to which environmental or physiological factors affect their expression and activity and what we still need to know if we are to exploit these molecules to enhance disease control in aquaculture.

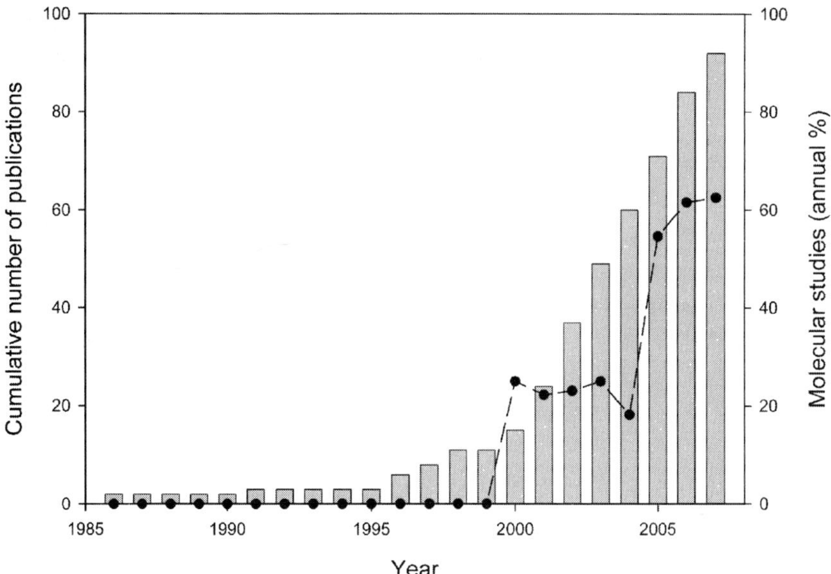

Fig. 8.1 Research on fish antimicrobial peptides over the last two decades. Bars show the cumulative number of publications regarding antimicrobial peptides from bony fish. The dashed line represents the percentage of papers reporting molecular studies within each year.

General Characteristics of Animal AMPs and their Role as Components of the Immune System

Antimicrobial peptides are low molecular weight cationic proteins, usually comprising less than 100 amino acids. They are products of single genes and kill microbes in a stoichiometric rather than enzymatic manner. In all the animal groups studied so far, AMPs are typically produced by epithelial cells, the liver, lymphoid tissue or phagocytes and may be expressed either constitutively or induced upon non-self challenge. Most AMPs kill their targets by depolarizing and permeabilizing the outer bacterial cell wall in various ways but others may disrupt cell metabolism or interfere with DNA synthesis (Devine and Hancock, 2002; Zasloff, 2002). Some synergize together or with lysozyme to attack bacteria in a multifarious manner (Patrzykat *et al.*, 2001; Concha *et al.*, 2004).

AMP's function in immunity is to disinfect the exposed surfaces such as the eye, gut and urogenital tract, but they also kill bacteria that may enter the body through injuries, preventing them from proliferating until they are sequestered and eliminated by the phagocytes and other components of the systemic immune system. Very few have cytotoxic effects on eukaryotic cells because they attack bacterial targets—namely the cytoplasmic membrane underlying the cell wall—that have a different composition in eukaryotic cells. They are, in effect, direct broad-spectrum natural antibiotics and their great value is that there are only few ways by which the bacteria can resist their effects (Devine and Hancock, 2002). Interestingly, however, in addition to their disinfecting properties, some cationic AMPs in mammals help modulate the inflammatory response by stimulating leucocyte chemotaxis, inhibiting or promoting chemokine and cytokine production and by regulating pro-inflammatory responses to microbial cell wall components (Bowdish *et al.*, 2005). A few have also been found to kill tumours or interrupt viral replication (Hancock *et al.*, 1995). Thus, they not only help the host control its microflora, but also protect it from the harmful effects of their cell wall components or products.

From the hundreds now characterized from multicellular organisms, it is clear that AMPs comprise a very diverse collection of proteins but most are cationic, hydrophobic and amphipathic, i.e., folded into a shape that segregates clusters of charged amino acids from hydrophobic ones. There is no formal and universally accepted classification system for AMPs as a whole, but for convenience they are usually divided into structurally different categories. These include: (1) linear peptides with an α-helix, (2) cysteine-rich molecules with two or more disulphide bonds that form β-sheets, (3) tailed loop peptides constrained by a disulphide bridge and (4) peptides with over representation of one or two amino acids, one of which is often proline (Boman, 1995; Devine and Hancock, 2002). These are all ribosomally encoded and are usually synthesized as prepropeptides that may undergo posttranslational modification, such as terminal amidation. The immature proteins may be stored in intracellular granules before being processed by proteolytic cleavage so as to liberate the active peptide. However, a few unusual AMPs are also known that are fragments of proteins with other known functions (Boman, 1995).

Many animals express a variety of AMP types, no doubt, to have an arsenal capable of dealing with a variety of microbial types. Whilst some AMP types are common across a variety of taxa, e.g., defensins and cathelicidins, others may possess types that are unique to that family or genus. Certainly, individual taxa tend to share similar types of AMPs perhaps because the AMPs have evolved and diversified along with their hosts. Thus, for example, different amphibian species tend to express similar groups of peptides to each other but are unlike those expressed by mammals or insects. Given that the teleosts constitute a hugely diverse and ancient group of vertebrates it is, therefore, likely that they will have their own unique clusters of AMPs as well as others well known in other vertebrates. Here we will describe the diversity of AMPs found so far in fish and consider their structure, phylogenetic relationships, bioactivities and patterns of expression.

Antimicrobial Peptides from Teleosts

Given that bony fish are the largest and most diverse group of vertebrates, comprising over 24,000 species (Hedges, 2002), it is somehow surprising that only *circa* 60 AMPs have been identified to date in some 30 teleost species (Table 8.1). Nevertheless, the relatively limited research done on fish AMPs has unveiled a remarkable diversity amongst these molecules, ranging from amphipathic α-helices to β-strands with complex intramolecular disulfide bond patterns. Bony fish also express a surprisingly high number of AMPs derived from proteins known to have other functions.

Pardaxins were the first antimicrobial peptides to be isolated from fish. These excitatory toxins with cytotoxic properties are secreted by flatfish of the genus, *Pardachirus* (Lazarovici *et al.*, 1986; Thompson *et al.*, 1986). Initially, these 33-residue peptides were isolated on the basis of their pore-forming properties and were thought to protect the fish against predation as they possess shark-repellent properties (Lazarovici *et al.*, 1986; Thompson *et al.*, 1986). The antimicrobial properties of pardaxin were established *circa* 10 years later by Oren and Shai (1996). In sodium dodecylphosphocholine micelles, pardaxin has an amphipathic hinge-helix-hinge-helix structure with separately clustered hydrophobic and hydrophilic regions (Porcelli *et al.*, 2004). The α-helical structure is common to many known animal AMPs (Zasloff, 2002) and its amphipathic nature is fundamental to disruption of bacterial cytoplasmic membranes, either by a 'barrel-stave' or 'carpet' mechanism (Huang, 2000).

Table 8.1 Alphabetical list of antimicrobial peptides identified to date from bony fish.

Peptide	*Species*	*References*
ApoA-I, -II	Common carp (*Cyprinus carpio*)	Concha *et al.* (2004)
CATH-1, -2	Rainbow trout (*Oncorhynchus mykiss*) Atlantic salmon (*Salmo salar*)	Chang *et al.* (2005) Chang *et al.* (2006)
Chrysophsins	Red seabream (*Pagrus major*)	Iijima *et al.* (2003)
DB-1, -2, -3	Zebrafish (*Danio rerio*) Orange-spotted grouper (*Epinephelus coioides*) Tiger pufferfish (*Takifugu rubripes*) Green-spotted pufferfish (*Tetraodon nigroviridis*)	Zou *et al.* (2007) Zou *et al.* (2007) Zou *et al.* (2007) Zou *et al.* (2007)
Dicentracin	European sea bass (*Dicentrarchus labrax*)	Salerno *et al.* (2007)
Epinecidin-1	Orange-spotted grouper (*E. coioides*)	Yin *et al.* (2006)
Hepcidins	Red sea bream (*P. major*) European sea bass (*D. labrax*) Zebrafish (*D. rerio*) Atlantic halibut (*Hippoglossus hippoglossus*) Blue catfish (*Ictalurus furcatus*) Channel catfish (*Ictalurus punctatus*) Japan sea perch (*Lateolabrax japonicus*) Hybrid striped bass (*Morone chrysops* × *M. saxatilis*) Mozambique tilapia (*Oreochromis mossambicus*) Japanese flounder (*Paralichthys olivaceus*) Winter flounder (*Pleuronectes americanus*) Atlantic salmon (*S. salar*) Turbot (*Scophthalmus maximus*)	Chen *et al.* (2005) Rodrigues *et al.* (2006) Shike *et al.* (2004) Park *et al.* (2005) Bao *et al.* (2005) Bao *et al.* (2005) Ren *et al.* (2006) Shike *et al.* (2002) Huang *et al.* (2007) Hirono *et al.* (2005) Douglas *et al.* (2003a) Martin *et al.* (2006) Chen *et al.* (2007)
Hipposin	Atlantic halibut (*H. hippoglossus*)	Birkemo *et al.* (2003)
Histone H1	Coho salmon (*Oncorhynchus kisutch*) Rainbow trout (*O. mykiss*) Atlantic salmon (*S. salar*)	Patrzykat *et al.* (2001) Noga *et al.* (2001) Richards *et al.* (2001)
Histone H2A	Rainbow trout (*O. mykiss*)	Fernandes *et al.* (2002)
Histone H2B	Atlantic cod (*Gadus morhua*) Hybrid striped bass (*M. saxatilis* × *M. chrysops*)	Bergsson *et al.* (2005) Noga *et al.* (2001)
HLPs	Channel catfish (*I. punctatus*)	Robinette *et al.* (1998)
HSDF-1	Coho salmon (*Oncorhynchus kisutch*)	Patrzykat *et al.* (2001)
Kenojeinin I	Fermented skate (*Raja kenojei*)	Cho *et al.* (2005)
LEAP-2	Blue catfish (*I. furcatus*) Channel catfish (*I. punctatus*) Rainbow trout (*O. mykiss*)	Bao *et al.* (2006) Bao *et al.* (2006) Zhang *et al.* (2004)
MAPP	Oriental weatherloach (*Misgurnus anguillicaudatus*)	Dong *et al.* (2002)
Misgurin	Oriental weatherloach (*M. anguillicaudatus*)	Park *et al.* (1997)
Oncorhyncin I	Rainbow trout (*O. mykiss*)	Smith *et al.* (2000)
Oncorhyncin II	Rainbow trout (*O. mykiss*)	Fernandes *et al.* (2004b)
Oncorhyncin III	Rainbow trout (*O. mykiss*)	Fernandes *et al.* (2003)
Parasin I	Amur catfish (*Parasilurus asotus*)	Park *et al.* (1998)

(Table 8.1 contd.)

(Table 8.1 contd.)

Pardaxins	Red Sea Moses sole (*Pardachirus marmoratus*) Peacock sole (*Pardachirus pavonicus*)	Lazarovici *et al.* (1986) Thompson *et al.* (1986)
Pleurocidins	Atlantic halibut (*H. hippoglossus*) American plaice (*Hippoglossoides platessoides*) Yellowtail flounder (*Limanda ferruginea*) Mud dab (*Limanda limanda*) Winter flounder (*P. americanus*)	Patrzykat *et al.* (2003) Patrzykat *et al.* (2003) Patrzykat *et al.* (2003) Brocal *et al.* (2006) Cole *et al.* (1997)
Piscidins	White bass (*Morone chrysops*) Striped bass (*Morone saxatilis*) Hybrid striped bass (*M. saxatilis* × *M. chrysops*)	Lauth *et al.* (2002) Lauth *et al.* (2002) Silphaduang and Noga (2001)
Ribosomal proteins	Atlantic cod (*G. morhua*) Rainbow trout (*O. mykiss*)	Bergsson *et al.* (2005) Fernandes and Smith (2002)
SAMP H1	Atlantic salmon (*S. salar*)	Luders *et al.* (2005)
SSAP	Rockfish (*Sebastes schlegeli*)	Kitani *et al.* (2007)

Another group of α-helical fish AMPs are the pleurocidins. They are 19-26 residue peptides found as multiple isoforms in several flatfish species, including the winter flounder, American plaice, white flounder and halibut (Patrzykat *et al.*, 2002; Douglas *et al.*, 2003b). In the winter flounder, pleurocidin is expressed in epithelial mucous cells of the skin, indicating that it plays a role in mucosal immunity (Cole *et al.*, 1997). Parasin is a 19-residue N-terminal histone-derived, α-helical peptide fragment produced by the amur catfish in response to epidermal injury (Park *et al.*, 1998). Other histone-derived AMPs isolated from various fish species include histone-derived fragment 1 (HSDF-1) (Patrzykat *et al.*, 2001), hipposin (Birkemo *et al.*, 2003), oncorhyncin II (Fernandes *et al.*, 2004b), oncorhyncin III (Fernandes *et al.*, 2003) and salmon antimicrobial peptide (SAMP) (Luders *et al.*, 2005) (Table 8.1, Fig. 8.2).

Piscidins are 22-residue α-helical AMPs first discovered in the gills (Silphaduang and Noga, 2001; Lauth *et al.*, 2002) and on the skin of bass (Lauth *et al.*, 2002). In common with pleurocidin and parasin, piscidins can adopt an amphipathic α-helical structure in hydrophobic environments that mimic cell membranes (Noga and Silphaduang, 2003; Campagna *et al.*, 2007). Similar piscidin-like peptides, chrysopsin, epinecidin and dicentracin, have been identified in the red sea-bream (Iijima *et al.*, 2003), orange-spotted grouper (Yin *et al.*, 2006) and European sea bass (Salerno *et al.*, 2007), respectively. In fact, there is

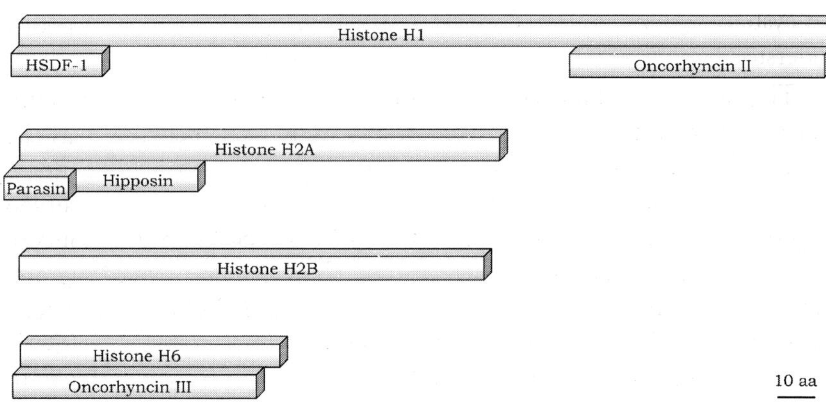

Fig. 8.2 Schematic representation of antimicrobial fish histones and histone-derived peptides. Histones H1, H2A and H2B from bony fish have well documented antibacterial properties. In addition, potent antimicrobial peptides are generated by proteolytic cleavage of these molecules, including HSDF-1 and oncorhyncin II from histone H1, hipposin and parasin from histone H2A and oncorhyncin III from the chromosomal protein H6. In this scale bar indicates 10 residues (*circa* 1 KDa).

growing evidence that piscidins are widespread amongst Perciformes, as Silphaduang *et al.* (2006) detected piscidin-like epitopes in members of the families Moronidae, Sciaenidae, Cichlidae, Siganidae and Belontidae.

Hepcidins (formerly termed LEAP-1 for liver-expressed antimicrobial peptide 1) are cysteine-rich antimicrobial peptides. This group of AMPs was first discovered in mammals as low molecular weight peptide antibiotics synthesized in the liver (Park *et al.*, 2001). The first fish hepcidin was found by Shike *et al.* (2002) in gill extracts of hybrid striped bass. To date, this remains the only fish hepcidin for which the primary structure has been obtained experimentally. In solution, this 21-residue peptide has two anti-parallel β-sheets and its 8 cysteines form four intramolecular disulfide bonds, a structure that is similar to its mammalian counterpart (Lauth *et al.*, 2005). Another liver-expressed antimicrobial peptide (LEAP-2) has been found in rainbow trout (Zhang *et al.*, 2004) and catfish (Bao *et al.*, 2006). It is a systemic AMP that contains a core structure with two disulfide bonds formed by cysteine residues in relative 1-3 and 2-4 positions (Krause *et al.*, 2003). However, not all cysteine-rich fish AMPs are hepcidins, as has been exemplified by an unusual cysteine-rich AMP isolated from loach (Dong *et al.*, 2002). This 94-residue peptide has no homology to known proteins, is anionic and *circa* 20% of its residues are cysteine (Dong *et al.*, 2002) but in other respects it remains poorly understood.

The cathelicidin family of AMPs are a large number of diverse proteins that are characterized by a conserved N-terminal region, known as the cathelin-like domain (Tomasinsig and Zanetti, 2005). All known cathelicidins share structural features, namely a highly conserved preproregion and four invariant cysteines in the cathelin-like domain, which are likely to be involved in two disulfide bonds (Tomasinsig and Zanetti, 2005). Two cathelicidin genes, CATH-1 and CATH-2, have been identified in rainbow trout and Atlantic salmon (Chang *et al.*, 2005, 2006). Like their mammalian counterparts, CATH-1 and CATH-2 are likely to be processed by elastase (Chang *et al.*, 2006).

A third group of cysteine-rich AMPs are the defensins. In vertebrates, this AMP family is large and subdivided into α-, β- and θ-defensins, according to the pattern of their 3 intramolecular disulfide bridges (Ganz, 2003a). Using a comparative genomics strategy, Zou and collaborators have recently identified classical defensins in zebrafish, tiger pufferfish, green-spotted pufferfish and orange-spotted grouper (Zou *et al.*, 2007). The putative primary structure of fish defensins suggests that they are homologous to mammalian β-defensins and that their secondary structure is likely to be composed of three β-strands with a conserved pattern of 3 intramolecular disulfide bonds (Zou *et al.*, 2007). In zebrafish, defensins are present as multiple copies clustered in the genome, similarly to their tetrapod counterparts.

Classification of Fish Antimicrobial Peptides

Despite the diversity amongst their primary structures, most fish AMPs can be grouped in five families, namely the cathelicidins, defensins, LEAPs, piscidins and histone-derived peptides (Table 8.2).

Cathelicidins have been found in mammalian species from various orders (Tomasinsig and Zanetti, 2005), chicken (Xiao *et al.*, 2006), bony fish (Chang *et al.*, 2005, 2006) and Atlantic hagfish (Uzzell *et al.*, 2003), confirming that cathelicidin genes have arisen early during chordate evolution. CATH-1 and CATH-2 from salmonids have a conserved gene structure comprising four exons and are closely related to hagfish HFIAP, rabbit p15S and guinea pig CAP11 cathelicidins (Chang *et al.*, 2006). The finding of β-defensin-like peptides in teleosts supports the hypothesis that

Table 8.2 Categorization of fish antimicrobial peptides in families, based on their homology, secondary structure and genomic organisation. Relevant references are listed on Table 8.1.

Family	*Structure*	*Activity*	*Exons*	*Members*
Cathelicidins	N/d	Gram-(+), Gram-(–)	4	CATH-1, -2
Defensins	Three β-strands with 3 disulfide bonds[a]	N/d	3	DB-1, -2, -3
LEAP	Two β-sheets with 2/4 disulfide bonds	Gram-(–), fungi	3	HepcidinLEAP-2
Piscidins	Amphipathic α-helix	Gram-(+), Gram-(–)	4[b]	Pleurocidin Piscidin Chrysophsin Epinecidin-1 Dicentracin
Histone-derived	Variable	Gram-(+), Gram-(–), fungi	N/d	Parasin HSDF-1 Oncorhyncin II, III Hipposin SAMP H1

[a]Secondary structure predicted by comparative modelling.
[b]Gene structure only determined for epinecidin-1.
N/d: not determined.

β-defensins may represent the ancestral form of the defensin family in vertebrates.

Most teleost species examined contain more than one hepcidin gene. Expansion of the hepcidin gene family is particularly noteworthy within the Percomorpha group, which includes Perciformes, Pleuronectiformes and Tetraodontiformes. Remarkably, we have recently identified at least 13 hepcidin genes clustered in six scaffolds of the tiger pufferfish genome (J.M.O. Fernandes, R. Sugamata, V.J. Smith and Y. Suzuki, unpubl. data), emphasizing how prominent these types of AMPs are in teleosts. Phylogenetic analysis of all the hepcidin-coding sequences identified in fish reveals the relationships between the various paralogues (Fig. 8.3), so it is likely that most tiger pufferfish hepcidins could have been generated by recent tandem gene duplication events. Interestingly, hepcidins from the more ancient fish groups (Salmoniformes, Cypriniformes and Siluriformes) cluster together with the mammalian genes (Fig. 8.3).

The structural similarity between hepcidin and LEAP-2 indicates that they belong to the same family, which collectively constitute liver-expressed antimicrobial peptides. For example, channel catfish hepcidin

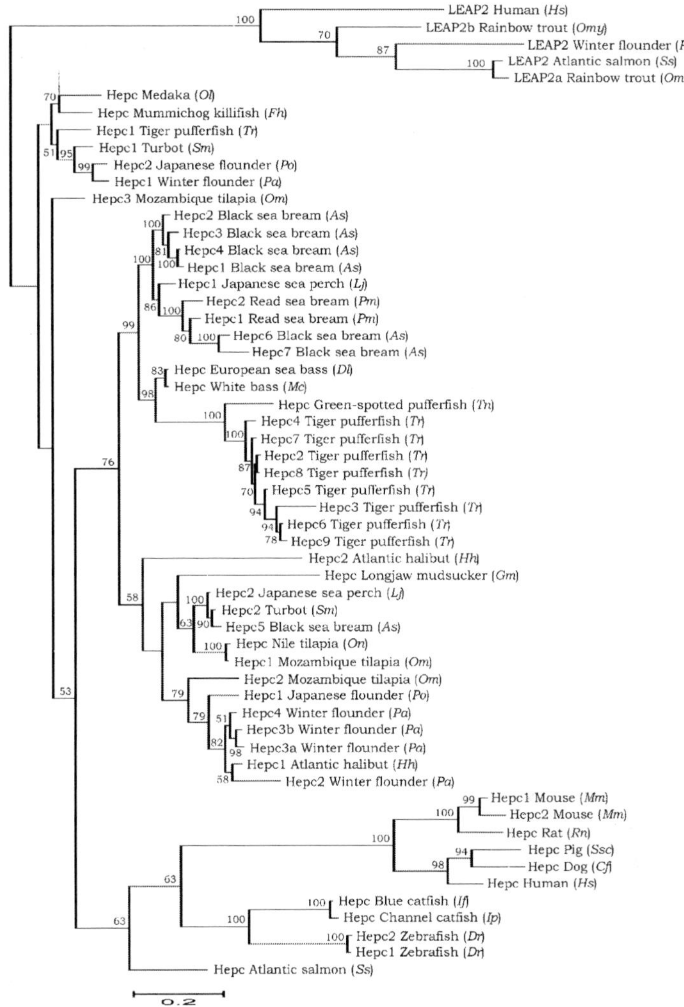

Fig. 8.3 Phylogram of hepcidins. This phylogenetic tree was reconstructed from a Bayesian inference of phylogeny conducted for 1,000,000 generations. Convergence was achieved after *circa* 150,000 generations. Bayesian posterior probabilities are represented as percentages at the tree nodes. Only values greater than 50% are shown. The liver-expressed antimicrobial peptide 2 (LEAP2) sequences were used as outgroup to root the tree. Abbreviations of binomial scientific names are as follows: *As*, *Acanthopagrus schlegelii*; *Cf*, *Canis familiaris*; *Dl*, *Dicentrarchus labrax*; *Dr*, *Danio rerio*; *Fh*, *Fundulus heteroclitus*; *Gm*, *Gillichthys mirabilis*; *Hh*, *Hippoglossus hippoglossus*; *Hs*, *Homo sapiens*; *If*, *Ictalurus furcatus*; *Ip*, *Ictalurus punctatus*; *Lj*, *Lateolabrax japonicus*; *Mc*, *Morone chrysops*; *Mm*, *Mus musculus*; *Ol*, *Oryzias latipes*; *Om*, *Oreochromis mossambicus*; *Omy*, *Oncorhynchus mykiss*; *On*, *Oreochromis niloticus*; *Pa*, *Pleuronectes americanus*; *Pm*, *Pagrus major*; *Po*, *Paralichthys olivaceus*; *Rn*, *Rattus norvegicus*; *Sm*, *Scophthalmus maximus*; *Ss*, *Salmo salar*; *Ssc*, *Sus scrofa*; *Tn*, *Tetraodon nigroviridis*; *Tr*, *Takifugu rubripes*.

(Bao *et al.*, 2005) and LEAP-2 (Bao *et al.*, 2006) have a similar 3 exon/2 intron structure and their coding sequences share approximately 44% identity at the nucleotide level.

A closer look at the genomic organization, amino acid sequence and secondary structure of pleurocidins, piscidins, chrysophsins, epinecidin-1 and dicentracin reveals striking similarities. Bayesian inference of phylogeny using their putative precursor sequences clearly shows that they share a common evolutionary origin (Fig. 8.4) and that the current nomenclature system is misleading. Pleurocidin WF3 from the winter flounder, for instance, is more closely related to moronecidin (piscidin)

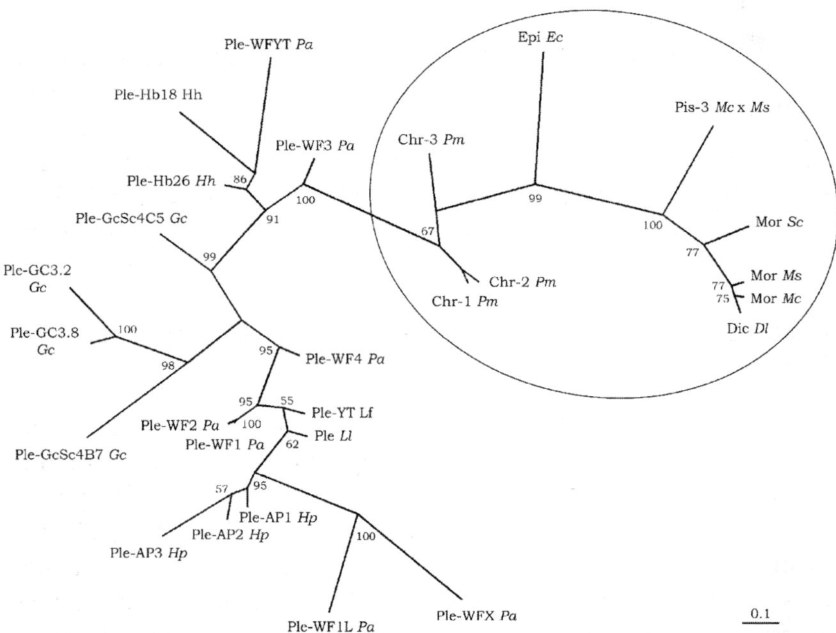

Fig. 8.4 Unrooted radiation tree of piscidins. A Bayesian phylogenetic analysis of pleurocidins (*Ple*), moronecidins (*Mor*), chrysopsins (*Chr*), piscidin (*Pis*), epinecidin (*Epi*) and dicentracin (*Dic*) was performed essentially as described in (Fernandes *et al.*, 2007a). The tree was reconstructed using an average mixed model of amino acid evolution and 1,000,000 generations. The reliability of this topology is supported by the generally high Bayesian posterior probabilities, calculated from the last 9,000 trees and shown as percentages at the tree nodes. Only values greater than 50% are indicated. Abbreviations of species names are as follows: *Dl*, *Dicentrarchus labrax* ; *Ec*, *Epinephelus coioides*; *Gc*, *Glyptocephalus cynoglossus*; *Hh*, *Hippoglossus hippoglossus*; *Hp*, *Hippoglossoides platessoides*; *Lf*, *Limanda ferruginea*; *Ll*, *Limanda limanda*; *Mc*, *Morone chrysops*; *Ms*, *Morone saxatilis*; *Pa*, *Pleuronectes americanus*; *Pm*, *Pagrus major*, *Sc*, *Siniperca chuatsi*.

from the Chinese perch than to pleurocidin WF1L from its own species (Fig. 8.4). We propose that pleurocidins, piscidins, chrysophsins, epinecidin-1 and dicentracin are members of the same family and that this family be designated the piscidins. Pleurocidin has recently been shown to be part of the cecropin superfamily of AMPs, which includes not only the insect and mammalian cecropins, but also the amphibian dermaseptins and insect ceratotoxins (Tamang and Saier, 2006). Hence, we deduce that piscidins are a family of ancient host defence peptides widespread across invertebrate and vertebrate taxa. It is not clear whether pardaxins are homologous to piscidins, as their primary structures are rather different. Without examining their precursor sequences and gene structure we are not confident at this point in time to include them within the piscidin family.

Some AMPs are produced by proteolytic cleavage of larger proteins of previously known function (Boman, 1995; Zasloff, 2002). In fish, most AMPs from this category isolated to date are derived from histones. Histone-derived peptides have been found in diverse groups of fish, including Salmoniformes, Siluriformes and Pleuronectiformes (Table 8.2). A few others include a 40S ribosomal protein, S30 (Fernandes and Smith, 2002) and apo-lipoproteins Apo-A1 and 2 (Concha *et al.*, 2004).

Effect of Microbial Infection or Immune Challenge on AMP Production

Antibacterial proteins produced by the skin or at mucosal surfaces are often expressed constitutively at levels which make isolation of the proteins from mucus extracts fairly easy (Lazarovici *et al.*, 1986; Thompson *et al.*, 1986; Cole *et al.*, 1997; Fernandes *et al.*, 2002, 2003, 2004b; Fernandes and Smith, 2002) but this is not always the case with systemic AMPs. With the internally synthesized AMPs, there may be no baseline expression until immune challenge, or else the levels are too low in naïve fish for assay led detection. Isolation may be further confounded by constitutive lysozymes, which are abundant in teleosts (Yano, 1996; Smith *et al.*, 2000; Fernandes *et al.*, 2004a). However, several systemic fish AMPs are upregulated by challenge with the non-self, such as injection of LPS, bacteria or vaccines, which increases their concentration to detectable levels and facilitates purification or finding of mRNA transcripts. Fish AMPs that strongly induced by immune challenge include cathelicidins

(Chang *et al.*, 2005, 2006) and hepcidins or hepcidin-like LEAPs (Shike *et al.*, 2002; Douglas *et al.*, 2003a; Zhang *et al.*, 2004; Bao *et al.*, 2005; Chen *et al.*, 2005; Lauth *et al.*, 2005; Rodrigues *et al.*, 2006; Huang *et al.*, 2007) and a BPI-like gene in channel catfish (Xu *et al.*, 2005). In tiger pufferfish, immunostimulation by subcutaneous injection of endotoxin (LPS) has a differential effect in expression levels of the various hepcidin genes (J.M.O. Fernandes, R. Sugamata, V.J. Smith and Y. Suzuki, unpublished), thus showing that the hepcidin paralogues are differentially regulated and might have distinct functions in immunity and iron homeostasis. Defensins are expressed—albeit at low levels—in untreated fish although this can up-regulated by immune stimulation (Zou *et al.*, 2007).

The response to induction may be strong and occur within hours. For example, in channel catfish, liver hepcidin expression may increase fourfold within 4 h, rising to over 20 fold in 48 h following bacterial injection (Hu *et al.*, 2007). However, upregulation in other organs may be less dramatic (Hu *et al.*, 2007). In channel catfish with anaemia—a disease commonly found in pond-raised individuals—the hepatic hepcidin transcript levels are approximately 14% of that of healthy specimens (Hu *et al.*, 2007), indicating that hepcidin is also involved in iron homeostasis in fish. Enhanced AMP expression and concomitant increased levels of mature proteins in the body no doubt help combat infection, as demonstrated by Jia *et al.* (2000), who reduced mortality of coho salmon from *Listonella (Vibrio) anguillarum* by pump injections of synthetic α-helical AMPs. Unfortunately, there is a dearth of data on AMP expression and survival of juveniles or adult fish to infection. Thus, while we view the role of antimicrobial peptides as key effectors in the innate defence armoury of fish, there is insufficient experimental evidence to support the notion that genetic modification of fish to increase AMP expression would be beneficial to disease control in aquaculture.

Antimicrobial Potency and Spectra of Activity

Most fish AMPs are potent antibiotics with a broad spectrum of activity at micromolar concentrations (Table 8.2). In fish, data for bioactivities are limited to those proteins that have been purified *de novo*, expressed recombinantly or made synthetically. As many recent reports of fish AMPs have come from data mining approaches, little is known about the spectra of activity or potency of the encoded proteins. As far as we are aware, only pleurocidin has been expressed *in vitro* by recombinant technology (Brocal *et al.*, 2006), although more studies have been performed with synthetic peptides. These include HSDF-1 and -2 (Patrzykat *et al.*, 2001), pleurocidin (Murray *et al.*, 2003), pardaxin (Porcelli *et al.*, 2004), piscidin (Chinchar *et al.*, 2004), hepcidin (Lauth *et al.*, 2005) and cathelicidin (Chang *et al.*, 2006). Comparing the potency of different peptides, wherever material is available is difficult, since various researchers have used different assays and the bacterial strains tested are not standardized. In general, however, most are active against both Gram positive and Gram negative bacteria. The mode of action of AMPs usually involves binding, insertion and disruption of the bacterial cell membrane via the 'carpet' or 'barrel-stave' mechanism proposed by Shai (1999) and Huang (2000). Several fish AMPs, namely pardaxin, show some degree of selectivity, which is related to their biophysical properties and to the composition of lipid moiety with which they interact (Thennarasu and Nagaraj, 1996; Porcelli *et al.*, 2004). Pardaxin has distinct haemolytic and antibacterial domains; the 11-residue carboxy-terminal tail is responsible for its non-selective effects against erythrocytes and bacteria, whereas its amino-terminal helix-hinge-helix accounts for its cytolytic activity against bacteria (Oren and Shai, 1996). Piscidins have broad-spectrum antibacterial, antifungal and antiparasitic activity against a number of fish and human pathogens, including MRSA strains of *Staphylococcus* (Silphaduang and Noga, 2001; Noga and Silphaduang, 2003). In addition, piscidins also have potent antiviral activity and they reduce the viral infectivity of channel catfish virus by 50% at a concentration of 4 μM (Chinchar *et al.*, 2004). Synthetic white bass hepcidin is only active against Gram-positive bacteria and filamentous fungi (Lauth *et al.*, 2005). It does not kill Gram-negative bacteria or yeast (Lauth *et al.*, 2005). Rainbow trout cathelicidins CATH-1 and CATH-2, produced by chemical synthesis, on the other hand, inhibit the growth of Gram-positive and Gram-negative bacteria but are only bactericidal against Gram-negative bacteria (Chang *et al.*, 2005). The potential antimicrobial activity of fish defensins is yet to be confirmed experimentally, but histones and histone-derived peptides have been shown to exhibit potent broad-spectrum antibacterial and antifungal activities (Park *et al.*, 1998; Robinette *et al.*, 1998; Richards *et al.*, 2001; Fernandes *et al.*, 2002, 2003, 2004b; Birkemo *et al.*, 2003; Bergsson *et al.*, 2005). The catfish HLPs, which are similar to histone H1 and histone H2B, additionally, have antiparasitic activity against young and mature trophonts (feeding stage) of the protozoan ectoparasite *Amyloodinium ocellatum*, the causative agent of amyloodiniosis (Noga *et al.*, 2001, 2002).

Synergistic activity between fish AMPs was first demonstrated in coho salmon for HSDF-1, a peptide derived from the N-terminus of histone H1. On its own, this synthetic 26-residue peptide is devoid of antimicrobial activity against the fish pathogens, *Aeromonas salmonicida* and *Listonella anguillarum*, even at concentrations in excess of 1 $mg{\cdot}ml^{-1}$, but in the presence of winter flounder pleurocidin or hen egg white lysozyme, it is highly active against these bacteria at concentrations as low as 16 $\mu g{\cdot}ml^{-1}$ (Patrzykat *et al.*, 2001). Reports of synergism between fish AMPs and other antimicrobial proteins are scarce but significant synergy has been observed between white bass hepcidin and moronecidin (Lauth *et al.*, 2005) and between a cationic peptide derived from the carboxy-terminus of the apolipoprotein A-I and hen egg white lysozyme (Concha *et al.*, 2004).

Sites of Tissue Expression

Essentially there are two main sites of AMP expression in fish. One is the boundary surface that interfaces with the outside environment, and the other is the blood and internal organs. Both face different types of microbial threat. The external surface, which includes not only the skin epithelia but also the eye, gill surface, gut and urogenital tract, has to deal with water-borne opportunists of a diverse type. The body organs, on the other hand, are largely protected from microbial attack by the external barriers; so they face threats from potential pathogens, many of which have developed strategies to avoid recognition and elimination by the host's adaptive immune system. It might be expected, therefore, that the mucosal surfaces and internal tissues might express their own suites of antimicrobial proteins.

Considering first the mucosal surfaces, it is striking that many appear to belong to the category of AMPs that are, or are derived from, larger proteins with other functions. They include histones, histone-like proteins (Park *et al.*, 1998; Robinette *et al.*, 1998; Noga *et al.*, 2001; Patrzykat *et al.*, 2001; Richards *et al.*, 2001; Cho *et al.*, 2002a; Fernandes *et al.*, 2002, 2003, 2004b; Birkemo *et al.*, 2003; Bergsson *et al.*, 2005), ribosomally derived proteins (Fernandes and Smith, 2002; Bergsson *et al.*, 2005), high density lipoprotein plus its principal apolipoproteins (Concha *et al.*, 2003, 2004), and an L-amino acid oxidase (Kitani *et al.*, 2007) (Table 8.1). These are not 'dedicated' classical AMPs but proteins—or protein fragments—present throughout the body that participate in a number of other important cellular or physiological processes. Furthermore, these types of proteins are not confined to teleosts but are present in nearly all multicellular animals. In fish mucosa, they are conspicuous because of their antimicrobial activities, although some are known contribute to defence in other vertebrate taxa, notably amphibian skin (Park *et al.*, 1996) and mouse macrophages (Hiemstra *et al.*, 1999). Histones themselves have been known for over forty years to have antibacterial properties in mammals (Hirsch, 1958). Since these types of molecules must occur in their intact form in every cell, one might expect that they also contribute to internal defence throughout the body. However, they do not seem to feature significantly in reports for blood or tissue AMPs from most multicellular organisms. Their dominance as antimicrobial effectors in fish mucosa might be because the fish skin offers unique opportunities for them to be exposed in an appropriate form to interact with and kill microorganisms *in vivo*. As yet, we do not know the full details of the underlying mechanisms, but in many ways fish skin, like that of amphibians, represents a 'special case' because it is not keratinized. Thus, the epithelial cells are very vulnerable to abrasion and physical damage. Shearing forces could allow organelles and nucleosome proteins to be released into the skin mucus, where they could be processed to a microbicidal form, a scenario that has been very elegantly demonstrated for parasin by Cho *et al.* (2002a). In this study, immunohistochemistry was used to show that unacetylated histone H2A (the precursor of parasin), together with procathepsin D is present in the cytoplasm of mucous gland cells of catfish. Upon wounding, a metalloprotease (matrix metalloproteinase 2) activates procathepsin D to cathepsin D (Cho *et al.*, 2002b) which, in turn, cleaves histone H2A (Cho *et al.*, 2002a). This enzymatic cascade reaction culminates in the production of parasin, which is then secreted to the mucosal layer that coats the skin (Cho *et al.*, 2002a). However, it is unclear whether other histone- or ribosomally derived proteins are generated in the skin in the same way.

Conspicuous as the above proteins are in fish skin, more conventional AMPs are certainly also expressed in the mucosal surface. The best understood to date are the pleuricidins and pardaxins (Thompson *et al.*,

1986; Cole *et al.*, 1997, 2000; Adermann *et al.*, 1998; Douglas *et al.*, 2003b; Patrzykat *et al.*, 2003) (Table 8.1). Pleuricidins are present in the mucin granules of the skin and the intestinal goblet cells and sometimes also in the gill filaments (Cole *et al.*, 1997, 2000; Douglas *et al.*, 2003b). Pardaxins are present in exocrine secretions from the epithelial glands of sole (Lazarovici *et al.*, 1986; Thompson *et al.*, 1986; Adermann *et al.*, 1998). Some piscidins also appear to occur in the skin epithelia of several fishes from the suborder Percoidei, including bass and Atlantic croaker (*Micropogonias undulatus*), although expression is not confined to the epithelial cells (Lauth *et al.*, 2002; Silphaduang *et al.*, 2006).

Of the AMPs expressed within the internal tissues of fish, the best studied are the hepcidin and the hepcidin-like proteins, including the LEAP peptides. In all species found to synthesize these, the main site of expression is the liver (Table 8.3). However, lower but significant transcript levels have also been found in spleen, gill or intestine (Bao *et al.*, 2005, 2006; Chen *et al.*, 2007), especially after immune stimulation, with detectable levels found to be expressed in stomach, trunk kidney, gut, blood, skin or gonad depending on species (Bao *et al.*, 2005, 2006; Chen *et al.*, 2007) (Table 8.3). Occasionally hepcidins or hepcidin-like transcripts have been reported for muscle but this may depend on the species and peptide type. Atlantic salmon, for example, appear to express one type of hepcidin (*sal-1*) in muscle but not another (*sal-2*) (Douglas *et al.*, 2003a). In turbot (*Scophthalmus maximus*), no expression seems to occur in muscle, although hepcidin genes are transcribed in a wide range of other tissues (Chen *et al.*, 2007). With respect to LEAP-2, both channel (*Ictalurus punctatus*) and blue (*I. furcatus*) catfishes express it constitutively in various tissues, including skin, intestine, head-kidney and muscle (Bao *et al.*, 2006), whereas in rainbow trout LEAP-2 is only expressed in the liver (Zhang *et al.*, 2004) (Table 8.3).

Cathelicidins, are another group of fish expressed by multiple tissues, but are predominant in lymphoid tissues. In rainbow trout, CATH-1 is only expressed upon bacterial challenge, whilst CATH-2 transcripts are constitutively present in spleen, head-kidney, intestine, gills and skin (Chang *et al.*, 2006). Similarly, β-defensins in pufferfish and grouper are expressed by several organs, usually the gill, gonad, gut, kidney, muscle, skin and spleen (Zou *et al.*, 2007), but it is not clear from the assays used for these analyses how much the signal detected in non-lymphoid organs

Table 8.3 Expression sites of the main AMPs in fish tissues.

AMP	*Type*	*Primary site*	*Secondary sites*	*Species*	*References*
Histones, histone-like and Histone-derived fragments	H2A	Skin epithelium		Channel catfish	Robinette *et al.* (1998) Noga *et al.* (2001)
	H2A (parasin 1)	Skin epithelium		Amur catfish	Park *et al.* (1998)
	H2A	Skin epithelium		Rainbow trout	Fernandes *et al.* (2002)
	HSD-F	Skin mucus	Blood	Coho salmon	Patrzykat *et al.* (2001)
	HLP (HLP1, 2 & 3)	Skin		Channel catfish	Robinette *et al.* (1998)
	H2B	Skin epithelium	Gill, spleen	Cod	Bergsson *et al.* (2005)
	H1 (oncorhyncin II)	Skin		Rainbow trout	Noga *et al.* (2001) Fernandes *et al.* (2004)
	H1	Skin	Blood	Coho salmon	Patrzykat *et al.* (2001)
	H1	Liver		Atlantic salmon	Richards *et al.* (2001)
	H1			Channel catfish	Noga *et al.* (2001)
	H6 (oncorhyncin III)	Skin		Rainbow trout	Fernandes *et al.* (2003)
	Hipposin	Skin		Atlantic halibut	Birkemo *et al.* (2003, 2004)
Pardaxins	P1, P2	Epithelial glands		Peacock sole	Thompson *et al.* (1986)
	P3	Exocrine secretion		Peacock sole	Zagorski *et al.* (1991)
	P4, P5	Skin		Red Moses sole	Lazarovici *et al.* (1986) Nagarajai (1996) Aldermann *et al.* (1998)
High-density lipoproteins	HDL	Skin	Blood	Carp	Concha *et al.* (2002)
HDL-derived apolipoproteins	ApoA-1, ApoA-2	Blood		Carp	Concha *et al.* (2004)

(Table 8.3 contd.)

(Table 8.3 contd.)

Piscidins and pleuricidins	Piscidin 1	Mast cells	Skin, gill, spleen, head-kidney	White bass	Silphaduang and Noga (2001)
	Piscidin 2, 3	Mast cells	Skin, gill, gut	Striped bass	Lauth *et al.* (2002) Silphaduang and Noga (2001)
	Dicentracin	Leucocytes		European sea bass	Salerno *et al.* (2007)
	Epinecidin-1	Leucocytes		Spotted grouper	Yin *et al.* (2006)
	Pleuricidin	Skin epithelium	Gill	Winter flounder	Cole *et al.* (1997) Patrzykat *et al.* (2002) Douglas *et al.* (2003)
	Chrysophsin	Gill		Red seabream	Iijima *et al.* (2003)
	Pleuricidin	Skin and epithelia		Winter flounder	Cole *et al.* (1997) Patrzykat *et al.* (2002) Douglas *et al.* (2003)
	Pleurocidin	Skin and epithelia		American plaice	Douglas *et al.* (2003)
	Pleurocidin	Skin and epithelia		American halibut	Patrzykat *et al.* (2002)
	Pleurocidin	Skin and epithelia		Mud dab	Brocal *et al.* (2006)
Hepcidins, hepcidin-like proteins	Hepcidin	Liver	Oesophagus, stomach	Winter flounder	Douglas *et al.* (2003)
	Sal-1	Liver	Blood, muscle	Atlantic salmon	Martin *et al.* (2006)
	Sal-2	Gill, skin		Atlantic salmon	Martin *et al.* (2006)
	Hepcidin	Liver	Skin heart, abdominal organs	Zebrafish	Shike *et al.* (2004)
	Hepcidin	Liver	Several tissues (not muscle)	Turbot	Chen *et al.* (2007)
	Hepcidin	Liver	Other tissues and brain	Red seabream	Chen *et al.* (2005)

(Table 8.3 contd.)

(Table 8.3 contd.)

	Hepcidin	Liver		Blue catfish	Hu *et al.* (2007)
	JF-1, JF-2	Liver	Other tissues	Japanese flounder	Hirono *et al.* (2005) Matsuyama *et al.* (2006)
	Hepcidin	Liver	Intestine, kidney, spleen	Tilapia	Huang *et al.* (2007)
	Hepcidin	Liver	Gills	Striped bass	Shike *et al.* (2002) Lauth *et al.* (2005)
	Hepcidin	Liver	Kidney, spleen	Halibut	Park *et al.* (2005)
	Hepcidin	Liver		Japanese sea perch	Ren *et al.* (2006)
	Hepcidin	Liver	Various tissues	European sea bass	Rodriques *et al.* (2006)
LEAPs	LEAP-2	Liver	Other organs (not brain)	Channel catfish	Bao *et al.* (2005)
	LEAP-2	Liver	Tissues (not brain or stomach)	Rainbow trout	Zhang *et al.* (2004)
Defensins	b-defensins	Various tissues		Zebrafish, Pufferfish	Zou *et al.* (2007)
Cathelidins	asCATH-1,	Head kidney		Atlantic salmon	Chang *et al.* (2006)
	as CATH-2	Head kidney	Gill, spleen	Atlantic salmon	Chang *et al.* (2006)
	rtCATH-2	Head kidney	Gill, intestine, skin, spleen	Rainbow trout	Chang *et al.* (2005)

can be attributed to AMPs expressed in the blood perfusing them. Pleurocidins and piscidins, in particular, show multi-tissue expression (Douglas *et al.*, 2003b; Silphaduang *et al.*, 2006) (Table 8.3) but as these proteins are abundant in the eosinophilic granular (or mast) cells (Silphaduang and Noga, 2001), the signals could come, at least in part, from the blood. With respect to blood-derived AMPs, piscidin-like molecules such as epinecidin-1 from the grouper (*E. coidoides*) and diacentrin from the European sea bass (*D. labrax*) have been reported for monocytes and granular leucocytes in the head-kidney, peripheral blood and peritoneal cavities (Yin *et al.*, 2006; Salerno *et al.*, 2007). However, crysophysin, a piscidin-like peptide from the red sea bream (*C. major*), has so far been isolated only from the gill epithelia (Iijima *et al.*, 2003). As regards the β-defensins, the zebrafish paralogues are differentially expressed in separate tissues: zfDB1 is preferentially expressed in skin, kidney, gill and muscle, whereas zfDB2 is only present in the gut and zfDB3 shows a ubiquitous pattern of expression (Zou *et al.*, 2007).

Other blood-borne antibacterial proteins include a 13 kDa protein isolated from the erythrocytes of *O. mykiss* (Fernandes and Smith, 2004) and the apolipoproteins, apoA-1 and apoA-2, purified from the plasma of carp, *C. carpio* (Concha *et al.*, 2004) (Table 8.3).

Changes in Expression during Development

With regard to the presence of AMPs during embryogenesis and maturation, it seems that some AMPs, mainly liver-expressed AMPs (hepcidins and hepcidin-like LEAPs) are present very early on in ontogeny of fish. Bao *et al.* (2005) have reported that catfish *(I. punctatus)* eggs express transcripts for hepcidin as early as 8 h post-fertilization, i.e., pre-organogenesis, and that expression continues and increases up to 17 days. Interestingly, the lowest levels of hepcidin transcription seem to occur immediately after hatching (Bao *et al.*, 2005). Another hepcidin-type AMP in eggs of catfish, LEAP-2, is not expressed until 3 days after fertilization, although it is doubtful that a functional protein is present in the eggs at this time as the transcript is spliced only on day 6 post hatching (Bao *et al.*, 2006). Thereafter, the mRNA for the mature protein is continuously expressed for at least a further 17 days, the length of the study period (Bao *et al.*, 2006). In turbot, increasing levels of hepcidin have been detected in embryos from 2 h post-fertilization until 95 h (the larval stage) and is most marked at 47 h when the tail buds form (Chen *et al.*, 2007). It should be noted that hepcidin and LEAP AMPs are not the only microbicidal molecules associated with juvenile fish. A study by Douglas *et al.* (2001) has found that pleurocidin transcripts are present in the juveniles of the Winter flounder (*P. americanus*) but become detectable only after at 13 days post-hatching.

It is interesting to note that peptides synthesized by the liver and other tissues in adults are already present pre-organogenesis in newly fertilized eggs. Certainly, eggs and embryos need protection from microbial colonization from the moment of release from the mother, as the aquatic environment is hostile and rich in microbes. One would, therefore, expect that to favour egg survival, either the mother would transfer antimicrobial proteins to them from her own body, or that the eggs would begin synthesizing their own antimicrobial arsenal as soon as possible. While IgM is known to be passively transferred to eggs by the mother (Mor and Avtalion, 1989; Takemura and Takano, 1997), less is known about maternal transfer of innate defence molecules, especially broad-spectrum bactericidal proteins. There are several opportunities for the female to endow her eggs with antibacterial molecules, either during development of the eggs in her ovary or when they pass down the oviduct to the outside. It is noteworthy that LEAP-2 is known to be expressed in ovary in catfish (Bao *et al.*, 2006), while hepcidin has been reported to be expressed in turbot gonad, albeit at levels lower than the liver (Chen *et al.*, 2007). It is clear that lysozyme may be transferred from the mother to her ova, as a study by Balfry and Iwama (2004) has shown that levels of lysozyme activity in the maternal kidney of coho salmon (*O. kisutch*) correlate with the levels of activity in her unfertilized eggs. Since the eggs then display increasing levels of lysozyme activities as they develop through alevins to first feeding fry, *de novo* synthesis of lysozyme must occur (Balfry and Iwama, 2004). It is plausible that AMPs, which are generally smaller, and therefore less 'costly' molecules to make than lysozyme, could also be available to the developing eggs in the same way. Lysozyme, whether maternally derived or synthesized by the eggs or hatchlings, is a powerful weapon against many Gram-positive bacteria, but is far less active against Gram-negatives or fungi. Working in synergy with AMPs, however, can maximize its antibacterial potency. This is clearly an interesting and important area for further investigation, but it would appear that the choice of model species is crucial, as differences in the levels of lysozyme activities at each developmental stage vary between species or strains (Balfry and Iwama, 2004). Thus, whilst AMPs are strong candidates for developmentally early defence molecules in fish because they are small, easily synthesized without mature organs or tissues and can diffuse rapidly to the egg surface, in reality the dynamics of expression and synthesis of bioactive proteins may be more complex.

Environmental Effects on AMP Expression and Activity

A smaller number of papers that examine the fish response to stress include some measurement of changes in antimicrobial activities of tissues or epidermal extracts, but very few have undertaken a thorough investigation of the environment. Environmental conditions are likely to have significant effects on fish, not only because factors such as temperature profoundly affect metabolism and growth (Johnston, 2003), embryonic development (Fernandes *et al.*, 2006; 2007b), behaviour (Wilson *et al.*, 2007) and functioning of the adaptive immune system of fish (Morvan *et al.*, 1998), but also because these animals are in such intimate contact with a medium that can vary dramatically in chemical composition. Thus, water-borne contaminants can easily cross the skin gills and other epithelial surfaces and disturb the homeostatic integrity, leading to a state of immune suppression (Dunier and Siwicki, 1993; Bly *et al.*, 1997; Arkoosh *et al.*, 1998).

Certainly, various physiological stresses such as handling are known to affect innate immune reactivity in fish but increased mucus production of the skin epithelium may give the appearance of enhanced antibacterial defence. Evidence for this has been given by Demers and Bayne (1997) in relation to lysozyme activity in rainbow trout, *O. mykiss*. More recent work has shown that mucus lysozyme levels may decrease in *O. mykiss* exposed to high levels of sewage although this is mainly in females and juveniles, which generally have higher levels of lysozyme in the mucosa than males (Hoeger *et al.*, 2005).

Surprisingly, few studies have actually measured AMP expression levels in fish exposed to pollutants, temperature change or other environmental disturbances. One study of temperature effects by Chinchar *et al.* (2004) found that the antiviral activity of four synthetic striped bass piscidins remains unchanged over a wide of water temperatures, being retained at 4°C as well as at 26°C. Fernandes *et al.*

(2004a) have also reported that muramidases from skin secretions of rainbow trout, *O. mykiss*, function at both 5°C and 20°C, but insufficient studies have been made to draw general conclusions about effects of temperature on the activity of fish AMPs. More importantly, we need to know not only how AMP potency is affected but also how temperature affects their expression at the gene level.

Apart from temperature, salinity is an important environmental factor for fish, especially those migrating between fresh and seawater for reproduction. Salt concentration is known to impair the activity of many animal AMPs by interfering with the interaction between the positively charged AMP molecules and the negatively charged bacterial surface. There is a dearth of information about how immune gene expression changes in fish during the transition from seawater to fresh, or vice versa, but functional studies of AMP activities *in vitro* indicate that some are sensitive to salt concentration but others are not. For example: in rainbow trout, *O. mykiss*, histone H2A is active, at least against the Gram-positive bacterium, *Planococcus citreus*, at NaCl concentrations ranging from 140 mM (freshwater levels) to 550 mM (seawater levels), but the MIC is 16-fold higher at 550 mM NaCl than at 140 mM, showing that some impairment of activity occurs at the higher salt levels (Fernandes *et al.*, 2002). Likewise oncorhyncin III, an N terminal fragment of non-chromosomal histone H6, shows weaker killing at seawater salt concentrations than in freshwater; in this case, the MIC is *circa* 8-fold higher at 550 mM NaCl than at 140 mM (Fernandes *et al.*, 2003). High salt has also been observed to completely inhibit the antibacterial activity of a ribosomal protein, L35, isolated from skin mucus of Atlantic cod, *G. morhua* (Bergsson *et al.*, 2005) but pleurocidin from flounder remains unaffected (Cole *et al.*, 2000; Douglas *et al.*, 2003b). As yet, we can only speculate that tolerance to high and low salt might be associated with migration between salt and freshwater as changes in AMP expression have yet to be tracked over the life history of anadromous species.

The influence of other cations on AMP activity have seldom been investigated, although Cole *et al.* (2000), have reported that 10 mM Mg^{2+} and Ca^{2+} impair antibacterial activity of pleurocidin from winter flounder. In contrast, increased NaCl concentrations, up to 150 mM, had no such effect (Cole *et al.*, 2000). Possibly, the divalent cations stabilize the bacterial cell membrane against depolarization, whereas Na^+ destabilizes it. Ions, such as NH_4^+ and Fe^{3+} may also affect antibacterial defence, but in unexpected ways. Increased levels of ammonia can occur under intense housing conditions, which are highly stressful for fish. Curiously, Robinette and Noga (2001) have found that while histone-like protein-1 is significantly depressed in epithelial skin scrapings taken from channel catfish under overcrowded conditions, overall antibacterial vigour of the skin secretions is actually elevated, indicating that other antimicrobial proteins might be stimulated by such stress.

The effect of iron levels on hepcidin expression is particularly interesting, as hepcidins appear to act as iron-regulating 'hormones' (Ganz, 2003b; Shi and Camus, 2006; Hu *et al.*, 2007). Elevated levels of iron in the diet have been found to increase transcript levels of hepcidin in sea bass (*D. labrax*) by approximately 30% compared to fish maintained on control or iron-deficient diets (Rodrigues *et al.*, 2006). Similar elevation in hepcidin expression under iron-overload conditions has also been noted in zebrafish (Fraenkel *et al.*, 2005), but in Japanese flounder, hepcidin-1 expression tends to decrease while hepcidin-2 remains unchanged (Hirono *et al.*, 2005). Indeed, in catfish, hepatic transcript levels of hepcidin correlate with serum iron concentrations and with the degree of saturation of transferrin (Hu *et al.*, 2007). In the natural environment, free iron is rapidly removed from solution by photosynthetic micro-algae; so it is unlikely than levels of iron would be a major soluble toxic threat to marine fish.

CONCLUSIONS

From the above review it is clear that in the last 10 years, our knowledge of teleost antimicrobial proteins has rapidly expanded, and nowadays an increasing number of AMPs are being discovered at a very rapid rate. Some, such as the hepcidins, cathelicidins and defensins, are well known in higher vertebrates, so their presence in fish demonstrates that they have been evolutionary conserved within the vertebrate line. However, the Teleosteii taxon contains over 24,000 species, so only a very small, and largely unrepresentative, number of fish species has been studied, most having been selected as experimental models for their commercial value. A great many more AMP types almost certainly exist. Until we have a more complete picture of the AMPs expressed within the fish group as a whole, it will be difficult to establish general trends and draw conclusions about AMP diversity and phylogeny in finfish. From the information available so far, however, it appears that fish AMPs show the same basic structural types as 'classical' AMPs (e.g., α-helices, and cys-rich proteins with disulphide bonds and) found in other groups.

Interestingly, whilst the application of molecular approaches, from PCR to EST studies and whole genome sequencing, have greatly improved our knowledge of fish AMP types, structure and expression, it has not provided as much information about the *in vivo* role of these molecules. Over expression and knockdown experiments using model fish species, such as the tiger pufferfish, zebrafish and medaka, would certainly shed light into this matter but to date, no such studies have been reported despite their feasibility. In fact, transgenic medaka carrying insect and porcine cecropin genes, previously integrated into their genomes, have been produced and the expression of cecropin transgenes conferred on the host was found to increase resistance to the fish pathogens *Pseudomonas fluorescens* and *Listonella anguillarum* (Sarmasik *et al.*, 2002). Apart from this pioneering study, we know very little about the spectra of activity of many fish AMPs and the extent to which the fish relies on them, as opposed to other innate defence strategies, in protection against infection is still largely conjectural. Certainly, there have been relatively few expression studies and of those published so far, most have focused on determining tissue sites of synthesis in adults. Additionally, we still need more studies that track AMP expression during early development and in response to experimental microbial challenge. Likewise, we are still largely ignorant about how fish AMPs interact with other phases of the immune system; how some exert their effects on their microbial targets; and how known pathogens evade their inhibitory or killing effects. It is likely that fish AMPs are not only powerful endogenous antibiotics with broad spectrum activity but they may also display roles, such as wound healing and regulation of inflammatory and immune responses. Important too is a clarification of how environmental conditions impact on AMP expression and activity. Such information could have great bearing not only for aquaculture production of eggs and young fry but also for protection and preservation of endangered wild stocks.

Finally, we have also yet to realise any potential benefits that fish AMP discovery and characterization might offer to biotechnology and biomedicine. Over the last decade, our knowledge about these molecules has massively improved but we are only just scratching the surface. There could well be some big surprises in store and many new features to be revealed about these remarkable, evolutionary ancient and potentially useful molecules in fish.

References

Adermann, K., M. Raida, Y. Paul, S. Abu-Raya, E. Bloch-Shilderman, P. Lazarovici, J. Hochman and H. Wellhoner. 1998. Isolation, characterization and synthesis of a novel paradaxin isoform. *FEBS Letters* 435: 173-177.

Arkoosh, M.R., E. Casillas, E. Clemons, A.N. Kagley, R. Olson and J. R. Stein. 1998. Effect of pollution on fish diseases: potential impact on salmonid populations. *Journal of Aquatic Animal Health* 10: 182-190.

Balfry, S.K. and G.K. Iwama. 2004. Observations on the inherent variability of measuring lysozyme activity in coho salmon (*Oncorhynchus kisutch*). *Comparative Biochemistry and Physiology* B138: 207-211.

Bao, B., E. Peatman, P. Li, C. He and Z. Liu. 2005. Catfish hepcidin gene is expressed in a wide range of tissues and exhibits tissue-specific upregulation after bacterial infection. *Developmental and Comparative Immunology* 29: 939-950.

Bao, B., E. Peatman, P. Xu, P. Li, H. Zeng, C. He and Z. Liu. 2006. The catfish liver-expressed antimicrobial peptide 2 (LEAP-2) gene is expressed in a wide range of tissues and developmentally regulated. *Molecular Immunology* 43: 367-377.

Bergsson, G., B. Agerberth, H. Jornvall and G.H. Gudmundsson. 2005. Isolation and identification of antimicrobial components from the epidermal mucus of Atlantic cod (*Gadus morhua*). *FEBS Journal* 272: 4960-4969.

Birkemo, G.A., T. Luders, O. Andersen, I.F. Nes and J. Nissen-Meyer. 2003. Hipposin, a histone-derived antimicrobial peptide in Atlantic halibut (*Hippoglossus hippoglossus* L.). *Biochimique et Biophysique Acta* 1646: 207-215.

Bly, J.E., S.M. Quiniou and L.W. Clem. 1997. Environmental effects on fish immune mechanisms. *Developmental Biology Standpoint* 90: 33-43.

Boman, H.G. 1995. Peptide antibiotics and their role in innate immunity. *Annual Review of Immunology* 13: 61-92.

Bowdish, D.M., D.J. Davidson, M.G. Scott and R.E. Hancock. 2005. Immunomodulatory activities of small host defence peptides. *Antimicrobial Agents and Chemotherapy* 49: 1727-1732.

Brocal, I., A. Falco, V. Mas, A. Rocha, L. Perez, J.M. Coll and A. Estepa. 2006. Stable expression of bioactive recombinant pleurocidin in a fish cell line. *Applied Microbiology and Biotechnology* 72: 1217-1228.

Campagna, S., N. Saint, G. Molle and A. Aumelas. 2007. Structure and mechanism of action of the antimicrobial peptide piscidin. *Biochemistry* 46: 1771-1778.

Chang, C.I., O. Pleguezuelos, Y.A. Zhang, J. Zou and C.J. Secombes. 2005. Identification of a novel cathelicidin gene in the rainbow trout, *Oncorhynchus mykiss*. *Infection and Immunity* 73: 5053-5064.

Chang, C.I., Y.A. Zhang, J. Zou, P. Nie and C.J. Secombes. 2006. Two cathelicidin genes are present in both rainbow trout (*Oncorhynchus mykiss*) and Atlantic salmon (*Salmo salar*). *Antimicrobial Agents and Chemotherapy* 50: 185-195.

Chen, S.L., M.Y. Xu, X.S. Ji, G.C. Yu and Y. Liu. 2005. Cloning, characterization, and expression analysis of hepcidin gene from red sea bream (*Chrysophrys major*). *Antimicrobial Agents and Chemotherapy* 49: 1608-1612.

Chen, S.L., W. Li, L. Meng, Z.X. Sha, Z.J. Wang and G.C. Ren. 2007. Molecular cloning and expression analysis of a hepcidin antimicrobial peptide gene from turbot (*Scophthalmus maximus*). *Fish and Shellfish Immunology* 22: 172-181.

Chinchar, V.G., L. Bryan, U. Silphadaung, E. Noga, D. Wade and L. Rollins-Smith. 2004. Inactivation of viruses infecting ectothermic animals by amphibian and piscine antimicrobial peptides. *Virology* 323: 268-275.

Cho, J.H., I.Y. Park, H.S. Kim, W.T. Lee, M.S. Kim and S.C. Kim. 2002a. Cathepsin D produces antimicrobial peptide parasin I from histone H2A in the skin mucosa of fish. *FASEB Journal* 16: 429-431.

Cho, J.H., I.Y. Park, M.S. Kim and S.C. Kim. 2002b. Matrix metalloproteinase 2 is involved in the regulation of the antimicrobial peptide parasin I production in catfish skin mucosa. *FEBS Letters* 531: 459-463.

Cho, S.H., B.D. Lee, H. An and J.B. Eun. 2005. Kenojeinin I, antimicrobial peptide isolated from the skin of the fermented skate, *Raja kenojei*. *Peptides* 26: 581-587.

Cole, A.M., P. Weis and G. Diamond. 1997. Isolation and characterization of pleurocidin, an antimicrobial peptide in the skin secretions of winter flounder. *Journal of Biological Chemistry* 272: 12008-12013.

Cole, A.M., R.O. Darouiche, D. Legarda, N. Connell and G. Diamond. 2000. Characterization of a fish antimicrobial peptide: gene expression, subcellular localization, and spectrum of activity. *Antimicrobial Agents and Chemotherapy* 44: 2039-2045.

Concha, M.I., S. Molina, C. Oyarzun, J. Villanueva and R. Amthauer. 2003. Local expression of apolipoprotein A-I gene and a possible role for HDL in primary defence in the carp skin. *Fish and Shellfish Immunology* 14: 259-273.

Concha, M.I., V.J. Smith, K. Castro, A. Bastias, A. Romero and R.J. Amthauer. 2004. Apolipoproteins A-I and A-II are potentially important effectors of innate immunity in the teleost fish *Cyprinus carpio*. *European Journal of Biochemistry* 271: 2984-2990.

Demers, N.E. and C.J. Bayne. 1997. The immediate effects of stress on hormones and plasma lysozyme in rainbow trout. *Developmental and Comparative Immunology* 21: 363-373.

Devine, D.A. and R.E. Hancock. 2002. Cationic peptides: Distribution and mechanisms of resistance. *Current Pharmacological Diseases* 8: 703-714.

Dong, X.Z., H.B. Xu, K.X. Huang, Q. Liou and J. Zhou. 2002. The preparation and characterization of an antimicrobial polypeptide from the loach, *Misgurnus anguillicaudatus*. *Protein Expression and Purification* 26: 235-242.

Douglas, S.E., J.W. Gallant, Z. Gong and C. Hew. 2001. Cloning and developmental expression of a family of pleurocidin-like antimicrobial peptides from winter flounder, *Pleuronectes americanus* (Walbaum). *Developmental and Comparative Immunology* 25: 137-147.

Douglas, S.E., J.W. Gallant, R.S. Liebscher, A. Dacanay and S.C. Tsoi. 2003a. Identification and expression analysis of hepcidin-like antimicrobial peptides in bony fish. *Developmental and Comparative Immunology* 27: 589-601.

Douglas, S.E., A. Patrzykat, J. Pytyck and J.W. Gallant. 2003b. Identification, structure and differential expression of novel pleurocidins clustered on the genome of the winter flounder, *Pseudopleuronectes americanus* (Walbaum). *European Journal of Biochemistry* 270: 3720-3730.

Dunier, M. and A.K. Siwicki. 1993. Effects of pesticides and other organic pollutants in the aquatic environment on immunity of fish: a review. *Fish and Shellfish Immunology* 3: 423-438.

Ellis, A.E. 1999. Immunity to bacteria in fish. *Fish and Shellfish Immunology* 9: 291-300.

Ellis, A.E. 2001. Innate host defense mechanisms of fish against viruses and bacteria. *Developmental and Comparative Immunology* 25: 827-839.

Fernandes, J.M.O. and V.J. Smith. 2002. A novel antimicrobial function for a ribosomal peptide from rainbow trout skin. *Biochemical and Biophysical Research Communications* 296: 167-171.

Fernandes, J.M.O. and V.J. Smith. 2004. Partial purification of antibacterial proteinaceous factors from erythrocytes of *Oncorhynchus mykiss*. *Fish and Shellfish Immunology* 16: 1-9.

Fernandes, J.M.O., G.D. Kemp, M.G. Molle and V.J. Smith. 2002. Anti-microbial properties of histone H2A from skin secretions of rainbow trout, *Oncorhynchus mykiss*. *Biochemical Journal* 368: 611-620.

Fernandes, J.M.O., N. Saint, G.D. Kemp and V.J. Smith. 2003. Oncorhyncin III: a potent antimicrobial peptide derived from the non-histone chromosomal protein H6 of rainbow trout, *Oncorhynchus mykiss*. *Biochemical Journal* 373: 621-628.

Fernandes, J.M.O., G.D. Kemp and V.J. Smith. 2004a. Two novel muramidases from skin mucosa of rainbow trout (*Oncorhynchus mykiss*). *Comparative Biochemistry and Physiology* B138: 53-64.

Fernandes, J.M.O., G. Molle, G.D. Kemp and V.J. Smith. 2004b. Isolation and characterisation of oncorhyncin II, a histone H1-derived antimicrobial peptide from skin secretions of rainbow trout, *Oncorhynchus mykiss*. *Developmental and Comparative Immunology* 28: 127-138.

Fernandes, J.M.O., M.G. MacKenzie, P.A. Wright, S.L. Steele, Y. Suzuki, J.R. Kinghorn and I.A. Johnston. 2006. Myogenin in model puffer fish species: comparative genomic analysis and thermal plasticity of expression during early development. *Comparative Biochemistry and Physiology* D1: 35-45.

Fernandes, J.M.O., J.R. Kinghorn and I.A. Johnston. 2007a. Characterization of two paralogous muscleblind-like genes from the tiger pufferfish (*Takifugu rubripes*). *Comparative Biochemistry and Physiology* B146: 180-186.

Fernandes, J.M.O., M.G. MacKenzie, J.R. Kinghorn and I.A. Johnston. 2007b. *FoxK1* splice variants show developmental stage-specific plasticity of expression with temperature. *Journal of Experimental Biology* 210: 3461-3472.

Fraenkel, P.G., D. Traver, A. Donovan, D. Zahrieh and L.I. Zon. 2005. Ferroportin1 is required for normal iron cycling in zebrafish. *Journal of Clinical Investigations* 115: 1532-1541.

Ganz, T. 2003a. Defensins: Antimicrobial peptides of innate immunity. *Nature Reviews Immunology* 3: 710-720.

Ganz, T. 2003b. Hepcidin, a key regulator of iron metabolism and mediator of anemia of inflammation. *Blood* 102: 783-788.

Hancock, R.E., T. Falla and M. Brown. 1995. Cationic bactericidal peptides. *Advances in Microbial Physiology* 37: 135-175.

Hedges, S.B. 2002. The origin and evolution of model organisms. *Nature Reviews Genetics* 3: 838-849.

Hiemstra, P.S., M.T. van den Barselaar, M. Roest, P. H. Nibbering and R. van Furth. 1999. Ubiquicidin, a novel murine microbicidal protein present in the cytosolic fraction of macrophages. *Journal of Leukocyte Biology* 66: 423-428.

Hirono, I., J.Y. Hwang, Y. Ono, T. Kurobe, T. Ohira, R. Nozaki and T. Aoki. 2005. Two different types of hepcidins from the Japanese flounder, *Paralichthys olivaceus*. *FEBS Journal* 272: 5257-5264.

Hirsch, J.G. 1958. Bactericidal action of histone. *Journal of Experimental Medicine* 108: 925-944.

Hoeger, B., B. Hitzfeld, B. Kollner, D.R. Dietrich and M.R. van den Heuvel. 2005. Sex and low-level sampling stress modify the impacts of sewage effluent on the rainbow trout (*Oncorhynchus mykiss*) immune system. *Aquatic Toxicology* 73: 79-90.

Hu, X., A.C. Camus, S. Aono, E.E. Morrison, J. Dennis, K.E. Nusbaum, R.L. Judd and J. Shi. 2007. Channel catfish hepcidin expression in infection and anemia. *Comparative Immunology and Microbial Infectious Diseases* 30: 55-69.

Huang, H.W. 2000. Action of antimicrobial peptides: two-state model. *Biochemistry* 39: 8347-8352.

Huang, P.-H., J.-Y. Chen and C.-M. Kuo. 2007. Three different hepcidins from tilapia, *Oreochromis mossambicus*: Analysis of their expressions and biological functions. *Molecular Immunology* 44: 1922-1934.

Hultmark, D., A. Engstrom, H. Bennich, R. Kapur and H.G. Boman. 1982. Insect immunity: isolation and structure of cecropin D and four minor antibacterial components from *Cecropia* pupae. *European Journal of Biochemistry* 127: 207-217.

Iijima, N., N. Tanimoto, Y. Emoto, Y. Morita, K. Uematsu, T. Murakami and T. Nakai. 2003. Purification and characterization of three isoforms of chrysophsin, a novel antimicrobial peptide in the gills of the red sea bream, *Chrysophrys major*. *European Journal of Biochemistry* 270: 675-686.

Jia, X., A. Patrzykat, R.H. Devlin, P.A. Ackerman, G.K. Iwama and R.E. Hancock. 2000. Antimicrobial peptides protect coho salmon from *Vibrio anguillarum* infections. *Applied Environmental Microbiology* 66: 1928-1932.

Johnston, I.A. 2003. Muscle metabolism and growth in Antarctic fishes (suborder Notothenioidei): evolution in a cold environment. *Comparative Biochemistry and Physiology* B136: 701-713.

Kitani, Y., C. Tsukamoto, G. Zhang, H. Nagai, M. Ishida, S. Ishizaki, K. Shimakura, K. Shiomi and Y. Nagashima. 2007. Identification of an antibacterial protein as L-amino acid oxidase in the skin mucus of rockfish *Sebastes schlegeli*. *FEBS Journal* 274: 125-136.

Krause, A., R. Sillard, B. Kleemeier, E. Kluver, E. Maronde, J.R. Conejo-Garcia, W.G. Forssmann, P. Schulz-Knappe, M.C. Nehls, F. Wattler, S. Wattler and K. Adermann. 2003. Isolation and biochemical characterization of LEAP-2, a novel blood peptide expressed in the liver. *Protein Science* 12: 143-152.

Lauth, X., H. Shike, J.C. Burns, M.E. Westerman, V.E. Ostland, J.M. Carlberg, J.C. Van Olst, V. Nizet, S.W. Taylor, C. Shimizu and P. Bulet. 2002. Discovery and characterization of two isoforms of moronecidin, a novel antimicrobial peptide from hybrid striped bass. *Journal of Biological Chemistry* 277: 5030-5039.

Lauth, X., J.J. Babon, J.A. Stannard, S. Singh, V. Nizet, J.M. Carlberg, V.E. Ostland, M.W. Pennington, R.S. Norton and M.E. Westerman. 2005. Bass hepcidin synthesis, solution structure, antimicrobial activities and synergism, and *in vivo* hepatic response to bacterial infections. *Journal of Biological Chemistry* 280: 9272-9282.

Lazarovici, P., N. Primor and L.M. Loew. 1986. Purification and pore-forming activity of two hydrophobic polypeptides from the secretion of the Red Sea Moses sole (*Pardachirus marmoratus*). *Journal of Biological Chemistry* 261: 16704-16713.

Luders, T., G.A. Birkemo, J. Nissen-Meyer, O. Andersen and I.F. Nes. 2005. Proline conformation-dependent antimicrobial activity of a proline-rich histone H1 N-terminal Peptide fragment isolated from the skin mucus of Atlantic salmon. *Antimicrobial Agents and Chemotherapy* 49: 2399-2406.

Martin, S.A., S.C. Blaney, D.F. Houlihan and C.J. Secombes. 2006. Transcriptome response following administration of a live bacterial vaccine in Atlantic salmon (*Salmo salar*). *Molecular Immunology* 43: 1900-1911.

Mor, A. and R.R. Avtalion. 1989. Transfer of antibody activity from immunised mother to embryo in tilapias. *Journal of Fish Biology* 37: 249-254.

Morvan, C.L., D. Troutaud and P. Deschaux. 1998. Differential effects of temperature on specific and nonspecific immune defences in fish. *Journal of Experimental Biology* 201: 165-168.

Murray, H.M., J.W. Gallant and S.E. Douglas. 2003. Cellular localization of pleurocidin gene expression and synthesis in winter flounder gill using immunohistochemistry and *in situ* hybridization. *Cell and Tissue Research* 312: 197-202.

Noga, E.J. and U. Silphaduang. 2003. Piscidins: A novel family of peptide antibiotics from fish. *Drug News Perspectives* 16: 87-92.

Noga, E.J., Z. Fan and U. Silphaduang. 2001. Histone-like proteins from fish are lethal to the parasitic dinoflagellate, *Amyloodinium ocellatum*. *Parasitology* 123: 57-65.

Noga, E.J., Z. Fan and U. Silphaduang. 2002. Host site of activity and cytological effects of histone-like proteins on the parasitic dinoflagellate, *Amyloodinium ocellatum*. *Diseases of Aquatic Organisms* 52: 207-215.

Oren, Z. and Y. Shai. 1996. A class of highly potent antibacterial peptides derived from pardaxin, a pore-forming peptide isolated from Moses sole fish *Pardachirus marmoratus*. *European Journal of Biochemistry* 237: 303-310.

Park, C.B., M.S. Kim and S.C. Kim. 1996. A novel antimicrobial peptide from *Bufo bufo* gargarizans. *Biochemical and Biophysical Research Communication* 218: 408-413.

Park, C.H., E.V. Valore, A.J. Waring and T. Ganz. 2001. Hepcidin, a urinary antimicrobial peptide synthesized in the liver. *Journal of Biological Chemistry* 276: 7806-7810.

Park, I.Y., C.B. Park, M.S. Kim and S.C. Kim. 1998. Parasin I, an antimicrobial peptide derived from histone H2A in the catfish, *Parasilurus asotus*. *FEBS Letters* 437: 258-262.

Park, C.B., J.H. Lee, I.Y. Park, M.S. Kim and S.C. Kim. 1997. A novel antimicrobial peptide from the loach, *Misgurnus anguillicaudatus*. *FEBS Letters* 411: 173-178.

Park, K.C., J.A. Osborne, S.C. Tsoi, L.L. Brown and S.C. Johnson. 2005. Expressed sequence tags analysis of Atlantic halibut (*Hippoglossus hippoglossus*) liver, kidney and spleen tissues following vaccination against *Vibrio anguillarum* and *Aeromonas salmonicida*. *Fish and Shellfish Immunology* 18: 393-415.

Patrzykat, A., L. Zhang, V. Mendoza, G.K. Iwama and R.E. Hancock. 2001. Synergy of histone-derived peptides of coho salmon with lysozyme and flounder pleurocidin. *Antimicrobial Agents and Chemotherapy* 45: 1337-1342.

Patrzykat, A., C.L. Friedrich, L. Zhang, V. Mendoza and R.E. Hancock. 2002. Sublethal concentrations of pleurocidin-derived antimicrobial peptides inhibit macromolecular synthesis in *Escherichia coli*. *Antimicrobial Agents and Chemotherapy* 46: 605-614.

Patrzykat, A., J.W. Gallant, J.K. Seo, J. Pytyck and S.E. Douglas. 2003. Novel antimicrobial peptides derived from flatfish genes. *Antimicrobial Agents and Chemotherapy* 47: 2464-2470.

Porcelli, F., B. Buck, D.K. Lee, K.J. Hallock, A. Ramamoorthy and G. Veglia. 2004. Structure and orientation of pardaxin determined by NMR experiments in model membranes. *Journal of Biological Chemistry* 279: 45815-45823.

Ren, H.L., K.J. Wang, H.L. Zhou and M. Yang. 2006. Cloning and organisation analysis of a hepcidin-like gene and cDNA from Japan sea bass, *Lateolabrax japonicus*. *Fish and Shellfish Immunology* 21: 221-227.

Richards, R.C., D.B. O'Neil, P. Thibault and K.V. Ewart. 2001. Histone H1: an antimicrobial protein of Atlantic salmon (*Salmo salar*). *Biochemical and Biophysical Research Communications* 284: 549-555.

Robinette, D., S. Wada, T. Arroll, M.G. Levy, W.L. Miller and E.J. Noga. 1998. Antimicrobial activity in the skin of the channel catfish *Ictalurus punctatus*: characterization of broad-spectrum histone-like antimicrobial proteins. *Cell and Molecular Life Sciences* 54: 467-475.

Robinette, D.W. and E.J. Noga. 2001. Histone-like protein: a novel method for measuring stress in fish. *Diseases of Aquatic Organisms* 44: 97-107.

Rodrigues, P.N., S. Vazquez-Dorado, J.V. Neves and J.M. Wilson. 2006. Dual function of fish hepcidin: response to experimental iron overload and bacterial infection in sea bass (*Dicentrarchus labrax*). *Developmental and Comparative Immunology* 30: 1156-1167.

Salerno, G., N. Parrinello, P. Roch and M. Cammarata. 2007. cDNA sequence and tissue expression of an antimicrobial peptide, dicentracin; a new component of the moronecidin family isolated from head kidney leukocytes of sea bass, *Dicentrarchus labrax*. *Comparative Biochemistry and Physiology* B146: 521-529.

Sarmasik, A., G. Warr and T.T. Chen. 2002. Production of transgenic medaka with increased resistance to bacterial pathogens. *Marine Biotechnology* (NY) 4: 310-322.

Selsted, M.E., S.S. Harwig, T. Ganz, J.W. Schilling and R.I. Lehrer. 1985. Primary structures of three human neutrophil defensins. *Journal of Clinical Investigations* 76: 1436-1439.

Shai, Y. 1999. Mechanism of the binding, insertion and destabilization of phospholipid bilayer membranes by alpha-helical antimicrobial and cell non-selective membrane-lytic peptides. *Biochimique et Biophysique Acta* 1462: 55-70.

Shi, J. and A.C. Camus. 2006. Hepcidins in amphibians and fishes: Antimicrobial peptides or iron-regulatory hormones? *Developmental and Comparative Immunology* 30: 746-755.

Silphaduang, U. and E.J. Noga. 2001. Peptide antibiotics in mast cells of fish. *Nature (London)* 414: 268-269.

Silphaduang, U., A. Colorni and E.J. Noga. 2006. Evidence for widespread distribution of piscidin antimicrobial peptides in teleost fish. *Developmental and Comparative Immunology* 72: 241-252.

Shike, H., C. Shimizu, X. Lauth and J.C. Burns. 2004. Organization and expression analysis of the zebrafish hepcidin gene, an antimicrobial peptide gene conserved among vertebrates. *Developmental and Comparative Immunology* 28: 747-754.

Shike, H., X. Lauth, M.E. Westerman, V.E. Ostland, J.M. Carlberg, J.C. Van Olst, C. Shimizu, P. Bulet and J.C. Burns. 2002. Bass hepcidin is a novel antimicrobial peptide induced by bacterial challenge. *European Journal of Biochemistry* 269: 2232-2237.

Smith, V.J., J.M.O. Fernandes, S.J. Jones, G.D. Kemp and M.F. Tatner. 2000. Antibacterial proteins in rainbow trout, *Oncorhynchus mykiss*. *Fish and Shellfish Immunology* 10: 243-260.

Takemura, T. and K. Takano. 1997. Transfer of maternal IgM to larvae of tilapia. *Fish and Shellfish Immunology* 7: 355-363.

Tamang, D.G. and M.H. Saier Jr. 2006. The cecropin superfamily of toxic peptides. *Journal of Molecular and Microbiological Biotechnology* 11: 94-103.

Thennarasu, S. and R. Nagaraj. 1996. Specific antimicrobial and hemolytic activities of 18-residue peptides derived from the amino terminal region of the toxin pardaxin. *Protein Engineering* 9: 1219-1224.

Thompson, S.A., K. Tachibana, K. Nakanishi and I. Kubota. 1986. Melittin-like peptides from the shark-repelling defense secretion of the sole *Pardachirus pavoninus*. *Science* 233: 341-343.

Tomasinsig, L. and M. Zanetti. 2005. The cathelicidins—structure, function and evolution. *Current Protein and Peptide Science* 6: 23-34.

Uzzell, T., E.D. Stolzenberg, A.E. Shinnar and M. Zasloff. 2003. Hagfish intestinal antimicrobial peptides are ancient cathelicidins. *Peptides* 24: 1655-1667.

Wilson, R.S., C.H. Condon and I.A. Johnston. 2007. Consequences of thermal acclimation for the mating behaviour and swimming performance of female mosquito fish. *Philosophical Transactions of the Royal Society of London* B 362: 2132-2139.

Xiao, Y., Y. Cai, Y.R. Bommineni, S.C. Fernando, O. Prakash, S.E. Gilliland and G. Zhang. 2006. Identification and functional characterization of three chicken cathelicidins with potent antimicrobial activity. *Journal of Biological Chemistry* 281: 2858-2867.

Xu, P., B. Bao, Q. He, E. Peatman, C. He and Z. Liu. 2005. Characterization and expression analysis of bactericidal permeability-increasing protein (BPI) antimicrobial peptide gene from channel catfish *Ictalurus punctatus*. *Developmental and Comparative Immunology* 29: 865-878.

Yano, T. 1996. The non-specific immune system: humoral defence. In: *The Fish Immune System: Organism, Pathogen and Environment*, G.K. Iwama and T. Nakanishi (eds.). Academic Press, San Diego, pp. 106-140.

Yin, Z.-X., W.-J. Chen, J.-H. Yan, J.-N. Yang, S.-M. Chan and J.-G. He. 2006. Cloning, expression and antimicrobial activity of an antimicrobial peptide, epinecidin-1, from the orange-spotted grouper, *Epinephelus coioides*. *Aquaculture* 253: 204-211.

Zasloff, M. 1987. Magainins, a class of antimicrobial peptides from *Xenopus* skin: isolation, characterization of two active forms, and partial cDNA sequence of a precursor. *Proceedings of the National Academy of Sciences of the United States of America* 84: 5449-5453.

Zasloff, M. 2002. Antimicrobial peptides of multicellular organisms. *Nature (London)* 415: 389-395.

Zhang, Y.A., J. Zou, C.I. Chang and C.J. Secombes. 2004. Discovery and characterization of two types of liver-expressed antimicrobial peptide 2 (LEAP-2) genes in rainbow trout. *Veterinary Immunology and Immunopathology* 101: 259-269.

Zou, J., C. Mercier, A. Koussounadis and C. Secombes. 2007. Discovery of multiple beta-defensin like homologues in teleost fish. *Molecular Immunology* 44: 638-647.

CHAPTER

9

Estrogens, Estrogen Receptors and Their Role as Immunoregulators in Fish

Luke R. Iwanowicz[1,*] and Christopher A. Ottinger[2]

INTRODUCTION

All vertebrates have mechanisms to protect themselves from pathogens. Likewise, all vertebrates possess a strategy to ensure individual reproductive success. While different classes of vertebrates approach these life-sustaining requirements differently, numerous commonalities exist. All vertebrates investigated to date possess an immune system comprising innate and adaptive branches. Immune responses are induced by numerous stimuli and are finely coordinated by intercellular associations of membrane complexes and signaling molecules. Due to cross talk between physiological systems, the immune status and, consequently, responsiveness to such stimuli are modulated by hormones, including

Authors' addresses: [1]USGS, Leetown Science Center, Aquatic Ecology Branch, Kearneysville, WV 25430, USA.
[2]USGS, Leetown Science Center, Fish Health Branch, Kearneysville, WV 25430, USA.
**Corresponding author*: E-mail: liwanowicz@usgs.gov

those critical to the reproductive system. As a result, general immunocompetence is influenced by the reproductive status. Fish are no exception to this vertebrate scenario. Surprisingly, communication between the immune system and reproductive endocrine system of this class of organisms is poorly characterized.

Fish immunologists and physiologists have recognized and experimentally addressed, to a limited extent, the communication network between the immune and endocrine systems. Such work has predominantly focused on neuropeptides (releasing hormones), polypeptides (prolactin, growth hormone and insulin growth factor-1) and the stress hormone cortisol (Maule and Schreck, 1991; Arnold and Rice, 1997; Narnaware *et al.*, 1997; Weyts *et al.*, 1999; Yada *et al.*, 1999, 2004, 2005; Esteban *et al.*, 2004). Cortisol, a steroid hormone, is perhaps the best-studied hormone effector molecule of the immune system. The role of cortisol as an immune modulator was first described in mammals, but immunomodulatory actions of this hormone have also been observed in fishes (Jaffe 1924; Selye, 1936a, b; Wendelaar Bonga, 1997). Similar to mammals, stressed fish exhibit elevated plasma cortisol levels and altered (generally suppressed) immune responses. The role of cortisol as an immune regulator has been the primary focus of fish immuno-endocrine research.

Steroid hormones, with the exception of cortisol, are generally not included in discussions of immunity. The intent of this chapter is to emphasize the importance of sex steroids (specifically estrogens) vis-a-vis the regulation of immune function. This will entail a brief introduction to steroids and fish immune function, an introduction to estrogen receptors, a brief summary on the effects of estrogens on the mammalian immune system and the current state of knowledge regarding estrogens and piscine immune function. In addition, the subject of endocrine disruption will be introduced and the potential of endocrine disruptors to affect normal immune function in fish considered.

STEROID HORMONES AND FISH IMMUNE FUNCTION

Steroid hormones are lipid-soluble, low molecular weight molecules that serve as cellular messengers in an intracrine, autocrine, paracrine or endocrine fashion. Transit of this class of hormones in the blood is often facilitated by non-covalent associations with binding proteins (i.e., sex hormone-binding globulin, cortisol-binding protein and serum albumin) that serve as circulating chaperones. Steroids may also circulate as free, unbound molecules. Given the lipophillic nature of steroids, they readily diffuse from the blood through the cell membrane and into the cytoplasm of both target and non-target cells. General cell signaling by this family of hormones involves cytoplasmic or nuclear receptors that bind to hormone response elements in the DNA and, subsequently, modulate transcriptional activity. Accumulated evidence also purports rapid, transcription-independent signaling initiated by steroid hormones (Bartholome *et al.*, 2004; Braun and Thomas, 2004; Mourot *et al.*, 2006). The potency of steroid hormones is similar to that of cytokines as they elicit their actions in target tissues at nanomolar concentrations (Turnbull and Rivier, 1999).

Steroid hormones are lumped into five major categories: progestens, mineralocorticoids, glucocorticoids, androgens and estrogens. These categories represent the receptor type targeted by these hormones. All steroid hormones and the related vitamin D are derived from the substrate cholesterol. Cholesterol is an endogenously synthesized 27-carbon molecule that is an integral component of cell membranes, and a required molecule for bile production and the metabolism of the fat soluble vitamins (A, D, E and K). Conversion of cholesterol into the 18-, 19-, and 21-carbon steroid hormones is initiated by the rate-limiting, irreversible cleavage of a 6-carbon residue from cholesterol and results in the production of pregnenolone. Subsequent enzyme-directed modifications, including aromatization, reduction or hydroxylation of the resulting product substrates, occur in specific cell-types and tissues to synthesize the intended hormone (Norman and Litwack, 1987). Steroid synthesis is highly conserved in vertebrates. For instance, the steroids cortisol, testosterone and estradiol synthesized in humans and mice are identical to the molecules produced by fishes. This is in stark contrast to the seemingly weak phylogenetic conservation of molecules such as cytokines and chemokines. Fishes have evolved converting enzymes that lead to hormones unique to this class of vertebrates (i.e., 11-ketotestosterone). However, as stated above, most of the functional steroids found in vertebrates are identical (Fig. 9.1).

Cortisol (or hydrocortisone) is perhaps the best-known and most biologically active natural glucocorticoid and is also a critical mediator of the hypothalmo-pituitary axis. While it is often colloquially referred to as the stress hormone, cortisol is an indispensable master regulator of

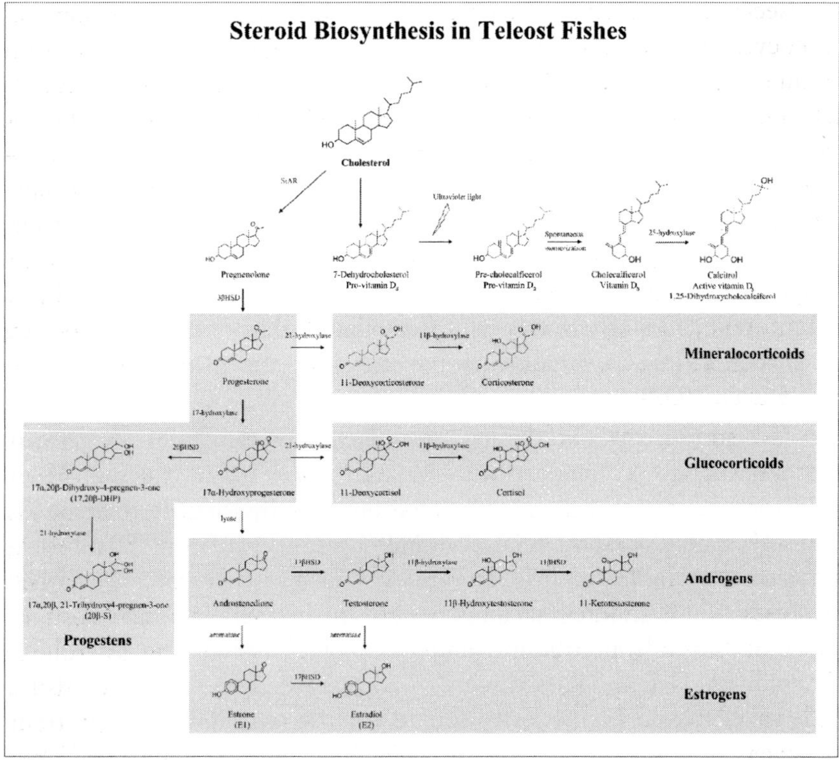

Fig. 9.1 Diagram of common steroid synthesis pathways in teleosts fish. Key enzymes for steroid biosynthesis include steroidogenic acute regulatory protein (StAR), 3β-hydroxysteroid dehydrogenase (3βHSD), 20β-hydroxysteroid dehydrogenase (20βHSD), 17β-hydroxysteroid hydrogenase (17βHSD), 11β-hydroxysteroid hydrogenase (11βHSD), 11β-hydrolase, 17-hydrolase, 21-hydrolase, lyase and aromatase.

metabolic, immunologic, and homeostatic processes. Glucocorticoid receptors (GR) are found in the cells of almost all vertebrate tissues including leucocytes (Stolte *et al.*, 2006). Elevated levels of plasma cortisol resulting from stress are commonly associated with suppressed immune function. On the other hand, physiological resting concentrations of cortisol are required for downregulation of the normal immune response. The intrinsic mechanisms of depressing immune function by cortisol are 'fail safes' that prevent hyperactive immune responses and excessive by-stander cell damage. Thus, while chronically elevated cortisol may negatively impact immune function, this hormone is essential for normal immune responses.

Receptors for corticosteroids have been detected in carp and salmonid leucocytes and circulating cortisol directly regulates GR expression (Maule and Schreck, 1991; Weyts *et al.*, 1998c; Stolte *et al.*, 2006). Elevated cortisol leads to distinct trafficking of GR-rich leucocyte subtypes from the circulation into lymphoid organs. Effects of cortisol on leucocyte viability are cell type-specific. For example, B-cells from carp are especially sensitive to cortisol, whereas thrombocytes and cells in the T-cell fraction are insensitive (Weyts *et al.*, 1998a). Cortisol also appears to prevent apoptosis in carp granulocytes (Weyts *et al.*, 1998b). While cortisol is a potent modulator of immune function, it has been reviewed in depth by others and will not be discussed further in this chapter (Weyts *et al.*, 1999; Stolte *et al.*, 2006).

Testosterone, the principle androgen in humans and mammals, is synthesized by both males and females. In the male, testosterone is synthesized in the leydig cells of the testes. In the female, testosterone is synthesized in the theca cells of follicles present in the ovary and is a required substrate for estradiol synthesis. This is true in mammals as well as fishes. While testosterone is known to regulate the synthesis of gonadotropins in the pituitary, the teleost specific 11-ketotestosterone is generally associated with spermatogenesis, secondary sex characteristics and male behavior (Kime, 1998). Binding affinity of the rainbow trout androgen receptor (ARα) for these androgens, however, is similar (Takeo and Yamashita, 2000).

Similar to cortisol, testosterone is immunosuppressive, although these actions are mediated via different receptor signaling pathways. Early evaluations of the modulatory effects of testosterone on the fish immune system were conducted by Slater and Schreck (1993). They pioneered the original characterization of androgen receptors in fish leucocytes utilizing binding assays and later described the physiological effects of this hormone related to receptor binding (Slater *et al.*, 1993; Slater *et al.*, 1995). Results from *in vitro* experiments designed to assess the effect of testosterone on the T-cell dependent antibody response utilizing trinitrophenyl lipopolysaccharide (TNP-LPS) demonstrated a time-dependent reduction in antibody producing cells due to cell death. Similarly Hou *et al.* (1999) demonstrated that the *in vivo* administration of testosterone or 11-ketotestosterone results in decreased plasma and mucous IgM in rainbow trout. Testosterone also impacts physiological responses of immunocytes involved in innate immune responses (Law *et al.*, 2001; Watanuki *et al.*, 2002).

Estrogens are the primary endogenous ligands of estrogen receptors. They are also the primary female sex steroids and potent orchestrators of cellular transcription during reproductive cycling. The granulosa cells of the ovary are the primary sites of estradiol synthesis in the female. Here testosterone, primarily produced in the theca cells, is converted into estradiol by aromatase (also known as estrogen synthase). Males synthesize physiologically relevant amounts of estrogen, but at considerably lower concentrations than the females. Extragonadal estrogen is produced in the brain, adipose tissue, bone and other peripheral tissues in mammals (Simpson *et al.*, 1999). Aromatase activity has also been detected in human peripheral blood leucocytes and blood cells of some fish (Zhang *et al.*, 2004a; Vottero *et al.*, 2006). We have also recently identified aromatase B expression in primary channel catfish leucocytes (Fig. 9.2). Tissue distribution of aromatase activity in fishes is similar to that in humans, but it is unlikely expressed in bone tissue (Piferrer and Blazquez, 2005). Estrogen synthesis in these extragonadal tissues is likely to exert local intracrine actions rather than system-wide endocrine actions (Simpson *et al.*, 1999).

Here it must be made clear that estrogen is not a single, specific hormone. Rather, estrogen refers to any of the three endogenous estrogens: estrone (E1), estradiol (E2) and estriol (E3). While estradiol is perhaps the most potent endogenous estrogen and is commonly the hormone implied by 'estrogen' these hormones have different affinities for estrogen receptor subtypes and therefore differential biological activities. Considering the intent and focus of this chapter, general actions of estrogens will not be included here. Rather, salient points germane to the immune system will be discussed throughout the following text.

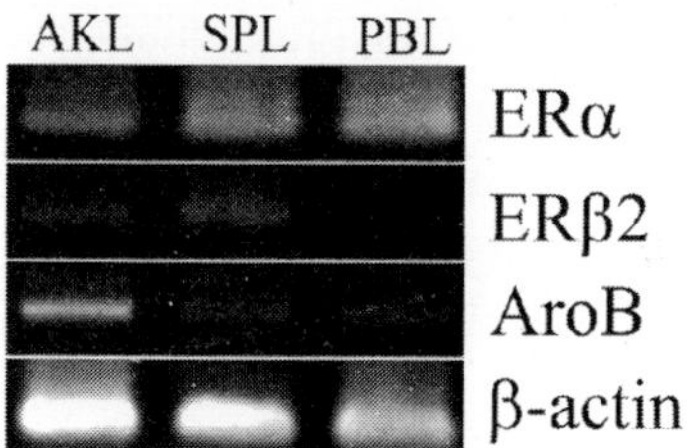

Fig. 9.2 ERα, ERβ2 and aromatase B (AroB) expression in primary channel catfish leucocytes. Samples are anterior kidney leucocytes (AKL) splenic leucocytes (SPL) and peripheral blood leucocytes (PBL).

ESTROGEN RECEPTORS

Estrogen receptors (ERs) made their evolutionary debut sometime during molluscan evolution, and are present in all known vertebrates (Thornton, 2001, 2003; Kajiwara *et al.*, 2006; Keay *et al.*, 2006). They are soluble, thermolabile, protease sensitive, modular, ligand-inducible transcription factors that belong to the nuclear receptor superfamily (Rollerova and Urbancikova, 2000). They are expressed in numerous tissue types and are involved in a multitude of regulatory processes including cell proliferation, differentiation, apoptosis and general cell signaling. While most often associated with the regulation of the reproductive system, ERs are functionally expressed and required for the normal function of the skeletal, nervous, respiratory, renal, circulatory and immune systems. Based on their almost ubiquitous tissue distribution, they clearly have an indispensable role in normal vertebrate homeostatic and general physiological processes. In the following section, we will highlight some of the general, but salient features of ERs. For a rigorous review of ER structure and function, please refer to Enmark and Gustaffson (1999) and Ascenzi *et al.* (2006).

Structurally, ERs are organized in five to six distinct, functional domains: a variable amino-terminus trans-activation domain (A/B), a highly conserved DNA-binding domain (C), a hypervariable hinge region (D), a well-conserved ligand-binding domain (E), and a variable C-terminal region (F) (Tsai and O'Malley, 1994; Rollerova and Urbancikova, 2000; Hewitt and Korach, 2002). The N-terminal A/B domain contains the activation function-1 (AF-1) region located in the A/B domain that is responsible for cell-specific ligand-independent transcriptional activation (Metzger *et al.*, 1995). The C domain is highly conserved across species and is responsible for specific binding to the estrogen response elements (EREs) of target genes that are characterized by palindromic inverted repeats (i.e., 5′-GGTCAnnnTGACC-3′) (Klinge, 2001). It also plays a pivotal role in receptor dimerization. The hinge domain (D) contains part of the nuclear localization signal, serves as a flexible 'bridge' between the C and E domains, and contains sites for acetylation and sumoylation post-translational modifications (McEwan, 2004; Sentis *et al.*, 2005; Huang *et al.*, 2006). The ligand-binding domain is often considered to consist of both the E and F domains. This region is required for homo- and hetero-dimerization, and contains the activation function-2 (AF-2) region that participates in ligand-dependant transcriptional activation. Additionally, part of the nuclear localization signal is located in the E domain. In the absence of ligand, the E domain is responsible for docking to heat shock proteins (Knoblauch and Garabedian, 1999; Gee and Katzenellenbogen, 2001). The AF-1 and AF-2 regions influence transcriptional activity as they serve as docking sites for co-activator and co-repressor proteins. Throughout the multiple domain structure are sites prone to post-translational modification including acetylation, glycosylation, myristolation, nitrosolation, palmitolation, phosphorylation, sumoylation and ubiquination that lead to functional changes (Ascenzi *et al.* 2006). While the domain structure illustrated here (Fig. 9.3) is the basic blueprint for all ERs, differences are noted across species in regards to the modular domains containing AF-1 and AF-2 (Xia *et al.*, 2000).

In mammals, two major ER subtypes, ERα (NR3A1) and ERβ (NR3A2), each encoded by a distinct gene have been described (Kuiper *et al.*, 1996; Mosselman *et al.*, 1996). They have distinct, yet partially

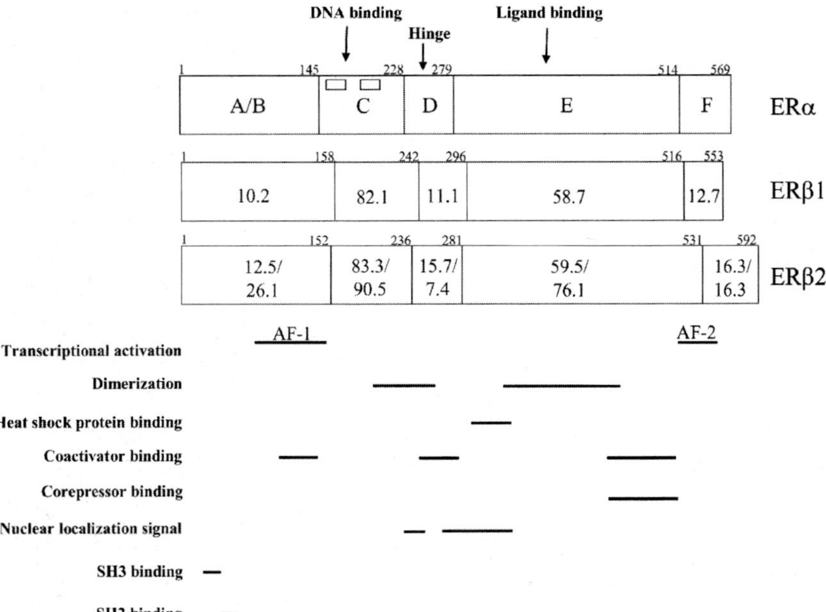

Fig. 9.3 Box diagrams of the *Danio rerio* (Zebrafish) estrogen receptor subtypes. Domain designations are primarily based on those described by Menuet *et al.* (2002) and protein alignments with other described fish ERs. Identification of functional sites is inferred from those described in human ERs or the results of Eukaryotic Linear Motif (ELM) pattern searches (Klinge, 2000).

overlapping, distributions in estrogen sensitive cell types and exhibit differential ligand-binding affinities and trans-activation properties. These differential expression profiles, ligand-binding properties and interactions with cofactors, in part, explain the pleiotrophic actions orchestrated by estrogens (Klinge, 2000; Moggs and Orphanides, 2001). Ligand-binding studies have shown that ERα has greater binding affinity for estradiol than ERβ (K_d = 0.06 and 0.24 nM respectively), yet ERβ tends to preferentially bind estriol and numerous phytoestrogens (Kuiper *et al.*, 1997; Escande *et al.*, 2006). Although ERs show similar DNA and ligand binding properties *in vitro*, ERβ is less potent than ERα at inducing transcription from the ERE-dependent signaling pathway. This is likely due to differences in the amino-terminus since ERβ lacks significant transcriptional capacity and capability of functional interaction with the carboxyl-terminus (McInerney *et al.*, 1998; Yi *et al.*, 2002). It has also been shown that the interaction of ERβ with EREs are independent of estradiol and are impaired by its amino-terminus (Huang *et al.*, 2005). The existence of functional splice variants resulting from alternative processing and exon usage for both ERα and ERβ should also be noted (Li *et al.*, 2003; Wang *et al.*, 2006). These non-traditional ER transcripts have received less attention than wild-type ERs, but are likely of great significance (Leung *et al.*, 2006).

Estrogen receptor function is best understood in mammals and has been elucidated primarily with the use of mutational analysis and knockout technology. Such work has been instrumental in determining active sites required for transcriptional activation, DNA-binding, ligand-binding, protein–protein interactions, and mechanisms of ligand-dependant as also ligand-independent receptor functions. Additionally, this research has led to insights regarding receptor trafficking and signaling. Not surprisingly, it appears that the original classical model of ER signaling is not the exclusive mechanism utilized by this receptor type. According to this classical model unliganded, monomeric ERs shuttle between the nucleus and cytoplasm. In the absence of the ligand, they associate with heat-shock proteins and other chaperones to form stable, protease-resistant complexes (Pratt and Toft, 1997; Knoblach and Garabedian, 1999). Upon ligand binding, these receptors undergo conformational changes leading to homo-or heterodimerization and the receptor–ligand complex is then transported to the nucleus. There the complex recruits nuclear cofactors and interacts with specific regulatory regions, EREs, of target genes to modulate transcription (Kumar and

Chambon, 1988; Beato and Sanchez-Pacheco, 1996; Torchia *et al.*, 1998; Hewitt and Korach, 2002). This mechanism of cell signaling, often termed 'the genomic pathway', is by far the best studied, accepted and characterized mode of estrogen signaling via ERs. However, it is not the exclusive signaling pathway exploited by these receptors.

Other models of ER function and signaling have since been developed and experimentally tested to explain the rapid, 'non-genomic' actions of estrogens. These models account for the approximately 33% of the genes in humans regulated by estrogen receptors that do not contain ERE-like sequences (O'Lone, 2004). Plasma membrane associated ERs that interact with ER antibodies or estrogen conjugates have been identified (Benten *et al.*, 1998, 2001; Watson *et al.*, 1999, 2002). Estrogen binding of these membrane receptors generally leads to rapid, cytoplamic calcium signaling. Complex localization and signaling processes at the plasma membrane are ascribed to the ER associating directly or indirectly with various scaffold proteins (caveolin), adapter proteins (shc and modulator of non-genomic activity of ER [MNAR]), tyrosine kinases (EGF, IGF receptors, and src), and G proteins (Filardo *et al.*, 2000; Kahlert *et al.*, 2000; Migliaccio *et al.*, 2002; Song *et al.* 2002, 2004; Wong *et al.*, 2002; Razandi *et al.*, 2003; Boonyaratanakornkit and Edwards, 2004; Evinger and Levin, 2005). Additionally, splice variants hERα46 and hERα36 have been shown to associate with the plasma membrane. The former is found throughout the cell but tends to associate with the plasma membrane in a palmitoylation-dependent manner. Estrogen signaling via hERα46 rapidly induces nitric oxide release via a phosphatidylinositol 3-kinase/Akt/endothelial nitric-oxide synthase pathway (Haynes *et al.*, 2000; Li *et al.*, 2003). The latter is a dominant-negative effector of both estrogen-dependent and estrogen-independent transactivation functions signaled through ERα and ERβ, and transduces membrane-initiated estrogen-dependent activation of the mitogen-activated protein kinase/extracellular signal-regulated kinase mitogenic signaling pathway (Wang *et al.* 2006). Full-length, wild type ERα has also been shown to associate with the plasma membrane as a dimer. This ER dimer activates ERK, cAMP, and phosphatidylinositol 3-kinase signaling upon estrogen binding, resulting from Gsα and Gqα activation (Razandi *et al.*, 2004).

Thomas *et al.* (2005) described an orphan receptor, GPR30, that is unrelated to nuclear estrogen receptors and has all the binding and signaling characteristics of a membrane-bound estrogen receptor. This discovery emphasizes the complexity of possible signaling by estradiol. It is likely that numerous cell types yet to be investigated may also transduce estrogen-specific cellular signals via undescribed pathways instead of, or in addition to classical ER pathways. The actions of estrogens are thus determined by the structure of the ligand, the ER subtype involved, the nature of the hormone-responsive gene promoter, and the character and balance of coactivators and corepressors that modulate the cellular response to the ER–ligand complex (Katzenellenbogen *et al.*, 2000). For a comprehensive review of alternative ER signaling, refer to Coleman and Smith (2001), Levin (2001), Sànchez *et al.* (2004), Simoncini *et al.* (2004), Björnström and Sjöberg (2005) and Acconcia and Kumar (2006).

Estrogen receptors have been described and functionally characterized in a number of teleosts primarily as the result of neuroendocrine and reproductive physiology research. Unlike humans and mice, many teleosts are known to express three distinct ER subtypes (Fig. 9.4). These subtypes are the products of different genes and are not simply the result of alternative RNA splicing, although alternative promoter usage and RNA processing of fish ERs clearly does occur (Pakdel *et al.*, 2000; Patino *et al.*, 2000; Greytak and Callard, 2006). These products are not true subtypes and their significance remains unknown. While the initial characterization of the recognized subtypes led to the assignments of ERs (α, β and γ), a revamping of fish ER designations has occurred in an effort to comply with *Danio rerio* nomenclature (Hawkins *et al.*, 2000; Menuet *et al.*, 2002; Halm *et al.*, 2004; Sabo-Attwood *et al.*, 2004; Filby and Tyler, 2005). Additionally, a second isoform of the ERα subtype has recently been identified in rainbow trout (Nagler *et al.*, 2007). Again, this isoform is the product of a specific gene, ERα2, and not a variant of ERα1. The presence of the second ERβ isoform is purportedly the result of a whole genome duplication event that occurred in Actinopterygians (ray finned fishes) following the evolutionary divergence of the Sarcopterygians (lobe-finned fishes) (Amores *et al.*, 1994; Nagler *et al.*, 2007). The second ERα isoform may, in fact, be the result of a salmonid-specific duplication of the ERα locus, as suggested by Nagler *et al.* (2007), and has not been identified in others teleosts to date. The fish ERs are now assigned ERα1 (NR3A1a), ERα2 (NR3A1b), ERβ1 (NR3A2a) and ERβ2 (NR3A2b). Thus, the ERβ subtype now consists of two isoforms ERβa (formerly ERγ) and ERβb (formerly ERβ).

Fish ERs have a similar domain structure (Fig. 9.3) as observed in mammals, have a high affinity for estrogen's and structurally related

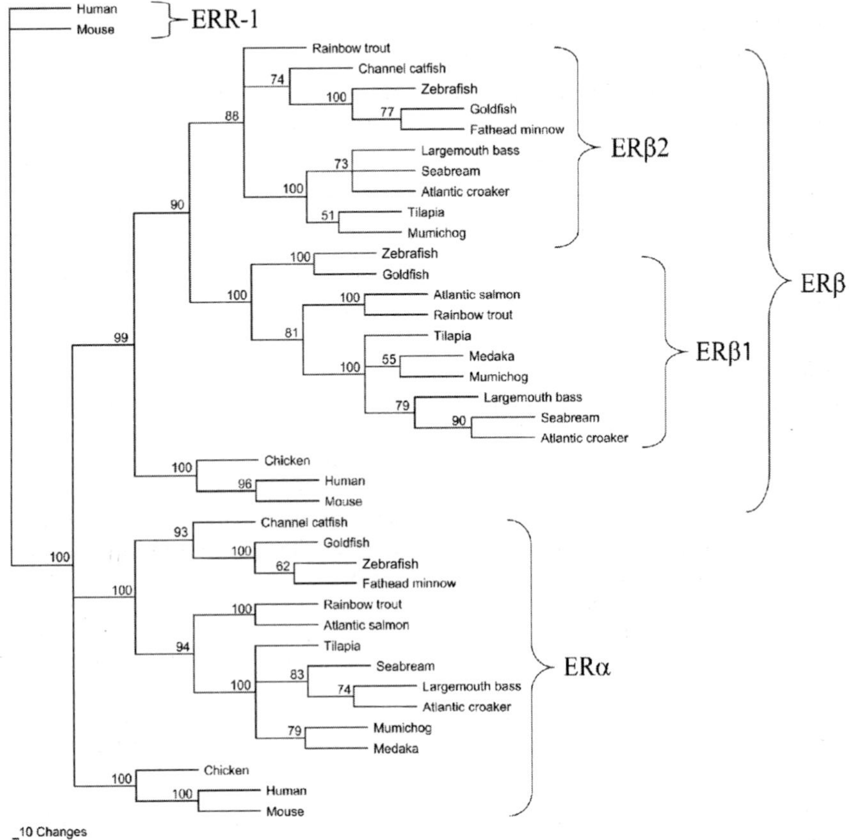

Fig. 9.4 Phylogram of the most parsimonious tree from a maximum parsimony analysis of fish, mammalian and avian estrogen receptor proteins. Maximum parsimony analysis was performed using PAUP*4.0b1. Gaps were treated as missing data and analysis utilized the Goloboff fit criterion (kappa = 2) using the heuristic search algorithm with 100 random additions of sequences and tree bisection-reconnection (TBR) branch swapping. Bootstrap values were calculated with 1000 replicates using the same heuristic search. Estrogen related receptor-1 (ERR-1) was set as an outgroup.

compounds, and also function as ligand-inducible transcription factors. In the A/B domain, there appears to be conservation of amino acids in the AF-1 region and retention of phosphorylation sites required for this activity. Additionally, amino acids of the E domain have been identified and localized within regions required for dimerization, estrogen binding and estrogen-dependent transactivation function (Xia *et al.*, 1999). Similar to mammalian ERs, the different subtypes also have differential binding preference for ligand and expression patterns are tissue dependant

(Table 9.1). Not surprisingly, expression is also dependent on season, sex and age (sexually mature or not) of the animal (Sabo-Attwood *et al.*, 2004). While it is unknown whether fish ERs exert their actions via 'non-genomic' pathways as observed in mammals, given the similarities between this well-conserved receptor system, it does seem likely. Also, the fact that fish have a more expansive repertoire of ERs than other known vertebrates lends the possibility of differences in ligand inducible signaling (and ligand affinity). Not all four ERs have been identified in all of the fish examined. This may reflect a lack of discovery, or that some species simply do not

Table 9.1 Relative binding affinity of human (h), Atlantic croaker (ac) and channel catfish (ccf) ERs—data complied from (a) Kuiper *et al.* (1997), (b) Hawkins and Thomas (2004), and (c) Gale *et al.* (2004). All experiments were performed separately with different concentrations of competitors, and RBA values were determined based on the IC50 values for each experiment. Direct comparisons between receptor sub-types or species should not be made in this table. This table is a general illustration of the unique differences in ligand binding to each ER sub-type. Binding is relative to E2 for each receptor. Ligand binding does not reflect receptor activation. Dissociation constants (K_d) must be considered to estimate the difference in affinity between receptors. The K_d values for each receptor are as follows: hERα = 0.06 nM, acERα = 0.33-0.40 nM, ccfERα = 0.47 nM, acERβa = 1.16 nM, hERβ = 0.24 nM, acERβb = 1.38 nM and ccfERβb = 0.21 nM. Lower K_d values indicate higher affinity.

Compound	*hERα*[a]	*acERα*[b]	*ccfERα*[c]	*acERβ1*[b]	*hERβ*[a]	*acERβ2*[b]	*ccfERβ2*[c]
Estradiol (E2)	100	100	100	100	100	100	100
Estrone (E1)	60	10	2.1	2.9	37	3.5	1.1
Estriol (E3)	14	3.9	1.5	9.8	21	1.7	3.0
Diethylstilbesterol	468	4898		96	295	315	
Hexestrol	302				234		
Dienestrol	223				404		
17α-Estradiol	58	9.6		2.1	11	1.6	
17α-Ethynylestradiol			409				108
Moxestrol	43	48		8.3	5	11	
4-OH-Tamoxifen	178	262		65	339	144	
Tamoxifen	7	25.4		1.0	6	4.8	
ICI 164,384	85	141		20	166	49	
ICI 182,780		706		36		324	
Genestein	5	2.4		9	36	18	
Coumestrol	94				185		
B-Zearalanol	16	97		4.6	14	3.6	
Bisphenol A	0.05				0.33		
Nonylphenol			0.78				0.01
Octylphenol			0.16				0.01
RU486				1.2			

have multiple, functional ER genes. As mentioned previously, the additional isoforms of the major ER subtypes are attributed to the genome duplication event that occurred in the Actinopterygians or specifically in the salmonid family (Nagler *et al.*, 2007). Other undiscovered family specific gene or genome duplications are also likely, given that fishes are the most species-rich class of vertebrates on the planet.

EFFECTS OF ESTROGEN ON MAMMALIAN IMMUNE FUNCTION

The notion that estrogens modulate immune function in humans has been speculated upon for over half a century (Nicol *et al.*, 1964; Kenny *et al.*, 1976). This is in part due to the fact that women have more vigorous humoral responses than men (Butterworth, 1967; Eidinger and Garrett, 1972). Likewise, there is an indisputable sex bias regarding the sex-based rates of autoimmune disease and responses to infection. Whether this is simply due to the immunosuppressive actions of androgens in men has been a point of intense debate, but numerous studies routinely determine that estrogen is the common factor in many autoimmune diseases (Holmdahl 1989; Carlsten *et al.*, 1991; Jansson and Holmdahl, 1998; Grimaldi *et al.*, 2002; Liu *et al.*, 2003).

Early research directed at identifying estrogen receptors in leucocytes suggested that some lymphocytes have few to no estrogen receptors. This research clearly identified estrogen binding in suppressor and cytotoxic T-cells, but not in T-helper cells (Cohen *et al.*, 1983; Stimson, 1988). However, since the advent of the reverse-transcriptase polymerase chain reaction (RT-PCR), ER transcripts have been identified in all of these T-cell populations (Suenaga *et al.*, 1998). This observation has led to speculation that some cell types synthesized ER mRNA that is not translated into functional protein. While this may be the case in some T-cells, functional estrogen receptor protein has been identified in most T-cells subtypes. Phiel *et al.* (2005) determined the relative expression of ERs in peripheral blood leucocytes (PBLs). Two populations of T-cells were examined and it was found that $CD4^+$ T-cells express relatively high levels of ERα mRNA compared to ERβ, while $CD8^+$ T-cells express low, but comparable levels of both ERs. Similar work by Benten *et al.* (1998) using splenic T-cells from female C57BL/10 mice identified plasma membrane-associated estrogen receptors. While this work did not employ PCR or ER antibodies, estradiol binding was identified on the surface of both $CD4^+$ and $CD8^+$ T-cells using an estradiol-BSA-FITC conjugate.

Sakazaki *et al.* (2002) have also demonstrated ERα in a fraction of CD3$^+$ T-cells via flow cytometry.

Cells of the B-cell lineage have also been demonstrated to express ERs. Suenaga *et al.* (1998) exploited the use of a B-cell line to demonstrate ER expression. Igarashi *et al.* (2001) took an *in vivo* approach that yielded very interesting data. They found that while B-cells expressed both ERs, expression is age and cell-stage dependent. Precursor B-cells isolated from the bone marrow of a congenic strain of mice do not express measurable amounts of either receptor in 5-day-old mice, they express only ERα at 3 weeks and both ERα and ERβ at 18 months. Neither ER could be detected in the fetal liver B-cell precursors. Work by Grimaldi (2002) demonstrates similar, but slightly different findings in regards to cell-stage specific differential ER expression. Here, they detected ERβ positive and low numbers of ERα positive B-cells at the pro/pre stage and only ERβ in immature bone marrow B-cells. Both ERα and ERβ positive cells were identified in splenic transitional, mature and germinal center-associated B-cells. Differences between these studies include the age (3-months old in the latter) and strain of mice employed, and method of receptor detection. Antibody detection via flow cytometry in addition to RT-PCR was utilized by Grimaldi (2002), thus, giving a more robust snapshot. Both studies show that expression of ERs is stage dependent in B-cells. Human PBLs have also been shown to express both ERs but a much higher level of ERβ (Benten *et al.*, 2002; Phiel *et al.*, 2005).

Estrogen receptors have been demonstrated in myeloid leucocytes as well. Early work in macrophages utilizing binding assays demonstrated the existence of two different estrogen receptors based on binding affinity (Gulshan *et al.*, 1990). Interestingly, this same work failed to detect ERs using monoclonal antibodies. This work suggested that estradiol binding might be by non-classical ERs. Using the same binding assay approach, estrogen receptors have been demonstrated in peripheral blood monocytes (Weusten, 1986). Interestingly, Benten *et al.* (2001) demonstrated sequesterable estrogen receptors in monocytes, which suggests the occurrence of non-traditional ERs. Thus, there is strong evidence that monocytes and macrophages possess proteins that bind estrogen and specifically respond to ligand-receptor engagement that are different than the ERs most commonly examined. However, authentic ERs are expressed in monocytes and macrophages (Khan *et al.*, 2005; Phiel *et al.*, 2005). Additionally, there is evidence that supports the assertion that ER subtype expression is cell-stage dependent in this lineage of cells similar to that reported in B-cells. In other words, monocytes tend to express ERβ while macrophages express ERα (Mor, 2003). While this may not be the case for all leucocytes, there is clearly a difference in the proportional expression of ERs. Estrogen receptors have also been described in dendritic cells, natural killer cells, platelets, neutrophils and eosinophils (Lee, 1982; Komi and Lassila, 2000; Curran *et al.*, 2001; Nealen *et al.*, 2001; Molero *et al.*, 2002; Nalbandian and Kovats, 2005b; Nalbandian *et al.*, 2005).

To summarize, leucocytes of numerous lineages express either ERα or ERβ, both or neither. Additionally, the relative proportion of these receptors may differ in the instance where both are expressed. The factors that dictate ER expression in these cells include age of the individual, stage of leucocyte development, site of cell residence in the case of tissue-associated leucocytes (microenvironment), and reproductive status of the individual. The intracellular environments of these different cells are also unique and likely to influence the transcriptional and non-transcriptional roles of ER signaling. Recent work has also offered insight that may explain early observations of estrogen binding in leucocytes with no detectable receptors (by antibody methods or RT-PCR). In other words, it is now known that peripheral blood mononuclear cells express wild-type full length ERs as well as a number of exon-deleted transcripts of both receptors. Western blotting with well-characterized monoclonal antibodies further indicates that some of these exon-deleted transcripts are translated into protein. Likewise neutrophils express transcripts and detectable ER proteins, that are never full-length, classical ERs (Stygar *et al.*, 2006). Given that leucocytes express ERs, they are likely direct targets of estrogens. Unraveling ligand-specific interactions will likely require intricate experimental designs that complement the complex transcriptional profiles of ER variant isoforms.

While ERs have clearly been demonstrated in leucocytes, the role of these receptors in relation to mammalian disease resistance has yet to be fully characterized. The influence of estrogens on the immune response directly or indirectly through leucocyte ERs has been investigated utilizing both *in vivo* and *in vitro* experimental systems. Despite valiant efforts to elucidate mechanisms of these effects *in vitro*, more potent effects of estrogen exposure are often observed using *in vivo* models. This observation should emphasize the complexity of estrogen signaling, the importance of the tissue microenvironment and suggests a critical role of estrogen induced factors by supportive cells (Jansson and Holmdahl, 1998).

Estrogen receptors are expressed in lymphoid and myeloid cells in addition to stromal and other supportive cells of the immune system. Functionally, they mediate physiological processes including immunocyte growth, hematopoiesis, differentiation, lymphocyte activation, T-helper cell (T_h) polarization and cytokine production (Deshpande *et al.*, 1997; Ito *et al.*, 2001; Carruba *et al.*, 2003; Maret *et al.*, 2003; Mor *et al.*, 2003; Lambert *et al.*, 2004; Salem, 2004; Nalbanian and Kovats, 2005b). Estrogens also affect both B and T-cell development as well as antigen-presenting-cell differentiation and homeostasis. In general, B, T and dendritic-cell development are suppressed with estrogen, but as noted earlier, such effects are strongly concentration dependent. Estrogens also affect chemotaxis, expression of adhesion molecules and chemokine receptors, matrix metalloprotease 9, morphological activation of macrophages, and synthesis of inducible nitric oxide synthase (iNOS) and neuronal nitric oxide synthase (nNOS) (Ito *et al.*, 1995; Molero *et al.*, 2002; You *et al.*, 2002, 2003; Janis *et al.*, 2004; Ghisletti *et al.*, 2005; Mo *et al.*, 2005). A list of immune-associated molecules modulated by estrogen is presented in Table 9.2. Supplemental literature regarding the effects of estrogens on specific leucocyte populations is given in Table 9.3. Other helpful reviews on this topic include Cutolo *et al.* (1995), Jansson and Holmdahl (1998), Druckmann (2001), Lang (2004), Obendorf and Patchev (2004) and Grimaldi *et al.* (2005).

As a rule of thumb, estrogens suppress cell-mediated, but augment humoral-based immunity in mice and humans (Smithson *et al.*, 1998; McMurray, 2001). However, it is becoming increasingly evident that this generalization is not dictated simply by the presence or absence of estrogen, but rather its concentration. That is, it appears that lower physiological concentrations of estrogen are stimulatory to the immune system while pharmacological doses adversely modulate cell-mediated immunity (Nalbandian and Kovats, 2005b). The relative physiological concentration of estradiol can be a critical determinant of leucocyte phenotype and, thus, the function.

Estrogen affects most leucocyte populations either directly via ERs, or indirectly by exerting their actions on accessory cells of the immune system. Direct impacts of estrogen on leucocytes or accessory cells can translate to indirect effects on other immunocytes via the modulatory actions orchestrated by cytokines and chemokines (Janis *et al.*, 2004; Sentman *et al.*, 2004; Mo *et al.*, 2005; Janele *et al.*, 2006). By general definition, cytokines are low molecular weight, polar proteins and glycoproteins secreted by immunocytes, and numerous other cell types

Table 9.2 Immune-associated molecules modulated by estradiol.

Sundstum *et al.* (1989) (a), Hamano *et al.* (1998) (b),
García-Durán *et al.* (1999) (c), Harris *et al.* (2000) (d),
McMurray *et al.* (2001) (e), Do *et al.* (2002) (f),
Matejuk *et al.* (2002) (g), Verdu *et al.* (2002) (h),
Kanda and Watanabe (2003) (i), Mor *et al.* (2003) (j),
Tomaszewskaa *et al.* (2003) (k), You *et al.* (2003) (l),
Chiang *et al.* (2004) (m), Gao *et al.* (2004) (n),
Miller *et al.* (2004) (o), Salem (2004) (p),
Sentman *et al.* (2004) (q), Vegeto *et al.* (2004) (r),
Zhang *et al.* (2004a) (s), Cutolo *et al.* (2005) (t),
Crane-Godreau *et al.* (2005) (u), Lambert *et al.* (2005) (v),
Mao *et al.* (2005) (w), Mo *et al.* (2005) (x),
Polanczyk *et al.* (2005) (y), Roberts *et al.* (2005) (z),
Sakazaki *et al.* (2005) (a1), Geraldes *et al.* (2006) (a2),
Pioli *et al.* (2006) (a3), and Shi *et al.* (2006) (a4),

Cytokines and cytokine receptors	*Cytokines and cytokine receptors*	*Other immune-related molecules*
IL-1β[d]	IL-8[a4]	B7-1, B7-2[s]
IL-2 IL-2R[e]		CD40, CD40L[s,a2]
IL-4[b,v]	MCP-1[g]	
IL-6[d]	MIP-1ß[g]	CTLA-4[g]
IL-7	MIP-2[d]	VCAM-1[o]
IL-10[p]	MIP3α[u]	ICAM-1[o]
IL-12[p]	MIP3β	P-selectin[o]
IL-13[h]	CINC-1, CINC-2β, CINC-3[o]	VEGF[g]
IL-15[z]		ECF
IL-18[g]	CCR1, CCR2, CCR5[g,x]	
	CXC10, CXC11[q]	
TNFα[p,t]		Metaloprotease 9[r]
IFNγ[t]	LT-β[g]	B-Defensin[a3]
TGFβ[n]		Complement C3[a]
APC and granulocyte associated	*Apoptosis/cell death*	
MHC-II[w]	Fas/FasL[j]	GATA-3[v]
iNOS[k,l,a1]	TRAIL[f]	FOXP3[y]
nNOS[c]	bcl-2	RANTES[g,i]
	shp-1	
Myeloperoxidase[m]		
Elastase[m]		
Superoxide[m]		

Table 9.3 Supplemental citation list of literature pertaining to the effects of estrogens on specific leukocyte populations.

B-cells	*T-cells*	*Professional APCs*	*Granulocytes*
Bynoe *et al.* (2000)	Deply *et al.* (2005)	Azenabor *et al.* (2004)	Abrahams *et al.* (2003)
Grimaldi *et al.* (2002)	Do *et al.* (2002)	Azenabor and Chaudhry (2003)	Bekesi *et al.*. (2000)
Grimaldi *et al.* (2006)	Erlandsson *et al.* (2001)	Bengtsson *et al.* (2004)	Chiang *et al.* (2004)
Masuzawa *et al.* (1994)	Herrera *et al.* (1992)	Benten *et al.* (2001)	Garcia-Duran *et al.* (1999)
Medina *et al.* (2000)	Kawashima *et al.* (1992)	Carruba *et al.* (2003)	Hamano *et al.* (1998)
Medina *et al.* (1993)	Lambert *et al.* (2005)	Chao *et al.* (2000)	Ito *et al.* (1995)
Paavonen *et al.* (1981)	Maret *et al.* (2003)	Chao *et al.* (1995)	Miller *et al.* (2004)
Peeva *et al.* (2005)	McMurray *et al.* (2001)	Cutolo *et al.* (2005)	Molero *et al.* (2002)
Smithson *et al.* (1998)	Mendelsohn *et al.* (1977)	Do *et al.* (2002)	Perez *et al.* (1996)
Smithson *et al.* (1995)	Neifeld and Tormey (1979)	Komi and Lassila (2000)	Ramos *et al.* (2000)
Sthoeger *et al.* (1988)	Okuyama *et al.* (1992)	Liu (2001)	Stefano *et al.* (2000)
Thurmond *et al.* (2000)	Polanczyk *et al.* (2005)	Mao *et al.* (2005)	Yu *et al.* (2006)
	Polanczyk *et al.* (2004a)	Matsuda *et al.* (1985)	
	Polanczyk *et al.* (2004b)	Mor *et al.* (2003)	
	Prieto and Rosenstein (2006)	Nalbandian and Kovats (2005a)	
	Rijhsinghani *et al.* (1996)	Nalbandian and Kovats (2005b)	
	Salem (2004)	Nalbandian *et al.* (2005)	
	Screpanti *et al.* (1991)	Paharkova-Vatchkova *et al.* (2004)	
	Staples *et al.* (1999)	Sakazaki *et al.* (2005)	
	Yellayi *et al.* (2000)	Salem *et al.* (1999)	
	Yron *et al.* (1991)	Stefano *et al.* (2003)	
	Zoller and Kersh (2006)	Thongngarm *et al.* (2003)	
		Tomaszewska *et al.* (2003)	
		Yang *et al.* (2006)	
		You *et al.* (2003)	
		Zhang *et al.* (2004a)	

that do not have an obvious role in immune function. Cytokines are critical effector molecules of both the innate and adaptive immune responses, and required for the intricate coordination of the immune system. Many of the cytokines and other immune related molecules affected by estradiol are included in Table 9.2. To date, little work has been done to identify EREs in all cytokine gene promoters, but EREs have been identified for some (O'Lone, 2004). Thus, expression of some cytokines is clearly directly influenced by estradiol. It has also been demonstrated that estradiol exerts inhibitory actions on the nuclear factor κB (NF-κB) signaling pathway via cytoplasmic ERs. The NF-κB family of transcription factors regulates numerous genes (including cytokines) that are essential for the development, maintenance and function of the innate and adaptive branches of the immune system (Kalaitzidis and Gilmore, 2005). Clearly, an inhibitory effect on this signaling pathway would have profound consequences on the immune response. The proposed mechanism of estrogen-mediated NF-κB inhibition is rapid and involves the interaction of ERα with the p85 subunit of phosphatidylinositol 3-kinase (PI3-kinase) in a ligand-dependent manner (Simoncini *et al.*, 2000, 2003). In macrophages, this rapid and persistent activation of PI3-kinase prevents the nuclear translocation of p65 and therefore prevents NF-κB induced gene transcription. In macrophages, this inhibition occurs without altering the Ikarose kinase κ-B (Iκ-B) degradation pathway (Ghisletti *et al.*, 2005). A similar estradiol-induced inhibitory effect of NF-κB is associated with an increase in Iκ-Bα protein levels in $CD4^+$ T-cells (McMurray *et al.*, 2001). In contrast, both ERα and ERβ have been shown to represses the translational activity of NF-κB in the presence or absence of ligand (Quaedackers *et al.*, 2001). It should be made clear that the non-classical effects of estradiol are likely to affect other signaling cascades. For instance, repression of the IL-6 gene by 17β-estradiol is mediated through the interaction of ERs with two transcription factors, NF-κB and C/EBPβ (Ray *et al.*, 1994; Stein and Yang, 1995). ERs located in the cytoplasm have been shown to efficiently induce transactivation of Stat-regulated promoters via non-genomic signaling (Björnström and Sjöberg, 2002). In any case, the modulatory actions of estrogens are diverse and exploit myriad cell-signaling networks. As an aside, but a point of interest, while the expression of cytokines IL-6 and TNFα is modulated by estradiol, these cytokines regulate the synthesis of estradiol in peripheral tissues. The activities of aromatase, and estradiol 17β-hydroxysteroid dehydrogenase are both increased by IL-6 and TNF-α (Purohit *et al.*, 2002). Thus, there appears to be a feedback loop in place between these cytokines and estradiol in peripheral tissues.

EFFECTS OF ESTROGEN ON IMMUNE FUNCTION IN FISH

To date, there is little published evidence conclusively demonstrating that fish leucocytes express estrogen receptors. This lack of evidence, however, is likely to reflect the lack of experiments intended to address the topic. Estrogen receptors are expressed in immune tissues including the spleen, anterior kidney and peripheral blood in some fish species (Xia *et al.*, 2000; Watanuki *et al.*, 2002; Wang *et al.*, 2005). These investigations, however, primarily focused on the general characterization of these receptors, and did not specifically isolate leucocytes. We have recently demonstrated the expression of both ERα and ERβ2 in these tissues using Histopague-1077 enriched fractions (Fig. 9.2). Interestingly, expression levels and patterns were dependent on the source organ. Tissue-associated leucocytes expressed both ER subtypes while only ERα was detected in PBLs. Earlier research during the original characterization of channel catfish ER reported a similar observation, that work demonstrated negligible or undetectable levels or ERβ expression in whole blood or anterior kidney (Xia *et al.*, 2000). Negligible levels of ERβ in the anterior kidney reported in that study may be a reflection of using whole tissue samples rather than enriched leucocytes as used in our laboratory. The dominant ER subtype expressed in mammalian PBLs is ERβ (Phiel *et al.*, 2005). Taken together, fish leucocytes do express at least two of the known ER subtypes and this expression is different than that observed in mammals. This is not entirely surprising, however, given the fact that the repertoire of available ERs in fish is different. Of importance is the fact that we have also recently demonstrated that that the expression of ERs in PBLs exposed to concanavalin A, lipopolysaccharide, or a mixed leucocyte culture *in vitro* is abolished during the first few days of activation. Functionally, this demonstrates the dynamic regulation of ERs that may render new proliferating leucocytes insensitive to estrogens (unpublished data).

There is no published data available on cell lineage-specific ER expression in fishes. Based on work in our laboratory utilizing long-term leucocyte cell lines from channel catfish, it is clear that ERs are differentially expressed in diverse leucocyte lineages (unpublished data). We have found that ERα is expressed in cell lines representing monocytes/ macrophages, T-cells and B-cells (Fig. 9.5). These cell lines are also positive for ERα via Western blotting. Expression of ERβ2 occurs in macrophage/ monocyte lines as well as a T-cell line. Expression of ERβ2 is low in the macrophage/moncocyte lines but is the dominant ER mRNA transcript in the T-cell line (Fig. 9.5). No ERβ2 has been detected in the cytotoxic T-cell line (32.15) (Miller *et al.*, 1998; Zhou *et al.*, 2001). General conclusions from this work indicate that all of these cells are likely to be sensitive to estrogens, and given the differential expression patterns of receptors, they would respond differently.

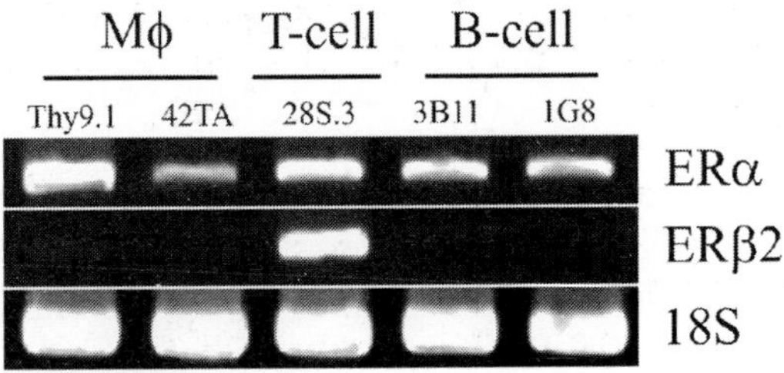

Fig. 9.5 Expression of ERα and ERβ2 in channel catfish long-term leucocyte cell lines. The Thy9.1 and 42TA cell lines are predominantly monocyte/macrophages, the 28S.3 cells are T-helper-like and the 3B11 and 1G8 cell lines are B-cells.

Experiments designed to specifically investigate the effects of estradiol on immune responses of fish are limited in number. Many of the speculated effects of estradiol on fish immune function are based primarily on observations of modulated immune function or humoral parameters during seasons of increased circulating estradiol. Wang and Belosevic (1994) conducted the earliest experiments designed to examine the effects of estradiol on immune function . *In vivo* exposure of goldfish to estradiol delivered via slow release implants led to an increased number of *Trypanosoma danilewski* following a challenge to this hemoflagellate. They also demonstrated that the *in vitro* proliferative response of goldfish primary PBLs induced by phorbol myristate acetate (PMA) and the calcium ionophore A23187 was suppressed following *in vitro* estradiol exposure in a dose-dependent manner (Wang and Belosevic, 1995). These authors also demonstrated that *in vivo* administration of estradiol led to a depressed PMA and calcium ionophore mitogenic response *in vitro* (Wang and Belosevic, 1995). Additional work by this group utilizing the goldfish kidney macrophage cell line (GMCL) demonstrated suppressive effects of *in vitro* administered estradiol on chemotaxis and phagocytosis, but not on nitric oxide production or the generation of superoxide (Wang and Belosevic, 1994). It should be noted that the concentrations of estradiol used in the cell line experiment (0.1–10 μM) reflect high- to super-physiological levels. However, Yamaguchi *et al.* (2001) did obtain similar results when using physiological concentrations of *in vitro* administered estradiol (0.1, 1, 10, 100 or 1000 nM) and primary leucocytes from carp. The effects of estradiol on common carp IgM secreting cells were examined by Saha *et al.* (2004). Nanomolar concentrations of *in vitro* administered estradiol do not appear to have an effect on IgM secretion from the carp PBLs, splenic or anterior kidney leucocytes. Estradiol also does not induce apoptosis in carp PBLs when administered *in vitro* (Saha *et al.*, 2002, 2004).

The reported effects of estrogen on fish immune function are important to consider. Law *et al.* (2001) found that estradiol had no effect on carp primary leucocyte phagocytosis. Watanuki *et al.* (2002) found that *in vivo* exposure of carp to estradiol leads to suppression of phagocytosis as well as super oxide and nitric oxide production in a dose-dependent manner. Although the effect on phagocytosis was similar, these results regarding the impact of estrogen on carp leucocyte super and nitric oxide production contrast with those reported for carp by Wang and Belosevic (1995) and Yamaguchi *et al.* (2001). The study by Watanuki *et al.* (2002) differs from those of Wang and Belosevic (1995) and Yamaguchi *et al.* (2001) in the sense that Watuanki *et al.* (2002) used *in vivo* as opposed to *in vitro* estrogen exposures. Hou *et al.* (1999) found that the *in vivo* administration of estradiol leads to decreased plasma and mucous IgM in rainbow trout. These results differ from those obtained by Saha *et al.* (2004) using *in vitro* exposure; however, different species of fish were examined and the immunoglobulin production was measured differently. Apparent contradictions arising from differences with *in vitro* or *in vivo* estrogen exposure are also common in the mammalian literature. The dichotomy of hormone-induced effects between *in vitro* and *in vivo* experiments is not surprising. Clearly, estradiol exerts actions in both experimental systems, but it is likely that the *in vivo* mechanisms are more complex with likely input from other tissue types. Additionally, elevated estradiol does not normally occur in the absence of testosterone. Exceptions include laboratory injection (or aqueous exposure studies with estradiol) and laboratory or environmental exposure to xenoestrogens. The collaborative signaling of androgen and estrogen may also explain some of these observed ambiguities. Season-dependent estradiol concentrations and associated receptor expression may also contribute to these differences. The potential impact of these seasonal differences, although hypothetically significant, is poorly understood.

ENDOCRINE DISRUPTION

During the mid to late 1990s, there has been increased awareness and concern regarding endocrine disrupting chemicals (EDCs; Sumpter, 1998). As defined by the World Health Organization (2002) an EDC is 'an exogenous substance or mixture that alters function(s) of the endocrine system and consequently produces adverse health effects in an intact organism, or its progeny, or (sub)populations'. These chemicals are virtually ubiquitous and have been identified in aquatic ecosystems across the world (Kime, 1998; Vos *et al.*, 2000; Noakson *et al.*, 2001; Gong *et al.*, 2003; Goksoyr, 2006). Due to the fact that hormones normally exert their physiological actions at nanomolar and even picomolar concentrations, environmental concentrations of EDCs that were once below detection limits are now known to inflict physiological insult. While most EDCs do not exhibit the same binding affinities as native ligands to their cognate receptor and higher concentrations are necessary for an observed biological effect, physiologically relevant concentrations of EDCs are found in many environmental aquatic systems (Petrovic *et al.*, 2002; Lintelmann *et al.*, 2003). While EDCs by definition may affect any of the hormone networks, discussion below will primarily focus on estrogenic EDCs (EEDCs). These compounds have also been shown to modulate immune responses (Ahmed, 2000; Ndebele *et al.*, 2004; Inadera, 2006).

Sources of EEDCs vary, depending on geographical location, but include both natural and anthropogenic origins. Natural sources include natural estrogens excreted by humans and animals that are introduced to aquatic systems via municipal wastewater systems and animal husbandry runoff (Finlay-Moore *et al.*, 2000; Herman and Mills, 2003; Hanselman *et al.*, 2004; Soto *et al.*, 2004). Additionally, plants and cyanobacteria produce phytoestrogens and mycoestrogens, respectively (Vlata *et al.*, 2006). Synthetic EEDCs include plasticizers, detergents, pharmaceuticals, personal care products, herbicides, pesticides, many of the legacy compounds (i.e., PCBs) and others. These compounds are introduced to aquatic systems via industrial and sewage discharges, active application and runoff, and atmospheric deposition. Unfortunately, many of the anthropogenic EEDCs such as the active compounds in some contraceptive pills are designed to be more physiologically active and stable than endogenous, natural estrogens. Clearly, such pharmaceutical compounds are likely to be more persistent in the environment while exerting physiological effects at very low concentrations. Additionally, it must be noted that these EEDCs are very rarely present in singularity. Rather complex mixtures of these chemicals are the realistic expectation. The EEDCs in such mixtures affect physiological systems in aquatic biota synergistically, or exert complex agonistic-antagonistic actions of unknown outcome.

Based on the broad definition of EEDCs and a general understanding of endocrine networks, the potential mechanisms and pathways available for EEDCs to 'short-circuit' the normal endocrine regulation are myriad. Perhaps the best-studied mechanism of endocrine disruption is hormone mimicry. In this case, chemicals structurally similar to the endogenous ligand bind as a functional agonist and activate receptors. Mimicry leads to the inappropriate induction of estrogen responsive genes and synthesis of proteins. During periods of high circulating estrogen concentrations, exposure to an EEDC may be of little consequence. However, if exposure occurs during a life-history stage or season when estrogen concentrations are low or undetectable, such exposure may have biologically profound consequences. Due to the conservation of endocrine systems across species, these structurally diverse EEDCs similarly induce gene expression mediated by ERs in various species (Matthews, 2002). However, mimicry need not result in inappropriate gene activation to have a significant impact. Endocrine disruption may also occur when a mimic binds a hormone receptor without inducing activation. In this case, the mimic serves as a functional antagonist and competes for receptor-binding sites with endogenous ligand. Consequently, normal transcription induced by the endogenous ligand is lessened or ablated by the competing disruptor due to reduced receptor availability. Similarly, competition for binding on circulating binding proteins may also occur. Additional mechanisms, which are not necessarily mutually exclusive, include modifying normal hormone metabolism (clearance), synthesis or receptor expression (Thibaut and Porte, 2004). Data also suggest that EDCs can function as hormone sensitizers by inhibiting histone deacetylase activity and stimulating mitogen-activated protein kinase activity, or have genome-wide effects by affecting DNA methylation thus altering gene expression (Hong *et al.*, 2006; Tabb and Blumberg, 2006). The result of the above actions is a disturbance in normal hormone physiology and, at the very least, exerts a physiological stress to maintain homeostasis.

Most EEDC associated reports primarily involve reproductive and developmental effects. However, a number of immune-associated effects have been documented. As an example, 4-nonylphenol (NP) is an estrogen mimic and is perhaps one of the best-known EEDCs. Recently NP has been shown to inhibit LPS-induced NO and TNFα production, which is attributed to an ER-dependent inhibition of NF-κB transactivation. This response is not associated with ERE directed transcription (You *et al.*, 2002). Others have shown immunological effects induced by other alkyphenols. For instance, p-n-nonylphenol suppresses T_{h1} development and enhances T_{h2} development. Exposure to p-n-octylphenol elicits similar effects, while NP and p-t-octylphenol have weaker effects. Interestingly using the same *in vitro* systems (isolated CD4+CD8+ thymocytes differentiated into T_{h1} and T_{h2} populations or purified naive CD4+ T-cells from DO11.10 T-cell receptor-transgenic and RAG-2-deficient mice differentiated into T_{h1} and T_{h2} populations) exclusive treatment with estradiol by itself fails to affect T_{h1}/T_{h2} development (Iwata *et al.*, 2004). Another EEDC, bisphenol A (BPA) has been shown to affect non-specific immunodefenses against non-pathogenic *Escherichia coli* (Sugita-Konshi *et al.*, 2003).

In the case of fish, NP and bisphenol A (BPA) affect the normal function of carp anterior kidney phagocytes at nanomolar concentrations *in vitro*. Specifically, BPA and NP exposure leads to an increased production of superoxide anions and a decrease in phagocytic activity (Gushiken, 2002). Pthalates have also been shown to negatively impact phagocytic cells of common carp (Watanuki, 2003). Interestingly, and of particular significance, early life-stage exposure to the EDCs o,p-DDE and Aroclor 1254 are known to induce long-term immunomodulation in salmonids (Milston *et al.*, 2003; Iwanowicz *et al.*, 2005). Thus, in addition to transient effects on immune function, exposure to contaminants and EEDCs during critical developmental windows may permanently affect normal life-long immune responses.

Before concluding, the nature of EEDCs must be partially clarified. In other words, while these chemicals exert their disruptive actions on normal estrogen signaling, some are promiscuous and affect additional cell signaling pathways. For instance, while NP is an ER agonist, it also regulates some genes in an ER-independent manner (Larkin *et al.*, 2002, 2003). It also has a weak affinity for the progesterone receptor, is a weak androgen receptor agonist, and affects CYP3A and CYP1A1 by signaling via the pregnane X and the arylhydrocarbon receptors (AhR) (Sohoni and Sumpter, 1998; Laws *et al.*, 2000; Meucci and Arukwe, 2006). Signaling through the AhR is perhaps one of the most recognized means of contaminant-induced immunotoxicity and, recently, it has been shown that some AhR agonists signal through the ER (Kerkvliet, 1995; Kerkvliet *et al.*, 2002; Abdelrahim *et al.*, 2006; Matthews and Gustafsson, 2006). Thus, traditional immunotoxicolgy may benefit from the newly recognized interplay between the ERα and AhR signaling pathways (Matthews *et al.*, 2005).

CONCLUSION

Estrogen receptors are clearly involved in the regulation of immune processes in mammals and fish. Research in this area is lagging in the instance of fish. However, there has been a resurgence of interest due to the issue of EEDCs'. It is now clear that fish leucocyte express estrogen receptors, but their significance in regard to immune function and disease resistance are relatively unknown. Seasonal regulation of estrogen receptors in fish is also unexplored. Given the evolutionary conservation of estrogens, their cognate receptors and cell signaling processes; it likely that many parallels can be adopted from the mammalian literature. Such work should provide a sound foundation upon which to build future research in this developing, interdisciplinary field of immuno-endocrinology.

At present, our knowledge of the specific effects and consequences of EEDCs on immune function in fish is limited. The effects of these contaminants on wild populations of fish are poorly studied, in part, due to the difficulty in studying such populations. Unlike cultured fish and laboratory populations of animals, numerous factors influence immune responses in wild fish and it is not possible to identify true 'control' populations. Additionally, wild populations are rarely exposed to a single contaminant. Thus, developing a better understanding of the effects of mixtures of such compounds on wild fish is critical to predicting the outcome of such exposure.

ACRONYMS

AF, activation function; AhR, arylhydrocarbon receptor; AKL, anterior kidney leucocytes; AKT, protein kinase B; APC, antigen presenting cell; AR, androgen receptor; AroB, aromatase B; BPA, bisphenol A, cAMP, cyclic adenosine monophosphate; CD, cluster of differentiation; CINC, cytokine-induced neutrophil chemoattractant; Con A, concanavalin A; CTLA, cytotoxic T lymphocyte antigen; ECF, eosinophilic chemotactic protein; EDC, endocrine disrupting chemical; EEDC, estrogenic endocrine disrupting chemical; ER, estrogen receptor; ERE, estrogen response element; ERK, extracellular signal-regulated kinase; FasL, Fas ligand; GATA, GATA binding protein; GMCL, goldfish macrophage cell line; GPR, G-protein coupled receptor; GR, glucocorticoid receptor; HPA, hypothalmo-pituitary-axis; ICAM, intracellular adhesion molecule; Ig, immunoglobulin; IFN, interferon; IκB, Ikaros κB; IL, interleukin; iNOS, inducible nitric oxide synthase; LPS, lipopolysaccharide; LT, lymphotoxin; MCP, monocyte chemotactic protein; MHC, major histocompatibility complex; MIP, macrophage inflammatory protein; NF-κB, nuclear factor-κB, nNOS, neuronal nitric oxide synthase; NO, nitric oxide; NP, 4-nonylphenol; PBL, peripheral blood leucocytes; PMA, phorbol 12-myristate 13-acetate; PMNL, polymorphonuclear leucocytes; RANTES, regulated on activation normal T-cell expressed and secreted; SPL, splenic leucocytes; TGF, transforming growth factor; TNF, tumor necrosis factor; TNP, trinitrophenyl; TRAIL, TNF-related apoptosis-inducing ligand; VCAM, vascular endothelial growth factor.

References

Abdelrahim, M., E. Ariazi, K. Kim, S. Khan, R. Barhoumi, R. Burghardt, S. Liu, D. Hill, R. Finnell, B. Wlodarczyk, V.C. Jordan and S. Safe. 2006. 3-Methylcholanthrene and other aryl hydrocarbon receptor agonists directly activate estrogen receptor alpha. *Cancer Research* 66: 2459-2467.

Abrahams, V.M., J.E. Collins, C.R. Wira, M.W. Fanger and G.R. Yeaman. 2003. Inhibition of human polymorphonuclear cell oxidative burst by 17-beta-estradiol and 2,3,7,8-tetrachlorodibenzo-p-dioxin. *American Journal of Reproductive Immunology* 50: 463-472.

Acconcia, F. and R. Kumar. 2006. Signaling regulation of genomic and nongenomic functions of estrogen receptors. *Cancer Letter* 238: 1-14.

Ahmed, S.A. 2000. The immune system as a potential target for environmental estrogens (endocrine disrupters): A new emerging field. *Toxicology* 150: 191-206.

Amores, A., A. Force, Y.L. Yan, L. Joly, C. Amemiya, A. Fritz, R.K. Ho, J. Langeland, V. Prince, Y.L. Wang, M. Westerfield, M. Ekker and J.H. Postlethwait. 1994. Zebrafish hox clusters and vertebrate genome evolution. *Science* 282: 1711-1714.

Ascenzi, P., A. Bocedi and M. Marino. 2006. Structure-function relationship of estrogen receptor α and β: Impact on human health. *Molecular Aspects of Medicine* 27: 299-402.

Arnold, R.E. and C.D. Rice. 1997. Channel catfish lymphocytes secrete ACTH in response to corticotropic releasing factor (CRF). *Developmental and Comparative Immunology* 21: 152.

Azenabor, A.A. and A.U. Chaudhry. 2003. 17 beta-Estradiol induces L-type Ca super(2+) channel activation and regulates redox function in macrophages. *Journal of Reproductive Immunology* 59: 17-28.

Azenabor, A.A., S. Yang, G. Job and O.O. Adedokun. 2004. Expression of iNOS gene in macrophages stimulated with 17beta-estradiol is regulated by free intracellular Ca^{2+} *Biochemical and Cellular Biology* 82: 381-390.

Bartholome, B., C.M. Spies, T. Gaber, S. Schuchmann, T. Berki, D. Kunkel, M. Bienert, A. Radbruch, G.R. Burmester, R. Lauster, A. Scheffold and F. Buttgereit. 2004. Membrane glucocorticoid receptors (mGCR) are expressed in normal human peripheral blood mononuclear cells and upregulated after *in vitro* stimulation and in patients with rheumatoid arthritis. *FASEB Journal* 18: 70-80.

Beato, M. and A. Sanchez-Pacheco. 1996. Interaction of steroid hormone receptors with the transcription initiation complex. *Endocrinology Reviews* 17: 587-609.

Bekesi, G., R. Kakucs, S. Varbiro, K. Racz, D. Sprintz, J. Feher and B. Szekacs. 2000. In vitro effects of different steroid hormones on superoxide anion production of human neutrophil granulocytes. *Steroids* 65: 889-894.

Bengtsson, A.K., E.J. Ryan, D. Giordano, D.M. Magaletti and E.A. Clark. 2004. 17β-Estradiol (E_2) modulates cytokine and chemokine expression in human monocyte-derived dendritic cells. *Blood* 104: 1404-1410.

Benten, W.P., C. Stephan and F. Wunderlich. 2002. B cells express intracellular but not surface receptors for testosterone and estradiol. *Steroids* 67: 647-654.

Benten, W.P.M., M. Lieberherr, G. Giese and F. Wunderlich. 1998. Estradiol binding to cell surfaces raises cytosolic free calcium in T cells. *FEBS Letters* 422: 349-359.

Benten, W.P.M., C. Stephan, M. Lieberherr and F. Wunderlich. 2001. Estradiol signaling via sequesterable surface receptors. *Endocrinology* 142: 1669-1677.

Björnström, L. and M. Sjöberg. 2002. Signal transducers and activators of transcription as downstream targets of nongenomic estrogen receptor actions. *Molecular Endocrinology* 16: 2202-2214.

Björnström, L. and M. Sjöberg. 2005. Mechanisms of estrogen receptor signaling: convergence of genomic and non-genomic actions on target genes. *Molecular Endocrinology* 19: 833-842.

Boonyaratanakornkit, V. and D.P. Edwards. 2004. Receptor mechanisms of rapid extranuclear signaling initiated by steroid hormones. *Essays in Biochemistry* 40: 105-120.

Braun, A.M. and P. Thomas. 2004. Biochemical characterization of a membrane androgen receptor in the ovary of the Atlantic croaker (*Micropogonias undulatus*). *Biology of Reproduction* 71: 146-155.

Butterworth, M., B. McClellan and M. Allansmith. 1967. Influence of sex on immunoglobulin levels. *Nature (London)* 214: 1224-1225.

Bynoe, M.S., C.M. Grimaldi and B. Diamond. 2000. Estrogen up-regulates Bcl-2 and blocks tolerance induction of naive B cells. *Proceedings of the National Academy of Sciences of the United States of America* 97: 2703-2708.

Carlsten, H., N. Nilsson, R. Jonsson and A. Tarkowski. 1991. Differential effects of oestrogen in murine lupus: Acceleration of glomerulonephritis and amelioration of T cell-mediated lesions. *Journal of Autoimmunity* 4: 845-856.

Carruba, G., P. D'Agostino, M. Miele, M. Calabro, C. Barbera, G. Di Bella, S. Milano, V. Ferlazzo, R. Caruso, M. La Rosa, L. Cocciadiferro, I. Campisi, L. Castagnetta and

E. Cillari. 2003. Estrogen regulates cytokine production and apoptosis in PMA-differentiated, macrophage-like U937 cells. *Journal of Cell Biochemistry* 90: 187-196.

Chao, T.-C., P.J. Van Alden, J.A. Greager and R.J. Walters. 1995. Steroid sex hormones regulate the release of tumor necrosis factor by macrophages. *Cellular Immunology* 160: 43-49.

Chao, T.-C., H.-H. Chao, M.-F. Chen, J.A. Greager and R.T. Walter. 2000. Female sex hormones modulate the function of LPS-treated macrophage. *American Journal of Reproductive Immunology* 44: 310-318.

Chiang, K., S. Parthasarathy and N. Santanam. 2004. Estrogen, neutrophils and oxidation. *Life Sciences* 75: 2425-2438.

Cohen, J.H.M., L. Danel, G. Cordier, S. Saez and J.-P. Revillard. 1983. Sex steroid receptors in peripheral T cells: absence of androgen receptors and restriction of estrogen receptors to OKT8-positive cells. *Journal of Immunology* 131: 2767-2771.

Coleman, K.M. and C.L. Smith. 2001. Intracellular signaling pathways: nongenomic actions of estrogens and ligand-independent activation of estrogen receptors. *Frontiers in Biosciences* 6: D1379-D1391.

Crane-Godreau, M.A. and C.R. Wira. 2005. Effects of estradiol on lipopolysaccharide and Pam3Cys stimulation of CCL20/macrophage inflammatory protein 3 alpha and tumor necrosis factor alpha production by uterine epithelial cells in culture. *Infection and Immunology* 73: 4231-4237.

Curran, E.M., L.J. Berghaus, N.J. Vernetti, A.J. Saporita, D.B. Lubahn and D.M. Estes. 2001. Natural killer cells express estrogen receptor-alpha and estrogen receptor-beta and can respond to estrogen via a non-estrogen receptor-alpha-mediated pathway. *Cellular Immunology* 214: 12-20.

Cutolo, M., A. Sulli, B. Seriolo, S. Accardo and A.T. Masi. 1995. Estrogens, the immune response and autoimmunity. *Clinical and Experimental Rheumatology* 13: 217-226.

Cutolo, M., S. Capellino, P. Montagna, P. Ghiorzo, A. Sulli and B. Villaggio. 2005. Sex hormone modulation of cell growth and apoptosis of the human monocytic/ macrophage cell line. *Arthritis Research and Therapy* 7: R1124-R1132.

Delpy, L., V. Douin-Echinard, L. Garidou, C. Bruand, A. Saoudi and J.C. Guery. 2005. Estrogen enhances susceptibility to experimental autoimmune myasthenia gravis by promoting type 1-polarized immune responses. *Journal of Immunology* 175: 5050-5057.

Deshpande, R., H. Khalili, R.G. Pergolizzi, S.D. Michael and M.D. Chang. 1997. Estradiol down-regulates LPS-induced cytokine production and NFkB activation in murine macrophages. *American Journal of Reproductive Immunology* 38: 46-54.

Do, Y., S. Ryu, M. Nagarkatti and P.S. Nagarkatti. 2002. Role of death receptor pathway in estradiol-induced T-cell apoptosis *in vivo*. *Toxicologic Science* 70: 63-72.

Druckmann, R. 2001. Review: female sex hormones, autoimmune diseases and immune response. *Gynecological Endocrinology* (Supplement) 6: 69-76.

Eidinger, D. and T.J. Garrett. 1972. Studies of the regulatory effects of the sex hormones on antibody formation and stem cell differentiation. *Journal of Experimental Medicine* 136: 1098-1116.

Enmark, E. and J.-A. Gustafsson. 1999. Oestrogen receptors—An overview. *Journal of Internal Medicine* 246: 133-138.

Erlandsson, M.C., C. Ohlsson, J.A. Gustafsson and H. Carlsten. 2001. Role of oestrogen receptors alpha and beta in immune organ development and in oestrogen-mediated effects on thymus. *Immunology* 103: 17-25.

Escande, A., A. Pillon, N. Servant, J. Cravedi , F. Larrea, P. Muhn, J. Nicolas, V. Cavaillès and P. Balaguer. 2006. Evaluation of ligand selectivity using reporter cell lines stably expressing estrogen receptor alpha or beta. *Biochemical Pharmacology* 71: 1459-1496.

Esteban , M.A., A. Rodríguez, A. García Ayala and J. Meseguer. 2004. Effects of high doses of cortisol on innate cellular immune response of sea bream (*Sparus aurata* L.). *General and Comparative Endocrinology* 137: 89-98.

Evinger, A.J. III and E.R. Levin. 2005. Requirements for estrogen receptor α membrane localization and function. *Steroids* 70: 361-363.

Filardo, E.J., J.A. Quinn, K.I. Bland and A.R. Frackelton. 2000. Estrogen-induced activation of Erk-1 and Erk-2 requires the G protein-coupled receptor homolog, gpr30, and occurs via transactivation of the epidermal growth factor receptor through release of HBEGF. *Molecular Endocrinology* 14: 1649-1660.

Filby, A.L. and C.R. Tyler. 2005. Molecular characterization of estrogen receptors 1, 2a, and 2b and their tissue and ontogenic expression profiles in fathead minnow (*Pimephales promelas*). *Biology of Reproduction* 73: 648-662.

Finlay-Moore, O., P.G. Hartel and M.L. Cabrera. 2000. 17β-estradiol and testosterone in soil and runoff from grasslands amended with broiler litter. *Journal of Environmental Quality* 29: 1604-1611.

Gale, W.L., R. Patino and A.G. Maule. 2004. Interaction of xenobiotics with estrogen receptors alpha and beta and a putative plasma sex hormone-binding globulin from channel catfish (*Ictalurus punctatus*). *General and Comparative Endocrinology* 136: 338-345.

Gao, Y, W.P. Qian, K. Dark, G. Toraldo, A.S. Lin, R.E. Guldberg, R.A. Flavell, M.N. Weitzmann and R. Pacifici. 2004. Estrogen prevents bone loss through transforming growth factor beta signaling in T cells. *Proceedings of the National Academy of Sciences of the United States of America* 101: 16618-16623.

Garcia-Duran, M., T. de Frutos, J. Diaz-Recasens, G. Garcia-Galvez, A. Jimenez, M. Monton, J. Farre, L. Sanchez de Miguel, F. Gonzalez-Fernandez, M.D. Arriero, L. Rico, R. Garcia, S. Casado and A. Lopez-Farre. 1999. Estrogen stimulates neuronal nitric oxide synthase protein expression in human neutrophils. *Circulation Research* 85: 1020-1026.

Gee, A.C. and J.A. Katzenellenbogen. 2001. Probing conformational changes in the estrogen receptor: Evidence for a partially unfolded intermediate facilitating ligand binding and release. *Molecular Endocrinology* 15: 421-428.

Geraldes, P., S. Gagnon, S. Hadjadj, Y. Merhi, M.G. Sirois, I. Cloutier and J.F. Tanguay. 2006. Estradiol blocks the induction of CD40 and CD40L expression on endothelial cells and prevents neutrophil adhesion: An ERalpha-mediated pathway. *Cardiovascular Research* 71: 566-573.

Ghisletti, S., C. Meda, A. Maggi and E. Vegeto. 2005. 17beta estradiol inhibits inflammatory gene expression by controlling NF-kB intracellular localization. *Molecular Cell Biology* 25: 2957-2968.

Goksoyr, A. 2006. Endocrine disruptors in the marine environment: mechanisms of toxicity and their influence on reproductive processes in fish. *Journal of Toxicology and Environmental Health* A69: 175-184.

Gong, Y., H.S. Chin, L.S. Lim, C.J. Loy, J.P. Obbard and E.L. Yong. 2003. Clustering of sex hormone disruptors in Singapore's marine environment. *Environmental Health Perspectives*111: 1448-1453.

Greytak, S.R. and C.V. Callard. 2006. Cloning of three estrogen receptors (ER) from killifish (*Fundulus heteroclitus*): Differences in populations from polluted and reference environments. *General and Comparative Endocrinology* 150: 174-188.

Grimaldi, C.M., J. Cleary, A.S. Dagtas, D. Moussai and B. Diamond. 2002. Estrogen alters thresholds for B cell apoptosis and activation. *Journal of Clinical Investigation* 109: 1625-1633.

Grimaldi, C.M., L. Hill, X. Xu, E. Peeva and B. Diamond. 2005. Hormonal modulation of B cell development and repertoire selection. *Molecular Immunology* 42: 811-820.

Grimaldi, C.M., V. Jeganathan and B. Diamond. 2006. Hormonal regulation of B cell development: 17 beta-estradiol impairs negative selection of high-affinity DNA-reactive B cells at more than one developmental checkpoint. *Journal of Immunology* 176: 2703-2710.

Gulshan, S., A.B. McCruden and W.H. Stimson. 1990. Oestrogen receptors in macrophages. *Scandinavian Journal of Immunology* 6: 691-697.

Gushiken, Y., H. Watanuki and M. Sakai. 2002. *In vitro* effect of carp phagocytic cells by bisphenol A and nonylphenol. *Fisheries Science* 68: 178-183.

Halm, S., G. Martinez-Rodriguez, L. Rodriguez, F. Prat, C.C. Mylonas, M. Carrillo and S. Zanuy. 2004. Cloning, characterization, and expression of three oestrogen receptors (ERalpha, ERbeta1 and ERbeta2) in the European sea bass, *Dicentrarchus labrax*. *Molecular Cell Endocrinology* 223: 63-75.

Hamano, N., N. Terada, K. Maesako, T. Numata and A. Konno. 1998. Effect of sex hormones on eosinophilic inflammation in nasal mucosa. *Allergy Asthma Proceedings* 19: 263-269.

Hamano, N., N. Terada, K. Maesako, G. Hohki, T. Ito, T. Yamashita and A. Konno. 1998. Effect of female hormones on the production of IL-4 and IL-13 from peripheral blood mononuclear cells. *Acta Oto-Laryngologica* 118: 27-31.

Hanselman, T.A., D.A. Graetz and A.C. Wilkie. 2004. Comparison of three enzyme immunoassays for measuring 17beta-estradiol in flushed dairy manure wastewater. *Journal of Environmental Quality* 33: 1919-1923.

Harris, M.T., R.S. Feldberg, K.M. Lau, N.H. Lazarus and D.E. Cochrane. 2000. Expression of pro-inflammatory genes during estrogen-induced inflammation of the rat prostate. *Prostate* 44: 19-25.

Hawkins, M.B. and P. Thomas. 2004. The unusual binding properties of the third distinct teleost estrogen receptor subtype ERßa are accompanied by highly conserved amino acid changes in the ligand binding domain. *Endocrinology* 145: 2968-2977.

Hawkins, M.B., J.W. Thornton, D. Crews, JK. Skipper, A. Dotte and P. Thomas. 2000. Identification of a third distinct estrogen receptor and reclassification of estrogen receptors in teleosts. *Proceedings of the National Academy of Science of the United States of America* 97: 10751-10756.

Haynes, M.P., D. Sinha, K.S. Russell, M. Collinge, D. Fulton, M. Morales-Ruiz, W.C. Sessa and J.R. Bender. 2000. Membrane estrogen receptor engagement activates endothelial nitric oxide synthase via the PI3-Kinase–Akt pathway in human endothelial cells. *Circulation Research* 87: 677-682.

Herman, J.S. and A.L. Mills. 2003. Biological and hydrogeological interactions affect the persistence of 17β-estradiol in an agricultural watershed. *Geobiology* 1: 141-151.

Herrera, L.A., R. Montero, J.M. Leon-Cazares, E. Rojas, M.E. Gonsebatt and P. Ostrosky-Wegman. 1992. Effects of progesterone and estradiol on the proliferation of phytohemagglutinin-stimulated human lymphocytes. *Mutation Research* 270: 211-218.

Hewitt, S.C. and K.S. Korach. 2002. Estrogen receptors: structure, mechanisms and function. *Reviews in Endocrine Metabolic Diseases* 3: 193-200.

Holmdahl, R. 1989. Estrogen exaggerates lupus but suppresses T-cell-dependent autoimmune disease. *Journal of Autoimmunity* 2: 651-656.

Hong, E.J., S.H. Park, K.C. Choi, P.C. Leung and E.B Jeung. 2006. Identification of estrogen-regulated genes by microarray analysis of the uterus of immature rats exposed to endocrine disrupting chemicals. *Reproductive Biology and Endocrinology* 4: 49-59.

Hou, Y.Y., Y. Suzuki and K. Aida. 1999. Effects of steroid hormones on immunoglobulin M (IgM) in rainbow trout, *Oncorhynchus mykiss*. *Fish Physiology and Biochemistry* 20: 155-162.

Huang, J., X. Li., C.A. Maguire, R. Hilf, R.A. Bambara and M. Muyan. 2005. Binding of estrogen receptor b to estrogen response element in situ is independent of estradiol and impaired by its amino terminus. *Molecular Endocrinology* 19: 2696-2712.

Huang, J., X. Li, T. Qiao, R. Bambara, R. Hilf and M. Muyan. 2006. A tale of two estrogen receptors (ERs): how differential ER-estrogen responsive element interactions contribute to subtype-specific transcriptional responses. *Nuclear Receptor Signal* 4: 15.

Igarashi, H., T. Kouro, T. Yokota, P.C. Comp and P.W. Kincade. 2001. Age and stage dependency of estrogen receptor expression by lymphocyte precursors. *Proceedings of the National Academy of Sciences of the United States of America* 98: 15131-15136.

Iguchi, T., H. Watanabe and Y. Katsu. 2001. Developmental effects of estrogenic agents on mice, fish, and frogs: A mini-review. *Hormone and Behaviour* 40: 248-251.

Inadera, H. 2006. The immune system as a target for environmental chemicals: Xenoestrogens and other compounds. *Toxicological Letters* 164: 191-206.

Ito, I., T. Hayashi, K. Yamida, M. Kuzuya, M. Naito and A. Iguchi. 1995. Physiological concentration of estradiol inhibits polymorphonuclear leukocyte chemotaxis via a receptor-mediated system. *Life Sciences* 56: 2247-2253.

Ito, A., B.F. Bebo, A. Matejuk, A. Zamora, M. Silverman, A. Fyfe-Johnson and H. Offner. 2001. Estrogen treatment down-regulates TNF-α production and reduces the severity of experimental autoimmune encephalomyelitis in cytokine knockout mice. *Journal of Immunology* 167: 542-552.

Iwanowicz, L.R., D.T. Lerner, V.S. Blazer and S.D. McCormick. 2005. Aqueous exposure to Aroclor 1254 modulates the mitogenic response of Atlantic salmon anterior kidney T-cells: indications of short- and long-term immunomodulation. *Aquatic Toxicology* 72: 305-314.

Iwata, M., Y. Eshima, H. Kagechika and H. Miyaura. 2004. The endocrine disruptors nonylphenol and octylphenol exert direct effects on T cells to suppress Th1 development and enhance Th2 development. *Immunology Letters* 94: 135-139.

Jaffe, H.L. 1924. The influence of the suprarenal gland on the thymus. III. Stimulation of the growth of the thymus gland following double suprarenalectomy in young rats. *Journal of Experimental Medicine* 40: 753-760.

Janele, D., T. Lang, S. Capellino, M. Cutolo, J.A. Da Silva and R.H. Straub. 2006. Effects of testosterone, 17β-Estradiol, and downstream estrogens on cytokine secretion from human leukocytes in the presence and absence of cortisol. *Annals of the New York of Sciences* 1069: 168-182.

Janis, K., J. Hoeltke, M. Nazareth, P. Fanti, K. Poppenberg and S.M. Aronica. 2004. Estrogen decreases expression of chemokine receptors and suppresses chemokine bioactivity in murine monocytes. *American Journal of Reproductive Immunology* 51: 22-31.

Jansson, L. and R. Holmdahl. 1998. Estrogen-mediated immunosuppression in autoimmune diseases. *Inflammation Research* 47: 290-301.

Kajiwara, M., S. Kuraku, T. Kurokawa, K. Kato, S. Toda, H. Hiros, S. Takahashi, Y. Shibata, T. Iguchi, T. Matsumoto, T. Miyata, T. Miura and Y. Takahashi. 2006. Tissue preferential expression of estrogen receptor gene in the marine snail, *Thais clavigera*. .*General and Comparative Endocrinology* 148: 315-326.

Kahlert, S., S. Nuedling, M. van Eickels, H. Vetter, R. Meyer and C. Grohe. 2000. Estrogen receptor rapidly activates the IGF-1 receptor pathway. *Journal of Biological Chemistry* 275: 18447-18453.

Kalaitzidis, D. and T.D. Gilmore. 2005. Transcription factor cross-talk: the estrogen receptor and NF-kappaβ. *Trends in Endocrinology and Metabolism* 16: 46-52.

Kanda, N. and S. Watanabe. 2003. 17beta-estradiol inhibits the production of RANTES in human keratinocytes. *Journal of Investigative Dermatology* 120: 420-427.

Katzenellenbogen, B.S., I. Choi, R. Delage-Mourroux, T.R. Ediger, P.G.V. Martini, M. Montano, J. Sun, K. Weis and J.A. Katzenellenbogen. 2000. Molecular mechanisms of estrogen action: selective ligands and receptor pharmacology. *Journal of Steroid Biochemistry and Molecular Biology* 74: 279-285.

Keay, J., J.T. Bridgham and J.W. Thornton. 2006. The *Octopus vulgaris* estrogen receptor is a constitutive transcriptional activator: evolutionary and functional implications. *Endocrinology* 147: 3861-3869.

Kenny, J.F., P.C. Pangburn and G. Trail. 1976. Effect of estradiol on immune competence: in vivo and in vitro studies. *Infection and Immunity* 13: 448-456.

Khan, K.N., M. Hideaki, F. Akira, K. Michio, S. Ichiro, M. Toshifumi and I. Tadayuki. 2005. Estrogen and progesterone receptor expression in macrophages and regulation of hepatocyte growth factor by ovarian steroids in women with endometriosis. *Human Reproduction* 20: 2004-2013.

Kime, D.E. 1998. *Endocrine Disruption in Fish.* Kluwer Academic Publishers, Boston.

Klinge, C.M. 2000. Estrogen receptor interaction with co-activators and co-repressors. *Steroids* 65: 227-251.

Klinge, C.M. 2001. Estrogen receptor interaction with estrogen response elements. *Nucleic Acids Research* 29: 2905-2919.

Kerkvliet, N.I. 1995. Immunologic effects of chlorinated dibenzo-p-dioxins. *Environmental Health Perspective* 103 (Supplement 9): 47-53.

Kerkvliet, N.I., D.M. Shepherd and L. Baecher-Steppan. 2002. T lymphocytes are direct, aryl hydrocarbon receptor (AhR)-dependent targets of 2,3,7,8-tetrachlorodibenzo-p-dioxin (TCDD): AhR expression in both CD4+ and CD8+ T cells is necessary for full suppression of a cytotoxic T lymphocyte response by TCDD. *Toxicology and Applied Pharmacology* 185: 146-152.

Knoblauch, R. and M.J. Garabedian. 1999. Role of Hsp90-associated cochaperone p23 in estrogen receptor signal transduction. *Molecular Cell Biology* 19: 3748-3759.

Komi, J. and A. Lassila. 2000. Nonsteroidal anti-estrogens inhibit the functional differentiation of human monocyte-derived dendritic cells. *Blood* 95: 2875-2882.

Kuiper, G.G., E. Enmark, M. Pelto-Huikko, S. Nilsson and J.A. Gustafsson. 1996. Cloning of a novel estrogen receptor expressed in rat prostate and ovary. *Proceedings of the National Academy of Sciences of the United States of America* 93: 5925-5930.

Kuiper, G.G., B. Carlsson, K. Grandien, E. Enmark, J. Haggbald, S. Nillson and J.A. Gustafsson. 1997. Comparison of the ligand binding specificity and transcript tissue distribution of estrogen receptors α and β. *Endocrinology* 138: 863-870.

Kuiper, G.G., J.G. Lemmen, B. Carlsson, J.C. Corton, S.H. Safe, P.T. van der Saag, B. van der Burg and J.A. Gustafsson. 1998. Interaction of estrogenic chemicals and phytoestrogens with estrogen receptor beta. *Endocrinology* 139: 4252-4263.

Kumar, V. and P. Chambon. 1988. The estrogen receptor binds tightly to its responsive element as a ligand-induced homodimer. *Cell* 55: 145-156.

Lambert, K.C., E.M. Curran, B.M. Judy, D.B. Lubahn and D.M. Estes. 2004. Estrogen receptor-alpha deficiency promotes increased TNF-alpha secretion and bacterial killing by murine macrophages in response to microbial stimuli in vitro. *Journal of Leukocyte Biology* 75: 1166–1172.

Lambert, K.C., E.M. Curran, B.M. Judy, G.N. Milligan, D.B. Lubahn and D.M. Estes. 2005. Estrogen receptor α (ERα) deficiency in macrophages results in increased stimulation of CD4+ T cells while 17β-estradiol acts through ER α to increase IL- 4 and GATA-3 expression in CD4+ T cells independent of antigen presentation. *Journal of Immunology* 175: 5716-5723.

Lang, T.L. 2004. Oestrogen as an immunomodulator. *Clinical Immunology* 113: 224-230.

Larkin, P., T. Sabo-Attwood, J. Kelso and N.D. Denslow. 2002. Gene expression analysis of largemouth bass exposed to estradiol, nonylphenol, and p,p'-DDE. *Comparative Biochemistry and Physiology* B133: 543-557.

Larkin, P., T. Sabo-Attwood, J. Kelso and N.D. Denslow. 2003. Analysis of gene expression profiles in largemouth bass exposed to 17-beta-estradiol and to anthropogenic contaminants that behave as estrogens. *Ecotoxicology* 12: 463-468.

Law, W., W. Chen, Y. Song, S. Dufour and C. Chang. 2001. Differential *in vitro* suppressive effects of steroids on leukocyte phagocytosis in two teleosts, tilapia and common carp. *General and Comparative Endocrinology* 121: 163-172.

Laws, S.C., S.A. Carey, J.M. Ferrell, G.J. Bodman and R.L. Cooper. 2000. Estrogenic activity of octylphenol, nonylphenol, bisphenol A and methoxychlor in rats. *Toxicology Science* 54: 154-167.

Lee, S.H. 1982. Uterine epithelial and eosinophil estrogen receptors in rats during the estrous cycle. *Histochemical Cell Biology* 74: 443-452.

Leung, Y.K., P. Mak, S. Hassan and S.M. Ho. 2006. Estrogen receptor (ER)-beta isoforms: a key to understanding ER-beta signaling. *Proceedings of the National Academy of Sciences of the United States of America* 103: 13162-13167.

Levin, E.R. 2001. Cell localization, physiology, and nongenomic actions of estrogen receptors. *Journal of Applied Physiology* 91: 1860-1867.

Li, L., M.P. Haynes and J.R. Bender. 2003. Plasma membrane localization and function of the estrogen receptor α variant (ER46) in human endothelial cells. *Proceedings of the National Academy of Sciences of the United States of America* 100: 4807-4812.

Lintelmann, J., A. Katayyama, N. Kurihara, L. Shore and A. Wenzel. 2003. Endocrine disruptors in the environment. *Pure and Applied Chemistry* 75: 631-681.

Liu, H., K.K. Loo, K. Palaszynski, J. Ashouri, D.B. Lubahn and R.R. Voskuhl. 2003. Estrogen receptor a mediates estrogen's immune protection in autoimmune disease. *Journal of Immunology* 171: 6936-6940.

Liu, Y.J. 2001. Dendritic cell subsets and lineages, and their functions in innate and adaptive immunity. *Cell* 1: 106-259.

Mao, A., V. Paharkova-Vatchkova, J. Hardy, M.M. Miller and S. Kovats. 2005. Estrogen selectively promotes the differentiation of dendritic cells with characteristics of Langerhans cells. *Journal of Immunology* 175: 5146-5151.

Maret, A., J.D. Courdert, L. Garidou, G. Foucras, P. Gourdy, A. Krust, S. Dupont, P. Chambon, P. Druet, F. Bayard and J.C. Guery. 2003. Estradiol enhances primary antigen-specific CD4 T cell responses and Th1 development *in vivo*. Essential role of estrogen receptor expression in hematopoietic cells. *European Journal of Immunology* 33: 512-521.

Masuzawa, T., C. Miyaura, Y. Onoe, K. Kusano, H. Ohta, S. Nozawa and T. Suda. 1994. Estrogen deficiency stimulates B lymphopoiesis in mouse bone marrow. *Journal of Clinical Investigations* 94: 1090-1097.

Matsuda, H., K. Okuda, K. Fukui and Y. Kamata. 1985. Inhibitory effect of estradiol-13 beta and progesterone on bactericidal activity in uteri of rabbits infected with *Escherichia coli*. *Infection and Immunity* 48: 652-657.

Matejuk, A., J. Dwyer, A. Zamora, A.A. Vandenbark and H. Offner. 2002. Evaluation of the effects of 17ß-Estradiol (17ß-E2) on gene expression in experimental autoimmune encephalomyelitis using DNA microarray. *Endocrinology* 143: 313-319.

Matthews, J. and J.A. Gustafsson. 2006. Estrogen receptor and aryl hydrocarbon receptor signaling pathways. *Nuclear Receptor Signal* 4: 16.

Matthews, J., B. Wihlen, J. Thomsen and J.A. Gustafsson. 2005. Aryl hydrocarbon receptor-mediated transcription: ligand-dependent recruitment of estrogen receptor alpha to 2,3,7,8-tetrachlorodibenzo-p-dioxin-responsive promoters. *Molecular Cell Biology* 25: 5317-5328 .

Matthews, J.B., K.C. Fertuck, T. Celius, Y.-W. Huang, C.J. Fong and T.R. Zacharewski. 2002. Ability of structurally diverse natural products and synthetic chemicals to induce gene expression mediated by estrogen receptors from various species. *Journal of Steroid Chemistry* 82: 181-194.

Maule, A.G. and C.B. Schreck. 1991. Stress and cortisol treatment changed affinity and number of glucocorticoid receptors in leukocytes and gill of coho salmon. *General and Comparative Endocrinology* 84: 83-93.

McEwan, I.J., 2004. Sex, drugs and gene expression: signaling by members of the nuclear receptor superfamily. *Essays in Biochemistry* 40: 1-10.

McMurray, M.C. 2001. Estrogen, prolactin, and autoimmunity: actions and interactions. *Internal Immunopharmacology* 1: 995-1008.

McMurray, R.W., K. Ndebele, K.J. Hardy and J.K. Jenkins. 2001. 17-beta-estradiol suppresses IL-2 and IL-2 receptor. *Cytokine* 14: 324-333.

McInerney, E.M., K.E. Weis, J. Sun, S. Mosselman and B.S. Katzenellenbogen. 1998. Transcription activation by the human estrogen receptor subtype β (ER β) studied with ER β and ER α receptor chimeras. *Endocrinology* 139: 4513-4522.

Medina, K.L., G. Smithson and P.W. Kincade. 1993. Suppression of B lymphopoiesis during normal pregnancy. *Journal of Experimental Medicine* 178: 1507-1515.

Medina, K.L., A. Strasser and P.W. Kincade. 2000. Estrogen influences the differentiation, proliferation, and survival of early B-lineage precursors. *Blood* 95: 2059-2067.

Mendelsohn, J., M.M. Multer and J.L. Bernheim. 1977. Inhibition of human lymphocyte stimulation by steroid hormones: cytokinetic mechanisms. *Clinical and Experimental Immunology* 27: 127-134.

Menuet, A., E. Pellegrini, I. Anglade, O. Blaise, V. Laudet, O. Kah and F. Pakdel. 2002. Molecular characterization of three estrogen receptor forms in zebrafish: binding characteristics, transactivation properties, and tissue distributions. *Biology of Reproduction* 66: 1881-1892.

Metzger, D., S. Als, J.M. Bornert and P. Chambon. 1995. Characterization of the amino-terminal transcriptional activation function of the human estrogen receptor in animal and yeast cells. *Journal of Biological Chemistry* 270: 9535-9542.

Meucci, V. and A. Arukwe. 2006. The xenoestrogen 4-nonylphenol modulates hepatic gene expression of pregnane X receptor, aryl hydrocarbon receptor, CYP3A and CYP1A1 in juvenile Atlantic salmon (*Salmo salar*). *Comparative Biochemistry and Physiology* C142: 142-150.

Migliaccio, A., G. Castoria, M. Di Domenico, A. de Falco, A. Bilancio, M. Lombardi, D. Bottero, L. Varricchio, M. Nanayakkara, A. Rotondi and F. Auricchio. 2002. Sex steroid hormones act as growth factors. *Journal of Steroid Biochemistry and Molecular Biology* 83: 31-35.

Miller, A.P., W. Feng, D. Xing, N.M. Weathington, J.E. Blalock, Y.F. Chen and S. Oparil. 2004. Estrogen modulates inflammatory mediator expression and neutrophil chemotaxis in injured arteries. *Circulation* 110: 1664-1669.

Miller, N.M., M. Wilson, E. Bengten, T. Stuge, G. Warr and W. Clem. 1998. Functional and molecular characterization of teleost leukocytes. *Immunological Reviews* 166: 187-197.

Milston, R.H., M.S. Fitzpatrick, A.T. Vella, S. Clements, D. Gundersen, G. Feist, T.L. Crippen, J. Leong and C.B. Schreck. 2003. Short-term exposure of Chinook salmon (*Oncorhynchus tshawytscha*) to o,p-DDE or DMSO during early life-history stages causes long-term humoral immunosuppression. *Environmental Health Perspective* 111: 1601-1607.

Mo, R., J. Chen, A. Grolleau-Julius, H.S. Murphy, B.C. Richardson and R.L. Yung. 2005. Estrogen regulates CCR gene expression and function in T lymphocytes. *Journal of Immunology* 174: 6023-6029.

Moggs, J.G. and G. Orphanides. 2001. Estrogen receptors: orchestrators of pleiotrophic cellular responses. *EMBO Reports* 21: 775-781.

Molero, L., M. Garcia-Duran, J. Diaz-Recasens, L. Rico, S. Casado and A. Lopez-Farre. 2002. Expression of estrogen receptor subtypes and neuronal nitric oxide synthase in neutrophils from women and men: regulation by estrogen. *Cardiovascular Research* 56: 4-7.

Mosselman, S., J. Polman and J. Dijkema. 1996. ERβ: Identification and characterization of a novel human estrogen receptor. *FEBS Letters* 392: 49-53.

Mor, G., E. Sapi, V.M. Abrahams, T. Rutherford, J. Song, X. Hao, S. Muzaffuar and F. Kohen. 2003. Interaction of estrogen receptors with the Fas ligand promoter in human monocytes. *Journal of Immunology* 170: 114-122.

Mourot, B., T. Nguyen, A. Fostier and J. Bobe. 2006. Two unrelated putative membrane-bound progestin receptors, progesterone membrane receptor component 1 (PGMRC1) and membrane progestin receptor (mPR) beta, are expressed in the rainbow trout oocyte and exhibit similar ovarian expression patterns. *Reproductive Biology and Endocrinology* 4: 6-19.

Nagler, J.J., T. Cavileer, J. Sullivan, D.G. Cyr and C. Rexroad III. 2007. The complete nuclear estrogen receptor family in the rainbow trout: Discovery of the novel ERα2 and both ERβ isoforms. *Gene*. (In Press).

Nalbandian, G. and S. Kovats. 2005a. Estrogen, immunity and autoimmune disease. *Current Medicinal Chemistry—Immunology, Endocrine and Metabolic Agents* 5: 85-91.

Nalbandian, G. and S. Kovats. 2005b. Understanding sex biases in immunity: effects of estrogen on the differentiation and function of antigen-presenting cells. *Immunological Research* 31: 91-106.

Nalbandian, G., V. Paharkova-Vatchkova, A. Mao, S. Nale and S. Kovats. 2005. The selective estrogen receptor modulators, tamoxifen and raloxifene, impair dendritic cell differentiation and activation. *Journal of Immunology* 175: 2666-2675.

Narnaware, Y.K., S.P. Kelly and N.Y.S. Woo. 1997. Effect of injected growth hormone on phagocytosis in silver sea bream (*Sparus sarba*) adapted to hyper- and hypo-osmotic salinities. *Fish and Shellfish Immunology* 7: 515-517.

Ndebele, K., P.B. Tchounwou and R.W. McMurray. 2004. Coumestrol, bisphenol-A, DDT, and TCDD modulation of interleukin-2 expression in activated CD+4 Jurkat T cells. *International Journal of Environmental Research and Public Health* 1: 3-11.

Nealen, M.L., K.V. Vijayan, E. Bolton and P.F. Bray. 2001. Human platelets contain a glycosylated oestrogen receptor beta. *Circulation Research* 88: 438-442.

Neifeld, J.P. and D.C. Tormey. 1979. Effects of steroid hormones on phytohemagglutinin-stimulated human peripheral blood lymphocytes. *Transplant* 27: 309-314.

Nicol, T., D.L.J. Bilbey, L.M. Charles, J.L. Cordingley and B. Vernon-Roberts. 1964. Oestrogen: The natural stimulant of body defence. *Journal of Endocrinology* 30: 277-291.

Noaksson, E., U. Tjarnlund, A.T. Bosveld and L. Balk. 2001. Evidence for endocrine disruption in perch (*Perca fluviatilis*) and roach (*Rutilus rutilus*) in a remote Swedish lake in the vicinity of a public refuse dump. *Toxicology and Applied Pharmacology* 174: 160-176.

Norman, A.W. and G. Litwack (eds.). 1987. *Hormones*. Academic Press, Orlando.

Obendorf, M. and V.K. Patchev. 2004. Interactions of sex steroids with mechanisms of inflammation. *Current Drug Targets—Inflammation and Allergy* 3: 425-433.

Okuyama, R., T. Abo, S. Seki, T. Ohteki, K. Sugiura and A. Kusumi. 1992. Estrogen administration activates extrathymic T cell differentiation in the liver. *Journal of Experimental Medicine* 175: 661-669.

O'Lone, R., M.C. Frith, E.K. Karlsson and U. Hansen. 2004. Genomic targets of nuclear estrogen receptors. *Molecular Endocrinology* 18: 1859-1875.

Paavonen, T., L.C. Andersson and H. Aldercreutz. 1981. Sex hormone regulation of in vitro immune response. Estradiol enhances human B cell maturation via inhibition of suppressor T cells in pokeweed mitogen-stimulated cultures. *Journal of Experimental Medicine* 154: 1935-1945.

Paharkova-Vatchkova, V., R. Maldonado and S. Kovats. 2004. Estrogen preferentially promotes the differentiation of $CD11c^{+}$ $CD11b^{intermediate}$ dendritic cells from bone marrow precursors. *Journal of Immunology* 172: 1426-1436.

Pakdel, F., R. Métivier, G. Flouriot and Y. Valotaire. 2000. Two estrogen receptor (ER) isoforms with different estrogen dependencies are generated from the trout ER gene. *Endocrinology* 141: 571-580.

Patino, R., Z. Xia, W.L. Gale, C. Wu, A.G. Maule and X. Chang. 2000. Novel transcripts of the estrogen receptor α gene in channel catfish. *General and Comparative Endocrinology* 120: 314-325.

Peeva, E., J. Venkatesh and B. Diamond. 2005. Tamoxifen blocks estrogen-induced B cell maturation but not survival. *Journal of Immunology* 175: 1415-1423.

Perez, M.C., E.E. Furth, P.D. Matzumura and C.R. Lyttle. 1996. Role of eosinophils in uterine responses to estrogen. *Biology of Reproduction* 54: 249-254.

Petrovic, M., M. Solé, M.J. López de Alda and D. Barceló. 2002. Endocrine disruptors in sewage treatment plants, receiving river waters, and sediments: integration of chemical analysis and biological effects on feral carp. *Environment and Toxicological Chemistry* 21: 2146-2156.

Phiel, K.L., R.A. Henderson, S.J. Adelman and M.M. Elloso. 2005. Differential estrogen receptor gene expression in human peripheral blood mononuclear cell populations. *Immunological Letters* 97: 107-113.

Piferrer, F. and M. Blazquez. 2005. Aromatase distribution and regulation in fish. *Fish Physiology and Biochemistry* 31: 215-226.

Pfeifer, R.W. and R.M. Patterson. 1985. Modulation of nonspecific cell-mediated growth inhibition by estrogen metabolites. *Immunopharmacology* 10: 127-135.

Pioli, P.A., L.K. Weaver, T.M. Schaefer, J.A. Wright, C.R. Wira and P.M. Guyre. 2006. Lipopolysaccharide-induced IL-1 beta production by human uterine macrophages up-regulates uterine epithelial cell expression of human beta-defensin 2. *Journal of Immunology* 176: 6647-6655.

Polanczyk, M.J., C. Hopke, J. Huan, A.A. Vandenbark and H. Offner. 2005. Enhanced FoxP3 expression and Treg cell function in pregnant and estrogen-treated mice. *Journal of Neuroimmunology* 170: 85-92.

Polanczyk, M.J., R.E. Jones, S. Subramanian, M. Afentoulis, C. Rich, M. Zakroczymski, P. Cookes, A.A. Vanderbark and H. Offner. 2004a. T lymphocytes do not directly mediate the protective effect of estrogen on experimental autoimmune encephalomyelitis. *American Journal of Pathology* 165: 2069-2077.

Polanczyk, M.J., B.D. Carson, S. Subramanian, M. Afentoulis, A.A. Vandenbark, S.F. Ziegler and H. Offner. 2004b. Cutting edge: estrogen drives expansion of the CD4+CD25+ regulatory T cell compartment. *Journal of Immunology* 173: 2227-2230.

Pratt, W.B. and D.O. Toft. 1997. Steroid receptor interaction with heat shock protein and immunophillin chaperones. *Endocrine Reviews* 18: 306-360.

Prieto, G.A. and Y. Rosenstein. 2006. Oestradiol potentiates the suppressive function of human CD4 CD25 regulatory T cells by promoting their proliferation. *Immunology* 118: 58-65.

Purohit, A., S.P. Newman and M.J. Reed. 2002. The role of cytokines in regulating estrogen synthesis: implications for the etiology of breast cancer. *Breast Cancer Research* 4: 65-69.

Quaedackers, M.E., C.E. Van Den Brink, S. Wissink, R.H. Schreurs, J.A. Gustafsson, P.T. Van Der Saag and B.B. Van Der Burg. 2001. 4-hydroxytamoxifen trans-represses nuclear factor-kappa B activity in human osteoblastic U2-OS cells through estrogen receptor (ER)alpha, and not through ER beta. *Endocrinology* 142: 1156-1166.

Ramos, J.G., J. Varayoud, L. Kass, H. Rodriguez, M. Munoz de Toro, G.S. Montes and E.H. Luque. 2000. Estrogen and progesterone modulation of eosinophilic infiltration of the rat uterine cervix. *Steroids* 65: 409-414.

Ray, A., K.E. Prefontaine and P. Ray. 1994. Down-modulation of interleukin-6 gene expression by 17 beta-estradiol in the absence of high affinity DNA binding by the estrogen receptor. *Journal of Biological Chemistry* 269: 12940-12946.

Raznandi, M., A. Pedram, I. Merchenthaler, G.L. Greene and E.R. Levin. 2004. Plasma membrane estrogen receptors exist and functions as dimers. *Molecular Endocrinology* 18: 2854-2865.

Raznandi, M., G. Alton, A. Pedram, S. Ghonshani, P. Webb and E.R. Levin. 2003. Identification of a structural determinant necessary for the localization and function of the estrogen receptor α at the plasma membrane. *Molecular Cell Biology* 23: 1633-1646.

Rijhsinghani, A.G., K. Thompson, S.K. Bhatia and T.J. Waldschmidt. 1996. Estrogen blocks early T cell development in the thymus. *American Journal of Reproductive Immunology* 36: 269-277.

Roberts, M., X. Luo and N. Chegini. 2005. Differential regulation of interleukins IL-13 and IL-15 by ovarian steroids, TNF-alpha and TGF-beta in human endometrial epithelial and stromal cells. *Molecular Human Reproduction* 11: 751-760.

Rollerova, E. and M. Urbancikova. 2000. Intracellular estrogen receptors, their characterization and function (review). *Endocrine Regulation* 34: 203-218.

Sabo-Attwood, T., K.J. Kroll and N.D. Denslow. 2004. Differential expression of largemouth bass (*Micropterus salmoides*) estrogen receptor isotypes alpha, beta, and gamma by estradiol. *Molecular Cell Biology* 218: 107-118.

Saha, N.R., T. Usami and Y. Suzuki. 2002. Seasonal changes in the immune activities of common carp (*Cyprinus carpio*). *Fish Physiology and Biochemistry* 26: 379-387.

Saha, N.R., T. Usami and Y. Suzuki. 2004. In vitro effects of steroid hormones on IgM-secreting cells and IgM secretions in common carp (*Cyprinus carpio*). *Fish and Shellfish Immunology* 17: 149-158.

Sakazaki, H., H. Ueno and K. Nakamuro. 2002. Estrogen receptor α in mouse splenic lymphocytes: Possible involvement in immunity. *Toxicology Letters* 133: 221-229.

Sakazaki, H., R. Ido, H. Ueno and K. Nakamuro. 2005. 17β-estradiol primes elicitation of inducible nitric oxide synthase expression by lipopolysaccharide and interferon-Γ in mouse macrophage cell line J774.1. *Journal of Health Sciences* 51: 62-69.

Salem, M.L. 2004. Estrogen, a double-edged sword: modulation of TH1- and TH2-mediated inflammations by differential regulation of TH1/TH2 cytokine production. *Current Drug Targets—Inflammation and Allergy* 3: 97-104.

Sanchez, R., D. Nguyen, W. Rocha, J.H. White and S. Mader. 2004. Diversity in the mechanisms of gene regulation by estrogen receptors. *Bioessays* 24: 244-254.

Screpanti, I., D. Meco, S. Morrone, A. Gulino, B.J. Mathieson and L. Frati. 1991. In vivo modulation of the distribution of thymocyte subsets: Effects of estrogen on the expression of different T cell receptor V beta gene families in CD4-, CD8-thymocytes. *Cellular Immunology* 134: 414-426.

Selye, H. 1936a. A syndrome produced by diverse nocuous agents. *Nature (London)* 138: 32.

Selye, H. 1936b. Thymus and adrenals in the response of the organism to injuries and intoxications. *British Journal of Experimental Pathology* 17: 234-248.

Sentis, S., M. Le Romancer, C. Bianchin, M.-C. Rostan and L. Corbo. 2005. Sumoylation of the estrogen receptor α hinger region regulates transcriptional activity. *Molecular Endocrinology* 19: 2671-2684.

Sentman, C.L., S.K. Meadows, C.R. Wira and M. Eriksson. 2004. Recruitment of uterine NK cells: induction of CXC chemokine ligands 10 and 11 in human endometrium by estradiol and progesterone. *Journal of Immunology* 173: 6760-6766.

Shi, T.L., X.Z. Luo, X.Y. Zhu, K.Q. Hua, Y. Zhu and D.J. Li. 2006. Effects of combined 17β-estradiol with TCDD on secretion of chemokine IL-8 and expression of its receptor CXCR1 in endometriotic focus-associated cells in co-culture. *Human Reproduction* 21: 870-879.

Simoncini, T., E. Rabkin and J.K. Liao. 2003. Molecular basis of cell membrane estrogen receptor interaction with phosphatidylinositol 3-kinase in endothelial cells. *Arteriosclerosis, Thrombosis and Vascular Biology* 23: 198-203.

Simoncini, T., A. Hafezi-Moghadam, D.P. Brazil, K. Ley, W.W. Chin and J.K. Liao. 2000. Interaction of oestrogen receptor with the regulatory subunit of phosphatidylinositol-e-OH kinase. *Nature (London)* 407: 538-541.

Simoncini, T., P. Mannella, L. Fornari, A. Caruso, G. Varone and A.R. Genazzani. 2004. Genomic and non-genomic effects of estrogens on endothelial cells. *Steroids* 69: 537-542.

Simpson, E., G. Rubin, C. Clyne, K. Robertson, L. O'Donnell and M. Jones. 1999. Local estrogen biosynthesis in males and females. *Endocrine-Related Cancer* 6: 131-137.

Slater, C.H. and C.B. Schreck. 1993. Testosterone alters the immune response of chinook salmon (*Oncorhynchus tshawytscha*). *General and Comparative Endocrinology* 89: 291-298.

Slater, C.H. and C.B. Schreck. 1997. Physiological levels of testosterone kill salmonid leukocytes in vitro. *General and Comparative Endocrinology* 106: 113-119.

Slater, C.H., M.S. Fitzpatrick and C.B. Schreck. 1995. Characterization of an androgen receptor in salmonid lymphocytes: Possible link to androgen induced immunosuppression. *General and Comparative Endocrinology* 100: 218-225.

Smithson, G., K. Medina, I. Ponting and P.W. Kincade. 1995. Estrogen suppresses stromal cell-dependent lymphopoiesis in culture. *Journal of Endocrinology* 155: 3409-3417.

Smithson, G., J.F. Couse, D.B. Lubahn, K.S. Korack and P.W. Kincade. 1998. The role of estrogen receptors and androgen receptors in sex steroid regulation of B lymphopoiesis. *Journal of Immunology* 161: 27-34.

Sohoni, P. and J.P. Sumpter. 1998. Several environmental oestrogens are also anti-androgens. *Journal of Endocrinology* 158: 327-339.

Song, R.X., R.A. McPherson, L. Adam, Y. Bao, M. Shupnik, R. Kumar and R.J. Santen. 2002. Linkage of rapid estrogen action to MAPK activation by ERα-Shc association and Shc pathway activation. *Molecular Endocrinology* 16: 116-127.

Song, R.X., C.J. Barnes, Z. Zhang, Y. Bao, R. Kumar and R.J. Santen. 2004. The role of Shc and insulin-like growth factor 1 receptor in mediating the translocation of estrogen receptor to the plasma membrane. *Proceedings of the National Academy of Sciences of the United States of America* 101: 2076-2081.

Soto, A.M., J.M. Calabro, N.J. Prechtl, A.Y. Yau, E.F. Orlando, A. Daxenberger, A.S. Kolok, L.J. Guillette Jr., B. le Bizec, I.G. Lange and C. Sonnenschein. 2004. Androgenic and estrogenic activity in water bodies receiving cattle feedlot effluent in eastern Nebraska, USA. *Environmental Health Perspectives* 112: 346-352.

Staples, J.E., T.A. Gasiewicz, N.C. Fiore, D.B. Lubahn, K.S. Korach and A.E. Silverstone. 1999. Estrogen receptor alpha is necessary in thymic development and estradiol-induced thymic alterations. *Journal of Immunology* 163: 4168-4174.

Stefano, G.B., P. Cadet, C. Brenton, Y. Goumon, V. Prevot, J.P. Dessaint, J.C. Beauvillain, A.S. Roumier, I. Welters and M. Salzet. 2000. Estradiol-stimulated nitric oxide release in human granulocytes is dependant on intracellular calcium transients: evidence of a cell surface estrogen receptor. *Blood* 95: 3951-3958.

Stefano, G.B., V. Prevot, J.-C. Beauvillain, C. Fimaini, I. Welters, P. Cadet, C. Brenton, J. Pestel, M. Salzet and T.V. Bilfinger. 2003. Estradiol coupling to human monocyte nitric oxide release is dependent on intracellular calcium transients: evidence for an estrogen surface receptor. *Journal of Immunology* 163: 3758-3763.

Stein, B. and M.X. Yang. 1995. Repression of the interleukin-6 promoter by estrogen receptor is mediated by NF-kappa B and C/EBP beta. *Molecular Cell Biology* 15: 4971-4979.

Sthoeger, Z.M., N. Chiorazzi and R.G. Lahita. 1988. Regularion of the immune response by sex hormones. I. *In vitro* effects of estradiol and testosterone on pokeweed mitogen-induced human B cell differentiation. *Journal of Immunology* 141: 91-98.

Stimson, W.H. 1988. Oestrogen and human T lymphocytes: Presence of specific receptors in the T suppressor/cytotoxic subset. *Scandinavian Journal of Immunology* 28: 345-350.

Stolte, E.H., B.M.L. Verburg van Kemenade, H.F.G. Savelkoul and G. Flik. 2006. Evolution of glucocorticoid receptors with different glucocorticoid sensitivity. *Journal of Endocrinology* 190: 17-28.

Stygar, D., P. Westlund, H. Eriksson and L. Sahlin. 2006. Identification of wild type and variants of oestrogen receptors in polymorphonuclear and mononuclear leukocytes. *Clinical Endocrinology* 64: 74-81.

Suenaga, R., M.J. Evans, K. Mitamura, V. Rider and N.I. Abdou. 1998. Peripheral blood T cells and monocytes and B cell lines derived from patients with lupus express estrogen receptor transcripts similar to those of normal cells. *Journal of Rheumatology* 25: 1305-1312.

Sugita-Konishi, Y., S. Shimura, T. Nishikawa, F. Sunaga, H. Naito and Y. Suzuki. 2003. Effect of Bisphenol A on non-specific immunodefenses against non-pathogenic *Escherichia coli*. *Toxicology Letters* 136: 217-227.

Sumpter, J.P. 1998. Xenoendorine disrupters—environmental impacts. *Toxicology Letters* 102-103: 337-342.

Sundstrom, S.A., B.S. Komm, H. Ponce-de-Leon, Z. Yi, C. Teuscher and C.R. Lyttle. 1989. Estrogen regulation of tissue-specific expression of complement C3. *Journal of Biological Chemistry* 264: 16941-16947.

Tabb, M.M. and B. Blumberg. 2006. New modes of action for endocrine-disrupting chemicals. *Molecular Endocrinology* 20: 475-482.

Takeo, J. and S. Yamashita. 2000. Rainbow trout androgen receptor-alpha fails to distinguish between any of the natural androgens tested in transactivation assay, not just 11-ketotestosterone and testosterone. *General and Comparative Endocrinology* 117: 200-206.

Thibaut, R. and C. Porte. 2004. Effects of endocrine disrupters on sex steroid synthesis and metabolism pathways in fish. *Journal of Steroid Biochemistry and Molecular Biology* 92: 485-494.

Thomas, P., Y. Pang, E.J. Filardo and J. Dong. 2005. Identity of an estrogen membrane receptor coupled to a G protein in human breast cancer cells. *Endocrinology* 146: 624-632.

Thongngarm, T., J.K. Jenkins, K. Ndebele and R.W. McMurray. 2003. Estrogen and progesterone modulate monocyte cell cycle progression and apoptosis. *American Journal of Reproductive Immunology* 49: 129-138.

Thornton, J.W. 2001. Evolution of vertebrate steroid receptors from an ancestral estrogen receptor by ligand exploitation and serial genome expansions. *Proceedings of the National Academy of Sciences of the United States of America* 98: 5671-5676.

Thornton, J.W. 2003. Resurrecting the ancestral steroid receptor: ancient origin of estrogen signaling. *Science* 301: 1714-1717.

Thurmond, T.S., F.G. Murante, J.E. Staples, A.E. Silverstone, K.S. Korach and T.A. Gasiewicz. 2000. Role of the estrogen receptor alpha in hematopoetic stem cell development and B lymphocyte maturation in the male mouse. *Endocrinology* 141: 2309-2318.

Tomaszewskaa, A., I. Guevara, T. Wilczokb and A. Dembiska-Kiea. 2003. 17β-Estradiol- and Lipopolysaccharide-induced changes in nitric oxide, tumor necrosis factor-α and vascular endothelial growth factor release from RAW 264.7 macrophages. *Gynaecology and Obstetrics Investigations* 56: 152-159.

Torchia, J., C. Glass and M.G. Rosenfeld. 1998. Co-activators and co-repressors in the integration of transcriptional responses. *Current Opinion in Cell Biology* 10: 373-383.

Tsai, M.J. and B.W. O'Malley. 1994. Molecular mechanisms of action of steroid/thyroid receptor superfamily members. *Annual Review of Biochemistry* 63: 451-486.

Turnbull, A.V. and C.L. Rivier. 1999. Regulation of the hypothalamic-pituitary-adrenal axis by cytokines: actions and mechanisms of action. *Physiological Reviews* 79: 1-71.

Vegeto, E., S. Ghisletti, C. Meda, S. Etteri, S. Belcredito and A. Maggi. 2004. Regulation of the lipopolysaccharide signal transduction pathway by 17beta-estradiol in macrophage cells. *Journal of Steroid Biochemistry and Molecular Biology* 91: 59-66.

Vegeto, E., S. Belcredito, S. Etteri, S. Ghisletti, A. Brusadelli, C. Meda, A. Krust, S. Dupont, P. Clana, P. Chambon and A. Maggi. 2003. Estrogen receptor-α mediated the brain anti-inflammatory activity of estradiol. *Proceedings of the National Academy of Sciences of the United States of America* 100: 9614-9619.

Verdu, E.F., Y. Deng, P. Bercik and S.M. Collins. 2002. Modulatory effects of estrogen in two murine models of experimental colitis. *American Journal of Physiology and Gastrointestinal Liver Physiology* 283: G27-G36.

Vlata, Z., F. Porichis, G. Tzanakakis, A. Tsatsakis and E. Krambovitis. 2006. A study of zearalenone cytotoxicity on human peripheral blood mononuclear cells. *Toxicology Letters* 165: 274-281.

Vos, J.G., E. Dybing, H.A. Greim, O. Ladefoged, C. Lambre, J.V. Tarazona, I. Brandt and A.D. Vethaak. 2000. Health effects of endocrine-disrupting chemicals on wildlife, with special reference to the European situation. *Critical Reviews in Toxicology* 30: 71-133.

Vottero, A., V. Rochira, M. Capelletti, I. Viani, L. Zirilli, T.M. Neri, C. Carani, S. Bernasconi and L. Ghizzoni. 2006. Aromatase is differentially expressed in peripheral blood leukocytes from children, and adult female and male subjects. *European Journal of Endocrinology* 154: 425-431.

Wang, Z.Y., W. Zhang, P. Shen, B.W. Loggie, Y. Chang and T.F. Deuel. 2006. A variant of estrogen receptor-α, hER-α36: Transduction of estrogen- and antiestrogen-dependent membrane-initiated mitogenic signaling. *Proceedings of the National Academy of Sciences of the United States of America* 103: 9063-9068.

Wang, D.S., B. Senthilkumaran, C.C. Sudhakumari, F. Sakai, M. Matsuda, T. Kobayashi, M. Yoshikuni and Y. Nagahama. 2005. Molecular cloning, gene expression and characterization of the third estrogen receptor of the Nile tilapia, *Oreochromis niloticus*. *Fish Physiology and Biochemistry* 31: 255-266.

Wang, R. and M. Belosevic. 1994. Estradiol increases susceptibility of goldfish to *Trypanosoma danilewski*. *Developmental and Comparative Immunology*18: 377-387.

Wang, R. and M. Belosevic. 1995. The *in vitro* effects of estradiol and cortosol on the function of a long-term goldfish macrophage cell line. *Developmental and Comparative Immunology* 19: 327-336.

Watanuki, H., T. Yamaguchi and M. Sakai. 2002. Suppression in function of phagocytic cells in common carp *Cyprinus carpio* L. injected with estradiol, progesterone or 11-ketotestosterone. *Comparative Biochemistry and Physiology* C132: 407-413.

Watanuki, H., Y. Gushiken and M. Sakai. 2003. *In vitro* modulation of common carp (*Cyprinus carpio* L.) phagocytic cells by Di-n-butyl phthalate and Di-2-ethylhexyl phthalate. *Aquatic Toxicology* 63: 119-126.

Watson, C.S., C.H. Campbell and B. Gametchu. 1999. Membrane estrogen receptors on rat pituitary tumor cells: Immunoidentification and responses to estradiol and xenoestrogens. *Experimental Physiology* 84: 1013-1022.

Watson, C.S., C.H. Campbell and B. Gametchu. 2002. The dynamic and elusive membrane estrogen receptor-α. *Steroids* 67: 429-437.

Wendelaar Bonga, S.E. 1997. The stress response in fish. *Physiological Reviews* 77: 591-625.

Weusten, J.J., M.A. Blankenstein, F.H. Gmelig-Meyling, H.J. Schuurman, L. Kater and J.H. Thijssen. 1986. Presence of oestrogen receptors in human blood mononuclear cells and thymocytes. *Acta Encocrinologia* 112: 409-414.

Weyts, F.A., G. Flik, J.H. Rombout and B.M. Verburg van Kemenade. 1998a. Cortisol induces apoptosis in activated B-cells, but not other lymphoid cells of the common carp, *Cyprinus carpio* L. *Developmental and Comparative Immunology* 22: 551-562.

Weyts, F.A., G. Flik and B.M. Verburg van Kemenade. 1998b. Cortisol inhibits apoptosis in carp neutrophilic granulocytes. *Developmental and Comparative Immunology* 22: 563-572.

Weyts, F.A., B.M. Verburg van Kemenade and G. Flik. 1998c. Characterization of glucocorticoid receptors in peripheral blood leukocytes in carp, *Cyprinus carpio* L. *General and Comparative Endocrinology* 111: 1-8.

Weyts, F.A., N. Cohen, G. Flik and B.M. Verburg van Kemenade. 1999. Interactions between the immune system and the hypothalamo-pituitary-interrenal axis in fish. *Fish and Shellfish Immunology* 9: 1-20.

Wong, C.W., C. McNally, E. Nickbarg, B.S. Komm and B.J. Cheskis. 2002. Estrogen receptor-interacting protein that modulates its nongenomic activity-crosstalk with Src/Erk phosphorylation cascade. *Proceedings of the National Academy of Sciences of the United States of America* 99: 14783-14788.

World Health Organization. 2002. IPCS, Global assessment of the state-of-the-science of endocrine disruptors. WHO/PCS/EDC/02.2

Xia, Z., R. Patino, W.L. Gale, A.G. Maule and L.D. Densmore. 1999. Cloning, *in vitro* expression, and novel phylogenetic classification of a channel catfish estrogen receptor. *General and Comparative Endocrinology* 113: 360-368.

Xia, Z., W.L. Gale, X. Chang, D. Langenau, R. Patino, A.G. Maule and L.D. Densmore. 2000. Phylogenetic sequence analysis, recombinant expression, and tissue distribution of channel catfish estrogen receptor β. *General and Comparative Endocrinology* 118: 139-149.

Yada, T., M. Nagae, S. Moriyama and T. Azuma. 1999. Effects of prolactin and growth hormone on plasma immunoglobulin M levels of hypophysectomized rainbow trout, *Oncorhynchus mykiss*. *General and Comparative Endocrinology* 115: 46-52.

Yada, T., I. Misumi , K. Muto , T. Azuma and C.B. Schreck. 2004. Effects of prolactin and growth hormone on proliferation and survival of cultured trout leucocytes. *General and Comparative Endocrinology* 136: 298-306.

Yada, T., K. Muto, T. Azuma, S. Hyodo and C.B. Schreck. 2005. Cortisol stimulates growth hormone gene expression in rainbow trout leucocytes in vitro. *General and Comparative Endocrinology* 142: 248-255.

Yamaguchi, T., H. Watanuki and M. Sakai. 2001. Effects of estradiol, progesterone and testosterone on the function of carp, *Cyprinus carpio*, phagocytes in vitro. *Comparative Biochemistry and Physiology* C129: 49-55.

Yang, L., Y. Hu and Y. Hou. 2006. Effects of 17beta-estradiol on the maturation, nuclear factor kappa B p65 and functions of murine spleen CD11c-positive dendritic cells. *Molecular Immunology* 43: 357-366.

Yellayi, S., C. Teuscher, J.A. Woods, T.H. Welsh, K.S. Tung, M. Nakai, C. Rosenfeld, D.B. Lubahn and P.S. Cooke. 2000. Normal development of thymus in male and female mice requires estrogen/estrogen receptor-alpha signaling pathway. *Endocrine* 12: 207-213.

Yi, P., M.D. Driscoll, J. Huang, S. Bhagat, R. Hilf, R.A. Bambara and M. Muyan. 2002. The effects of estrogen-responsive element- and ligand-induced structural changes on the recruitment of cofactors and transcriptional responses by ER α and ER β. *Molecular Endocrinology* 16: 674-693.

You, H.J., J.Y. Kim and H.G. Jeong. 2003. 17 β-estradiol increases inducible nitric oxide synthase expression in macrophages. *Biochemical and Biophysical Research Communication* 303: 1129-1134.

You, H.J., C.Y. Choi, Y.J. Jeon, Y.C. Chung, S.K. Kang, K.S. Hahm and H.G. Jeong. 2002. Suppression of inducible nitric oxide synthase and tumor necrosis factor-alpha expression by 4-nonylphenol in macrophages. *Biochemical and Biophysical Research Communication* 294: 753-759.

Yron, I., A. Langer, T. Weinstein, E. Sahar, Y. Lidor, Y. Pardo, I. Katz, L. Shohat, Y. Kalechman and J. Ovadia. 1991. Effect of sex hormones on human T cell activation by concanavalin A. *Natural Immunity and Cell Growth Regulation* 10: 32-44.

Yu, H.P., T. Shimizu, Y.C. Hsieh, T. Suzuki, M.A. Choudhry, M.G. Schwacha and I.H. Chaudry. 2006. Tissue-specific expression of estrogen receptors and their role in the regulation of neutrophil infiltration in various organs following trauma-hemorrhage. *Journal of Leukocyte Biology* 79: 963-970.

Zhang, Q.H., Y.Z. Hu, J. Cao, Y.Q. Zhong, Y.F. Zhao and Q.B. Mei. 2004a. Estrogen influences the differentiation, maturation and function of dendritic cells in rats with experimental autoimmune encephalomyelitis. *Acta Pharmacologia Sinica* 25: 508-513.

Zhang, Y., W. Zhang, L. Zhang, T. Zhu, J. Tian, X. Li and H. Lin. 2004b. Two distinct cytochrome P450 aromatases in the orange-spotted grouper (*Epinephelus coiloides*): cDNA cloning and differential mRNA expression. *Journal of Steroid Biochemistry and Molecular Biology* 92: 39-50.

Zhou, H., T.B. Stuge, N.W. Miller, E. Bengten, J.P. Naftel, J.M. Beranke, V.G. Chinchar, L.W. Clem and M. Wilson. 2001. Heterogeneity in channel catfish CTL with respect to target recognition and cytotoxic mechanisms employed. *Journal of Immunology* 167: 1325-1332.

Zoller, A.L. and G.J. Kersh. 2006. Estrogen induces thymic atrophy by eliminating early thymic progenitors and inhibiting proliferation of beta-selected thymocytes. *Journal of Immunology* 176: 7371-7388.

CHAPTER

10

Immune Response of Fish to Eukaryotic Parasites

Dave Hoole

FISH AS AN ENVIRONMENT FOR EUKARYOTIC PARASITES

Although the origin of the relationship between two organisms may be shaded by the mists of time, the consequences of this relationship have a direct impact, not only on the physiology of each partner, but also—when one partner is of economical importance—the scientific effort placed into understanding this relationship. This is perhaps most applicable when considering the origins of the interactions between parasites and their hosts. It would be somewhat remiss when considering the physiological interactions between a host and its parasite fauna not to consider when and how this relationship first originated. Such a consideration has direct implications not only for the adaptations of the parasite to a parasitic mode of life, but also how the host might adapt its behaviour or physiology, e.g., immune response, to evade or accommodate the symbiotic animal. It

Author's address: Keele University, Keele, Staffordshire, ST5 5BG, UK.
E-mail: d.hoole@biol.keele.ac.uk; d.hool@keele.ac.uk

would be inappropriate to enter the debate to correctly define a parasite. This has been extensively reviewed by several learned authorities (see Brooks and McLennan, 1993, for review). In the context of this chapter, the definition of the parasitic mode of life as described by MacInnis in 1976 will be adopted, viz., 'one partner, the parasite, of a pair of interacting species, is dependent upon a minimum of one gene or its products from the other interacting species, defined as the host, for its survival'. Parasitism has originated independently in several taxa. Indeed, in some taxa, it is thought to have arisen on separate occasions and may be an important component within that taxon. The evolution of prospective hosts has been inevitably associated with the evolution of parasitism. Studies on fossil records (see Conway Morris, 1981 for review) have revealed that the origin and radiation of the metazoan phyla may have occurred in the Vendian Period approximately 680 million years ago. The first jawless fish originated within the Cambrian Period, 580 million years ago, and fossil records suggest that potentially parasitic organisms of fish, e.g., pre-acanthocephalans also occurred during this geological era. Cressey and Patterson (1973) described some well-preserved parasitic copepods in the gill chambers of Cretaceous fish from Brazil. The relationship between parasites and their fish hosts is, in geological time, a long-standing interaction. Observations on the interactions between fish and other aquatic animals have also revealed a trend towards a parasitic way of life in some groups. For example, Williams and Bunkley-Williams (1994) noted that some larval and adult crabs can feed on the gill filaments of fish caught in traps on the sea floor, and Brusca (1981) proposed that cymothoid isopods, which are obligate fish parasites, probably originated from facultative parasitic Cymothoidae.

Since the relationship between fish and their parasites has evolved over several geological eras, it is perhaps not surprising to find that there are few organs within the fish that are not considered suitable for residence by a parasite. The external covering of fish is an ideal location for the attachment and growth of organisms. Several small invertebrates can attach to the surface of larger animals, a phenomenon termed phorsey, which aids in the dispersal of the smaller organism. This may have been a stage in the development of an ectoparasitic mode of life and certainly the intimate contact between the fish and its water environment does mean that the external surfaces provide an easily assessible location for parasites. Although the surface of a fish can be colonized by numerous commensals, these usually do little harm unless they are present in large numbers or if the fish is experiencing stressful conditions. The scales, which cover the surface of the majority of fish, vary considerably, but all of them form an effective physical barrier to the penetration of several potential parasites. An epidermis that contains mucous glands overlies the scales, which are located in the dermis. The mucus produced serves a variety of functions, including protection against infection and an osmoregulatory role. However, the presence of mucus, epidermal cells and blood also provides a ready source of nutrients for parasites. Perhaps the most vulnerable external feature of the fish for infection is the gills. Arising from each gill arch are filaments from which protrude numerous lamellae. Whilst these serve as the major site for gaseous exchange their structure, which comprises a thin epithelial layer, supportive pillar cells, mucous cells and a very efficient and extensive blood supply, means that they are also an easy source of nutrients. In addition, although they are subjected to extensive water currents, parasites can evade the worst of these by attaching at specific locations within the various gill arches. The route of infection of endoparasites is primarily either via penetration of the external surfaces of the fish or through consumption. The latter has resulted in the evolution of parasitic stages that can withstand the adverse conditions that may be encountered in the alimentary tract of the fish host. The adoption of a parasitic strategy has meant that intestinal infections, e.g., cestodes, acanthocephalans utilize an intermediate host which serves as prey for the fish.

IMMUNE RESPONSES TO SURFACE PARASITES

The ciliate, *Ichthyophthirius multifiliis*, is a common pathogenic parasite that has been observed infesting freshwater fish inhabiting tropical, subtropical and temperate regions of the world. This wide host and geographical range has meant that extensive records exist, which indicate that the parasite causes significant mortalities and leads to heavy economic losses in aquaculture and the ornamental trade. Control of the disease, termed ichthyophthiriasis (ich) or white spot, is financially desirable, which has resulted in considerable efforts for finding control strategies. Chemical treatments added to water, whilst giving some control, have several drawbacks in that they may be stressful to fish, have an ecological impact on the environment, can affect human health, and are expensive. One of the main problems associated with the external application of chemicals is that the *I. multifiliis* is, in fact, an internal parasite. This has meant there has been considerable interest in the possibility of producing a vaccine against this parasite, and hence there have been extensive studies on the host/pathogen interactions. The parasite has a life cycle consisting of an infective theront that penetrates the epidermis of the host fish by utilizing a range of enzymes, e.g., hyaluronidase, acid phosphatase and non-specific esterases (Kozel, 1980), possibly released from the mucocysts. The mature trophont exits the host and is released into the water where it transforms into a reproductive cyst, the tomont. After repeated divisions, the infective theronts are released. This direct life cycle means that infections can rapidly accumulate within a fish population with individuals showing characteristic white spots that can reach 1mm in diameter.

The ability of a fish to develop protection against white spot has been known for a considerable time and was probably first noted by Buschkiel in 1910. However, investigations to elucidate the possible mechanisms involved in this protection were not extensively carried out until the 1970s. Protection attributed to acquired immunity was demonstrated in several fish species for example carp (Hines and Spira, 1973a, b; Houghton and Matthews, 1986), channel catfish (Dickerson *et al.*, 1984), rainbow trout (Wahli and Meier, 1985) and tilapia (Subasinghe and Sommerville, 1986). However, the protection invoked appears to be dependent on several factors such as the antigen source. For example, Burkart *et al.* (1990) using intra-peritoneal injections of live theronts and formalin-fixed trophonts, successfully immunized channel catfish but could not repeat the success utilizing a preparation of cilia, formalin fixed theronts, or lysates of theronts. In contrast, earlier studies such as those carried out by Goven *et al.* (1980), revealed that cilia fractions obtained from *I. multifiliis* successfully immunized channel catfish when given intra-peritoneally. Indeed, this has been achieved in a range of fish species using the related ciliate, *Tetrahymena thermophilia* (Goven *et al.*, 1981; Dickerson *et al.*, 1984; Houghton *et al.*, 1992). Perhaps one of the most significant contributions to the studies on the immunological interactions between *I. multifiliis* and its host fish was made by Cross and Matthews in 1992 and relates to how the parasite might respond to any immune response evoked. These researchers noted that theronts entered the epidermis of immune carp at levels comparable to naïve control fish. However, in contrast to the latter, in immunized individuals, approximately 79% of the parasites had exited the fish and entered the external environment usually within 2 hours of penetration. Houghton and Matthews subsequently revealed in 1993 that parasites could survive for up to 4 days in the immune host.

are these more invasive / virulent?

E ict grown under acidic conditions

Fe starved or loaded

w excess tryptophan

grown on fish lysate

grown w sublethal amounts of H_2O_2

Why are Gambusia infected by E ict?

1) infects broadly like tarda just not enough investigation done

2) catfish & Gambusia share in lacking some defense common to other fish

3) cat & Gamb share something exploited by Ict –

long term plasma cells survive in some niche in the kidney

some fish Ig have redox diversity, so same Ig has different disulfide-bonded forms

see how serum drawn from fish, esp Gambusia or goldfish or zebra (as a small fish)
then could look at complement killing & SDS-PAGE of Ig

p166 fish tranferrin - may be cleaved to generate inflammatory and/or antibacterial peptides

Sea bass is a developing model

COX-2 is major player in cell activation in innate immun
good for RT-PCR a target for aspirin

p194 refs on serum total protein, serum Ig, IgM, serum lysozyme, hyperoxygenated water, cortisol le

p197 immersion vaccination discussion

p198 blood taken from 5 g fish by cutting lethal tail, pooling from multiple fish

[could determine total body load over time of rif^R E ict using Elm-r
& grind whole body extract

p247 histone-derived AMPs (antimicrobial peptides) - because cationic?
odd protein to have other functions since so conserved

→ can human or murine histones be processed into AMPs?
p257 one mouse ref

p265 typical freshwater [NaCl] = 140 mM, saltwater = 550 mM

p279 cortisol is a glucocorticoid, steroid hormone

should scan skin slime for free fatty acids,

iron chelation, lysozyme activity, antimicrobial peptides

(1)

- big dif is most fish are not homoeothermic

- how many & what fish genomes done?

IL-1β identified in most fish, also TNF-α

any Gambusia cell lines, like MP?

rIL-1β co-admin w/ Aeromonas or Vibrio killed vaccine enhanced Ab response

rIL-1β protected trout vs Aeromonas after 2d but not after 7d

p10
inflamm screen in trout of IL-1β, TNF-α, & IL-8

p13
IL-12 made by leukocytes in resp to bacteria infection, early - hrs

unclear if std mammal TH1 vs TH2 responses are present in fish

IgM in fish is tetrameric, most prevalent Ab - also monomeric mIgM

p81
no class switching, but does have AID (zebrafish)

receptor editing seen

IgD - coexpressed w/ IgM having same V region

another locus makes IgT (own V&J) and IgZ

affinity maturation more limited (so boosts should have small effect)
10-100 fold not like humans of 10^4

p96
no bone marrow. hematopoesis in kidneys, esp anterior

cortisol is a major immune regulator

Kidney acts like spleen - traps bacteria, particles

fish - may have plasma cells AND plasmablasts ← might be developmental stage towards plasma cells
to make Ig

It is the mechanisms by which this immune protection is bought about and its efficiency which has concentrated the minds of several fish immunologists and parasitologists over the last two decades. For example, several workers (Hines and Spira, 1974b; Subasinghe and Sommerville, 1986; Clark *et al.*, 1988) have established that immobilization of theronts occurs, indeed immobilization and agglutination form the basis of several immunological assays for *I. multifiliis* (Wahli and Meier, 1985, Houghton and Matthews, 1986), although their role in the immune protection invoked in the epidermis has not been clearly established.

It is perhaps not surprising given the size, activity and location of the theront within the dermis, a layer which contains an extensive capillary network, that specific antibodies against *I. multifiliis* have been detected in several fish species (Wahli *et al.*, 1986; Cross and Matthews, 1993a,b). Whilst monoclonal antibodies have been used by Lin and Dickerson (1992) to confer passive immunity against *I. multifiliis*, Dickerson in later studies (Clark *et al.*, 1996) demonstrated that this immobilization monoclonal was effective against theronts only *in vitro*; in the *in vivo* situation the antibody caused rapid exit of the parasite from the fish, but immobilization was not observed. Over recent years, there have been several studies into the role of systemically and locally produced antibodies in the immune protection against *I. multifiliis*. Several authors, e.g., Thorburn and Jansson (1988) and Margarinos *et al.* (1994), have noted that serum antibody levels did not correlate with immune protection evoked by immunization against a range of pathogens. In addition, studies by Houghton and Matthews (1990) have revealed that immune carp, which had been immunosuppressed by injection with corticosteroids, were no longer protected against Ich, although their circulating antibody levels still remained high. These observations, together with previous studies (Lobb and Clem, 1981; Lobb, 1987; Rombout *et al.*, 1993) that have shown that cutaneous antibodies are involved in immune protection in fish, have led to increased interest in the local immune response in the skin of fish infected with *I. multifiliis*. Xu *et al.* (2002) have shown that theronts were immobilized and agglutinated when exposed *in vitro* for 1-24 h to excised skin obtained from immune channel catfish, *Ictalurus punctatus*. Immunofluorescence and Western blot analysis revealed that the extracellular substance in the culture medium was a specific antibody against Ich. Since the skin had been extensively washed, they proposed that the antibody had been synthesized locally. Further studies by Xu and

Klesius (2002) revealed that theronts exposed to culture medium obtained from excised immune skin were smaller in size than those exposed to culture medium obtained from naïve fish skin, and the infectivity of these parasites was significantly reduced. Antibodies at high concentrations or at lower dilutions over longer exposure times were considered to have induced the effects on the parasite. In a further development, Xu and Klesius (2003) suggested that when immune catfish cohabit with non-immune fish, the former may protect the latter from infection with Ich. Emzyme-linked immunosorbent assays revealed that anti-Ich antibodies occurred in the water containing immune fish. A deeper relationship with immunization and antibody levels has also been recently explored. Dalgaard *et al.* (2002) noted that rainbow trout immunized with sonicated trophonts by intra-peritoneal injection had lower infection levels than fish immunized by immersion with sonicated trophonts or naïve controls. However, there was no difference in serum antibody levels between the treatments. In contrast, Xu *et al.* (2004) noted that in channel catfish protection occurred in fish immunized with live theronts by immersion and intra-peritoneal injection, and with sonicated trophonts by i.p. injection. In addition, there was a positive correlation between cutaneous anti-Ich antibodies and host survival in immunized fish.

The antigens associated with the presence of immobilization antibodies, the so-called immobilization antigens (i-antigens), were identified by Lin and Dickerson (1992) as 48 and 60 KDa proteins structurally associated with the cilia membrane. The antigens belong to a GPI-anchored surface glycoproteins (Dickerson *et al.*, 1989; Clark *et al.*, 2001), which are analogous to i-antigens found in the free-living ciliates, *Tetrahymena* and *Paramecium*. There has been much controversy as to whether administration of ciliates such as *Tetrahymena pyriformis* provides protection against Ich (for example, see Clark *et al.* 1988; Burkart *et al.*, 1990; Ling *et al.*, 1993). However, it would appear that, whilst the antigens analogous to Ich i antigens found in *Tetrahymena* and *Paramecium* can undergo change due to the effects of temperature and osmolarity (Clark and Forney, 2003), i antigens in Ich appear to be more stable and do not undergo switching. In 1993, Dickerson and co-workers noted that *I. multifiliis* occurred in different serotypes and that each serotype elicited the production of a unique antibody that only immobilized the serotype that generated it and not a heterologous serotype. This may have wide-ranging implications as recent investigations by Swennes *et al.* (2006) have revealed that different serotypes may have different virulence in channel catfish. In recent studies carried out by Xu *et al.* (2006), it was noted that whilst antisera obtained from immunized channel catfish did not immobilize heterologous theronts, the antisera did react with proteins from sonicated heterologous theronts. The authors suggested that the antiserum is reacting with non-immobilizing antigens. Indeed, Wang *et al.* (2002) utilized two strains of *I. multifiliis* (NY1 and G5) and concluded that active immunity in channel catfish in response to a natural infection with Ich was not serotype-specific. The mechanism by which antibodies induce death of the parasite has recently been elucidated by studies carried out by Xu and co-workers. In 2005, Xu and co-workers noted that cutaneous antibody in channel catfish may induce apoptosis in Ich theronts, and in 2006 these authors also noted that the apoptosis is positively correlated with the expression of Fas, a known receptor for the induction of apoptosis, on the surface of Ich theronts.

This extensive work on the involvement of antibodies in the immune protection of fish against Ich has somewhat distracted attention from the possible role of other components of the immune response. Several histological studies, e.g., Hines and Spira (1974a), Ventura and Paperna (1985), Cross and Matthews (1993a,b) and Cross (1994) have noted that in fish infected with *I. multifiliis*, there was an infiltration and migration of leucocytes into the site of infection. Indeed, as previously stated, a possible cellular involvement in immune protection is indicated, as immune suppression using corticosteroids made immune carp susceptible to infection, even though antibody levels remained high (Houghton and Matthews, 1990). Graves *et al.* (1985) suggested that non-specific cytotoxic cells were involved in the immunity of channel catfish to *I. multifiliis*. Molecular studies carried out by Sigh and co-workers (2004a) have shown that several pro-inflammatory cytokines, i.e., IL-1β, TNF-α, IL-8, IL-1RII are expressed in skin, head kidney and spleen of rainbow trout infected with *I. multifiliis*. Recent studies by Gonzalez *et al.* (2007a) have also revealed a significant upregulation in infected skin of the genes that encode for the chemokine CXCa, chemokine receptor CRXR1 and the pro-inflammatory interleukins IL-1β and TNFα. Buchmann and co-workers speculate that the presence of these cytokines could, in part, explain the histological observations made by previous authors. For example, IL-8, a CXC chemokine, is known to be a potent attractant of mammalian neutrophils, and Cross (1994) noted an influx of neutrophils into the skin of Ich-infected fish. In addition, they proposed that the skin was an important source for pro-inflammatory molecules. Further molecular studies carried out by Sigh *et al.* (2004b) also revealed that there was an induction of C3 transcription levels in the skin of rainbow trout, which supported the authors' previous observations (Buchmann *et al.* 1999) that complement was involved in the immune response against Ich. This has been recently supported by molecular studies carried out by Gonzalez *et al.* (2007b,c) which have revealed that a significant upregulation of the gene encoding for the complement component factor B/C2-A and serum amyloid A in Ich-infected skin. In addition, Buchmann *et al.* (1999) also noted that MHC II and IgM expression increased in the skin and head-kidney of the infected rainbow trout. The observation that inducible nitric oxide expression was only increased briefly in the skin during infection led Buchmann and co-workers to postulate that—as the immune response against Ich appeared to be antibody based—it was mediated through a T_H2 route rather than a T_H1 route, although the authors acknowledge that as yet there is no clear evidence for the presence of a T_H1/T_H2 dichotomy in fish.

The surface of fish can also serve as a suitable medium for metazoan parasites several of which, e.g., monogeneans and crustaceans, have become recognized as important pathogens in fish farming. Monogeneans, which can be either oviparous or viviparous, have a single host—the fish—within their direct life cycle and, as such, numbers can rapidly increase within a fish population. The physical interaction between the parasite and the host surface can be the focus of the immune response. Monogeneans mechanically disrupt the epidermis of their host by the insertion of hooks or hooklets located on an attachment organ, the opisthaptor. In addition, the movement of the parasite utilizing an anterior adhesive organ means that attachment points are the possible sites of immunological reactions. The mechanism of feeding may also produce an immunological response; for example, the monopisthocotyleans are thought to browse on the epithelia of the fish and thus consume mucus and epidermal cells, whilst the polyopisthocotyleans consume host blood and are, therefore, exposed to the immunological factors, e.g., complement, antibodies contained therein. Indeed, Buchmann and Bresciani (1998) suggested that site selection of gyrodactylids on salmonids might be partly governed by the immune response of the skin.

Several studies have revealed that infections of monogeneans may induce immune protection in fish, for example, Molnar (1971) cyprinids against dactylogyrids; Slotved and Buchmann (1993) pseudodactylogyrids on eels, and gyrodactylids in a range of fish species (Scott and Robinson, 1984; Richards and Chubb, 1996; Cable *et al.*, 2000). Several studies have also revealed that the immune response generated is sensitive to treatments with immunosuppressants, e.g., gyrodactylids on salmonids. Recent studies carried out by Harris *et al.* (2000) and Olfsdottir *et al.* (2003) have shown that the administration of corticosteriods such as hydrocortisone and dexamethasone can affect the host specificity of monogeneans. Whilst an immunosuppressive route could mediate this effect, Olafsdottir *et al.* (2003) also speculated that these steroids could increase the turnover of fish epithelial cells which provide the gyrodactylid with an accessible food source thus increasing the parasites' productivity. This aside, several studies have been carried out to elucidate the role of both humoral and cellular components in the immune response of fish to monogenetic flukes. Several investigations (Harris *et al.*, 1998; Rubio-Godoy *et al.*, 2004) have revealed that complement may be involved in the killing of mongeneans, e.g., *Gyrodactylus salaris* and *Discocotyle sagittata*. Furthermore, *in vitro* investigations by Buchmann and Bresciani (1999) have shown that complement-containing serum from rainbow trout produced damage to the worm tegument. Such studies supported the observations made by Buchmann in 1998, where factor C3 was bound to carbohydrate-rich epitopes of *Gyrodactylus derjavini*. It is interesting to note that Buchmann (2001) also observed that carbohydrate epitopes, possibly containing mannose, galactose and lactose, in the glycocalyx of this parasite, reacted with lectins in rainbow trout. These *in vitro* investigations, which indicate the possible role of complement in the immune response to gyrodactylids, were substantiated *in vivo* by Bakke *et al.* (2000), who noted an increase in complement factors in Atlantic salmon infected with these parasites. The possible role of specific humoral factors, i.e., antibodies, in the immune response against monogeneans has generated much controversy. Several investigations have found no association between infection and the generation of antibodies. For example, Thoney and Burreson (1988) did not observe any binding of immunoglobulin from *Leiostomus xanthurus* to homogenates of *Heteraxinoides xanthophilis*, Bondad-Reantaso *et al.* (1995) failed to detect an increase in antibody levels in flounder infected with *Neobenedenia girellae,* and more recently, Buchmann (1998) could not find specific antibodies in rainbow trout infected with *G. derjavini*. In contrast, Wang *et al.* (1997) noted that antibody from tiger puffer reacted with *Heterobothrium okamotoi*. Recent studies have also confirmed an association between infection with monogeneans and the production of an antibody response in fish. Rubio-Godoy *et al.* (2003) noted that antibody titres were significantly higher in rainbow trout infected with *Discocotyle sagittata* rather than in naïve fish, although there was no correlation between the intensity of infection and antibody titres. In addition, in an interesting turn of events, Hatanaka *et al.* (2005) have utilized the knowledge acquired regarding the ciliary antigens of *I. multifiliis* and have revealed that a surface glycoprotein on *Neobenedenia girellae* acts as an agglutination/immobilization antigen which is capable of inducing an immune response in *Paralichthys olivaceus*. This 8 kDa glycoprotein was thought to be an intergral component of the membrane fraction of the cilia of the oncomiracidia.

Whilst there have been extensive studies on the humoral immune response to mongenean infections in fish there have been, in comparison, few detailed studies on the role of the cellular component. Several studies have revealed that infection with monogeneans is associated with cellular infiltration and hyperplasia at the infection site (see Buchmann and Lindenstrom, 2002 for review), and that epidermal mucous cells may play a role in the protective response, e.g., Wells and Cone (1990), Buchmann and Bresciani (1998). Indeed, further detailed studies by Buchmann and co-workers (Buchmann and Bresciani, 1999; Sigh and Buchmann, 2000) revealed that putative mast cells and macrophages may be involved in the immune response of fish to surface parasites such as gyrodactylids.

The recent application of molecular biological techniques has enabled a more detailed understanding of the local immune response and the role of various cytokines to be elucidated. Lindenstrom *et al.* (2003) noted that in the skin of rainbow trout infected with G. *derjavini,* there was an increase in the expression of IL-1β1 and IL-1β2 isoforms, and speculated that the increase in expression of IL-1R1II during the course of the primary infection may have been involved with the downregulation of IL-1. These studies were continued and Lindenstrom *et al.* (2004) revealed that in the same host-parasite system there was an increase expression of TNF-α1 and, to a lesser extent, TNF-α2, TGF-β, iNOS and COX-2, whilst there were parasite related changes in the expression of IL-8 and the two cell markers, TCRβ and MHC IIβ. The authors then extended the hypothesis initially proposed by Buchmann (1999) that infections of gyrodactylids lead to an increase in the local production of IL-1β that induces mucus secretion, activates macrophages and increases the expression of a number of NF-κB-dependent genes. Activation and degranulation of mast cells, perhaps associated with TNF-α, may further increase the inflammatory process, leading to the generation of reactive oxygen and nitrogen species and the secretion of complement. Indeed, Lindenstrom *et al.* (2006) proposed that an increase in IL-1β expression might lead to an increase in mucus production that would increase the susceptibility of a fish to infection, as the mucus may improve parasite propagation. In addition, Chaves *et al.* (2006) noted that in pompano (*Trachinotus marginatus*) infected with *Bicotylophora trachinoti* that there was an increase in the number of leucocytes and phagocytosis in splenic and pronephric cells, although the latter was depressed later in the infection. The authors suggested that this monogenean species induces a biphasic response with an initial stimulation of the non-specific defence mechanisms followed by a depression.

In contrast to the number of research studies on external protozoan and small metazoan parasites of fish, few investigations have been carried out on the immunological interactions between the large metazoans, e.g., crustaceans, leeches, lampreys that infest the surface of fish. Perhaps the exception is the parasitic copepods that, due to their economic importance in fish cultured in marine and brackish waters, have been the focus of attention over several years. The most important group of parasitic copepods are commonly referred to as sea lice, and includes two major genera, *Lepeophtheirus* and *Caligus*. Of all the sea lice species, *Lepeophtheirus salmonis*, the salmon louse, has been studied in most detail, although several members of the *Caligus* i.e., *Caligus elongate*, *C. epidemicus*, *C. orientalis*, *C. punctatus* and *C. rogercressey*, because of their disease potential in aquaculture, have also been investigated. Sea lice, like the monogeneans, have a direct life cycle, which means that under favourable conditions, numbers occurring on a fish can increase rapidly. Like their free-living relatives, sea lice have a typical complex copepod life cycle. This comprises two free-living planktonic naupliar stages and a free-living infectious copepodid stage. Parasitic stages, which can be found attached to the fish, include four to six chalimus stages and one or two pre-adult stages prior to the adult stage.

The parasitic stages of sea lice feed on a range of host nutrient sources including mucus, skin and blood. These feeding activities, together with those associated with attachment, can cause the development of lesions which, in heavy infections, cause disease, and on some occasions fish death. The form and intensity of the pathological responses induced varies, depending upon the parasitic stage and species, and also the host species infected. These responses have been extensively reviewed by Johnson and Fast (2004). Mild inflammatory responses have been noted in Atlantic salmon infested with *L. salmonis*. In such instances, small haemorrhages and epidermal hyperplasia occur; however, where the epidermis is breached and the underlying dermis is exposed, inflammatory infiltrates consisting of neutrophils and lymphocytes have been observed. In contrast, infections on coho salmon can induce acute inflammation comprising intense epidermal hyperplasia and infiltration of neutrophils, macrophages and lymphocytes. Indeed, in some instances, partial or complete encapsulation of the copepod can also occur. It would appear that these intense pathological responses induced by sea lice may be limited to a number of host species such as coho salmon or restricted to some host/sea lice associations, e.g., *Lepeophtheirus pectoralis* on flounder, *Platichthys flesus*. Mild inflammatory responses to the attachment of sea lice have also been noted in *Caligus* sp. infections. The lack of an intense host response may indicate that the sea lice suppress the immune response of the host. However, even low numbers of the pre-adult stage of *L. salmonis* can induce subtle changes in the host, for example, increased necrosis and/or apoptosis in epithelial and chloride cells, and loss of epithelial cells (Nolan *et al.*, 1999).

In contrast to the wealth of information on the immune response of fish to *I. multifiliis*, studies on the immune response generated by the naturally or experimentally acquired infections of crustacean parasites are limited. Johnson and Fast (2004) have suggested that investigations have been restricted by the absence of infection protocols that mimic acquisition of infections in the wild. In addition, studies have been focussed on the drive to produce an effective vaccine; hence, antigens are usually applied in association with vaccine trials. This aside, several studies have investigated the importance of acquired immunity against crustacean infections and the possible role of antibodies within the immune response. Investigations have revealed that parasitic crustaceans may induce an acquired immune response. For example, Johnson (1993) suggested that the increase in the production of eggs of *L. salmonis* in mature coho salmon was possibly related to a reduction in immune function, and Woo and Shariff (1990) also noted that the reproductive output of *Lernaea cyprinacea* on the kissing gourami, *Helostoma temmincki*, was reduced on the infected fish. In contrast, other studies have revealed that acquired immunity is not evoked in other crustacean/host systems, for example, *Lernaea minuta* infections on Javanese carp, *Puntius gonionotus*. The role of antibodies in the immune response of fish to crustacean infections is also controversial. Whilst Thoney and Burreson (1988) failed to detect antibodies in the spot chocker, *Leiostomus xanthurus*, infected with the tissue-dwelling copepod, *Lernaeenicus radiatus*, Grayson *et al.* (1991) did observe specific antibodies in Atlantic salmon naturally infected with *L. salmonis*. Interestingly, specific antibodies were not detected in naturally infected rainbow trout. It is not known whether antibodies are protective. In contrast, several studies have demonstrated a humoral antibody response in fish immunized with extracts of crustacean parasites, for example rainbow trout immunized with extracts of *Argulus foliaceus* by Ruane *et al.* (1995) and Atlantic salmon exposed to extracts of *C. elongates* and *L. salmonis* by Reilly and Mulcahy (1993).

The inflammation associated with crustacean infections on fish has recently been studied utilizing molecular biological techniques. Fast *et al.* (2006) noted an increase expression of several immune-related genes i.e. IL-1β, TNFα-like cytokine, MH II, TGFβ-like in the head-kidney of Atlantic salmon infected with *L. salmonis* and Walker *et al.* (2004) reported work carried out by Haond and Wiegertjes that revealed that there was an increased expression of the cytokines, IL-1 and TNFα, in the skin of carp infected with *Argulus japonicus*. Walker *et al.* (2004) also suggested that these pro-inflammatory cytokines were associated with the leucocyte response to crustacean infections, indeed Haond and Wiegertjes also noted an increase in granulocytes and monocytes in the blood of infected fish. However, the role of the leucocyte component of the immune response against parasitic crustaceans has to be questioned. Mustafa *et al.* (2000) noted that there was no increase in the respiratory burst and phagocytic activity of macrophages isolated from the head-kidney of Atlantic salmon infected with *L. salmonis*.

IMMUNE RESPONSES TO INTERNAL PARASITES

Once a parasite has entered the fish's body, it is open to the full force of the immune response of the host plus the challenge of adapting to a greater range of physiological parameters.

Migration to the final organ of infection is mediated through a variety of routes, for example, the blood system. This might be considered as a particularly immunologically hostile environment and yet there are several protozoan and metazoan parasites that reside permanently within the blood. Kinetoplastid flagellates are extracellular parasites that live in the vascular system of a wide variety of fishes. In European cyprinids, two species predominate, *Trypanosoma carassii* (syn. *T. danilewskyi*) and *Trypanoplasma borreli*, which are transmitted by leeches. An initial rise in parasitaemia followed by a decline in parasite numbers would indicate the presence of a partially successful immune response. Several studies have revealed that prominent pathological effects associated with *Typanoplasma* infections include occlusion of capillaries, local haemorrhages and infiltration of inflammatory cells to sites of infection. Investigations carried out by Bunnajirakul *et al.* (2000) have revealed that blood vessels associated with the kidney, liver, heart and intestine are the prime sites for infection and cellular infiltration. Similar reactions have also been noted by Dykova and Lom (1979) and Lom *et al.* (1986) in goldfish infected with *T. borreli*, and salmonids infected with *Cryptobia salmositica* (Woo and Poynton, 1995). The effects of *T. borreli* are, however, not limited to the vascular system, as ultrastructural studies by Rudat *et al.* (2000) have revealed that subcellular changes occur within the renal tubule epithelial cells. Parasitaemia patterns associated with kinetoplastid infections in fish, together with observations that fish control acute infection of *T. carassii* and *T. borreli* and may be resistant to infections of these parasites obtained from chronic infections (e.g., Overath *et al.*, 1999), have led several authorities to suggest that parasite-specific antibodies may be important in controlling infections. Indeed, these antibodies have been detected by, for example, Lom and Dykova (1992), Jones *et al.* (1993), Wiegertjes *et al.* (1995, 1996) and Chin *et al.* (2004) in infected fish and are capable of transferring resistance to naïve cyprinids (Woo, 1981; Wiegertjes *et al.*, 1995; Overath *et al.*, 1999). Recent studies carried out by Bienek *et al.* (2002) have demonstrated that immunization of goldfish with excretory/secretory (ES) products of *T. danilewskyi* increased resistance to challenge infections, particularly when administered in conjunction with Freund's complete adjuvant. Further studies by Bienek *et al.* (2006) identified that α- and β-tubulin were major components of the ES products that induced the protective immune response in goldfish. Tubulin molecules are cytoskeletal proteins, which polymerise to form microtubules, and it is possible that antibodies in the infected fish inhibit new microtubule formation and destabilize their structure thus affecting parasite structure, motility and division. Indeed, exposure of *T. danilewskyi in vitro* to anti-recombinant α-tubulin IgG generated in rabbits caused a dramatic decline in parasite numbers. The relationship between kinetoplastid infections and the antibody response may not however be simplistic, as Wiegertjes *et al.* (1995) proposed that the genetic status of the carp host affects the response induced. Some strains of carp were found to be highly susceptible to infection with *T. borreli* that resulted in 100% mortality. Parasite-specific antibodies were not detected in these carp. The reason for this lack of antibody response is unclear. Wiegertjes *et al.* (1995) suggested that a general B-cell anergy occurred in these fish. In recent investigations carried out by Scharsack *et al.* (2004), however, it was noted that peripheral blood leucocytes from both susceptible and resistant carp responded to mitogenic stimuli, although the lymphoid cells from these two groups of fish differed in the spectrum of immunomodulatory mediators produced. These authors suggested the differences in susceptibility of carp to *T. borreli* might be based on a strong polyclonal lymphocyte activation, which is not downregulated by inhibitory signals. Where antibodies are produced, they may activate the complement system, as has been observed in another haemoflagellate, *Cryptobia salmositica*, in *Oncorhynchus mykiss* and *Salvelinus fontinalis* (Li and Woo, 1995; Ardelli and Woo, 1997; Feng and Woo, 1997). These latter authors showed that parasites were also lysed by activation of complement via the alternative pathway, which has been noted to occur against *T. danilewskyi* by Plouffe and Belosevic (2004). Studies carried out by Saeij *et al.* (2003a, b) have recently confirmed that *T. borreli* is lysed by antibody mediated complement activity. These workers showed that a heat-liable fraction of *T. borreli* and CpG motifs in the parasite DNA induced nitric oxide production and increased expression of TNFα, IL-1β, C3, serum amyloid A and α2-macroglobulin. They concluded that *T. borreli* is highly immunogenic without being very antigenic. Scharsack and co-workers have attempted to characterize the role of nitric oxide in the immune response of carp to *T. borreli*. *In vitro* and *in vivo* studies (Scharsack *et al.*, 2003a) revealed that whilst the parasite was capable of inducing high levels of nitric oxide this was not apparently harmful to the parasite. Indeed, in further investigations by Scharsack *et al.* (2003b), it was shown that nitric oxide and reactive oxygen species production occurred in parasite stimulated head-kidney leucocytes. *T. borreli* was, however, not sensitive to activated neutrophils. It was suggested that the parasite might interfere with the immunoregulatory signals produced by this leucocyte type. The role of macrophages in the immune response to haemoflagellates in fish is also problematic. For whilst Sypek and Burreson (1983) and Li and Woo (1995) noted that fish macrophages were associated with phagocytosis of blood flagellates, studies by Scharsack *et al.* (2003b) using flow cytometry found no evidence of phagocytosis of *T. borreli*, which was further substantiated by Saeij *et al.* (2003a), who could not detect phagocytosis of the parasite in the spleen, liver and pronephros of heavily infected carp. In addition, the application of a macrophage depletion model confirmed that peritoneal macrophages had a minor effect on the resistance of carp to haemoflagellates. In an interesting development on the interaction between blood flagellates and carp macrophages, Wiegertjes and co-workers (Joerink *et al.*, 2004; Wiegertjes and Joerink, 2004; Wiegertjes *et al.*, 2005) proposed that *T. borreli* may be a useful parasite model to elucidate the functional activities of Th1 and Th2 cells in fish whose responses may be mediated through different cytokine spectrum and the associated inducible nitric oxide synthase and arginase macrophage activity, respectively.

In addition to protozoan infections of the fish's vascular system, this organ also plays host to a limited number of metazoan parasites. However, in contrast to mammalian hosts where these parasites have been extensively studied due to their pathological effects on humans, for example in the case of schistosomiasis, very little is known about their 'fishy' equivalents, i.e., the Sanguinicolidae. This family of parasitic blood flukes comprises at least 60 species that inhabit the blood system of a range of freshwater and marine fish. The most economically important genus is the *Sanguinicola*, of which *S. inermis*—which infects carp—has been the subject of extensive research. This parasite species has a typical digenean life cycle comprising a snail intermediate host and a fish definitive host, which is infected by penetration of the cercarial stage. The adult flukes usually reside in the heart, bulbus arteriosus, ventral aorta and branchial vessels, while shed eggs can be entrapped in several organs, particularly the liver, kidney and gills. Miracidia are released into the water body from eggs. It is, therefore, perhaps not surprising that extensive pathology has been associated with entrapped eggs, escaping miracidia and pentrating cercarial stages (Kirk and Lewis, 1994; Richards *et al.*, 1994a). Infection has also been associated with an alteration in the cellular composition of the immune organs, i.e., spleen and pronephros of carp (Richards *et al.*, 1994b), although these effects may also be affected by the water quality (Schuwerack *et al.*, 2001, 2003; Hoole *et al*, 2003). Indeed, the level and nature of the immune response generated is determined by a range of parameters including duration of infection (Richards *et al.*, 1994a,b) and temperature (Richards *et al.*, 1996a). *In vitro* studies by Richards *et al.* (1996a, b) have shown that cercarial and adult stages of *S. inermis* induce polarization of carp leucocytes and proliferation of lymphocytes. Recent studies carried out by Roberts *et al.* (2005) have also indicated that the immune response generated may be dependent on the parasite stage and route of exposure. Cercariae increased complement activity in experimentally infected and injected carp, although serum antibody levels were increased only in the injected fish. Exposure to infection did not appear to stimulate antibody production which may be associated with the apparent increase in apoptotic levels in pronephric cells in infected fish (Hoole *et al.*, 2003). The role of serum components within the immune response, at least to some stages of the *S. inermis*, was also questioned by Richards *et al.* (1996c). *In vitro* studies revealed that the presence of carp serum did not increase the number of pronephric leucocytes and the damage they induced in the cercarial stage of this parasite. Interestingly, these leucocytes only attached to adult flukes that had died in culture and a few cells were found attached to live parasites. This is perhaps not surprising, as the mammalian equivalent of this blood fluke, i.e., the schistosomes has devised a range of strategies for evading the immune response in the immunologically hostile blood environment. However, it would appear that an effective immune response is generated in carp against *S. inermis* as the recovery of adult flukes in a challenge infection given 8 months after the primary was significantly less than in the initial infection. Based on the extensive pathological response to the penetrating cercarial stage and entrapped eggs, it is tempting to speculate that these stages may be of prime importance in stimulating the immune response. This too is perhaps not surprising since several studies have suggested that an immune response may be generated by intracellular and extracellular parasites of the musculature or other tissues and cavities of the fish.

Parasites that have a histiozoic and/or coelozoic habitat include larval and adult stages of cestodes, digenetic trematodes and nematodes as well as Microsporidia and Myxosporea. However, extensive studies on their association with the immune response have been restricted to a few parasitic species that have an economical or ecological impact. The histiozoic niche can be the permanent site of residence of several parasitic species or may be a site of transient parasite migration. It would, however, appear that both permanent and transient association primarily results in a similar response that comprises a cellular accumulation at the site of infection or along the migratory pathway. Numerous histological examinations have shown that a parasite is surrounded by a cellular infiltrate comprising, for example, monocytes/macrophages, lymphocytes, mast cell equivalents (eosinophilic granular cells), neutrophils and non-specific cytotoxic cells (NCC). Several studies have attempted to elucidate the immunological components and mechanisms involved in this cell accumulation, and in particular, the recruitment of leucocytes to the site of infection, the attachment of leucocytes to the pathogen, and the mechanisms of killing. Perhaps the most extensively studied migratory metazoan in fish is the diplostomule stage of the digenetic trematode, *Diplostomum spathaceum*. This post-cercarial stage of the parasite penetrates the surface of its fish host and transforms into the diplostomule and is, thus, exposed to the immune system of the fish for a window of approximately 24 hours, during which time it migrates to the relatively immunologically privileged site in the lens of the eye. On a primary challenge, 60-85% of the penetrating cercariae fail to establish in the eye (Whyte *et al.*, 1991). Several *in vitro* studies have been carried out to elucidate the role of the immune system in this parasite loss. Whyte *et al.* (1988) noted that diplostomules incubated in the presence of normal trout sera or trout anti-diplostomule sera had an increased mortality rate compared to parasites cultured with foetal calf serum, and even though the cytotoxic effect was reduced when the serum was heat inactivated, it was speculated that complement activated by the alternate pathway may have a role against the migrating parasite. Indeed, Wood and Matthews (1987) also suggested that complement might be active against the cercariae of *Cryptocotyle lingua* in the mullet, *Chelon labrosus*. Several researchers (e.g., Whyte *et al.*, 1990) have noted that the immunization of rainbow trout with temperature killed or sonicated cercariae, conferred protection against diplostomid infection, and specific antibody occurred 6 weeks post-immunization. Further studies by Whyte *et al.* in 1989 revealed that cercariae, diplostomule and metacercarial stages of *D. spathaceum* may share common surface antigens, although serum antibody titre is not positively correlated with the number of metacercariae in the eye, which may suggest a possible cell mediated involvement in immune protection. Indeed, *in vitro* studies by these authors showed that activated macrophages had enhanced killing activity against diplostomules, particularly when the parasite was opsonized with specific trout antibody. The parasite elicited a respiratory burst in these leucocytes, thus, increasing superoxide anion and hydrogen peroxide production. However, normal macrophages produced less hydrogen peroxide than was required to kill the parasite *in vitro* (Chappell *et al.*, 1994). Recent studies carried out by Kalbe and Kurtz (2006) have revealed that genetic status and gender of the fish host may affect the immune response to *Diplostomum*. Pronephric cells from stickleback populations obtained from a lake situation infected by *D. pseudospathaceum* showed significantly higher respiratory burst activity than fish from a river location, or indeed their hybrids. In addition, females showed a substantially greater activity in this immune parameter compared to male hosts.

This pattern of immune response is not limited to the migration of parasites, and similar responses have also been noted to both protozoan and metazoan parasites resident in organs or tissues. For example, Leiro *et al.* (2004) noted that the ciliate *Philasterides dicentrarchi* did not affect the phagocytic activity of leucocytes isolated from the peritoneal cavity of turbot, *Scophthalmus maximus*, although serum from naturally infected fish increased phagocytic activity, possibly by activation of the complement system. Over 16 years ago, Sharp and co-workers noted that metacestodes of the tapeworm, *Diphyllobothrium dendriticum*, were encapsulated on the peritoneal surface of the stomach and pyloric caeca or within the musculature of the stomach of rainbow trout, and that parasite produced substances that were chemoattractive to trout leucocytes (Sharp *et al.*, 1989, 1991a, 1992). Attachment of these cells to the parasite was antibody- and complement-mediated, and the macrophages produced both hydrogen peroxide and superoxide anions when exposed to worm-conditioned medium or extracts of the tapeworm (Sharp *et al.*, 1991b). Such responses are not limited to metazoan infection, as recent studies on the immunological interactions between the *Tetracapsula bryosalmonae* and its fish host have revealed. This myxozoan parasite formerly designated PKX, is the causative organism of the immunopathological condition 'proliferative kidney disease (PKD)' of salmonid fish, where it induces a hyperplasic response in the lymphoid tissue (Morris *et al.*, 2005). The presence of a humoral immune response to PKD is well established (e.g., Hedrick *et al.*, 1993; Saulnier and de Kinkelin, 1996). However, the role of the cellular component of the immune response is debated. Recent studies carried out by Chilmonczyk and co-workers (Chilmonczyk and Monge, 1999; Chilmonczyk *et al.*, 2002) have shown that lymphocytes constitute a major component of the leucocyte response to the parasite and undergo extensive cell proliferation. In contrast, the respiratory burst and phagocytosis were depressed in infected rainbow trout. This is somewhat at odds with the ultrastructural observations made by MacConnell *et al.* (1989), who suggested that macrophages were the main cell in the kidney response.

In contrast to the numerous parasite species that have evolved a histiozioc habitat, there are relatively few that are truly coelozoic. Perhaps the most significant, at least in terms of size, and thus perhaps immunological insult, is the metacestode stage of the pseudophyllidean tapeworms that reside in the body cavity of several fish species. The two cestodes that have received the greatest attention are *Ligula intestinalis* that infects a range of cyprinid fish, and *Schistocephalus solidus*, which primarily infests the stickleback, *Gasterosteus aculeatus*. Infections of *L.intestinalis* are quite impressive as the weight of parasite tissue can, in some instances, exceed the weight of the host. In this respect, it is quite an unusual host/parasite system. The parasite elicits an intense cellular response, which in roach, *Rutilus rutilus*, comprises macrophages, neutrophils, monocytes, vacuolated granulocytes and lymphocytes (Hoole and Arme, 1982, 1983a,b). This leucocyte involvement in the immune response is reflected by quantitative and qualitative changes that occur in the leucocyte component of the pronephros and spleen of infected fish (Taylor and Hoole, 1989). Further studies by these workers have also revealed that *Ligula* induces migration of pronephric leucocytes (Taylor and Hoole, 1993), proliferation of lymphocytes (Taylor and Hoole, 1994) and killing by cyprinid leucoytes as monitored using chemiluminescence (Taylor and Hoole, 1995). Indeed, ultrastructural studies by Hoole and Arme (1986) revealed that the microthrix border of *Ligula* was phagocytosed by macrophages and neutrophils. In addition, these researchers also speculated that complement (Hoole and Arme, 1986) and possibly C-reactive protein (Hoole and Arme, 1988) are involved in the immune response of roach to *Ligula*. Further studies carried out by Williams and Hoole (1992, 1995) revealed that antibody producing cells in the spleen and pronephros increased in number in fish exposed to parasite antigens, and that antibody occurred on the surface of the metacestode, suggesting that an antibody-mediated cell cytotoxic response may occur to these large parasites.

Perhaps the ancestral coelozoic site for parasitic infection could be considered as the alimentary canal as it as been proposed that the relationship between the parasite and its host may—in evolutionary terms—have been mediated through the accidental acquisition of organisms via the oral route (Wakelin, 1996). This has led to the presence of numerous commensal species being located within the gut of many vertebrates, perhaps highlighted by the ruminants. Surprisingly, the alimentary tract, in relative terms compared to the other organs that have immune capabilities, has only recently been extensively investigated. Several reviews (e.g., Zapata *et al.*, 1996; Press and Evensen, 1999) have shown that the gut of fish has an immune role and contains gut-associated leucocytes, e.g., macrophages, lymphocytes, granulocytes and plasma cells, and humoral components, e.g., Ig. This immune function appears to be present from an early age, for example, in recent studies carried out by Huttenhuis *et al.* (2006), it was shown that putative T lymphocytes appeared in the intestine of *Cyprinus carpio* 3 days post-fertilization, whilst monocytes/macrophages occurred in the laminar propria of the intestine 7 days post-fertilization. Whilst those intestinal infections which induce severe local and/or systemic pathology have received some attention, it is somewhat surprising that endemic parasites within the gut of fish have not been more extensively studied, given the recent interest in the use of orally delivered probiotics and vaccines (e.g., Irie *et al.*, 2003; Kim and Austin, 2006) and the possibility that intestinal parasites may affect their efficacy. It has been known for many years that parasitic infection can induce a range of pathological responses within the intestinal wall, although the form and intensity of these responses may vary, dependent upon the intensity of the relationship between the parasite and host tissue. Studies on intestinal protozoa *Goussia carpelli* by Hemmer *et al.* (1998) have revealed that during the merogonic and gamogonic development of the parasite, there is severe damage to the epithelium of the infected gut of its carp host. However, the regenerative capacity of this organ may explain the mild clinical symptoms associated with this disease. In contrast, in other infections such as the myxosporeans, e.g., *Myxidium fugu* and *Leptotheca fugu* which infect the gut of the tiger puffer, *Takifugu rubripes*, the epithelium becomes detached from the lamina propria and cell debris containing macrophages accumulates between the two layers (Tun *et al.*, 2002). In addition, there is infiltration of macrophages and lymphocytes into the infected sites and, on some occasions, apparently dead parasites have been noted. These pathological responses do not only appear to be restricted to protozoan infections, as there are numerous studies that have shown that metazoan infections can also induce significant changes within the gut wall. Although, as with protozoans, the response invoked may be related to the intensity of infection and the association between the parasite and its host. For example, several studies have revealed that cestodes, such as *Bothriocephalus* species and *Khawia sinensis,* invoke mucus production, haemorrhages, cellular necrosis, epithelial cell detachment and an increase in leucocytes in the mucosa and submucosa not only at the site of attachment of the scolex to the gut wall (Fig. 10.1), but also in the surrounding area (Scott and Grizzle, 1979; Hoole and Nisan, 1994; Morley and Hoole, 1995). Indeed, Hoole and co-workers noted that these leucocytes were also capable of attaching to the parasite tegument. In contrast, Molnar *et al.* (2003) observed that in carp infected with *Atractolytocestus huronsis*, while nuclei in infected areas exhibited karyorrhexis, inflammatory changes did not occur. In contrast to the majority of tapeworms, the acanthocephalans, the so-called thorny headed worms, have a particularly invasive mechanism of attachment where the attachment organ penetrates deeply through the host gut wall and, in doing so, induces extensive damage (Taraschewski, 1989; Dezfuli, 1991; Dezfuli *et al.*, 1998, 2002). Indeed, recent studies on one species of pathogenic acanthocephalan, *Pomphorhynchus laevis*, in the intestine of *Leuciscus cephalus* have revealed that the form of damage also relates to the

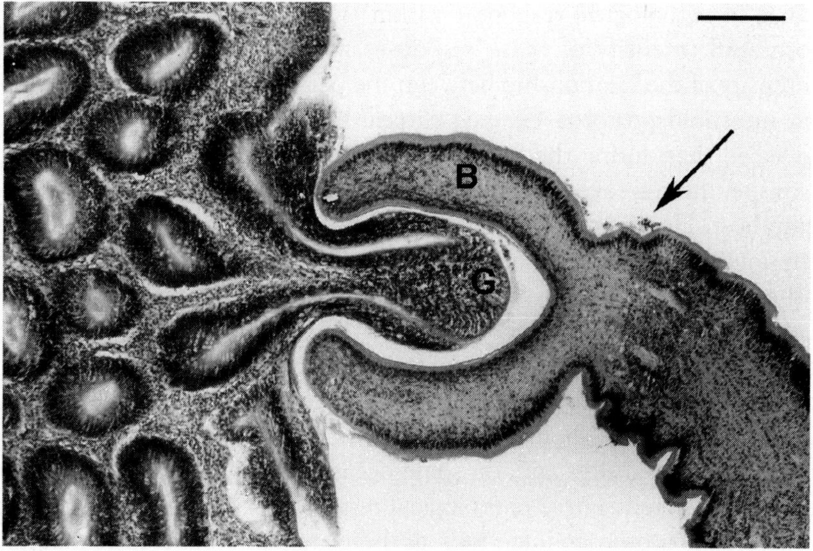

Fig. 10.1 *Bothriocephalus acheilognathi* (B) attached to the intestinal wall (G) of carp. Note presence of host leucocytes on parasite surface (arrow). Scale bar 100 µm. (Copyrighted image reproduced by kind permission of C. Williams, Environment Agency, UK.)

area and depth of penetration of the parasites' attachment bulb and proboscis. Intestinal infections also affect the physiological status of the alimentary tract and there are now several well-documented cases of both tapeworms and acanthocephalans affecting the host-gut neuroendocrine system. For example, infections of trout with the tapeworms *Eubothrium crassum* and *Cyathocephalus truncatus*, and the acanthocephalan *Pomphorhynchus laevis* induce a decrease in the number of bombesin-, gastrin-releasing peptide and glucagon-like endocrine cells, and an increase in the density of cells containing cholecystokinin-8- and gastrin-like substances (Dezfuli *et al.*, 2003; Bosi *et al.*, 2005). Given the recent interest in the interactions with the endocrine and immune systems this is an area that requires further elucidation.

The effects of enteric parasites are, however, not restricted to local inflammatory response, and several studies have revealed that systemic changes are also induced. These changes are reflected in both the innate and acquired cellular and humoral components of the immune response and have been ascertained using a variety of techniques. For example, Steinhagen *et al.* (1997) noted that there was an increase in peripheral granulocyte numbers on days 11 and 14 post-exposure of carp to *Goussia carpelli.* Steinhagen also noted in 1997b that this parasite increased the number of immunoglobulin-positive cells in the pronephros and intestine of carp, and that the peak of the response correlated to the stage of oocyst formation by the parasite, and was not affected by low temperatures. He also showed that the parasite induced an enhanced phagocytic activity in pronephric cells during the period of merogonic and gamogonic development of the parasite, and that in carp with gamogonic and sporogonic stages, there was an increase in the cytotoxic activity of activated macrophages. Parasite molecules also enhanced the opsoninization of non-activated macrophages (Steinhagen and Hespe, 1997). The possible role of a cellular component in the killing of intestinal protozoans has also been suggested by studies carried out on *Enteromyxum leei* infections of gilthead sea bream (*Sparus aurata*) where an increase in serum peroxidase levels occurred after 10 days' infection (Cuesta *et al.*, 2006). These effects on the systemic responses are not only limited to protozoan infections, although it is true that, in contrast, there have been fewer studies on the immunological interactions between enteric metazoans and their fish hosts. Studies by Nie and Hoole (1996, 2000) did, however, reveal that in *in vitro* assays containing pronephric cells from carp, antigens from *Bothriocephalus acheilognathi* induced proliferation of lymphocytes and stimulated migration primarily of neutrophils and eosinophils. In addition, in carp injected with *Bothriocephalus* extracts, there was an increase in both antibody producing cells in the pronephros and specific antibody serum levels (Nie and Hoole, 1999). However, this increase was not as great in naturally infected fish. Recent studies on the host-parasite interactions between fish and enteric parasites have concentrated on the involvement of cytokines. Investigations carried out by Cuesta *et al.* (2006) revealed an upregulation in expression of the pro-inflammatory cytokine, IL-1β, in the pronephros of gilthead sea bream exposed to *Enteromyxum leei*, in contrast there was no clear difference in the expression of TNFα in infected compared to uninfected fish. Clearly, more research is required to ascertain the involvement of cytokines in both the systemic and local immunity to intestinal parasites in fish.

PARASITES AS STRESSORS: EFFECTS ON HOST IMMUNITY

Details considered so far have been related to the direct effects of parasitic infection on the immune status of the fish host. It should not be forgotten, however, that eukaryotic parasites, either by their population size as in the case of protozoans, individual size, for example, cestodes and crustaceans, or the pathology invoked, can be considered as stressors, and as such can affect the immune status of the host, mediated through the stress response. As in higher vertebrates, the stress response in fish involves hormonal, physiological and behavioural changes. The hormonal changes, which are indicative of the stress response, include the production of catecholamines and corticosteroids (Wendelaar Bonga, 1997), and whilst the details of this response are beyond the remit of this chapter, it is worth considering how parasites might affect the stress hormonal balance within the host and thus the immune status. It is somewhat surprising that the relationship between the parasite, host stress response and immune status has not been extensively studied, since there may be implications relating to the control of infection in general and the efficacy of vaccination protocols. There have, however, been several studies in a range of vertebrates which have linked the presence of parasitic infection with an increase in levels of corticosteroids, for example, the nematode, *Ostertagia ostertagi* in the cow, *Bos taurus* (Fleming, 1998). Recently, however, the relationship between several eukaryotic parasites and the stress hormonal status of their fish hosts has come under the spotlight. Whilst there are reports that parasitic infections do not affect the plasma cortisol levels of their hosts, for example, gnathiid isopods in the coral reef fish, *Hemigymnus melapterus* (Grutter and Pankhurst, 2000), these are outweighed by the presence of several investigations which have indicated a link between parasitic infection and an increase in cortisol levels. Studies on several external parasites have revealed that the parasite need not penetrate the host body in order to induce a stress response. For example, Stoltze and Buchmann (2001) noted an increase in cortisol levels in rainbow trout infected with *Gyrodactylus derjavini*, and Poole *et al.* (2000) observed a similar response in sea trout (*Salmo trutta*) parasitized by *Lepeophtheirus salmonis*. Walker *et al.* (2004) speculated that these effects of external parasites were probably mediated through disturbance of the hydromineral balance of the fish host, resulting from damage and pathology associated with infection. The association between cortisol and external infections has also been confirmed by the administration of corticosteroids to the host. For example, Harris *et al.* (2000) implanted hydrocortisone acetate into Artic charr (*Salvelinus alpinus*) and brook trout (S. *frontalis*) and obtained a significant increase in the population of *Gyrodactylus salaris*, implying the immune status of the fish may have been compromised. The association between cortisol and parasites is not only linked to external infections, as Sures *et al.* (2001) observed that the nematode *Anguillicola crassus* induced an increase in plasma cortisol levels in the European eel, *Anguilla anguilla* which was correlated with the development of the larval parasite and the appearance of the adult worm.

Whilst there have been extensive studies which have revealed that cortisol affects the immune status of a fish, there have been limited investigations which have established a clear route between parasitic infection, the stress response evoked and a decrease in the immune status of the fish host. In fact, it is only recently that Saeij *et al.* (2003a) proposed a link between stress induced by handling in carp, cortisol levels, the host immune response and parasitic infection. Handling apparently increased susceptibility to the protozoan parasite, *Trypanoplasma borreli*. Indeed, cortisol suppressed parasite-induced expression of a range of immune parameters such as inducible nitric oxide, serum amyloid A, tumour necrosis factor-α, interleukin-1β.

ENVIRONMENTAL EFFECTS ON IMMUNE RESPONSES: IMPLICATIONS FOR PARASITIC INFECTIONS

In the early 1990s, there was increased interest in the possible use of parasites as monitors of pollution (e.g., Poulin, 1992) in both the marine and freshwater habitats. These proposals were based partly on the numerous field studies which had revealed that the prevalence and intensity of both ecto- and endoparasites was affected by water quality, primarily the presence of pollution (Khan and Thulin, 1991). Whilst this effect could be mediated by direct effects of the pollutants on the free-living and parasitic stages, and/or the intermediate host, the possibility that the pollutant, either directly or indirectly via a stress response, may affect immune status, and thus host susceptibility was also proposed.

There have been numerous studies which have revealed that the immune status of vertebrates is affected by environmental conditions. Indeed, the changes in immune parameters in fish, were proposed by Wester *et al.* (1994) to be, in themselves, possible biomarkers for pollution. Several investigations have revealed that those features which have been highlighted above as being associated with the immune response of hosts to parasitic infections, are affected by the presence of pollutants. For example, over 20 years ago, it was noted that migration of fish leucocytes and activation of lymphocytes were affected by such pollutants as aromatic hydrocarbons, organophosphates and cadmium (see Hoole, 1997, for a review). However, there are few studies which link the deterioration of water quality with alteration in immune status of lower vertebrates and a subsequent increase in parasitic infection. Studies in amphibians have, however, shown that such a link may exist. For example, Linzey *et al.* (2003) noted that there was a decrease in splenic lymphocyte blastogenesis in marine toads, *Bufo marinus*, exposed to cadmium pollution, and that this may be related to the increase in incidence of *Mesocoelium monas*. In addition, tadpoles of *Rana sylvatica* exposed to a range of pesticides exhibited a decrease in circulating eosinophils and an increase in the prevalence of metacercariae of *Ribeiro* sp. and *Telorchis* sp. (Kiesecker, 2002). The limited studies in fish have also indicated a direct link between pollution, the immune response and parasitic infection. Scherwerack *et al.* (2001, 2003) noted that proliferation of pronephric lymphocytes of carp increased with mitogen stimulation when the cells had been obtained from fish infected with the trematode, *Sanguinicola inermis* and exposed to ammonia or cadmium. In addition, in fish exposed to heavy metals, lymphocyte blastogenesis induced by the cercarial extracts was greater when compared to cells isolated from unexposed infected fish. It would appear, however, that these responses, as well as effects on the immune organs, were dependent on whether the pollutant preceded or came after the parasitic infection. The combined effects of the pollutant and parasitization on immune status of the fish have also been studied by Jacobson *et al.* (2003). These workers revealed that Chinook salmon exposed to polychlorinated biphenyl (PCB) and infected with the trematode, *Nanophyetus salmincola*, had a low primary haemolytic plaque-forming response in the pronephros. The effects of simultaneous parasitic and pollutant exposures on antibody responses, however, appear to be inconclusive. For example, Sures and Knopf (2004) noted that in eels infected with *Anguillicola crassus* and exposed to cadmium, there was a negligible effect on antibody production.

THE ELUSIVE PARASITE: EVASION AND SUPPRESSION OF THE IMMUNE RESPONSE?

Parasitization perhaps represents one of the stongest forces for the evolution of host organisms. The co-evolution of both host and parasite is, in many respects, similar to an evolutionary arms race involving the immune response of the host on one side and the evasion and/or suppression strategies of the parasite on the other. This battle is perhaps most eloquently explained in the 'Red Queen Hypothesis' proposed by van Valen in 1973. Both parasite and host are, therefore, under considerable selection pressure to evolve and adapt to their conjoined existence. Thus, by definition, if a parasite exists, it must have evolved strategies to withstand the immune response of the host. Whilst these strategies have been the subject of extensive investigation in medically important parasites (see Wakelin, 1996, for a review), their existence in parasites of fish is primarily based on speculation. Whilst many workers have acknowledged that such strategies must occur in fish parasites, detailed investigations are lacking (Fig. 10.2).

As indicated above, the fact that a parasite may act as a stressor and thus affect corticosteroid levels in the host could be interpreted as an immune suppression strategy due to the suppressive effects of cortisol on several immune parameters (Schreck, 1996). In addition to this indirect effect of parasitization on the downregulation of the immune response there is evidence that fish parasites might directly suppress and/or evade

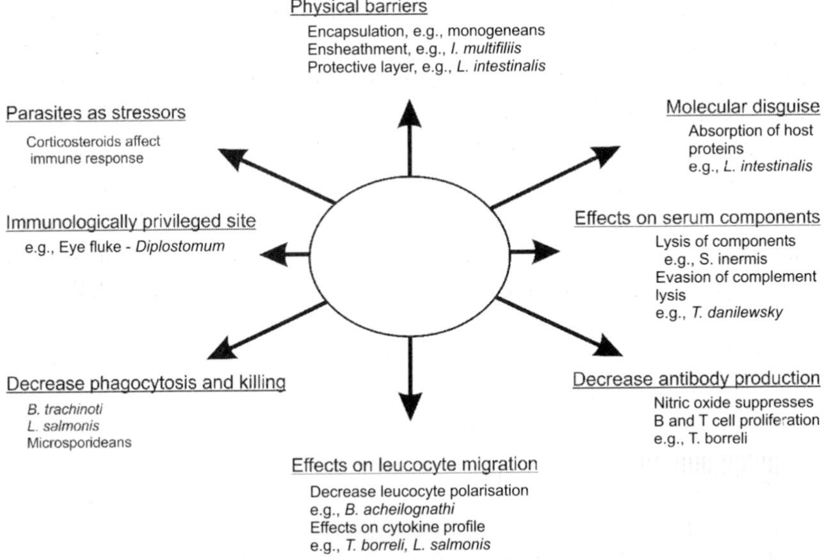

Fig. 10.2 Examples of speculative immune evasion and immune suppression strategies adopted of parasites infecting fish.

the immune response of a fish. These strategies may either involve physical barriers, infection site choice and/or effects on both the humoral and cellular components of the immune response.

The physical barriers present can be either host or parasite derived. The interactions between monogenean parasites and the external surface of their host are complex and varied. Whilst some parasitic species, such as *Pseudodactylogyrus anguillae*, induce a hyperplastic response in the gill of their eel host which eventually leads to a decline in parasite numbers others, such as *P. bini*, appear to revel in the tissue reaction evoked and are relatively unaffected by the response (Buchmann and Lindenstom, 2002). It would appear that the gill is particularly effective in partly embedding monogeneans, as a tissue reaction has been described in several host-parasite interactions, for example *Lepomis macrochirus* on *Cleidodiscus robustus* (Thune and Rogers, 1981) and *Callorhynchicola multitesticulatus* on *Callorhynchus millii* (Llewellyn and Simmons, 1984). Such an encapsulation could, as suggested by Buchmann and Lindenstrom (2002), create a shelter for the parasite. The concept of a 'protective shelter' has also been proposed by Cross (1993) for the gelatinous material which ensheathed the immobilized *Ichthyophthirius multifiliis*. Such a barrier was thought to prevent excessively large quantities of specific antibodies binding to the cilia membrane. Physical barriers may not only result from the host response, but it is also feasible that the complex nature and structure of several parasite surfaces may assist in evasion of the immune response. This is particularly highlighted in the case of the tapeworms, which comprise a large microthrix border. Hoole and Arme (1982) noted that the plerocercoid stage of the cestode, *Ligula intestinalis*, which resides in the body cavity of several species of cyprinid fish, contains an unusually extensive microthrix border of 10 μm in depth. Ultrastructural observations revealed that this border might prevent penetration of host leucocytes to the more vunerable tegumental surface (Fig. 10.3).

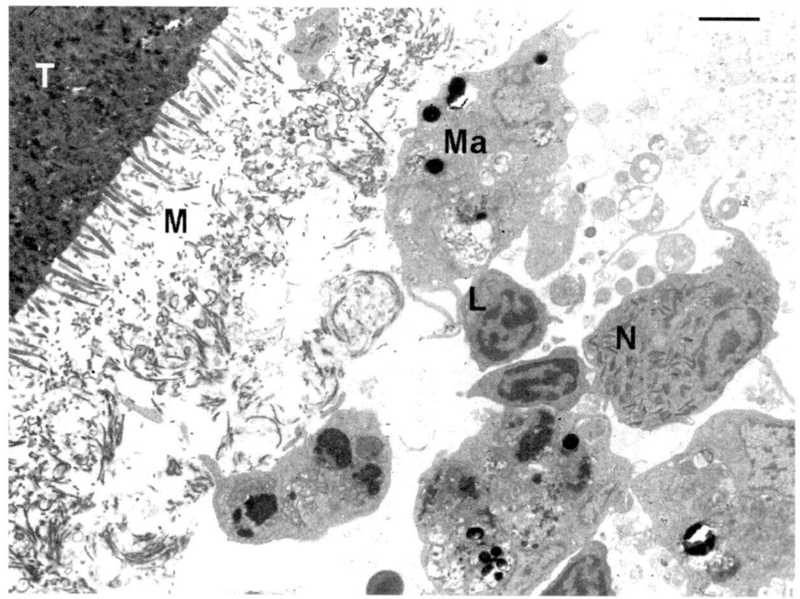

Fig. 10.3 Host reaction of roach (*Rutilus rutilus*) on the surface of the plerocercoid of *Ligula intestinalis.* Macrophages (Ma), lymphocytes (L) and neutrophils (N) appear to be incapable of pentrating the extensive microthrix border (M) and thus obtaining access to the tegument of the parasite (T). Scale bar 2 μm.

The surface membrane of a parasite may also be associated with immune evasion as several metazoans which parasitize mammals, for example, the trematode, *Schistosoma mansoni* and the nematodes *Brugia pahangi* and *Wuchereria bancrofti* have the ability to acquire or mimic host molecules and incorporate them into their outer surface (Sher *et al.*, 1978; Premaratne *et al.*, 1989; Kar *et al.*, 1993, respectively). Host molecules thus disguise the parasite against immune attack. Although there is no direct evidence that such a strategy occurs in the host-parasite interaction in fish, circumstantial evidence does suggest that it may be possible. Hoole and Arme (1983) utilized a transplantation protocol incorporating the plerocercoid of *Ligula intestinalis* and two cyprinid hosts, *Rutilus rutilus* and *Gobio gobio*, to propose that the parasite might acquire host proteins onto its surface that may be involved in disguise. In addition, Hoole and co-workers (Roberts *et al.*, 2005) detected 'host-like' molecules on the surface of the blood fluke, *Sanguinicola inermis*. There have been several studies in which fish immunoglobulins have been noted on the surface of parasites, although the possibility that such molecules may be 'blocking antibodies', as have been noted in parasites occurring in mammals (Wakelin, 1996), has not been considered. There have, however, been a limited number of studies that have indicated that parasites in fish may evade the humoral immune responses evoked.

Studies have revealed that serum factors, e.g., complement and antibody, mediate adherence of fish leucocytes to metazoan parasites (Hoole and Arme, 1986; Sharp *et al.*, 1991b). The absence of such a reaction to the adult stage of *Sanguinicola inermis* led Richards *et al.* (1996) to propose that the parasite may produce substances which breakdown host serum components. There have been several reports suggesting that the alternative complement pathway may kill gyrodactylid monogeneans. This, together with the fact that a number of isoforms of the complement proteins are known to exist in teleosts (Holland and Lambris, 2002; Boshra, Li and Sunyer, 2006), led Buchmann and Lindenstrom (2002) to propose that monogenans lacking the particular epitopes to activate these complement isoforms may evade the lysis and remain attached to the host surface. The differential effects of both the alternative and classical system of complement activation on parasite survival has been noted in infections of the protozoan, *Trypanoplasma borreli* in carp. Saeij *et al.* (2003b) observed that, at least *in vitro*, the parasite was lysed by the classical complement pathway but not by the alternative pathway. Wiegertjes *et al.* (2005) suggested that this differential activation of the complement system might be related to properties of the surface coat of some flagellate protozoans. The fish trypanosome, *Trypanosoma danilewsky,* has a surface coat comparable to non-Salivarian trypanosomes whose carbohydrate moiety of the mucins comprises sialic acid. This is transferred by trans-sialidase from the host glycoconjugates to the parasite surface. These 'pathogen-associated molecular patterns' appear to provide a protection against the initial attack by the complement system via the alternative pathway, although they may be required for antibody-mediated lysis. Wiegertjes and co-workers (1995) also suggested that the *T. borreli* induced mortality in carp was related to the lack of antibody production and further speculated in 2002 (Saeij *et al.*, 2002) that this may be associated with an increase in production of nitric oxide in the pronephros which could suppress antibody production by suppressing B and T cell proliferation. They likened it to 'bystander lymphocyte autotoxicity' suggested by Eisenstein *et al.* in 1994. Indeed, Scharsack *et al.* (2003a) also noted that *T. borreli*-induced NO production and significantly inhibited the proliferative response of mitogen-stimulated carp pronephros and peripheral blood lymphocytes. Such observations suggest that the cellular component of the immune response of fish may directly, or indirectly, be subject to parasite-induced suppression, or that the parasite may evade the cellular response provoked.

There have been several reports which suggest that factors produced by helminth parasites can influence the immune response of fish. For example, extracts from the plerocercoid of *Ligula intestinalis* suppressed the mitogen-induced proliferative response of splenic lymphocytes from roach (Taylor and Hoole, 1994) and temperature-induced migration, i.e., polarization of roach leucocytes (Taylor and Hoole, 1993). In addition, pronephric granulocytes from carp, either naturally or experimentally exposed to antigens of the cestode, *Bothriocephalus acheilognathi* exhibit a decrease in the polarization response to extracts of the parasite when compared to cells obtained from naïve fish (Nie and Hoole, 2000). How migration of fish leucocytes is downregulated by parasitization is unkown although Saeij *et al.* (2003b) did suggest an interesting hypothesis that the reduction in the IL-8 receptor homologue (CXCR2) in the head-kidney of carp infected with *T. borreli* might interfere with migration of neutrophils to the site of infection. Extensive studies carried out on *Lepeophtheirus salmonis* have suggested that this large crustacean parasite also produces immunomodulatory substances. Fast *et al.* (2005) noted that PGE_2, a component of *L. salmonis* secretions, inhibited expression of several immune-related genes, e.g., IL-8β, COX-2, MH class I and II, and suggested that this substance may prevent leucocyte recruitment at the feeding site and presentation of parasite antigens to T cells.

It has also been proposed that parasites may mediate the phagocytic and killing activity of fish leucocytes. Chaves *et al.* (2006) noted that the phagocytic response of spleen and pronephric cells of *Trachinotus marginatus* infected with the monogenean *Bicotylophora trachinoti* were first stimulated by the infection, but was later depressed. They suggested that this biphasic reponse might be the result of immuno-suppressive factors produced by the parasite. A similar biphasic response was noted in *Gasterosteus aculeatus* infected with the tapeworm, *Schistocephalus solidus*, where pronephric leucocytes isolated from infected fish 44 and 48 days' post-infection failed to respond to parasite antigens (Scharsack *et al.*, 2004). It has also been clearly established that secretions of *L. salmonis* produce a range of enzymes with proteolytic activity which may reduce phagocytic and respiratory burst responses (see Johnson and Fast, 2004, for review). This inhibition of killing activity of fish leucocytes is not limited the metazoan parasites. Studies on infections by intracellular parasites, such as microsporideans, partially suppress the respiratory burst in turbot phagocytes (Leiro *et al.*, 2001), which suggests that parasites may produce ROS-scavengers such as catalase, superoxide dismutase, or glutathione peroxidase (Lobo-da-Cunha and Azevedo, 1993; Kwon *et al.*, 2002).

The concept of an immunologically privileged site where a parasite can spend its life unperterbed by the host immune response is perhaps a parasite view of nirvhana, which can never be effectively achieved. There are, however, some organs where the lack of an intense immune response may aid parasite survival. At high intensities, it has been noted that gyrodactylid monogeneans aggregate on the cornea of the eye, and the 30% of the infection can occur on less than 1% of the total fish surface. Sigh and Buchmann (2000) proposed that these observations, together with a poor mast and mucous cell complement in the corneal surface, may mean that the eye is an 'immunological refugium' for monogeneans.

One must surely marvel at the enigmatic parasite!

THE FUTURE: OVERCOMING THE LIMITATIONS

Although significant advances have been made over the last two decades in the knowledge of the immunological interactions between the immune response of fish and parasitic infection, these have been relatively minor when compared to the advances made on this interaction with prokaryotic infections. One of the main reasons for this relates to the difficulty in establishing parasite life cycles in the laboratory. To undertake controlled infections, repeatable laboratory maintenance of infective stages must be obtained. This is perhaps highlighted by the fact that success in the studies of the immune response of fish to eukaryotic parsites has been primarily obtained with selected protozoan parasites, e.g., *Trypanoplasma borreli*, *Ichthyophthirius multifiliis*. When one considers the metazoan parasites, real significant advances have only been made on the external parasites, e.g. *Gyrodactylus* spp, *Lepeophtheirus salmonis* which have a direct life cycle that lacks an intermediate host. Although there have been isolated studies on endoparasitic metazoan parasites, e.g., *Sanguinicola, Diplostomum, Bothriocephalus* the difficulty in maintaining more than one host and producing significant numbers of parasite not only to maintain the life cycle but carry out experimental infections have usually meant that such studies have not been continued. One area that has not been fully explored is the *in vitro* culture and maintenance of fish parasites, perhaps because this, in many ways, is more difficult than establishing indirect parasite life cycles in the laboratory.

Perhaps what is required to overcome these problems is the application of technology that will allow numerous parameters to be monitored during a parasitic infection. Recent advances in molecular biology may allow this. The application of post-genomic techniques such as microarray analysis offers exciting and detailed insights to studies on several physiological processes in fish. Studies by Cossins *et al.* (2006) on carp have revealed numerous changes in the gene profile in fish that have been environmentally challenged. Such an approach has highlighted a number of candidate genes which because they are up- or downregulated and are targets fot further studies. Such techniques have recently been applied to carp infected with *I. multifiliis* and have revealed that the skin is an important site of expression of immune proteins in infected fish (Gonzalez *et al.*, 2007), and in Japanese flounder (*Paralichthys olivaceus*) infected with the mongenean *Neoheterobothrium hirame* where several important immune genes are uprgulated in peripheral blood leucocytes (Matsuyama *et al.*, 2007). Such insights allow more target investigations to be carried out on the association between the parasite and the cellular and humoral components of the immune response of fish.

Until our knowledge of how the immune response of the fish attempts to eradicate the parasitic infection is significantly advanced, mechanisms utilized by the parasite to evade these responses can only remain mere speculation, having serious consequences for contol scenarios. Until that time the parasite, as it has done since it first formed an association with its host in the Cambrian Period, must remain an enigma.

References

Ardelli, B.F. and P.T.K. Woo. 1997. Protective antibodies and anamnestic response in *Salvelinus fontinalis* and innate resistance of *Salvelinus namaycush* to the hemoflagellate. *Journal of Parasitology* 83: 943-946.

Bakke, T.A., A. Soleng, H. Lunde and P.D. Harris. 2000. Resistance mechanisms in *Salmo salar* stocks infected with *Gyrodactylus salaris*. *Acta Parasitologica* 45: 272.

Bienek, D.R., D.A. Plouffe, G.F. Wiegertjes and M. Belosevic. 2002. Immunization of goldfish with excretory/secretory molecules of *Trypanosoma danilewskyi* confers protection against infection. *Developmental and Comparative Immunology* 26: 649-657.

Bondad-Reantaso, M.G., K. Ogawa, T. Yoshinaga and H. Wakabayashi. 1995. Acquired protection against *Neobenedenia girellae* in Japanese flounder. *Fish Pathology* 30: 233-238.

Boshra, H., J. Li and J.O. Sunyer. 2006. Recent advances on the complement system of teleost fish. *Fish and Shellfish Immunology* 20: 239-262.

Bosi, G., A.P. Shinn, L. Giara, E. Simoni, F. Pironi and B.S. Dezfuli. 2005. Changes in the neuromodulators of the diffuse endocrine system of the alimentary canal of farmed rainbow trout, *Oncorhynchus mykiss* (Walbaum), naturally infected with *Eubothrium crassum* (Cestoda). *Journal of Fish Diseases* 28: 703-711.

Brooks, D.R. and D.A. McLennan. 1993. *Parascript: Parasites and the Language of Evolution*. Smithersonian Institution Press, Washington.

Brusca, R.C. 1981. A monograph on the Isopoda Cymothoidae (Crustacea) of the eastern Pacific. *Zoological Journal of the Linnean Society* 73: 117-199.

Buchmann, K. 1998. Binding and lethal effect of complement from *Oncorhynchus mykiss* on *Gyrodactylus derjavini* (Platyhelminthes, Monogenea). *Diseases of Aquatic Organisms* 32: 195-200.

Buchmann, K. 1999. Immune mechanisms in fish skin against monogenean infections—a model. *Folia Parasitologica (Praha)* 46: 1-9.

Buchmann, K. 2001. Lectins in fish skin: Do they play a role in monogenean-fish interactions? *Journal of Helminthology* 75: 227-232.

Buchmann, K. and J. Bresciani. 1998. Microenvironment of *Gyrodactylus derjavini* on rainbow trout *Oncorhynchus mykiss*: association between mucous cell density in the skin and site selection. *Parasitology Research* 84: 17-24.

Buchmann, K. and J. Bresciani. 1999. Rainbow trout leukocyte activity: Influence on the ectoparasitic monogenean, *Gyrodactylus derjavini*. *Diseases of Aquatic Organisms* 35: 13-22.

Buchmann, K. and T. Lindenstrom. 2002. Interactions between monogenean parasites and their fish hosts. *International Journal for Parasitology* 32: 309-319.

Buchmann, K., T. Lindenstrom and J. Sigh. 1999. Partial cross protection against *Ichthyophthirius multifiliis* in *Gyrodactylus derjavini* immunized trout. *Journal of Helminthology* 73: 189-195.

Bunnajirakul, S., D. Steinhagen, U. Hetzel, W. Korting and W. Drommer. 2000. A study of sequential histopathology of *Trypanoplasma borreli* (Protozoa: Kinetoplastida) in susceptible common carp *Cyprinus carpio*. *Diseases of Aquatic Organisms* 39: 221-229.

Burkart, M.A., T.G. Clark and H.W. Dickerson. 1990. Immunization of channel catfish, *Ictalurus punctatus* Rafinesque, against *Ichthyophthirius multifiliis* (Fouquet): killed verus live vaccine. *Journal of Fish Biology* 13: 401-410.

Buschkile, A.L. 1910. Beitrage zur Kenntnis des *Ichthyophthirius multifiliis* Fouquet. *Archiv fur Protistenkunde* 21: 61-102.

Cable, J. P.D. Harris and T.A. Bakke. 2000. Population growth of *Gyrodactylus salaris* on resistant and susceptible salmon. *Acta Parasitologica* 45: 272.

Chappell, L.H., L.J. Hardie and C.J. Secombes. 1994. Diplostomiasis: The disease and host-parasite interactions. In: *Parasitic Diseases of Fish,* A.W. Pike and J.W. Lewis (eds.). Samara Publishers, Tresaith, UK, pp. 59-86.

Chaves, I.S., R. Luvizzotto-Santos, L.A.N. Sampaio, A. Bianchini and P.E. Martinez. 2006. Immune adaptive response by *Bicotylophora trachinoti* (Monogenea: Diclidophoridae) infestation in pompano *Trachinotus marginatus* (Perciformes: Carangidae). *Fish and Shellfish Immunology* 21: 242-250.

Chilomonczyk, S. and D. Monge. 1999. Flow cytometry as a tool for assessment of the fish cellular immune response to pathogens. *Fish and Shellfish Immunology* 9: 319-333.

Chilomonczyk, S., D. Monge and P. de Kinkelin. 2002. Proliferative kidney disease: Cellular aspects of the rainbow trout, *Oncorhynchus mykiss* (Walbuam), response to parasitic infection. *Journal of Fish Diseases* 25: 217-226.

Chin, A., B.D. Glebe and P.T.K. Woo. 2004. Humoral response and susceptibility of five full-sib families of Atlantic salmon, *Salmo salar* L., to the haemoflagellate, *Cryptobia salmositica. Journal of Fish Diseases* 27: 471-481.

Clark, T.G. and I.D. Forney. 2003. Free living and parasitic ciliates, In: *Antigenic Variation,* A.O. Craig and A. Scherf (eds.). Academic Press, Amsterdam, pp. 376-402.

Clark, T.G., H.A. Dickerson and R.C. Findley. 1988. Immune response of channel catfish to ciliary antigens of *Ichthyophthirius multifiliis. Developmental and Comparative Immunology* 12: 581-594.

Clark, T.G., T.L. Lin and H.W. Dickerson. 1996. Surface antigen cross-linking triggers forced exit of a protozoan parasite from its host. *Proceedings of the National Academy of Sciences of the United States of America* 93: 6825-6829.

Clark, T.G., Y. Gao, J. Gaertig, X. Wang and G. Cheng. 2001. The I-antigens of *Ichthyophthirius multifiliis* are GPI-anchored proteins. *Journal of Eukaryotic Microbiology* 48: 332-337.

Conway Morris, S. 1981. Parasites and the fossil record. *Parasitology* 82: 489-509.

Cossins, A., J. Fraser, M. Hughes and A. Gracey. 2006. Post-genomic approaches to understanding the mechanisms of environmentally induced phenotypicplasticity. *Journal of Experimental Biology* 209: 2328-2336.

Cressey, R. and C. Patterson. 1973. Fossil parasitic copepods from a Lower Cretaceous fish. *Science* 180: 1283-1285.

Cross, M.L. 1993. Antibody binding following exposure of live *Ichthyophthirius multifiliis* (Ciliophora) to serum from immune carp *Cyprinus carpio. Diseases of Aquatic Organisms* 17: 159-164.

Cross, M.L. 1994. Localized cellular responses to *Ichthyophthirius multifiliis*: protection and pathogenesis? *Parasitology Today* 10: 364-368.

Cross, M.L. and R.A. Matthews. 1992. Ichthyophthiriasis in carp, *Cyprinus carpio* L., fate of parasites in immunized fish. *Journal of Fish Diseases* 19: 497-507.

Cross, M.L. and R.A. Matthews. 1993a. Localized leucocyte response to *Ichthyophthirius multifiliis* establishment in immune carp *Cyprinus carpio*. *Veterinary Immunology and Immunopathology* 38: 341-358.

Cross, M.L. and R.A. Matthews. 1993b. *Ichthyophthirius multifiliis* Froquet (Ciliophora): The location of sites immunogenic to the host *Cyprinus carpio* (L.). *Fish and Shellfish Immunology* 3: 13-24.

Cuesta, A., P. Munoz, A. Rodriguez, I. Salinas, A. Sitja-Bobadilla, P. Alvarez-Pelliero, M.A. Esteban and J. Meseguer. 2006. Gilthead sea bream (*Sparus aurata* L.) innate defence against the parasite *Enteromyxum leei* (Myxozoa). *Parasitology* 132: 95-104.

Dalgaard, M., K. Buchmann and A. Li. 2002. Immunization of rainbow trout fry with *Ichthyophthirius multifiliis* sonicate: Protection of host and immunological changes. *Bulletin of the European Association of Fish Pathologists* 22: 288-297.

Dezfuli, B.S. 1991. Histopathology in *Leuciscus cephalus* (Pisces: Cyprinidae) resulting from infection with *Pomphorhynchus laevis* (Acanthocephala). *Parasitologia* 33: 137-145.

Dezfuli, B.S., S. Capuano and M. Manera. 1998. A description of rodlet cells from the alimentary canal of *Anguilla anguilla* and their relationship to parasitic helminthes. *Journal of Fish Biology* 53: 1084-1095.

Dezfuli, B.S., L. Giari, S. Arrighi, C. Domeneghini and G. Bosi. 2003. Influence of enteric helminthes on the distribution of intestinal endocrine cells belonging to the diffuse endocrine system in brown trout, *Salmo trutta* L. *Journal of Fish Diseases* 26: 155-166.

Dezfuli, B.S., L. Giari, E. Simoni, G. Bosi and M. Manera. 2002. Histopathology, immunohistochemistry and ultrastucture of the intestine of *Leuciscus cephalus* (L.) naturally infected with *Pomphorhynchus laevis* (Acanthocephala). *Journal of Fish Diseases* 25: 7-14.

Dickerson, H.W., T.G. Clark and R.C. Findly. 1989. *Ichthyophthirius multifiliis* has membrane-associated immobilization antigens. *Journal of Protozoology* 36: 159-164.

Dickerson, H.W., T.G. Clark and A.A. Leff. 1993. Serotypic variation among isolates of *Ichthyophthirius multifiliis* based on immobilization. *Journal of Eukaryotic Microbiology* 40: 816-820.

Dickerson, H.W., J. Brown, D.L. Dawe and J.B. Gratzek. 1984. *Tetrahymena pyriformis* as a protective antigen against *Ichthyophthirius multifiliis* infection: Comparisons between isolates and ciliary preparation. *Journal of Fish Biology* 24: 523-529.

Dykova, I. and J. Lom. 1979. Histopathological changes in *Trypanosoma danilewskyi* Leveran & Mesnil, 1904 and *Trypanoplasma borreli* Leveran & Mesnil, 1902 infections of goldfish, *Carassius auratus* (L.). *Journal of Fish Diseases* 2: 381-390.

Eisenstein, T.K., D. Huang, J.J. Meissler and B. Al-Ramadi. 1994. Macrophage nitric oxide mediates immunosuppression in infectious inflammation. *Immunobiology* 191: 441-447.

Fast, M.D., N.W. Ross and S.C. Johnson. 2005. Prostoglandin E_2 modulation of gene expression in an Atlantic salmon (*Salmo salar*) macrophage-like cell line (SHK-1). *Developmental and Comparative Immunology* 29: 951-963.

Fast, M.D., D.M. Muise, R.E. Easy, N.W. Ross and S.C. Johnson. 2006. The effects of *Lepeophtheirus salmonis* infections on the stress response and immunological status of Atlantic salmon (*Salmo salar*). *Fish and Shellfish Immunology* 21: 228-241.

Feng, S. and P.T.K. Woo. 1997. Complement fixing antibody production in thymectomized *Oncorhynchus mykiss*, vaccinated against or infected with the pathogenic haemoflagellate *Cryptobia salmositica*. *Folia Parasitologica* 44: 188-194.

Fleming, M.W. 1998. Experimental inoculations with *Ostertagia ostertagi* or exposure to artificial illumination alters peripheral cortisol in dairy calves (*Bos taurus*). *Comparative Biochemistry and Physiology* A119: 315-319.

Gonzalez, S.F., K. Buchmann and M.E. Nielsen. 2007a. Real-time gene expression analysis in carp (*Cyprinus carpio* L.) skin: Inflammatory responses caused by the ectoparasite *Ichthyophthirius multifiliis*. *Fish and Shellfish Immunology* 22: 641-650.

Gonzalez, S.F., K. Buchmann and M.E. Nielsen. 2007b. Complement expression in common carp *(Cyprinus carpio L.)* during infection with *Ichthyophthirius multifiliis*. *Developmental and Comparative Immunology* 31: 576-586.

Gonzalez, S.F., K. Buchmann and M.E. Nielsen. 2007c. *Ichthyophthirius multifiliis* infection induces massive up-regulation of serum amyloid A in carp *(Cyprinus carpio)*. *Veterinary Immunology and Immunopathology* 115: 172-178.

Gonzalez, S.F., N. Chatziandreou, M.E. Nielsen, W.Z. Li, J. Rogers, R. Taylor, Y. Santos and A. Cossins. 2007d. Cutaneous immune responses in the common carp detected using transcript analysis. *Molecular Immunology* 44: 1664-1679.

Goven, B.A., D.L. Dawe and J.B. Gratzek. 1980. Protection of channel catfish *Ictalurus punctatus* against *Ichthyophthirius multifiliis* by immunization. *Journal of Fish Biology* 17: 311-316.

Goven, B.A., D.L. Dawe and J.B. Gratzek. 1981. Protection of channel catfish (*Ictalurus punctatus*) against *Ichthyophthirius multifiliis* Fouquet by immunization with varying doses of *Tetrahymena pyriformis* (Lwoff) cilia. *Aquaculture* 23: 269-273.

Graves, S.S., D.L. Evans and D.L. Dawe. 1985. Antiprotozoan activity of nonspecific cytotoxic cells (NCC) from channel catfish (*Ictalurus punctatus*). *Journal of Immunology* 134: 78-85.

Grayson, T.H., P.G. Jenkins, A.B. Wrathmell and J.E. Harris. 1991. Serum responses to salmon louse, *Lepeophtheirus salmonis* (Kroyer, 1838), in naturally infected salmonids and immunized rainbow trout (*Oncorhynchus mykiss*, Walbaum), and rabbits. *Fish and Shellfish Immunology* 1: 141-155.

Grutter, A.S. and N.W. Pankhurst. 2000. The effects of capture, handling, confinement and ectoparasite load on plasma levels of cortisol, glucose and lactate in the coral reef fish *Hemigymnus melapterus*. *Journal of Fish Biology* 57: 391-401.

Harris, P.D., A. Soleng and T.A. Bakke. 1998. Killing of *Gyrodactylus salaris* (Platyhelminthes: Monogenea) mediated by host complement. *Parasitology* 117: 137-143.

Harris, P.D., A. Soleng and T.A. Bakke. 2000. Increased susceptibility of salmonids to the monogenean *Gyrodactylus salaris* following administration of hydrocortisone acetate. *Parasitology* 120: 57-64.

Hatanaka, A., N. Umeda, S. Yamashita and N. Hirazawa. 2005. A small ciliary surface glycoprotein of the monogenean parasite *Neobenedenia girellae* acts as an agglutination/immobilization antigen and induces an immune response in the Japanese flounder *Paralichthys olivaceus*. *Parasitology* 131: 591-600.

Hedrick, R.P., E. MacConnell and P. de Kinkelin. 1993. Proliferative kidney disease of salmonid fish. In: *Annual Review of Fish Diseases,* M. Faisal and F.M. Hetrick (eds.). Elsevier Science, Oxford, UK, Vol. 3, pp. 277-290.

Hemmer, N., D. Steinhagen, W. Drommer and W. Korting. 1998. Changes of intestinal epithelial tunover in carp *Cyprinus carpio* infected with *Goussia carpelli* (Protozoa: Apicomplexa). *Diseases of Aquatic Organisms* 34: 39-44.

Hines, R.S. and D.T. Spira. 1974a. Ichthyophthiriasis in mirror carp *Cyprinus carpio* (L.) III Pathology. *Journal of Fish Biology* 6: 189-196.

Hines, R.S. and D.T. Spira. 1974b. Ichthyophthiriasis in the mirror carp *Cyprinus carpio* L., V Acquired immunity. *Journal of Fish Biology* 6: 373-378.

Holland, M.C.H. and J.D. Lambris. 2002. The complement system in teleosts. *Fish and Shellfish Immunology* 12: 399-420.

Hoole, D. 1997. The effects of pollutants on the immune reponse of fish: Implications for helminth parasites. *Parasitologia* 39: 219-225.

Hoole, D. and C. Arme. 1982. Ultrastructural studies on the cellular response of roach, *Rutilus rutilus*, to the plerocercoid larva of the pseudophyllidean cestode, *Ligula intestinalis*. *Journal of Fish Diseases* 5: 131-144.

Hoole, D. and C. Arme. 1983a. *Ligula intestinalis* (Cestoda: Pseudophyllidea): An ultrastructural study on the cellular response of roach fry, *Rutilus rutilus*. *International Journal for Parasitology* 13: 359-363.

Hoole, D. and C. Arme. 1983b. Ultrastructural studies on the cellular response of fish hosts following experimental infection with the plerocercoid of *Ligula intestinalis* (Cestoda: Pseudophyllidea). *Parasitology* 87: 139-149.

Hoole, D. and C. Arme. 1986. The role of serum in leucocyte adherence to the plerocercoid of *Ligula intestinalis* (Cestoda: Pseudophyllidea). *Parasitology* 92: 413-424.

Hoole, D. and C. Arme. 1988. *Ligula intestinalis*: Phosphorylcholine inhibition of leucocyte adherence. *Diseases of Aquatic Organisms* 5: 29-33.

Hoole, D. and H. Nisan. 1994. Ultrastructural studies on the intestinal response of carp to the pseudophyllidean tapeworm, *Bothriocephalus acheilognathi*. *Journal of Fish Diseases* 17: 623-629.

Hoole, D., J.W. Lewis, P-M.M. Schuwerack, C. Chakravarthy, A.K. Shrive, T.G Greenhough and J.R. Cartwright. 2003. Inflammatory interactions in fish exposed to pollutants and parasites: A role for apoptosis and C reactive protein. *Parasitology* 126: S71-S85.

Houghton, G. and R.A. Matthews. 1986. Immunosuppression of carp (*Cyprinus carpio* L.) to Ichthyophthiriasis using the corticosteroid triamcinolone acetonide. *Veterinary Immunology and Immunopathology* 12: 413-419.

Houghton, G. and R.A. Matthews. 1990. Immunosuppression in juvenile carp, *Cyprinus carpio* L.: The effects of the corticosteroids trimcinolone acetonide and hydrocortisone 21- hemisuccinate (cortisol) on acquired immunity and the humoral antibody response to *Ichthyophthirius multifiliis*. *Journal of Fish Diseases* 13: 269-280.

Houghton, G. and R.A. Matthews. 1993. *Ichthyophthirius multifiliis* Fouquet: Survival within immune juvenile carp, *Cyprinus carpio* L. *Fish and Shellfish Immunology* 16: 301-312.

Houghton, G., L.J. Healey and R.A. Matthews. 1992. The cellular proliferative response, humoral antibody response and cross reactivity studies of *Tetrahymena pyriformis* with *Ichthyophthirius multifiliis* Fouquet in juvenile carp (*Cyprinus carpio* L.). *Developmental and Comparative Immunology* 16: 301-312.

Huttenhuis, H.B.T., N. Romano, C. Van Osterhoud, A.J. Taverne-Thiele, L. Mastrolia, W.B. Van Muiswinkel and J.H.W.M. Rombout. 2006. The ontogeny of the mucosal immune cells in common carp (*Cyprinus carpio* L.). *Anatomy and Embryology* 211: 19-29.

Irie, T., S. Watarai and H. Kodama. 2003. Humoral immune response of carp (*Cyprinus carpio*) induced by oral immunization with liposome-entrapped antigen. *Developmental and Comparative Immunology* 27: 413-421.

Jacobson, K.C., M.R. Arkoosh, A.N. Kagley, E.R. Clemons, T.K. Collier and E. Casillas. 2003. Cumulative effects of natural and anthropogenic stress on immune function and disease resistance in juvenile Chinook Salmon. *Journal of Aquatic Animal Health* 15: 1-12.

Joernik, M., J.P.J. Saeij, J.L. Strafford, M. Belosevic and G.F. Wiegertjes. 2004. Animal models for the study of innate immunity: Protozoan infections in fish. In: *Host-Parasite Interactions,* G.F. Wiegertjes and G. Flik (eds.). SEB Symposium Series, BIOS Scientific Publications, Abingdon, UK, Vol. 55, pp. 107-129.

Johnson, S.C. 1993. A comparison of development and growth rates of *Lepophtheirus salmonis* (Copepoda: Caligidae) on naïve Atlantic (*Salmo salar*) and Chinook (*Oncorhynchus tshawytscha*) salmon. In: *Pathogens of Wild and Farmed Fish: Sea Lice,* G.A. Boxshall and D. Defaye (eds.). Ellis Horwood, Chichester, UK, pp. 68-80.

Johnson, S.C. and M.D. Fast. 2004. Interactions between sea lice and their hosts. In: *Host-Parasite Interactions,* G.F. Wiegertjes and G. Flik (eds.). SEB Symposium Series BIOS Scientific Publishers, Abington, UK, Vol. 55, pp. 131-159.

Jones, S.R.M., M. Palmen and W.B. van Muiswinkel. 1993. The effects of inoculum route and dose on the immune response of common carp, *Cyprinus carpio* to the blood parasite, *Trypanoplasma borreli. Veterinary Immunology and Immunopathology* 36: 369-378.

Kalbe, M. and J. Kurtz. 2006. Local differences in immunocompetence reflect resistance of sticklebacks against the eye fluke *Diplostomum pseudospathaceum. Parasitology* 132: 105-116.

Kar, S.K., J. Mania, C.I. Baldwin and D.A. Denham. 1993. The sheath of the microfilaria of *Wuchereria bancrofti* has albumin and immunoglobulin on its surface. *Parasite Immunology* 15: 297-300.

Khan, R.A. and J. Thulin. 1991. Influence of pollution on parasites of aquatic animals. *Advances in Parasitology* 30: 199-238.

Kiesecker, J.M. 2002. Synergism between trematode infection and pesticide exposure: A link to amphibian limb deformities in nature? *Proceedings of the National Academy of Sciences of the United States of America* 99: 9900-9904.

Kim, D-H. and B. Austin. 2006. Innate immune responses in rainbow trout (*Oncorhynchus mykiss*, Walbaum) induced by probiotics. *Fish and Shellfish Immunology* 21: 513-524.

Kirk, R.S. and J.W. Lewis. 1994. Sanguinicoliasis in cyprinid fish in the UK. In: *Parasitic Diseases of Fish*, A.W. Pike and J.W. Lewis (eds.). Samara Publishers, Tresaith, UK, pp. 101-117.

Kozel, T.R. 1980. Contact and penetration of *Ictalurus punctatus* by *Ichthyophthirius multifiliis*. *Proceedings of the American Society of Parastilogists* 55: 201.

Kwon, S.R., C.S. Kim, J.K. Chung, H.H. Lee and K.H. Kim. 2002. Inhibition of chemiluminescent response of olive flounder *Paralichthyes olivaceus* phagocytes by the sucticociliate parasite *Uronema marinum*. *Diseases of Aquatic Organisms* 52: 119-122.

Leiro, J., R. Iglesias, A. Parama, M.L. Sanmartin and F.M. Ubeira. 2001. Effect of *Tetramicra brevifilum* (Microspora) infection on respiratory-burst responses of turbot (*Scophthalmus maximus* L.) phagocytes. *Fish and Shellfish Immunology* 11: 639-652.

Leiro, J., J.A. Arranz, R. Iglesias, F.M. Ubeira and M.L. Sanmartin. 2004. Effects of the histiophagous ciliate *Philasterides dicentrarchi* on turbot phagocyte responses. *Fish and Shellfish Immunology* 17: 27-39.

Li, S. and P.T.K. Woo. 1995. Efficiacy of a live *Cryptobia salmositica* vaccine, and the mechanisms of protection in vaccinated rainbow trout, *Oncorhynchus mykiss*, against cryptobiosis. *Veterinary Immunology and Immunopathology* 48: 343-353.

Lin, T.L. and H.W. Dickerson. 1992. Purification and partial characterization of immobilization antigens from *Ichthyophthirius multifiliis*. *Journal of Protozoology* 39: 457-463.

Lindenstrom, T., K. Buchmann and C.J. Secombes. 2003. *Gyrodactylus derjavini* infection elicits IL-1β expression in rainbow trout skin. *Fish and Shellfish Immunology* 15: 107-115.

Lindenstrom, T., C.J. Secombes and K. Buchmann. 2004. Expression of immune response genes in rainbow trout skin induced by *Gyrodactylus derjavini* infections. *Veterinary Immunology and Immunopathology* 97: 137-148.

Lindenstrom, T., J. Sigh, M.B. Dalgaard and K. Buchmann. 2006. Skin expression of IL-1β in East Atlantic salmon, *Salmo salar*, L., highly susceptible to *Gyrodactylus salaris* infection is enhanced compared to a low susceptibility Baltic stock. *Journal of Fish Diseases* 29: 123-128.

Ling, K.H., Y.M. Sin and T.J. Lam. 1994. Protection of goldfish against some common ectoparasitic protozoans using *Ichthyophthirius multifiliis* and *Tetrahymena pyriformis* for vaccination. *Aquaculture* 116: 303-314.

Linzey, D.W., J. Burroughs, L. Hudson, M. Marini, J. Robertson, J.P. Bacon, M. Nagarkatti and P.S. Nagarkitti. 2003. Role of environmental pollutants on immune functions, parasitic infections and limb malformations in marine toads and whistling frogs from Bermuda. *International Journal of Environmental Health Research* 13: 21-30.

Llewellyn, J. and J.E. Simmons. 1984. The attachment of the monogenean parasite, *Callorhyncicola multitesticulatus* to the gills of its holocephalan host *Callorhynchus millii*. *International Journal for Parasitology* 14: 191-196.

Lobb, C.J. 1987. Secretory immunity induced in catfish *Ictalurus punctatus* following bath immunization. *Developmental and Comparative Immunology* 11: 727-738.

Lobb. C.J. and L.W. Clem. 1981. The metabolic relationships of immunoglobulins in fish serum, cutaneous mucus and bile. *Journal of Immunology* 132: 1525-1529

Lobo-da-Cunha, A. and C. Azevedo. 1993. Ultrastructural and cytochemical identification of peroxisomes in the ciliate *Ichthyophthirius multifiliis*. *Journal of Eukaryotic Microbiology* 40: 169-171.

Lom, J. and I. Dykova. 1992. Protozoan parasites of fishes. *Developments in Aquaculture and Fisheries Science* 26: 1-315.

Lom, J., I. Dykova and B. Machackova. 1986. Experimental evidence of pathogenicity of *Trypanoplasma borreli* and *Trypanosoma danilewski* for carp fingerlings. *Bulletin of the European Association of Fish Pathologists* 6: 87-88.

MacConnell, E., C.E. Smith, R.P. Hedrick and C.A. Speer. 1989. Cellular inflammatory response of rainbow trout to the protozoan parasite that causes proliferative kidney disease. *Journal of Aquatic Animal Health* 1: 108-118.

MacInnis, A.J. 1976. How parasites find their host: Some thoughts on the inception of host-parasite integration. In: *Ecological Aspects of Parasitology*, C.R. Kennedy (ed.). North-Holland, Amsterdam, pp. 3-20.

Magarinos, B., J.L. Romalde, Y. Santos, J.F. Casal and A.E. Toranzo. 1994. Vaccination trials on glthead sea bream (*Sparus aurata*) against *Pasteurella piscicida*. *Aquaculture* 120: 201-208.

Matsuyama, T., A. Fujiwara, C. Nakayasu, T. Kamaishi, N. Oseko, N. Tsutsumi, I. Hirono and T. Aoki. 2007. Microarray analysis of gene expression in Japanese flounder *Paralichthys olivaceus* leucocytes during monogenean parasite *Neoheterobothrium hirame* infection. *Diseases of Aquatic Organisms* 75: 79-83.

Molnar, K. 1971. Studies on gill parasitosis of grass-carp (*Ctenopharyngodon idella*) caused by *Dactylogyrus lamellatus* Achmerow, 1952 II: Epizootiology. *Acta Veterinary Academy of Science, Hungary* 21: 361-375.

Molnar, K., G. Majoros, G. Csaba and C. Szekely. 2003. Pathology of *Atractolytocestus huronenesis* Anthony, 1958 (Cestoda, Caryophyllaeidae) in Hungarian pond-farmed common carp. *Acta Parasitologica* 48: 222-228.

Morris, D.J., H.W. Ferguson and A. Adams. 2005. Severe, chronic proliferative kidney disease (PKD) induced in rainbow trout *Oncorhynchus mykiss* held at a constant 18°C. *Diseases of Aquatic Organisms* 66: 221-226.

Morley, N.J. and D. Hoole. 1995. Ultrastructural studies on the host-parasite interface between *Khawia sinensis* (Cestoda: Caryophyllidea) and carp *Cyprinus carpio*. *Diseases of Aquatic Organisms* 23: 93-99.

Mustafa, A., C. MacWilliams, N. Fernandez, K. Matchett, G.A. Conboy and J.F. Burka. 2000. Effects of sea lice (*Lepeophtheirus salmonis* Kroyer, 1837) infestation on macrophage functions in Atlantic salmon (*Salmo salar* L.). *Fish and Shellfish Immunology* 10: 47-59.

Nie, P. and D. Hoole. 1999. Antibody response of carp, *Cyprinus carpio* to the cestode, *Bothriocephalus acheilognathi*. *Parasitology* 118: 635-639.

Nie, P. and D. Hoole. 2000. Effects of *Bothriocephalus acheilognathi* (Cestoda: Pseudophyllidea) on the polarization response of pronephric leucocytes of carp, *Cyprinus carpio*. *Journal of Helminthology* 74: 353-357.

Nie, P., D. Hoole and C. Arme. 1996. Proliferation of pronephric lymphocytes of carp, *Cyprinus carpio* L. induced by extracts of *Bothriocephalus acheilognathi* Yamaguti, 1934 (Cestoda). *Journal of Helminthology* 70: 127-131.

Nolan, D.T., P. Reilly and S.E. Wendelaar Bonga. 1999. Infection with low numbers of the sea louse *Lepeoptheirus salmonis* (Kroyer) induces stress-related effects in post-smolt Atlantic salmon (*Salmo salar* L.). *Canadian Journal of Fisheries and Aquatic Sciences* 56: 947-959.

Olafsdottir, S.H., H.P.O. Lassen and K. Buchmann. 2003. Labile resistance of Atlantic salmon, *Salmo salar* L., to infections with *Gyrodactylus derjavini* Mikailov, 1975: Implications for host specificity. *Journal of Fish Diseases* 26: 51-54.

Overath, P., J. Haag, M.G. Mameza and A. Lischke. 1999. Freshwater fish trypanosomes: Definition of two types, host control by antibodies and lack of antigenic variation. *Parasitology* 119: 591-601.

Plouffe, D.A. and M. Belosevic. 2004. Enzyme treatment of *Trypanosoma danilewskyi* (Leveran & Mesnil) increases its susceptibility to lysis by the alternative complement pathway of goldfish, *Carassius auratus* (L.). *Journal of Fish Diseases* 27: 277-285.

Plouffe, D.A. and M. Belosevic. 2006. Antibodies that recognize α- and β-tubulin inhibit *in vitro* growth of the fish parasite *Trypanosoma danilewskyi*, Laveran and Mesnil, 1904. *Developmental and Comparative Immunology* 30: 685-697.

Poole, W.R., D. Nolan and O. Tully. 2000. Modelling the effects of capture and sea lice (*Lepeophtheirus salmonis* [Kroyer]) infestation on the cortisol stress response in trout. *Aquaculture Research* 31 : 835-841.

Poulin, R. 1992. Toxic pollution and parasitism in freshwater fish. *Parasitology Research* 8: 58-61.

Premaratne, U.N., R.M.E. Parkhouse and D.A. Denham. 1989. Microfilariae of *Brugia pahangi* in the blood of cats have variable levels of feline IgG on the sheaths. *Journal of Parasitology* 75: 320-322.

Press, C. McL. and O. Evensen. 1999. The morphology of the immune system in teleost fishes. *Fish and Shellfish Immunology* 9: 309-318.

Rambout, H.W.M., N. Taverne, M. van de Kamp and A.J. Taverne-Thiele. 1993. Differences in mucus and serum immunoglobulin in carp (*Cyprinus carpio* L.). *Developmental and Comparative Immunology* 17: 309-317.

Reilly, P. and M.F. Mulcahy. 1993. Humoral antibody response in Atlantic salmon (*Salmo salar* L.) immunised with extracts derived from the ectoparasitic caligid copepods, *Caligus elongates* (Nordmann, 1832) and *Lepeophtheirus salmonis* (Kroyer, 1838). *Fish and Shellfish Immunology* 3: 59-70.

Richards, D.T., D. Hoole., J.W. Lewis, E. Ewens and C. Arme. 1994a. Ultrastructural observations on the cellular response of carp (*Cyprinus carpio* L.) to eggs of the blood fluke, *Sanguinicola inermis* Plehn, 1905 (Digenea: Sanguinicolidae). *Journal of Fish Diseases* 17: 439-446

Richards, D.T., D. Hoole, J.W. Lewis, E. Ewens and C. Arme. 1994b. Changes in the cellular composition of the spleen and pronephros of carp *Cyprinus carpio* infected with the blood fluke *Sanguinicola inermis* (Trematoda: Sanguinicolidae). *Diseases of Aquatic Organisms* 19: 173-179.

Richards, D.R., D. Hoole, J.W. Lewis, E. Ewens and C. Arme. 1996a. Stimulation of carp *Cyprinus carpio* lymphocytes by the blood fluke *Sanguinicola inermis* (Trematoda: Sanguinicolidae). *Diseases of Aquatic Organisms* 25: 87-93.

Richards, D.R., D. Hoole, J.W. Lewis, E. Ewens and C. Arme. 1996b. *In vitro* polarization of carp leucocytes in response to the blood fluke *Sanguinicola inermis* Plehn, 1905 (Trematoda: Sanguinicolidae). *Parasitology* 112: 509-513.

Richards, D.R., D. Hoole, J.W. Lewis, E. Ewens and C. Arme. 1996c. Adherence of carp leucocytes to the adult and cercarial stages of the blood fluke (*Sanguinicola inermis*) Plehn, 1905 (Trematoda: Sanguinicolidae). *Journal of Helminthology* 70: 63-67.

Richards, G.R. and J.C. Chubb. 1996. Host responses to initial and challenge infections, following treatment of *Gyrodactylus bullatarudis* and *G. turnbulli* (Monogenea) on the guppy (*Poecilia reticulate*). *Parasitology Research* 82: 242-247.

Roberts, M.L., J.W. Lewis, G.F. Wiegertjes and D. Hoole. 2005. Interaction between the blood fluke *Sanguinicola inermis* and humoral components of the immune response of carp, *Cyprinus carpio. Parasitology* 131: 261-271.

Rombout, J.H.W.M., A.J. Tavernethiele and M.I. Villena. 1993. The gut-associated lymphoid-tissue (GALT) of carp *(Cyprinus carpio L.)—an immunocytochemical analysis. Developmental and Comparative Immunology* 17: 55-66.

Ruane, N., T.K. McCarthy and P. Reilly. 1995. Antibody response to crustacean ectoparasites immunized with *Argulus foliaceus* L. antigen extract. *Journal of Fish Diseases* 18: 529-537.

Rubio-Godoy, M., R. Porter and R.C. Tinsley. 2004. Evidence of complement-mediated killing of *Discocotyle sagittata* (Platyhelminthes, Monogenea) oncomiracidia. *Fish and Shellfish Immunology* 17: 95-193.

Rubio-Godoy, M., J. Sigh, K. Buchmann and R.C. Tinsley. 2003. Antibodies against *Discocotyle sagittata* (Monogenea) in farmed trout. *Diseases of Aquatic Organisms* 56: 181-184.

Rudat, S., D. Steinhagen, U. Hetzel, W. Drommer and W. Korting. 2000. Cytopathological observations on renal tubule epithelium cells in common carp *Cyprinus carpio* under *Trypanoplasma borreli* (Protozoa: Kinetoplastida) infection. *Diseases of Aquatic Organisms* 40: 203-209.

Saeij, J.P.J., W.B. van Muiswinkel, A. Groeneveld and G.F. Wiegertjes. 2002. Immune modulation by fish kinetoplastid parasites: A role for nitric oxide. *Parasitology* 124: 77-86.

Saeij, J.P.J., L.B.M. Verburg-van Kemenade, W.B. van Muiswinkel and G.F. Wiegertjes. 2003a. Daily handling stress reduces resistance of carp to *Trypanoplasma borreli: In vitro* modulatory effects of cortisol on leukocyte function and apoptosis. *Developmental and Comparative Immunology* 27: 233-245.

Saeij, J.P.J., B.J. de Vries and G.F. Wiegertjes. 2003b. The immune response of carp *Trypanoplasma borreli*: Kinetics of immune gene expression and polyclonal lymphocyte activation. *Developmental and Comparative Immunology* 27: 859-874.

Saeij, J.P.J., A. Groeneveld, N. van Rooijen, O.L.M. Haenen and G.F. Wiegertjes. 2003c. Minor effect of depletion of resident macrophages from peritoneal cavity on resistance of common carp *Cyprinus carpio* to blood flagellates. *Diseases of Aquatic Organisms* 57: 67-75.

Saulnier, D. and P. de Kinkelin. 1996. Antigenic and biochemical study of PKX, the myxosporean causative agent of proliferative kidney disease of salmonid fish. *Diseases of Aquatic Organisms* 27: 103-114.

Scharsack, J.P., D. Steinhagen, C. Kleczka, J.O. Schmidt, W. Korting, R.D. Michael, W. Leibold and H.J. Schuberth. 2003a. The haemoflagellate *Trypanoplasma borreli* induces the production of nitric oxide, which is associated with modulation of carp (*Cyprinus carpio* L.) leucocyte functions. *Fish and Shellfish Immunology* 14: 207-222.

Scharsack, J.P., D. Steinhagen, C. Kleczka, J.O. Schmidt, W. Korting, R.D. Michael, W. Leibold and H.J. Schuberth. 2003b. Head-kidney neutrophils of carp (*Cyprinus carpio* L.) are functionally modulated by the haemoflagellate *Trypanoplasma borreli*. *Fish and Shellfish Immunology* 14: 389-403.

Scharsack, J.P., M. Kalbe, R. Derner, J. Kurtz and M. Milinski. 2004. Modulation of granulocyte responses in three-spined stickleback *Gasterosteus aculeatus* infected with the tapeworm *Schistocephalus solidus*. *Diseases of Aquatic Organisms* 59: 141-150.

Schuwerack, P-M.M., J.W. Lewis and D. Hoole. 2003. Cadmium-induced cellular and immunological changes in *Cyprinus carpio* infected with the blood fluke *Sanguinicola inermis*. *Journal of Helminthology* 77: 341-350.

Schuwerack, P.M.M., J.W. Lewis, D. Hoole and N. Morley. 2001. Ammonia-induced cellular and immunological changes in juvenile *Cyprinus carpio* infected with the blood fluke, *Sanguinicola inermis*. *Parasitology* 122: 339-345.

Schreck, C.B. 1996. Immunomodulation: Endogenous factors. In: *The Fish Immune System: Organisms, Pathogen and Environment*, G.K. Iwama and T. Nakanishi (eds.). Academic Press, London, pp. 311-337.

Scott, A.L. and J.M. Grizzle. 1979. Pathology of cyprinid fishes caused by *Bothriocephalus gowkongensis* Yea, 1955 (Cestoda: Pseudophyllidea). *Journal of Fish Diseases* 2: 69-73.

Scott, M.E. and M.A. Robinson. 1984. Challenge infections of *Gyrodactylus bullatarudis* (Monogenea) on guppies (*Poecilia reticulata*) following treatment. *Journal of Fish Biology* 24: 581-586.

Sharp, G.J.E., A.W. Pike and C.J. Secombes. 1989. The immune response of wild rainbow trout, *Salmo gairdneri* Richardson, to naturally acquired plerocercoid infections of *Diphyllobothrium dendriticum* (Nitzsch, 1824) and *D. ditremum* (Creplin, 1825). *Journal of Fish Biology* 35: 781-794.

Sharp, G.J.E., A.W. Pike and C.J. Secombes. 1991a. Leucocyte migration in rainbow trout (*Oncorhynchus mykiss* [Walbaum]): Optimization of migration conditions and responses to host and pathogen (*Diphyllobothrium dendriticum* [Nitzsch, 1824]) derived chemoattractants. *Developmental and Comparative Immunology* 15: 295-305.

Sharp, G.J.E., A.W. Pike and C.J. Secombes. 1991b. Rainbow trout (*Oncorhynchus mykiss* [Walbaum, 1792]) leucocyte interactions with metacestode stages of *Diphyllobothrium dendriticum* (Nitzsch, 1824), (Cestoda, Pseudophyllidea). *Fish and Shellfish Immunology* 1: 195-211.

Sharp, G.J.E., A.W. Pike and C.J. Secombes. 1992. Sequential development of the immune response in rainbow trout (*Oncorhynchus mykiss* [Walbaum]) to experimental plerocercoid infections of *Diphyllobothrium dendriticum* (Nitzsch, 1824). *Parasitology* 104: 169-178.

Sher, A., B.F. Hall and M.A. Vadas. 1978. Acquistion of murine major histocompatability complex gene products by schistomula of *Schistosoma mansoni*. *Journal of Experimental Medicine* 148: 46-50.

Sigh, J. and K. Buchmann. 2000. Associations between epidermal thionin-positive cells and skin parasitic infections in brown trout *Salmo trutta*. *Diseases of Aquatic Organisms* 41: 135-139.

Sigh, J., T. Lindenstrom and K. Buchmann. 2004a. Expression of pro-inflammatory cytokines in rainbow trout (*Oncorhynchus mykiss*) during an infection with *Ichthyophthirius multifiliis*. *Fish and Shellfish Immunology* 17: 75-86.

Sigh, J., T. Lindenstrom and K. Buchmann. 2004b. The parasite ciliate *Ichthyophthirius multifiliis* induces expression of immune relevant genes in rainbow trout, *Oncorhynchus mykiss* (Walbaum). *Journal of Fish Diseases* 27: 409-417.

Slotved, H.C. and K. Buchmann. 1993. Acquired resistance of the eel *Anguilla anguilla* L. to challenge infections with gill monogeneans. *Journal of Fish Diseases* 16: 585-591.

Steinhagen, D. 1997. Temperature modulation of the response of Ig-positive cells to *Goussia carpelli* (Protozoa: Apicomplexa) infections in carp, *Cyprinus carpio* L. *Journal of Parasitology* 83: 434-439.

Steinhagen, D. and K. Hespe. 1997. Carp coccidiosis: activity of phagocytic cells from common carp infected with *Goussia carpelli*. *Diseases of Aquatic Organisms* 31: 155-159.

Steinhagen, D., B. Oesterreich and W. Korting. 1997. Carp coccidiosis: clinical and haematological observations of carp infected with *Goussia carpelli*. *Diseases of Aquatic Organisms* 30: 137-143.

Stoltze, K. and K. Buchmann. 2001. Effect of *Gyrodactylus derjavini* infections on cortisol production in rainbow trout fry. *Journal of Helminthology* 75: 291-294.

Sures, B. and K. Knopf. 2004. Individual and conbined effects of cadmium and 3,3′, 4,4′5-pentachlorobiphenyl (PCB 126) on the humoral immune response in European eel (*Anguilla anguilla*) experimentally infected with larvae of *Anguillicola crassus* (Nematoda). *Parasitology* 128: 445-454.

Sures, B., K. Knopf and W. Kloas. 2001. Induction of stress by the swimbladder nematode *Anguillicola crassus* in European eels, *Anguilla anguilla*, after repeat experimental infection. *Parasitology* 123: 179-184.

Subasinghe, R.P. and C. Sommerville. 1986. An experimental study into the possible protection of fry from *Ichthyophthirius multifiliis* Fouquet. In: *The First Asian Fisheries Forum*, J.L. Maclean, L.B. Dizon and I.V. Hosillos (eds.). Asian Fisheries Society, Manila, Philippines, pp. 279-282.

Swennes, A.G., J.G. Noe, C. Findly and H.W. Dickerson. 2006. Differences in the virulence between two serotypes of *Ichthyophthirius multifiliis*. *Diseases of Aquatic Organisms* 69: 227-232.

Sypek, J.P. and E.M. Burreson. 1983. Influence of temperature on the immune response of juvenile summer flounder, *Paralichthys dentatus*, and its role in the elimination of *Trypanoplasma bullocki* infections. *Developmental and Comparative Immunology* 7: 277-286.

Taraschewski, H. 1989. *Acanthocephalus anguillae* in intra- and extraintestinal positions in experimentally infected juveniles of goldfish and carp and in stickleback. *Journal of Parasitology* 75: 108-118.

Taylor, M.J. and D. Hoole. 1989. *Ligula intestinalis* (Cestoda: Pseudophyllidea): Plerocercoid induced changes in the spleen and pronephros of roach (*Rutilus rutilus*) and gudgeon (*Gobio gobio*). *Journal of Fish Biology* 34: 583-596.

Taylor, M.J. and D. Hoole. 1993. *Ligula intestinalis* (Cestoda: Pseudophyllidea): Polarization of cyprinid leucocytes as an indicator of host and parasite derived chemoattractants. *Parasitology* 107: 433-440.

Taylor, M.J. and D. Hoole. 1994. Modulation of fish lymphocyte proliferation by extracts and isolated proteinase inhibitors of *Ligula intestinalis* (Cestoda). *Fish and Shellfish Immunology* 4: 221-230.

Taylor, M.J. and D. Hoole. 1995. The chemiluminescence of cyprinid leucocytes in response to zymosan and extracts of *Ligula intestinalis* (Cestoda). *Fish and Shellfish Immunology* 5: 191-198.

Thoney, D.A. and E.M. Burreson. 1988. Lack of specific humoral antibody response in *Leiostomus xanthurus* (Pisces: Serranidae) to parasitic copepods and monogeneans. *Journal of Parasitology* 74: 191-194.

Thorburn, M.A. and E.L.K. Jansson. 1988. The effects of booster vaccination and fish size on survival and antibody production following *Vibrio* infection of bath-vaccinated rainbow trout, *Salmo gairdneri*. *Aquaculture* 71: 285-291.

Thune, R.L. and W.A. Rogers. 1981. Gill lesions in bluegill, *Lepomis macrochirus* Rafinesque, infested with *Cleidodiscus robustus* Mueller, 1934 (Monogenea: Dactylogyridae). *Journal of Fish Diseases* 4: 277-280.

Tun, T., K. Ogawa and H. Wakabayashi. 2002. Pathological changes induced by three myxosporeans in the intestine of cultured tiger puffer, *Takifugu rubripes* (Temminck and Schlegal). *Journal of Fish Diseases* 25: 63-72.

Van Valen, L. 1973. A new evolutionary law. *Evolutionary Theory* 1: 1-30.

Ventura, M.T. and I. Paperna. 1985. Histopathology of *Ichthyophthirius multifiliis* infections in fishes. *Journal of Fish Biology* 27: 185-203.

Wahli, T. and W. Meier. 1985. Ichthyophthiriasis in trout: Investigations of natural defence mechanisms. In: *Fish and Shellfish Pathology*, A. Ellis (ed.). Academic Press, London, pp. 347-352.

Wahli, T., W. Meier and K. Pfister. 1986. Ascorbic acid induced immune-mediated decrease in mortality in *Ichthyophthirius multifiliis* infected rainbow trout (*Salmo gairdneri*). *Acta Tropica* 43: 287-289.

Wakelin, D. 1996. *Immunity to Parasites: How Parasitic Infections are Controlled*. Cambridge University Press, Cambridge, pp. 122-145.

Walker, P.D., G. Flik and S.E. Wendelaar Bonga. 2004. The biology of parasites from the genus *Argulus* and a review of the interactions with its host. In: *Host-Parasite Interactions*, G.F. Wiegertjes and G. Flik (eds.). SEB Symposium Series, BIOS Scientific Publishers, Abington, UK, Vol. 55, pp. 107-129.

Wang, R., J-H. Kim, M. Sameshima and K. Ogawa. 1997. Dection of antibodies against the monogenean *Heterobothrium okamotoi* in tiger puffer by ELISA. *Fish Pathology* 32: 179-180.

Wang, X., T.G. Clark, J. Noe and H.W. Dickerson. 2002. Immunisation of channel catfish, *Ictalurus punctatus*, with *Ichthyophthirius multifiliis* immobilisation antigens elicits serotype-specific prtotection. *Fish and Shellfish Immunology* 13: 337-350.

Wells, P.R. and D.K. Cone. 1990. Experimental studies on the effect of *Gyrodactylus colemanensis* and *G. salmonis* (Monogenea) on the density of mucous cells in the epidermis of fry of *Oncorhynchus mykiss*. *Journal of Fish Biology* 37: 599-603.

Wendelaar Bonga, S.E. 1997. The stress response in fish. *Physiological Reviews* 77: 591-625.

Wester, P.W., A.D. Vethaak and W.B. van Muiswinkel. 1994. Fish as biomarkers in immunotoxicology. *Toxicology* 86: 213-232.

Whyte, S.K., L.H. Chappell and C.J. Secombes. 1988. *In vitro* transformation of *Diplostomum spathaceum* (Digenea) cercariae and short-term maintenance of post-penetration larvae *in vitro*. *Journal of Helminthology* 62: 293-302.

Whyte, S.K., L.H. Chappell and C.J. Secombes. 1989. Cytotoxic reactions of rainbow trout, *Salmo gairdneri* Richardson, macrophages for larvae of the eye fluke *Diplostomum spathaceum* (Digenea). *Journal of Fish Biology* 35: 333-345.

Whyte, S.K., L.H. Chappell and C.J. Secombes. 1990. Protection of rainbow trout *Oncorhynchus mykiss* (Richardson), against *Diplostomum spathaceum* (Digenea): Role of specific antibody and activated macrophages. *Journal of Fish Diseases* 13: 281-291.

Whyte, S.K., C.J. Secombes and L.H. Chappell. 1991. Studies on the infectivity of *Diplostomum spathaceum* in rainbow trout (*Oncorhynchus mykiss*). *Journal of Helminthology* 65: 169-178.

Wiegertjes, G.F. and M. Joerink. 2004. Macrophage polarization in the immune response to parasites. *Bulletin of the European Association of Fish Pathologists* 24: 5-10.

Wiegertjes, G.F., A. Groeneveld and W.B. van Muiswinkel. 1995. Genetic variation in susceptibility to *Trypanoplasma borreli* infection in common carp (*Cyprinus carpio* L.). *Veterinary Immunology and Immunopathology* 47: 153-161.

Wiegertjes, G.F., A.B. Bongers, P. Voorthuis, B. Zandieh Doulabi, A. Groeneveld, W.B. van Muiswinkel and R.J. Stet. 1996. Characterisation of isogenic carp (*Cyprinus carpio* L.) lines with a genetically determined high and low antibody production. *Animal Genetics* 27: 313-319.

Wiegertjes, G.F, M. Forlenza, M. Joerink and J.P. Scharsack. 2005. Parasite infections revisited. *Developmental and Comparative Immunology* 29: 749-758.

Williams, E.H. Jr. and L. Bunkley-Williams. 1994. Four cases of unusual crustacean-fish associations and comments on parasitic processes. *Journal of Aquatic Animal Health* 6: 202-208.

Williams, M.A. and D. Hoole. 1992. Studies on the kinetics and specificity of the antibody response of roach, *Rutilus rutilus*, to *Ligula intestinalis*. *Diseases of Aquatic Organisms* 12: 83-89.

Williams, M.A. and D. Hoole. 1995. Immunolabelling of putative fish host molecules on the tegumental surface of *Ligula intestinalis* (Cestoda: Pseudophyllidea). *International Journal for Parasitology* 25: 249-256.

Woo, P.T.K. 1981. Acquired immunity against *Trypanosoma danilewskyi* in goldfish, *Carassius auratus*. *Parasitology* 83: 343-346.

Woo, P.T.K. and S.L. Poynton. 1995. Diplomonadida, kinetoplastida and amoebida (Phylum, Sarcomastigophora). In: *Fish Diseases and Disorders*, P.T.K Woo (ed.). CAB International, Wallingford, UK, pp. 27-96.

Woo, P.T.K. and M. Shariff. 1990. *Lernaea cyprinacea* L. (Copeoda: Caligidae) in *Helostoma temmincki* Cuvier & Valenciennes: The dynamics of resistance in recovered and naïve fish. *Journal of Fish Diseases* 13: 485-493.

Wood, B.P. and R.A. Matthews. 1987. The immune response of the thick-lipped grey mullet, *Chelon labrosus* (Risso, 1826), to metacercarial infections of *Crytocotyle lingua* (Creplin, 1825). *Journal of Fish Biology* A31: 175-183.

Xu, D-H. and P.H. Klesius. 2002. Antibody mediated immune response against *Ichthyophthirius multifiliis* using excised skin from channel catfish, *Ictalurus punctatus* (Rafinesque), immune to *Ichthyophthirius*. *Journal of Fish Diseases* 25: 299-306.

Xu, D-H. and P.H. Klesius. 2003. Protective effect of cutaneous antibody produced by channel catfish, *Ictalurus punctatus* (Rafinesquue), immune to *Ichthyophthirius multifiliis* Froquet on cohabited non-immune catfish. *Journal of Fish Diseases* 26: 287-291.

Xu, D-H., P.H. Klesius and R.A. Shelby. 2002. Cutaneous antibodies in excised skin from channel catfish, *Ictalurus punctatus* Rafineque, immune to *Ichthyophthirius multifiliis*. *Journal of Fish Diseases* 25: 45-52.

Xu, D-H., P.H. Klesius and R.A. Shelby. 2004. Immune responses and host protection of channel catfish, *Ictalurus punctatus* (Rafinesque), against *Ichthyophthirius multifiliis* after immunization with livetheronts and sonicated trophonts. *Journal of Fish Diseases* 27: 135-141.

Xu, D-H., P.H. Klesius and C.A. Shoemaker. 2005. Cutaneous antibodies from channel catfish, *Ictalurus punctatus* Rafinesque, immune to *Ichthyophthirius multifiliis* (Ich) may induce apoptosis of Ich theronts. *Journal of Fish Diseases* 28: 213-220.

Xu, D-H., P.H. Klesius and V.S. Panangala. 2006a. Induced cross-protection in channel catfish, *Ictalurus punctatus* (Rafinesque), against different immobilization serotypes of *Ichthyophthirius multifiliis*. *Journal of Fish Diseases* 29: 131-138.

Xu, D-H., P.H. Klesius and C.A. Shoemaker. 2006b. Apoptosis in *Ichthyophthirius multifiliis* is associated with expression of the Fas receptor of theronts. *Journal of Fish Diseases* 29: 225-232.

Zapata, A.G., A. Chiba and A. Varas. 1996. Cells and tissues of the immune system of fish. In: *The Fish Immune System: Organism, Pathogen and Environment*, G.K. Iwama and T. Nakanishi (eds.). Academic Press, London, UK, pp. 1-62.

Index

About the Editors

Giacomo Zaccone if Professor of Comparative Anatomy and also a teacher of Comparative Endocrinology at the University of Messina. He is a reviewer for many authoritative journals including the Anatomical Record, Journal of Comparative Neurology, Comparative Biochemistry and Physiology, Journal of Fish Biology and Fish Physiology book series. He is editor of International Journal of Biochemical Science and board member of Acta Histochemica. He has authored several invited chapters published by Academic Press, Wiley-Liss and Science Publishers covering neuroepithelial gill chemoreceptor cells and their associated autonomic innervation and biology of fish integument. He has coedited a book on Airway Chemoreceptors in the Vertebrates (with E. Cutz, D. Adriaensen and C. Nurse) from Science Publishers and a topical issue (with L. Sundin, C. Olsson and M. Jonz) on nervous control in visceral organs in fishes from Acta Histochemica, Elsevier. He also coedited the Fish Endocrinology book/with M. Reinecke from Science Publishers. His interest is now focused onto NO/NOS regulatory system and neuropeptides in the lung and heart of the primitive fishes.

Alfonsa García-Ayala obtained her PhD from the University of Murcia, Spain. She did post-doctoral work at the University of Utrecht working on reproductive endocrinology. She is now a teacher of Embryology and Principles of Applied Cell Biology and Immunology at the Cell Biology and Histology Department of the University of Murcia. She is Vice-Dean and Teaching Quality Co-ordinator of the Faculty of Biology of the University. Her research focuses on the role of leukocytes and cytokines in reproductive functions and on the effects of endocrine disruptors on innate immunity of teleosts.

José Meseguer Peñalver is Professor of Cell Biology, Head of the Fish Immunology research group and Dean of the Faculty of Biology at the University of Murcia, Spain. He is reviewer for many scientific journals. He represents the University of Murcia in the AQUA-TNET "Socrates Thematic Network Aquaculture, Fisheries and Aquatics Resource Management". His research is mainly focused on fish innate immunity and immune modulators, fish cytokines and the interactions between leukocytes and reproductive functions.

B.G. Kapoor was formerly Professor and Head of Zoology in Jodhpur University (India). He has co-edited several books including Sensory Biology of Jawed Fishes – New Insights, 2001 (with T.J. Hara); Fish Adaptations, 2003 (with A.L. Val); Catfishes, 2003 (with G. Arratia, M. Chardon and R. Diogo); Fish Chemosenses, 2005 (with Klaus Reutter); Communication in Fishes, 2006 (with F. Ladich, S.P. Collin and P. Moller); Fish Endocrinology, 2006 (with Manfred Reinecke and Giacomo Zaccone); Fish Cytogenetics, 2007 (with E. Pisano, C. Ozouf-Costaz and F. Foresti); Fish Respiration and Environment, 2007 (with Marisa N. Fernandes, Francisco T. Rantin and Mogens L. Glass); Fish Osmoregulation, 2007 (with Bernardo Baldisserotto and Juan Miguel Mancera); Fish Reproduction, 2008 (with Maria João Rocha and Augustine Arukwe); Fish Life in Special Environments, 2008 (with Philippe Sébert and D.W. Onyango); Fish Larval Physiology, 2008 (with R.N. Finn); Fish Diseases, 2008 (with J.C. Eiras, Helmut Segner and Thomas Wahli); Fish Behaviour, 2008 (with Carin Magnhagen, Victoria A. Braithwaite and Elisabet Forsgren); Feeding and Digestive Functions of Fishes, 2008 (with J.E.P. Cyrino and D.P. Bureau). All the foregoing titles are from Science Publishers, Enfield, NH, USA. Dr. Kapoor has also co-edited The Senses of Fish: Adaptations for the Reception of Natural Stimuli, 2004 (with G. von der Emde and J. Mogdans); and co-authored Ichthyology Handbook, 2004 (with B. Khanna), both from Springer, Heidelberg. Also, he has been a contributor in books from Academic Press, London (1969, 1975 and 2001). His E-mail is: bhagatgopal.kapoor@rediffmail.com